D0152346

# ARTIFICIAL INTELLIGENCE AND NATURAL MAN

# ARTIFICIAL INTELLIGENCE AND NATURAL MAN

*Second Edition, Expanded*

*Margaret A. Boden*

PROFESSOR OF PHILOSOPHY AND PSYCHOLOGY
UNIVERSITY OF SUSSEX

*Basic Books, Inc., Publishers*

NEW YORK

Library of Congress Cataloging-in-Publication Data

Boden, Margaret A.
Artificial intelligence and natural man.

Bibliography: p. 501
Includes index.
1. Artificial intelligence. 2. Thought and thinking.
I. Title.
Q335.B56 1987 006.3 86–47739
ISBN 0–465–00456–3 (pbk.)


Printed in the United States of America
Designed by Vincent Torre
87 88 89 90 HC 9 8 7 6 5 4 3 2 1

For Ruskin and Jehane

# CONTENTS

## Part III

## LANGUAGE AND UNDERSTANDING



## Part IV

## THE VISUAL WORLD



## Part V

## NEW THOUGHTS FROM OLD

## *Part* VI

## THE RELEVANCE OF ARTIFICIAL INTELLIGENCE

# PREFACE TO SECOND EDITION

ARTIFICIAL INTELLIGENCE has grown dramatically since this book first appeared: there are new researchers—and research groups—throughout the industrialized world. Its visibility has increased also, for the media have now discovered it.

But *growth,* whether visible or invisible, can take place without *development.* The recent explosion of funding and publicity is due to commercial and political factors, not to intellectual advances within the field. The central problems of artificial intelligence, and the theoretical basis of its achievements, have remained essentially the same. Naturally, there has been some advance, but most of it is in technological efficiency, not basic scientific understanding.

Despite the passage of time, then, the main message of this book is still timely. The book is not a research text on the nitty-gritty of programming, nor a "list" of the most up-to-date programs around. Indeed, two opening chapters (Chapters 2 and 3) describe a program (a simulation of neurosis) that was already out-of-date when I started writing the first edition in the early 1970s. I chose to discuss it nevertheless—and would do so again today—because its simplicity, and its obvious "stupidity," help introduce two wide-ranging questions: What sort of thing is a computer program? and What difficulties face those who try to make programs as powerful as human thinking? These questions are what this book is about, and they are hardly nearer an answer now than they were a decade ago.

To be sure, if I were to write this book "from scratch" now, it would not be exactly the same. Some recent programs would be included in addition to—occasionally, instead of—the ones originally chosen. And many projects cited in the first edition as exploratory research have been further developed, so could now be described in more detail.

In particular, I would now devote an entire chapter to low-level vision, instead of discussing it within the chapter on psychology (Chapter 13). Low-level vision is the area in which there has been the most significant *theoretical* advance. I would include an account of the current exploratory research on parallel processing ("connectionist") systems. I would allow more space to "production rules," because of their use in commercially available expert systems like those mentioned in Chapter 15. And in view of the current interest in logic programming, I would add PROLOG to the programming languages discussed in Chapter 12. (I discuss the first three of these topics at length, though from the psychologist's—not the technologist's—point of view, in Boden [in preparation].)

Comparatively minor adjustments would be made also. For example, my discussion of creativity (Chapter 11) would now include programs on heuristic exploration, which have been used to help plan experiments in genetic engineering and design three-dimensional "chips." Chapter 10 would refer to a recent theoretical analysis of learning programs that highlights their computational similarities and differences. Further work on text analysis and story understanding has been done by the workers mentioned in Chapters 7 and 11. And in Chapter 7, too, I would discuss attempts to program *conversation,* which previously I mentioned only in the notes. But none of this new material would invalidate the essential points made in the original version.

For this second edition, then, I have not altered the main text. Instead, I have added a Postscript chapter, and a separate Bibliography listing significant recent work. The Postscript explains the reasons for the recent upsurge of publicity, indicates the areas of theoretical novelty, identifies the main current research issues, and suggests what advances we can—or cannot—expect in the future.

M.A.B.

*Brighton, Sussex, April 1986*

# PREFACE TO FIRST EDITION

ARTIFICIAL INTELLIGENCE is not the study of computers, but of intelligence in thought and action. Computers are its tools, because its theories are expressed as computer programs that enable machines to do things that would require intelligence if done by people.

This book describes artificial intelligence in a way that stresses its human relevance. I have used plain language as far as possible, and have entirely avoided mathematical and formal symbolisms. No specific expertise is presupposed, although readers with psychological or philosophical interests will find many points relating to already familiar issues. In particular, I have not assumed any previous acquaintance with artificial intelligence, nor even any knowledge of programming. So that the reader may gradually develop a sense of what a program is and what a computer can do with it, I have tried to describe programs that do interesting things in a way that allows an understanding of these matters to deepen progressively throughout the book. Later chapters build on earlier ones in the sense that they may continue discussion of examples introduced previously, so the book is best read as a whole, from beginning to end.

I have selected for discussion a number of computer programs likely to be of interest to readers with psychological or philosophical concerns but no programming experience. These programs throw light on the nature of human personality, belief, language and communication, perception, learning, creativity, and problem solving. So that they may provide a starting point from which to progress into detailed study of the programming literature, I have given guidance to further primary sources in the notes. And so that work in artificial intelligence may be related to its wider human context, I have indicated some of the relevant psychological and philosophical literature.

Above all, I have tried to convey a sense of the relevance of artificial intelligence to the understanding of natural man. Contrary to what most people assume, this field of research has a potential for counteracting the dehumanizing influence of natural science, for suggesting solutions to many traditional problems in the philosophy of mind, and for illuminating the hidden complexities of human thinking and personal psychology. The common view that machine research must tend to display us humiliatingly to ourselves as "mere clockwork" is false. The more widely this is realized, the less of a threat will artificial intelligence present to humane conceptions of society.

M.A.B.

*Brighton, Sussex*
*August 1976*

# ACKNOWLEDGMENTS

I AM deeply grateful to Aaron Sloman for his careful reading of the draft manuscript, and for many conversations on related topics. Helpful comments were provided also by Roy Edgley and Marc Eisenstadt. Much of the book was written during a year's sabbatical leave, granted by the University of Sussex. I am grateful also to my children, to whom this book is dedicated, and who put up with me while I was writing it.

I would also like to thank the following publishers for permission to reprint excerpts from the titles listed below:

Terry Winograd, *Understanding Natural Language* (Academic Press; Edinburgh University Press), pp. 8–15.

Poem by Laurence Lerner from *A.R.T.H.U.R.: The Life and Opinions of a Digital Computer* (University of Massachusetts Press; Harvester Press), pp. 9–10.

Richard Wollheim, ed., *Freud: A Collection of Critical Essays* (Anchor; Doubleday), pp. 242–270.

# ACKNOWLEDGMENTS FOR THE SECOND EDITION

I AM grateful to Rudi Lutz for his comments on my Postscript chapter, and to Harold Cohen (of the University of California at San Diego, La Jolla, California) for permission to reproduce two of his computer-generated drawings (photo: Becky Cohen).

"Explain all that," said the Mock Turtle.

"No, no! The adventures first," said the Gryphon in an impatient tone: "explanations take such a dreadful time."

*Lewis Carroll*

# *Part* I

# INTRODUCTION

# 1

# What Is Artificial Intelligence?

ANYONE who mentions artificial intelligence in polite conversation can expect two perplexed reactions: "What?" and "So what?" (Try referring to it as "AI," and you will receive some very strange looks.) Many people have not yet heard of it; few have more than the sketchiest idea of what is involved in it; and most are deeply skeptical of its human value. Even professional conversation can lead to questioning and disagreement about the nature of this novel discipline, for there is no single definition of it that would be judged equally felicitous by all its practitioners.

One thing, however, is certain: artificial intelligence is not the study of computers. Computers are metallic machines of intrinsic interest to electronic engineers but not, as such, to many others. So if you are not enamored of tin cans, you need not fear to meet any in this book.

It would be somewhat more accurate to say that artificial intelligence is the study of computer programs. Indeed, many workers approach it from the context of programming science, being drawn primarily by the challenge it offers to programming techniques. And no one seriously concerned with it can avoid detailed reference to programs and programming.

Yet one leading figure in the field has been heard to retort impatiently to an enthusiastic colleague, "I'm not interested in programs!" This dismissive remark throws light not only on the implicit contrast between "polite" and "professional" conversation, but also on artificial intelligence itself. For many expert programmers, the activity of programming is strictly subordinate to a wider aim, such as "the development of a systematic theory of intellectual processes, wherever they may be found," or "the study of intelligence as computation."[1] Each of these has been offered as a definition, and each stresses intelligence in general rather than human thought in particular. Other conceptions of artificial intelligence link it more firmly to its human source, emphasizing its potential for generating precise formal models in theoretical psychology.[2]

Still other workers have only a secondary interest (if any) in explaining human thought, concentrating primarily on practical or technological problems such as programming an automatic speech recognizer, filing clerk, or medical diagnostician, or building a robot assembly-line worker or explorer of Mars. People working in a technological context sometimes describe artificial intelligence as "an engineering discipline,"[3] a description hardly calculated to appeal to those with primarily psychological interests.

Clearly, then, the concerns that individuals bring to this type of machine research vary greatly, the preferred definitions differing accordingly. Perhaps the least tendentious definition is M. L. Minsky's—"artificial intelligence is the science of making machines do things that would require intelligence if done by men"—which indicates the general nature of the field without specifying particular aims too closely.[4]

My own interests in artificial intelligence are biased toward its potential for counteracting the dehumanizing influence of natural science, for suggesting solutions to many traditional problems in the philosophy of mind, and for illuminating the hidden complexities of human thinking and personal psychology.

The new concept of "machine" provided by artificial intelligence is so much more powerful than familiar concepts of mechanism that the old metaphysical puzzle of how mind and body can possibly be related is largely resolved. Insights drawn from this source clarify the nature of human purpose, freedom, and moral choice, and help one to understand how it is possible for the character of a human life to depend upon the degree of self-knowledge enjoyed by the person concerned. Artificial intelligence, in short, cannot only acknowledge but can even elucidate the essentially subjective mental realities so stressed by humanist psychologists (as opposed to behaviorists or neurophysiologists).

Equally, it highlights the awesome complexity of the mind, for it leads to an appreciation of the enormous psychological subtlety of "simple" everyday achievements, such as chatting and choosing, of whose computational intricacies one is intuitively quite unaware. Not least, it can suggest specific hypotheses about the thought processes involved, which aid in the development of richly structured theories about the mind. In this way it helps us progress from psychological insight to psychological understanding.

By "artificial intelligence" I therefore mean the use of computer programs and programming techniques to cast light on the principles of intelligence in general and human thought in particular. In other words, I use the expression as a generic term to cover all machine research that is somehow relevant to human knowledge and psychology, irrespective of the declared motivation of the particular programmer concerned.

It follows that I make no basic distinction of principle between "artificial intelligence" and "computer simulation." There is admittedly a difference in emphasis between workers who try to make a machine do something, irrespective of how humans do it, and those who aim to write a program that is functionally equivalent to a psychological theory. In a computer simulation, every thought process posited by a certain psychological theory has a corresponding process specified in the program.[5] Computer simulations are thus directly parasitic upon some prior articulated theory about human psychology, whereas other programs are not.

But most programs depend, at least in part, on the programmer's intuitive notions (implicit theories) about how people function in comparable circumstances. It is not surprising, then, that the distinction between these two categories of machine research is becoming less clear and less relevant with increasing appreciation of the difficulties involved in "powerful" programming, that is, programming that enables the machine to function intelligently. Even such a strictly technological enterprise as designing a robot to reconnoiter the moon may turn out to involve simulation of the mammalian motivational system.[6] Consequently, there is little point in insisting that the labels "artificial intelligence" and "computer simulation" be used as hard and fast alternatives, and I normally intend the first in a wide sense so as to include the second.

Anyone who is not to regard an account of artificial intelligence as unintelligible gibberish must have at least a minimal sense of what a program is and what a computer can do with it. On the Gryphon's principle of "adventures before explanations," I have tried to describe programs that do interesting things in a way that allows a sense of these matters to develop progressively throughout the book, postponing the more abstract and technical discussions for as long as possible.

Examination of the wider social and philosophical implications of artificial intelligence is also postponed until the later chapters, so that these may be thought about in light of specific examples of the relevant research. We shall see then that, properly understood, artificial intelligence can *counteract* the dehumanizing influences of natural science that have been complained of by critics of the role of "scientism" in urban-industrial cultures. However, people who know nothing of the field are more likely to assume that it supports the crudely mechanistic conceptions of mankind that have been encouraged by scientific success in understanding and controlling the material world. We shall see also that the psychological terms used in describing programs are to be understood analogically, rather than literally, but that their use is justified provided that they are chosen with care by reference to the specific functional properties of the programs concerned.

I begin by describing a relatively simple program that in some respects resembles a neurotic person unable consciously to admit that she hates her father. That is, the program represents a psychological theory of neurosis. My account of the neurotic program is the most detailed description in the book. The lay reader who follows it carefully, step-by-step, will thereby develop a feeling for the way in which a (simple) program works, and especially for the rigor with which it must be expressed. Initial discussion of the many weaknesses of this comparatively stupid program leads to consideration in later chapters of much more intelligent programs, most of which are in some way germane to the question how one might overcome the undeniable crudities of the first example.

The neurotic program is outlined in Chapter 2 and explained and criticized in Chapter 3, and if you are firmly of the Gryphon's cast of mind you may prefer to turn immediately to it, returning to read the rest of this chapter later.

You may, however, share the Mock Turtle's wish to get his basic conceptual bearings as soon as possible, in which case some general introductory remarks would be helpful. In appreciating what a program is and what a computer can do with it, there are three basic points to be grasped. First, only a computer program can tell a computer what to do. Second, the telling has to be done in a language the machine can understand. And third, what a computer does is to combine specific symbols in varying ways by following the instructions expressed by the program.

The rest of this chapter comprises a preliminary explanation of these three points, each of which will become clearer as particular examples are discussed in the main body of the book.

## PROGRAMS TELL COMPUTERS WHAT TO DO

The first point is sometimes expressed by saying that a computer program, together with an appropriate machine, constitutes an "effective procedure."[7] This concept is fundamental to computer science. It denotes a set of rules (the program) unambiguously specifying certain processes, which processes can be carried out by a machine processor built in such a way as to accept these rules as instructions determining its operations.[8] In short, given the appropriate hardware, the program tells the machine what to do and the machine can be relied upon to do it.

But this does not mean that a program designed to answer a certain sort of question guarantees the right answer to it, nor that all programs rely on rigorous and infallible thinking. On the contrary, many programs use inferential procedures which (like human thought) work only *reasonably* well, *reasonably* often. In Chapter 4, for instance, a program is described which shows political prejudice—hardly the most reliable type of judgment. And in Chapter 9, programs are discussed which are subject to visual illusions or hallucinations. Indeed, we shall see that logical rigor and strict consistency may have to be deliberately sacrificed if one wishes to produce a really intelligent program.

Nor can the programmer always foresee every step the program will make. As Chapter 12 will show more fully, some programs are so written that certain decisions are left open, to be taken by the computer itself when the program is running, in light of the particular circumstances. To be sure, the manner in which these decisions will be taken has to be specified *at some level* by the programmer, since only a program can tell a computer what to do. But, as the next section will suggest, there are ways of writing programs which allow for a high degree of flexibility in the overall performance of the machine and that enable the programmer to ignore the most basic details of its thinking.

Accordingly, the common slogan, "A computer can only do what you tell it to do," may be misleading. If it is taken to mean that everything the computer does is done at the behest of the instructions in the program the slogan is, of course, true. But if it is taken to mean either that the programmer can foresee everything the program will do, or that the program will do *all and only* what the programmer intended it to do, then it is false.

In sum: whether or not a program is infallible, and whether or not every detail of its functioning has been explicitly represented or foreseen by the programmer, the program is a series of instructions that effectively specifies the information processing to be carried out by the machine.

## LANGUAGES THE COMPUTER UNDERSTANDS

The second point, that these instructions have to be written in a language that is intelligible to the machine, is a special case of the fact that if you ask someone to do something you had better use words that she understands. What constitutes understanding on the machine's part is the carrying out of appropriate processes, given the instructions.

At base, this is effected automatically, or "blindly," since the correlation between the symbols expressing the basic instructions and the corresponding machine processes is engineered by the machine's designer, and deliberately built into the hardware. But, as we shall see, the machine's understanding of higher level symbols may be a good deal more intelligent. For instance, if it (or you) were told to find the second noun phrase in the sentence "The cat sat on the mat," it (like you) would have to think rather harder than if it had been told, more specifically, to find the words *the mat*. Ideally, one would like to be able to communicate with computers in ordinary English, but—for reasons which will be examined later—this is not yet possible. It is possible, however, to have surprisingly sensible conversations drawing on a small subset of English words or restricted to a particular context: if you doubt this you should glance at the dialogues in Chapters 5 and 6.

For present purposes, the important point is that the computer has to be "taught" English, or any other high level language, by relating it to the language (or languages) it already knows. For any computer, like any person, will at a particular time understand only a limited range of symbols. Using that restricted language one can (as with a person) tell it things that enable it to understand things it could not understand before. For example, if a computer already understands the comparatively low level picture-description concepts of *line, point where lines meet, number*, and *angle of* $x°$, then it can be taught that an ARROW-shaped vertex is a point where three lines meet, with one of the angles bigger than 180°; Chapter 8 will show that the (higher level) concept of *arrow* is crucial for understanding many drawings of solid objects like *cubes* and *wedges* (higher level concepts yet).

Since a computer starts off knowing only its built-in machine language, whose dictates it must execute willy-nilly, it has to be given languages in which it can think about more things, more intelligently. How can this be done?

At the lowest level, which is the least interesting, is the *machine language* or *machine code*. The machine is constructed so that each instruction written in the machine code elicits a unique operation within the mechanism. Because of this close correspondence between code and mechanism, the machine codes of differently designed machines differ from each other, the structure of the code being dependent on the structure of the mechanism involved. For instance, because the basic active components of digital computers are normally in one of two stable states ("on" or "off"), existing machine codes are essentially binary in nature. That is, they use only two symbols—like the "0" and "1" of binary arithmetic. They can symbolize indefinitely many things (as can the 26 letters of the Roman alphabet) because they can be grouped together in indefinitely many ways to form distinct compound symbols.

At the level of the machine language, every single elementary operation that the machine is to carry out must be specified by a separate instruction—much as if one had to formulate a knitting pattern in terms of precisely detailed finger movements. These operations are few in number and simple in type: for instance, copying an item of information in a specified position in the storage register, shifting an item to another specified position, and counting the instructions executed so far. (The comparable operations basic to knitting include passing the wool *over* the needle, passing it *around* the needle, pushing the point of the right-hand needle into the *front* of the first stitch on the left-hand needle, pushing it into the *back* of the first stitch . . . and so on.) Consequently, the complete sequence of instructions, or program, required at this level is very long and cumbrous, even for relatively simple tasks.

This drawback of the machine language can be alleviated if the programmer groups the detailed instructions into *subroutines*: these are blocks of instructions defining a complex operation which may be required for many different problems, or many times in the course of one problem, and each subroutine can be called (activated) or referred to by means of a special code-name. A subroutine may or may not allow for some adaptablitiy to circumstances. This distinction too is paralleled in knitting patterns.

A knitting pattern may contain expressions like "st. st.," "K.," "P.," "K.1, P.1 rib," and "C6F," which respectively denote stocking stitch, a knit stitch, a purl stitch, a specific type of ribbing, and a particular type of cable stitch. The first three of these are straightforward shorthand for unvarying sequences of knitting operations. Thus a *knit* stitch starts with the wool lying behind the needle, and points the needle away from the knitter; a *purl* stitch has the wool lying in front of the needle, and points the needle towards the knitter; and *stocking stitch* is the complex operation of working alternate rows in *knit* (all stitches) and *purl* (all stitches). Knitting abbreviations denoting rib, cable, and so on, are explicitly defined in the initial section of the pattern (which precedes the list of instructions specific to the garment concerned), in terms of individual stitches and movements of needles. Only a few basic terms, such as *knit* and *purl,* are assumed to be already understood and so are not defined in the pattern.

Other knitting terms may involve different values (or *parameters*) according to circumstance. Moreover, the choice of value is sometimes made by the pattern designer, while at other times it is left (either entirely or with some guidance) to the discretion of the knitter as she is making the garment. For example, the definition of cable stitch involves the notions of knit, number (C6F differing from C4F), right needle, left needle, cable needle, forward movement of cable needle, and backward movement of cable needle. All this is spelled out in the "abbreviations" section at the beginning of the pattern. "C6F" will result in a wider cabling effect than "C4F." Since the width of individual cabled bands is something decided by the pattern designer, the knitter is given no choice, but merely interprets "C6F" and "C4F" so as to involve the specified number of stitches. By contrast, the width (size) of the garment itself is something the knitter may wish to control, so as to fit the person who is going to wear it. For this reason, knitting patterns include such instructions as : "Cast on 56 (60, 64, 68)," or "Work in st. st. for 2 (2½, 3, 3½) inches." The knitter is told which of these numbers (i.e. the first, the second, the third, or the fourth) corresponds to different body-sizes, and chooses one of the four options accordingly whenever she comes across this type of instruction in the pattern. Obviously, the use of variables (such as numbers, or terms like "main color" and "contrast color") in knitting patterns allows one to write more flexible designs than if one were to use subroutines containing no variable terms, or parameters.

Knitting patterns contain a number of other conventional devices which correspond to items used in computer programs. For instance, "fancy" knitting involves the repetition of an entire sequence of differing rows. The pattern may say something like this:

**Commence Diamond Pattern:**

** **1st row:** P.2 (1, 2, 1), [k.2, p.2] 2 (3, 3, 4) times, *s1.2F, p.1, k.2 from cable needle, s1.1B, k.2, p.1 from cable needle, p.2, rep. from * to last 8 (11, 12, 15) sts., [k.2, p.2] 1 (2, 2, 3) times, k.2, p.2 (1, 2, 1).

**2nd row:** . . . (*A similarly complex, but different, set of instructions.*)

**3rd row:** . . . (*Different again.*)

**4th row:** As 2nd row.

**5th row:** As 1st row.

**6th row:** . . . (*Different again.*) **

Repeat from ** to **, until work measures 4 (5, 6, 6) inches.

The phrase "Commence Diamond Pattern" is analogous to the *comments* inserted into programs by programmers; comments help the programmer to keep track of what is going on, but do not act as instructions for the machine itself. (We shall see in Chapter 10 that some programs can similarly make use of comments to help them understand what they are up to, and can use this knowledge to learn how to perform better.) "1st row," "2nd row," and so on are comparable to the *labels* attached by the programmer to particular lines of code, by means of which those lines can be quickly found (by program or programmer) from some faraway place in the program. "Repeat from ** to **" exemplifies the *iteration* of a subroutine. And "until work measures 4 (5, 6, 6) inches" implies reliance on a *monitoring routine* which will test the progress made so far at suitable intervals and switch to the next phase of instructions at the appropriate time. This monitoring routine is assumed to be already available to the knitter (though inch-measures are usually helpfully printed on the back of the pattern), but it may need to be freshly written for a computer. An instruction like "Work in st. st. until sleeve measures required length" offers no specific guidance as to what length may be suitable; but it makes clear that length is the relevant factor in deciding when to apply the stop-rule, and it cannot be obeyed unless the knitter can find a way to decide on the proper length.

You might find it amusing, and even illuminating, to read through a fancy pattern, comparing it to the computational concepts you come across in this book or in elementary programming exercises. Very few people doubt that they would be able to understand knitting patterns, abstract and technical though these are. But the same people who are rightly confident that they could master a knitting pattern, if they tried, are often unduly scared by the precision and abstractness of programming. Unfamiliar technical terms such as "subroutine," "monitor," "iteration," "parameter," and the like appear threatening to the uninitiated, whereas homely terms like "knit," "purl," and "rib" do not. The result is an unnecessary prejudice against programming and programs which misleads people into thinking that computational matters are intelligible only to mathematical geniuses.

There is a further analogy between patterns and programs that is relevant here. If one asks *why* pattern designers write "C6F" instead of giving the full cabling instructions every time cabling is required, the obvious answer is—to save space, so that the printed pattern will fit into a handbag and not need a briefcase. A less obvious, and more interesting, answer is that the designer (like the knitter) actually thinks in terms of these abbreviations as she describes the "details" of what she is doing, and could not proceed intelligently if she were not to do so. Certainly, she can laboriously translate into the language of "wool forward" or "wool backward" if she has to (perhaps to undo a mistake she has made, or to explain her skill to a child). But she cannot do this for large sections of the garment, or for the garment as a whole; if she tries to do so, she will be hopelessly lost.

For most people, including computer programmers, a program written in machine language is likewise difficult to cope with at the detailed level, and quite impossible to understand as a whole. One does not naturally think of information-processing procedures in such small-scale units, and it takes a great deal of effort to do so. Moreover, were one to think solely in these terms, one could not intelligibly express "the problem" being tackled by the program. Similarly, a knitting pattern that employed no abbreviations for the various subroutines and that even explicitly distinguished the finger movements involved in knit and purl (as *Teach Yourself Knitting* manuals have to commence by doing), would not give one any idea of the overall nature of the final product. In order to express this a language on a higher level is needed, a language that ignores the constituent details but represents instead the general contours and dimensions of the garment: girl's raglan-sleeved Fair Isle cardigan.

The higher languages used in artificial intelligence include the general purpose "programming languages" (such as LISP, SAIL, FORTRAN, PLANNER, CONNIVER QA4, POP-2, and POPLER), and special purpose languages embodying specialized knowledge relevant only to certain domains. PROGRAMMAR, for example, is restricted to grammatical matters, like finding the noun phrases in sentences, and its primitive units are instructions specifically concerned with parsing English. But many special purpose languages may be thought of as "dialects" of general purpose ones: thus PROGRAMMAR is closely modeled on LISP; and programs that are specifically concerned with vision, or with solving problems in organic chemistry, can have their own specialized vocabulary (connoting lines, points, regions, and vertices, or molecular structures of particular types) useful only to them, but written in a more generally useful language such as LISP.

Different programming languages are differentially appropriate to various problems. Subroutines for computing a square root, for locating a specified item on a list, or for finding an unnamed object that matches a general pattern (such as "the second noun phrase") without having to have a *specific* object (such as "the mat") in mind, are more easily written in one language than in another. Theoretical and practical advance in artificial intelligence depends on the development of increasingly powerful languages in which to write programs or (what comes to the same thing) in which to communicate with the machine.

Since initially the machine understands only machine code, all the programming languages have to be made intelligible to it in terms of this code. Correlatively, before a machine can actually execute a program it reads in a high level language, it has to tell itself what to do in terms of machine code instructions. The conversion of high level to low level language is done by special programs (either *compilers* or *interpreters*) that take as input a program in a high level language and produce as output a program in machine code.

Sometimes the conversion is comparable to a straightforward translation from one natural language to another: thus, a LISP compiler rewrites entire LISP programs in machine code, and effectively throws away the LISP version before the (translated) program is run. By contrast, interpreters keep hold of the original version of the program and execute it step by step; consequently, a LISP interpreter "understands" LISP in a more flexible way than does a LISP compiler, since the precise way in which it will execute a particular step can depend upon details of the immediate context that could not have been foreseen before the program was run. Sometimes the conversion is carried out in more than one stage: PLANNER procedures, for example, are interpreted into LISP, which is compiled as machine code. This has the advantage that the person who writes the conversion program for the new language can capitalize on conversion programs already in existence, so avoiding direct contact with the machine code: one can thus teach a computer a new programming language by relating it to one that the computer can already understand, without having to go back to square one.

The disadvantage of multiple-stage "translations" of this type—and, to some degree, of programming languages in general—is that the further one removes oneself from the machine code, the more likely one is to require of the machine unnecessarily complex operations that do not efficiently exploit its potential. Analogously, a knitter who was incapable of thinking about stitches in terms of finger movements would be unlikely to discover more efficient ways of achieving a certain result, or to invent a new stitch potentially within the repertoire of the human hand. She could, however, both understand conventional knitting patterns and create new ones expressing her own sartorial designs—and this ability is at least as important as the "time-and-motion" sort of skill that conserves the energy of one's fingers.

Those workers in computer science who can confidently cope with the machine code, and who can see the range of logical possibilities inherent in the basic operations available to a particular machine, sometimes bemoan their colleagues' unthinking reliance on high level programming languages. But lesser mortals—including many professional programmers, as well as the writer and readers of this book—may suitably be thankful for the existence of languages that enable one to understand and write programs while ignoring the mechanistic details of what goes on when a program is run.

The "shorthand" nature of programming languages, in which a single instruction corresponds to a complex series in machine code, is in principle useful to all programmers, whether human or not. In other words, a program for writing new programs could advantageously employ high level descriptions of the information processes available to it, and would write the new programs in a programming language rather than the machine code. Equally, a mechanical knitwear designer would save itself time and trouble by manipulating symbols like "rib" and "bobble" in preference to specifications of the equivalent finger movements. No mechanical Coco Chanel exists that can design an elegant knitted suit without human guidance; nor has automatic programming yet achieved comparable results, no complex and interesting programs having been produced by mechanical means. But a program to be described later can work out for itself how to get hold of a particular brick so as to move it into a box, and it does so in terms of the PLANNER vocabulary, wherein complicated procedures like clearing other bricks off the top of the one required can be symbolized directly. Similarly, a program that becomes more skilled by learning from its own mistakes relies crucially on its understanding of highly abstract notions, like "unsatisfied prerequisites" of goal-oriented action. The high level, abbreviational character of programming languages is relevant to powerful information processing in general, not merely to human thought.

Programming languages have a second feature that makes them useful to human beings, but that is of no concern to machines (at least, not yet). They typically employ symbols (such as English words or logical and mathematical signs) that are already familiar to us in richer semantic contexts than the programming context itself. They can thus activate a network of well-known associations on which the programmer can draw when writing (or the reader when reading) programs in these languages. Even though much of this familiar semantic content is strictly *in*effective —for it has not yet been made available to the machine—it may be heuristically or psychologically useful to the programmer working in the high level language. (Similarly, descriptive tags like "cable" and "bobble" help one to knit intelligently, even though they contribute nothing to the actual instructions.) For example, the programming language ALGOL is markedly similar to algebraic notation; not surprisingly, then, it is much easier to write sensible mathematical programs in ALGOL than in the binary notation of the machine language or any "nonalgebraic" programming language.

The early programming language IPL-V (which was practically at the machine code level) did not look like algebra, nor did it allow the importation of words borrowed from ordinary language: if one wished to write a weather-forecasting program in IPL-V one would have to denote "rain," "sun," and "rainbow" by distinct strings of numbers and letters, like B157. Consequently, one could not so readily locate (or write) the instruction telling the machine to forecast rainbows, given sun and rain simultaneously, as one could if one were allowed to write these words themselves into the program. Bearing in mind that the earliest versions of the neurotic program soon to be described were written in IPL-V, it is perhaps surprising that it is not even cruder than it is. Most currently used programming languages do allow for this introduction of familiar words into the program, and so are better suited than IPL-V to expressing the varied semantic and psychological relations that form the subject matter of everyday thought.

It may seem perplexing that a program supposed to represent a neurotic hatred of one's father could be written in the apparently meaningless character strings that comprise IPL-V instructions; nor does it help to be told that later versions of the program were written in a form of the superficially algebraic ALGOL. How can jumbles of numbers and letters, or algebraic formulae, have anything to say about hating one's father, neurotically or otherwise? This question raises the third of the three basic points mentioned earlier: that what a computer does is to manipulate symbols, according to the instructions in the program.

## COMPUTERS MANIPULATE SYMBOLS

It is essential to realize that a computer is not a mere "number cruncher," or supercalculating arithmetic machine, although this is how computers are commonly regarded by people having no familiarity with artificial intelligence. Computers do not crunch numbers; they manipulate symbols.

A symbol is an inherently meaningless cipher that becomes meaningful by having meaning assigned to it by a user, who thereafter interprets it in a particular way. Examples of symbols include road signs, maps, graphs, badges, drawings, spoken and written words in natural languages, hieroglyphics, alphabetic characters in various alphabets, "deaf and dumb" signs, and numerical digits—whether the I, V, X, . . . M of the Roman system, the Arabic 0, 1, 2, . . . 9 of ordinary decimal arithmetic, or the 0 and 1 of the less familiar binary code.

There need be no intrinsic similarity between a symbol and what it symbolizes. Sometimes the symbol is somehow "like" its significate, as the onomatopoeic "buzz" resembles the noise made by bees. And sometimes the likeness between sign and significate is rich and systematic, so that it can be exploited in using the symbolic representation as an aid to intelligent thinking about the thing symbolized. The clearest example of this type is a scale model, but other examples include engineers' blueprints, perspective drawings, and graphs and diagrams of various kinds. These points will be made less cryptically in due course: what is important here is that a symbol *need not* resemble what it represents. Any arbitrary sign may be chosen to symbolize a particular concept or thing.

It follows that the superficial form of a symbol says nothing definite about the nature of what it symbolizes. At most, it suggests matters that could relatively conveniently be represented by it (we have already seen that some programming languages are more readily amenable than others to symbolizing certain things). In particular, a superficially numerical code for expressing information need not be used for expressing numerical information. Just as numbers can be expressed in words, so matters usually expressed in words can be signified by numerals. One can write a symbol for a cat thus: "30120," instead of thus: "CAT." This apparently numerical symbol (which happens to be derived by reference to the order of the 26 letters in the Roman alphabet, but which could equally have been generated by some other method) has no numerical meaning in this instance, since by hypothesis it is interpreted to mean a furry, purry animal rather than the number thirty thousand, one hundred twenty.

The machine codes of computers, and even many programming languages, are superficially numerical in form. There are a number of reasons for this. The binary, on-off, nature of digital computers has already been remarked: although the machine code could be written as strings of "*" and "?," or of pink and purple dots, it is conventionally written as strings of "0" and "1," which unavoidably suggest the mathematical context of binary arithmetic. Second, as a matter of historical fact, programming techniques were originally developed with mathematical problems in mind, partly for pragmatic reasons (businessmen and missile scientists wanted their sums done for them) and partly because a binary code is obviously very convenient for doing test problems in binary arithmetic.

Third, the computer's relative ease in handling numbers makes it convenient to label specific locations in the memory store by numerical addresses, and these addresses feature in the most basic information processing directed by "nonnumerical" programming languages. For instance, if a programming language (such as LISP) allows one to write the word "CAT," the computer itself (by way of the LISP compiler) will silently assign a particular number to the place where "CAT" is stored, and this number will mediate its search for "CAT" in its dictionary ever afterward. The particular number assigned may be systematically generated or it may be chosen randomly.[9] For present purposes, the point is that the machine codes of computers, and also some programming languages, involve apparently numerical data and operations, in the sense that they are written in an apparently mathematical symbolism.

But we have already seen that superficially numerical symbols may be given a purely nonnumerical interpretation. It follows that computers can crunch numbers if specifically programmed so to do, but this is not their essential computational function. Digital computers, originally developed with mathematical problems in mind, are in fact general purpose symbol manipulating machines. It is up to the programmer to decide what interpretations can sensibly (that is, consistently) be put on the inherently meaningless ciphers of machine and programming languages. Devotees of James Bond should not be surprised to find that strings of numbers can be used to express instructions having nothing to do with arithmetic.

The terms "computer" and "computation" are themselves unfortunate, in view of their misleading arithmetical connotations. The definition of artificial intelligence previously cited—"the study of intelligence as computation"—does not imply that intelligence is really counting. Intelligence may be defined as the ability creatively to manipulate symbols, or process information, given the requirements of the task in hand. If the task is mathematical, then numerical information may need to be processed. But if the task is nonnumerical (or "semantic") in nature—for example, forecasting what a particular politician would say if asked to comment on American foreign policy, or planning how to build a steeple out of toy bricks when asked to do so—then the information that is coded and processed must be semantic information, irrespective of the superficial form of the symbols used in the information code. The force of the term "computation" in the quoted definition does not imply specifically numerical operations, but rather indicates the effective procedures by means of which the intelligent achievement can be generated.

In sum, artificial intelligence is the use of programs as tools in the study of intelligent processes, tools that help in the discovery of the thinking-procedures and epistemological structures employed by intelligent creatures. Even a preliminary understanding of what a program is, and what a computer can do with it, should suggest why programs are so useful for expressing theories about the representation and use of knowledge. Whether or not these theories are adequate to explain the intelligence of human beings—or chimps, or frogs—they help one to think clearly about what must be involved in intelligent behavior. It follows that artificial intelligence could not be dismissed as irrelevant to human psychology even if (as is sometimes claimed in a priori attacks on this type of research[10]) there were insuperable difficulties preventing truly comprehensive simulation of human abilities.

I shall say no more in defense of artificial intelligence now, for it would mean little or nothing to most readers: no one can discuss its general relevance sensibly before having learnt something about it. Not until much later, therefore, shall I discuss its wider philosophical implications—including the common fear that it is not merely irrelevant but actually inimical to human interests, and the common suspicion that "artificial intelligence" is a contradiction in terms. If the reader wishes meanwhile to place scare-quotes around the psychological vocabulary I use in describing programs, my purpose will not be affected.

Let us begin by assessing an early attempt to produce a programmed neurotic—or, if you prefer, a programmed "neurotic."

# *Part* II

# THE PERSONAL DIMENSION

# 2

# *Artificial Neurosis*

THE NEUROTIC PROGRAM I shall outline in this chapter suffers from repressed emotions like hatred for one's father, fantasies like belief in one's royal descent, twinges of conscience on disobeying a moral rule, and anxiety-ridden attitudes buttressed by strings of rationalizations that may bear very little relation to the truth.

Or, rather, it embodies theories representing clumsy approximations to these psychological phenomena. The program is based on Freud's theory, but to understand how it works you do not need to agree with (or even know much about) Freudian psychology. Whether or not one chooses to accept Freud's view in preference to alternative theories of neurosis, the importance of the program for present purposes is that it shows how one may attempt to express a psychological theory in computational form, and in what ways one may fail to capture the subtlety apparently possessed by the verbal version.

This theoretical model of neurosis was programmed relatively early, and is much simpler (and stupider) than most of the programs described in the rest of the book. Its author now acknowledges it to have been overambitious to a degree that was not fully realized at the time. You will not be amazed, therefore, to find that this artificial system parallels actual neurosis (and Freud's views about it) only to a very limited extent. What is perhaps more surprising is the way in which it can nonetheless help one to ask interesting questions about the psychological phenomena concerned, questions that would not have been posed in so clear a form (or, in some cases, perhaps not posed at all) if someone had not jumped in at the deep end and tried to produce a computer analogue of this aspect of the human mind.

In Chapter 3 I shall explain in detail just how the program does what it does, and shall also ask what it does not do that the human neurotic does. It will turn out that considering how this program might be radically improved raises deep questions concerning the inner structure of intelligent processes (artificial or not), questions that are relevant to the more complex—and more clever—programs examined later. By the time you have finished reading about those you will be in a better position to see what is wrong with the neurotic program, and what artificial intelligence workers would need to be able to do in order to strengthen it.

As a beginning, let us look at the broad outlines of what this program does, postponing detail and criticism to the following chapter.

## OUTLINE DESCRIPTION OF A NEUROTIC PROGRAM

K. M. Colby's "Simulation of a Neurotic Process," which is described in a series of papers from 1962 on,[1] represents a woman undergoing psychoanalysis, who believes that her father abandoned her but is unable consciously to accept that she hates him.

Her unconscious hatred may occasionally be entirely repressed, but more often appears in a less threatening form: perhaps as love for, judicious appreciation of, indifference to, or (at worst) rational antipathy to, him. These comparatively conflict-free sentiments are expressed by her saying such things, respectively, as "I love father," "I see faults in father," "I couldn't care less about father," or even "Yes, I dislike father—because he abandoned me long ago." At times, her animosity is deflected onto some innocent third party, so that she declares "I hate Ms. Smith." Sometimes her hatred of her father is projected onto him, as when she claims "Father hates me"; and, of course, his unfatherly behavior in abandoning her is cited in support of this claim.

So, to the extent that she recognizes her negative feelings toward him, she rationalizes them in terms of her father's betrayal. If challenged about the real or imagined nature of the betrayal, she gives reasons for her belief, such as "Mother says father abandoned me." At the same time she attempts to deny her hatred by giving reasons that count against it, perhaps pointing out that *she* would never abandon *him*. If her therapist suggests to her that she hates her father, she perceives this remark as threatening and resists it: that is, she not only answers "No!" to her analyst, but is unable to admit the aptness of the interpretation in her own conscious mind. At best, she accepts it "intellectually" rather than emotionally, and so derives no real benefit from it. As we shall see, each of these defensive (anxiety-reducing) maneuvers is represented in Colby's program.

Colby himself was for many years a practicing psychoanalyst, and the original impetus of his program was his desire to produce a clear theoretical model of the process of free association in psychotherapy. The earliest ("output") version of the program simulates the patient's spontaneous free association or introspection. Later ("input-output") versions simulate also the patient's response to the conversational and interpretative remarks of the analyst. Each sentence expressed (printed out) by the program should be seen as analogous perhaps to a single statement on the patient's part, but more likely to a whole series of remarks associated with a particular topic, or theme. Here is an example of the print-out generated by the input-output version:[2]

PROGRAM: FATHER IS INDIFFERENT TO ME.
MOTHER SAYS FATHER ABANDONED US.
FATHER PREFERRED SISTER.
I AVOID FATHER.
SISTER ADMIRED FATHER.
I FEEL TENSE NOW.

THERAPIST: YOU ADMIRE FATHER.

PROGRAM: NO.
I DO NOT ADMIRE FATHER.
I ADMIRE AL.
MOTHER OUGHT NOT TO ADMIRE AL.
MOTHER DETESTS FATHER.

Colby hopes that future developments may be of use in studying the (currently obscure) ways in which belief systems and defensive habits of thought can be modified by particular conceptual inputs—in other words, how a neurosis can be ameliorated or cured by some of the analyst's interventions and left unaffected or aggravated by others. Simulation of an individual person undergoing analysis, by feeding in her idiosyncratic thematic beliefs as expressed over hundreds of hours of therapy, might allow for preliminary "testing" of therapeutic strategies: if a certain input precipitates a crisis in the simulation, this may provide a warning practically appropriate to the specific person-to-person context. Colby also suggests that the program might eventually be used for initial training purposes: getting neophyte analysts to try out their word magic on a computer would have some distinct advantages over current practice.

Above all, Colby believes that his programming efforts, whether or not they ever achieve the sorts of therapeutic usefulness just listed, can considerably clarify the implicit assumptions and explicit theories that analysts and other psychologists bring to bear on the mental phenomena concerned.

The program is provided by Colby with a set of sentences representing some of the beliefs of a woman who was one of his patients in long-term analysis. As you might expect, *Stanford is in California* is not one of them, even though this doubtless was one of her beliefs. Examples of her more relevant beliefs that were included are *Father abandoned me, I am defective, I descend from royalty, I must not marry a poor man,* and *Mother is helpless.* (What makes a belief "relevant" to a neurotic hatred of one's father?)

Each belief represents a statement held to be true by the person concerned. But since the statements include examples like *I hate father* and *I love mother,* the term "beliefs" is here being used in a wide sense to include not merely "factual" beliefs (that I descend from royalty, for instance), but also what would more commonly be called "sentiments" or "emotions." And even the factual beliefs (true or false) included within the system have an emotional aspect to them, since they are considered to be emotionally significant to the person (in Freud's terms: *cathected*), rather than being coldly assented to as mere matters of fact (like Stanford's being in California).

Each belief accordingly has a number, or "charge," associated with it that reflects the degree of emotional importance it has within the mind in question. These charges vary during the running of the program in a way corresponding to what, according to psychoanalytic theory, goes on in the person's mind. In addition, there are five numbers ("monitors") representing the emotional states of anxiety, excitation, pleasure, self-esteem, and well-being. These also fluctuate according to circumstances. The emotional monitors make a difference to the thinking that goes on, for they influence the fate of individual beliefs and help to select the particular defense mechanism employed when the system experiences psychological conflict. Or, rather, they come into play when specifically *neurotic* conflict is experienced. (What is the difference between neurotic and other forms of psychological conflict?)

Since the self-activating version of the program represents free association rather than response (whether in thought or speech) to the analyst's remarks, there is no interaction between this version and the programmer: once it has started to run, the program continues "spontaneously" until it is stopped.

Like a person free-associating, the program attempts to express its beliefs. That is, it tries to print out the sentences stored in it: but, again like a neurotic person, it is not always successful. Beliefs that are too threatening cannot be directly expressed. Whether a belief is threatening or not depends partly on its own emotional charge (which can vary, as we shall see: but items like *I hate father,* for instance, always carry a fairly high negative charge), and partly on the overall emotional state of the system at the time; consequently, a belief like *I dislike father,* which in itself is mildly threatening, will be expressible if the overall well-being is high but not if the system has already reached a high level of anxiety. Similarly, a person will sometimes be able consciously to recognize her negative feelings, but sometimes she will not.

The program's first step is to choose a psychologically significant belief around which to form a "complex," or "pool," of beliefs to work with. The core belief is always a personal imperative or a moral injunction, like *I must not marry a poor man,* or *I must love people,* because the imperative *must* is assumed by Colby, as by Freud, to underlie all neurotic complexes. To form the pool of relevant beliefs, the program collects together all the topic-related sentences in a way to be explained later. Processing (thinking) now continues with respect to this pool alone; with respect to the sentences remaining latent in the background belief system, the program has a temporarily closed mind.

Next, a belief is chosen at random from the pool; this is called the "regnant" belief. The regnant is then tested against every other belief in the pool to see if it conflicts with any of them. For instance, *I hate father* conflicts with *I must love father.* (The criteria of "conflict" will be detailed in Chapter 3.) If no conflicting belief is found, or if the conflict involved is very slight, the regnant belief is expressed: that is, it is printed out. To prevent its being chosen again, and also to keep the overall number of beliefs in the pool constant, another belief (which is relevant to the one just expressed) is chosen from the background store and placed in the pool in its stead. Another random choice is then made, and the process is repeated.

But if a conflicting belief is found—and particularly if this belief is itself of high affective charge, as "superego" beliefs involving *must* or *ought* all are—a defensive routine is chosen from the set available to make the regnant less threatening.

The defense mechanisms in the program are symbol-manipulating routines that can change the beliefs, or their interrelations, in different ways. Thus there are eight "transforms" that change the meaning of an old belief in forming a new, distorted, derivative; and there are routines of isolation, denial, and rationalization that operate on interrelations between beliefs rather than beliefs taken singly. Distorted beliefs produced by a defense mechanism always have an emotional charge that is less than the charge on the original regnant—*how much* less depends on the particular mechanism, as we shall see. Consequently, the distorted belief may be expressible even though the original was not.

Expression of a belief (distorted or not) reduces its charge to zero, but if it is a distorted belief that is expressed, the original (repressed) belief remains in the system with some residual emotional charge and so can be chosen as a regnant again and again. New (distorted) beliefs are added to the memory, so that what the system believes (as well as what it actually expresses) changes over time. Moreover, new sources of conflict can arise, because the distorted belief is tested for conflict *only* against the current pool: the short-sighted neurotic policy of adjusting one's beliefs to avoid the current difficulty can lead to trouble if a newly accepted notion is inconsistent with beliefs temporarily latent in the mind.

## DEFENSE MECHANISMS THAT DISTORT BELIEFS

Colby's list of transforms, which change the meaning of an anxiety-ridden belief, is as follows:[3]

(1) DEFLECTION: Shift Object (Not Self)
(2) SUBSTITUTION: Cascade Verb
(3) DISPLACEMENT: Combine (1) and (2)
(4) NEUTRALIZATION: Neutralize Verb
(5) REVERSAL: Reverse Verb
(6) NEGATION: Insert *Not* Before Verb and Do (5)
(7) REFLECTION: Shift Object to Self
(8) PROJECTION: Switch Subject (Self) and Object (Not Self)

For the moment, these can best be understood by way of examples; *how* they work, given the way in which beliefs and concepts are represented in the data structure of the program, will be clarified in Chapter 3.

Let us suppose that the regnant belief randomly selected as a candidate for expression is *I hate father.* Since this conflicts with superego imperatives in the pool, it is too threatening to be directly expressed. Transform (1) would change this sentence to *I hate the boss,* or *I hate Ms. Smith.* Transform (2) relies on a dictionary of weak to strong "synonyms" of verbs, and in this case might give *I see faults in father.* Transform (3) could give *I see faults in Ms. Smith* as the final distorted version of the regnant. Both (1) and (3) would qualify as "displacement" in the Freudian sense. Transform (4) may give *I couldn't care less about father*; (5) gives *I love father*; and (6) gives *I do not love father.* These three transforms are analogous to what Freud called "reaction formation" of varying strengths. The use of (7) gives *I hate self,* and echoes introjection, or turning against the self. Finally, (8) gives *Father hates me,* a classic case of simple projection.

These being the alternative distorting transforms available to the program (ignoring, for the moment, RATIONALIZATION, ISOLATION, and DENIAL), how does it decide which transformation to express? Indeed, how does it decide to express any derivative at all, rather than entirely repressing the matter? And in the case of those transforms whose operation allows some latitude of choice, namely, the first four, how does the program select one application of the transform rather than another?

A trivial way of answering these questions would be to program a random operator (like that which selects beliefs from pools for possible expression) to activate and operate defense mechanisms. However, this would have little psychological interest or plausibility. For it is at least possible, perhaps even probable, that different techniques achieve defensive purposes to different degrees in the same or different situations. Crucial theoretical questions necessarily arise, then, as to the relative power of the various mechanisms in reducing anxiety and in effecting discharge of the underlying instinctual impulse that the person cannot consciously admit.

Consider transform (1), which Colby terms "DEFLECTION." The definition of this transform stipulates that the syntactic (and psychological) object of the belief sentence is to be altered. The new object must be chosen from a list of objects represented in the dictionary provided to the program. In Colby's simulation, only objects classed as persons by the dictionary are allowed as object substitutes in operating the transforms, because persons are the most commonly occurring objects of neurotic beliefs encountered in analysis. Let us suppose that there are only five persons listed in the dictionary: *father, mother, self, boss, Ms. Smith.* The new object cannot be *self,* since there is a universal constraint within the definition of transform (1) forbidding this; shifting to the self would in fact be a case of transform (7), REFLECTION. Shifting to *mother* would conflict with the highly charged *I must love mother,* and would improve the situation not at all. Shifting to *boss* or to *Ms. Smith* are both prima facie admissible.

If one considers a human being who unconsciously hates her father, it seems theoretically and phenomenologically plausible that greater satisfaction of the underlying impulse would be achieved by her consciously hating a male, dictatorial, high status authority figure than by her hating a gentle and obscure female. But the high anxiety level of boss-hate (based in these very attributes of the boss) may be too close for comfort to the even higher anxiety level of father-hate, so that the inoffensive Ms. Smith may be hated instead. In programming terms, if one wishes to avoid a random search through the list of persons, necessitating continual "affective experimentation" of ensuing anxiety levels, then specific constraints (corresponding to the theoretical point concerned) must be built into the program.

One might decide, for instance, to limit the search for a substitute to those persons having attributes (such as gender, age, status, personal habits, but *not* height or hair color) in common with the original object of the belief, with the caveat that not more than four common attributes are allowed. Colby's program (or, rather, Colby himself) solves this problem by use of a subroutine called FINDANALOG.[4]

This procedure computes the number (ignoring the nature) of properties two nouns have in common by referring to the data base. The data base, as we shall see, is an information matrix that codes (among other things) superordinate and subordinate categories for nouns; it provides the information, for example, that *woman* and *man* are each instances of the class *person,* and that *Ms. Smith* is an instance of *woman.* These attributes of nouns are also included as data for the program: thus *men* in general are viewed as *hostile,* whereas *Ms. Smith* is not. The strength of analogy required between the original and the substitute object is represented by a number specifying how many properties they should share. This number can be varied according to the specific psychological nature of the conflict situation. Thus FINDANALOG-2 (which tells FINDANALOG to look for *two* shared attributes) might be content with deflection to *Ms. Smith* in the previous example, whereas FINDANALOG-5 would not.

If no appropriate object can be found the program is still stuck with the inadmissible *I hate father,* so it then turns to consider a different transform entirely. But which one should it pick?

The nature of the transform selected depends on the degree of danger (anxiety) signaled by the DANGER monitor at the time, and also on whether or not SELF-ESTEEM is involved. A procedure called PICKTRANSFORM (whose functioning will be more fully described in Chapter 3) takes these monitors into account in deciding which transform to activate.

PICKTRANSFORM takes into account also the general psychological effectiveness of the different transforms. The example of hating the boss (as a deflective defense against hating one's father) shows that a defensive operation that is highly efficient in discharging the original instinctual impulse may be less satisfactory in reducing anxiety.

Colby commits himself (and therefore PICKTRANSFORM) to the theoretical position that this is generally true, for he describes the eightfold list of defensive transforms as rank-ordered (by way of their numerical labels (1) to (8)) from low to high according to their effectiveness in reducing anxiety, and from high to low in respect of their effectiveness in providing discharge of the charge on each belief. For example, (8) PROJECTION involves such gross distortion of meaning that it is highly efficient in reducing anxiety levels, but poor in reducing charges. (1) DEFLECTION, on the other hand, does little to reduce anxiety but is very satisfying in relieving instinctual tensions. In general, says Colby, "the need for discharge through expression drives the program down the list of transforms [that is: to one with a *lower* number as its label], while the need to reduce anxiety drives the program up [to one with a *higher* numerical label]. The sequence of derivatives created as output will represent a compromise formation between these processes."[5]

PICKTRANSFORM also keeps tabs on the actual success (in producing expressible beliefs) of the defensive measures used on particular occasions. Successful operation of a transform results in a greater probability of its being used on future occasions. Thus a preferred style of defense develops over a series of runs, depending on the idiosyncratic nature and processing history of the belief system currently being simulated. People, too, tend to use a preferred defensive style that has served them well in the past, as has been shown by one of the few elegant experiments on Freud's theory of defense.[6]

Distorted versions of beliefs always carry a lower charge than their original, so that expression of them may be possible. The more distorted the derivative, the lower its percentage of the original charge. Degree of distortion depends on which transform was responsible for producing the delusional derivative, DEFLECTION being the least and PROJECTION the most distortive. If the maximal distortion available fails to lower the charge below the anxiety threshold—something that can happen only when the charge on the regnant is exceptionally high—no output whatever is produced, and the program evasively changes the subject by (randomly) choosing a fresh regnant from the pool and trying to express that. In the latter case, the original regnant, having been totally repressed, remains latent in the pool with a continuing high charge that results in its being chosen again and again for processing—a computerized version of the neurotic's "repetition without remembering."

If the conflict is excessive (as measured by the WELL-BEING monitor), the whole pool is repressed, and a new pool is formed around a different core belief. Simultaneously, the charges of all the beliefs in the system are slightly increased (proportionate to the charge they are carrying already), so that the overall tension rises. Naturally, this leads the neurotic system into more and more difficulties: since the (unconsciously) experienced conflict is higher, there is more chance that the second pool may, in its turn, be repressed—which can only make matters worse for the next. In consequence, the beliefs may become progressively more distorted as the neurosis develops, without thereby effectively discharging the forbidden instinctual impulses that are causing the conflict. In the *output only mode* of the program there is no relief for this situation; in the *input mode*, as we shall see, carefully chosen remarks input by the therapist may trigger off ameliorative thinking in the model.

The varying affective tensions are (numerically) expressed from time to time in the printout: this parallels the nonverbal cues, such as tone of voice or facial and bodily movements, observed in the clinical situation and interpreted by the analyst as indicators of the person's current emotional state.

## LATER IMPROVEMENTS

The later versions of Colby's simulation of neurosis have some extra subtleties. For instance, they can incorporate compound beliefs whose subject or object is a belief, like *That mother married a poor man is not glamorous* and *I disapprove the fact that father does not respect women.* Also, they can attach different degrees of credibility and subjective certainty to individual beliefs, which affects the anxiety aroused when the belief is in conflict: beliefs that are only tentatively assented to in the first place are more readily repressed.

The differential credibility of beliefs is connected also with new "self-corrective" ways of *con*tending (rather than *de*fending) by facing conflicts and working on them in various ways. For example, a troublesome belief can be deliberately weakened in influence (through lowering its credibility) by finding reasons, and reasons for reasons, for disbelieving it: thus *I hate father* might be rendered less threatening by *I will never abandon father.* Even though the regnant belief is retained, it is simultaneously "denied" to such an extent that it may cause little psychological disturbance. This complex type of denial is more flexible, and potentially more stable, than those that rely on simple negation or on total repression of anxiety-ridden beliefs. Conversely, the program can find reasons, and reasons for reasons, for believing something, so that *Father abandoned me* is supported by *Mother says father abandoned me,* and *Father hates me* and *I dislike father* are each rationalized by *Father abandoned me.*

In programming these extra features into the mind of his mechanical neurotic, Colby incorporated some of the ideas of the social psychologist R. P. Abelson, whose work will be described in Chapter 4. In the early 1960s Abelson produced a simulation of affectively influenced thought, modeling the ways in which emotionally charged attitudes may be highly resistant to change, even when confronted with prima facie conflicting evidence. Abelson's work derives from the "cognitive balance" theorists such as Fritz Heider rather than Freud, but his early program involved (in addition to credibility measures) a type of rationalization and of denial that are commonly dubbed "Freudian" or "neurotic," particularly when they occur unconsciously.

This theoretical similarity and interchange between Colby's program and others should come as no surprise to those who take seriously Freud's remark, "Psychoanalysis is not a specialized branch of medicine. . . . Psychoanalysis is a part of psychology; not of medical psychology in the old sense, not of the psychology of morbid processes, but simply of psychology."[7] Indeed, although the defense mechanisms are often thought of as characteristically neurotic, it follows from Freud's definition of them as "the techniques which the ego makes use of in conflicts that may lead to neurosis"[8] that mechanisms of defense are employed by "normal" as well as "neurotic" personalities, the former being the more successful in controlling conflict by means of them.

In short—as Gertrude Stein might have said, but doubtless didn't—a belief system is a belief system, is a belief system. The processes involved in cognitive organization and transformation are unlikely to be totally different in cases of psychopathology and normal psychology, and a number of interesting questions arise if one asks how the same general principles of mental structure (the same basic program) should be differentially "set" or adjusted so as to result in different forms of thought. Abelson's most recent work illustrates this point, as we shall see. In the present context, what is important is the general likelihood that insights gained in connection with one program may be relevant to another, even when the main theoretical impetus derives from sources as different as Heider and Freud (or, for that matter, from language and vision).

Colby's neurotic program, in its later versions, is not limited to the self-activated (free-association) mode, for as well as producing output it is able to respond to input. It thus gives a closer analogy to the actual therapeutic situation. The sentence functioning as input is not added as such to the belief system. (Would you instantly accept anything your analyst said to you?) Rather, the system is searched for a topic-relevant belief, which is examined for agreement or disagreement with the input sentence and is also used as the core of a pool of beliefs (as described above).

Consequently, the effect of the input sentence is selectively to activate specific belief structures already present in the model, even though it is not itself accepted into the system of beliefs. Similarly, a remark about policemen's pay, or the duty owed to one's parents, will have very different effects on the inner thoughts of a "neutral" listener and one with a neurotic preoccupation with authority figures, whether the said remark is overhead in the café or endured in the consulting room. (A paranoid person may react to the remark in yet another way, if policemen figure in the psychotic delusion: we shall see in Chapter 5, when we shall discuss Colby's more recent work, that this applies to paranoid programs also.)

Whereas remarks about filial duty—such as "People ought to respect their father"—are likely to raise the inner tensions of a person neurotically hating her father, others may prove more soothing, even if they are not assimilated as such into the mind. For example, the type of denial described above could be triggered off by the therapist's asking "Would you abandon your father?" This would elicit the relevant belief *I will never abandon father*, which would lead to a lessening of conflict because it would be recognized as a reason against believing *I hate father*, and so would decrease that statement's credibility.

If an analyst boldly suggests that a person actually hates her father, she will reply "Yes," "No," or "Maybe." If the associated conflict has been sufficiently reduced (perhaps by bearing in mind how badly her father behaved in abandoning her, and how unjustly he despises all women), the answer "Yes" may constitute a genuine recognition on her part of the aptness of the interpretation. But "Yes" may alternatively represent her merely "intellectual" acceptance of it, in functional isolation from all other relevant beliefs: Colby's program has an ISOLATION routine that cuts down the extent to which a belief can be compared to others bearing on it. The reply "No" can also differ in significance, for it may or may not derive from defensive resistance on the person's part, and so may or may not be reliable.

Colby's program in the *input-output mode* similarly replies *Yes*, *No*, or *Maybe* in the course of its thinking about the analyst's remarks. And a *No* cannot always be taken at its face value. The answer *No* may be defensively produced even though the previous processing found that *I hate father* is indeed assented to at some level of the program. It is crucial to realize that this *No* does not represent a mere face-saving lie to avoid the embarrassment of admitting one's faults to the analyst. Rather, it represents a failure at the conscious level of the mind to recognize information concurrently taken into account at unconscious levels. As Colby himself puts it, "If the program attempts to interrogate itself about its own information, it cannot express directly some of its most highly charged beliefs and it receives as answers distorted derivatives of those beliefs."[9] Defensive thought processes, in short, lead to loss of information and misrepresentation of beliefs at certain points within the system itself.

The important everyday distinction between a lie and a neurotic inability to admit unpleasant truths to oneself is not actually made within Colby's program (which always expresses any belief that is not defensively repressed), raising the general question of how adequate Colby's program is as a theoretical model of the human mind. Complaining about the program's inability to lie would be as irrelevant as complaining that it cannot dance the tango. The program is intended as a theoretical model of neurosis, not as a theory of every aspect of human psychology. (This is not to say that no program could lie, for one reason or another; what more do you think Colby's program would need to be able to do, in order to have the ability to lie?)

The extent to which Colby's program succeeds in capturing neurotic thought processes can best be assessed by first inquiring just *how* it does what it does. Many of its inadequacies can be traced to specific features of the way in which its knowledge is represented in its data structure and used by its subroutines. We shall find that it could in principle be radically improved by providing it with ways of representing and using knowledge that are required for intelligent processes in general, and that are therefore relevant to programs concerned with, for example, language, vision, and learning as well as the more obviously "personal" issues discussed by Colby and Abelson.

# 3

# *Function and Failure in the Neurotic Program*

TO ASK how intelligent K. M. Colby's program is, is to ask what it knows and what it does with what it knows: what actually goes on when it selects a core belief, forms a pool, chooses a regnant, tries to express it, runs into conflict, and defensively transforms it into or associates it with another belief?

Knowledge (or belief: for present purposes the distinction may be ignored) is embodied in a program in two ways: in the memory, or data base, and in the procedures that operate on or by reference to the data. One of the strengths of artificial intelligence as a way of thinking about thinking is that it forces one to consider the dynamic aspects of intelligence. A functioning program is a theory that is intended more as a movie of the mind than as a portrait of it, and the programmer must specify precisely how successive frames are brought into being. Many verbal theories, like those of Freud himself, likewise are attempts to model the movement of thought. But it is all too easy, when theorizing in verbal terms, to imagine that one has made matters explicit that in fact one has not. Since only a program can tell a computer what to do, the programmer's largely intuitive psychological theory must be expressed in computational terms. It is not enough to say that a certain progression of thoughts can happen: the program must represent *how* it can happen.

The specific way in which the subroutines that "do the thinking" use the information in the data structure depends upon the way in which the information is represented there. Different programming languages are suited to different ways of storing information and doing things with it, much as different traps catch different animals in different ways.

For instance, the earliest versions of Colby's program were written in IPL-V, a "list-processing" language that expresses data in the form of lists (and hierarchically structured lists of lists . . .). The program's thinking operations were then IPL-V subroutines, that is, ways of searching through, comparing, adding symbols to, and deleting symbols from, lists. By contrast, the later versions (which I shall describe in this chapter) are written in a form of ALGOL, a language in which data are stored as "arrays" made up of rows and columns. Procedures in ALGOL locate and compare symbols within arrays (whether entire rows and columns or the intersection of a given row and column), and form new combinations of symbols by reference to the information coded in the various arrays stored in the data base.

It is not necessary to wrestle with ALGOL, which would arouse possibly painful memories of school algebra, in order to understand how Colby's program functions. The important point for our purposes is that, largely because of the thinking (and remembering) facilities provided by this programming language, Colby chose to represent his patient's hatred of her father in a manner that can be visualized as an interrelated system of two-dimensional tables, or arrays.

If you have never done any programming, you might take pencil and paper and try to mimic the functioning of Colby's program as I describe it in the next two sections. This should help to give you a sense of how a program works, and in particular of the clarity with which its thought processes must be expressed. None of the other programs I discuss will be described so fully as this one, so the chore need not be repeated.

Please bear in mind, while reading this chapter, that the neurotic program is markedly less intelligent than most of these other programs. You should not let its stupidity put you off artificial intelligence before you have considered them. Rather, you should ask yourself *in what ways* it is stupid, and *what sorts* of thought processes would need to be provided to make it less so. These questions are the topic of the third section.

## DATA USED BY THE NEUROTIC PROGRAM

The knowledge that is explicitly represented in Colby's program is stored in four arrays: the Belief Matrix, the Dictionary, the Substitute Matrix, and the Reason Matrix. These contain cross-references to each other, so

FIGURE 3.1

*Belief Matrix of Neurotic Program*

| 1<br>Row No. | 2<br>Subj. Modi-fier | 3<br>Sub-ject | 4<br>Verb Modi-fier | 5<br>Tense* | 6<br>Verb | 7<br>Obj. Modi-fier | 8<br>Ob-ject | 9<br>Imp. Level | 10<br>Com-plex-ity | 11<br>Fixed Chge. | 12<br>Resid-ual Chge. | 13<br>Credi-bility | 14<br>Row in Reason Matrix |
|---|---|---|---|---|---|---|---|---|---|---|---|---|---|
| 1. | 0 | 1 | 0 | 0 | 2 | 0 | 1 | −2 | 0 | 7000 | 6000 | 6000 | 0 |
| 2. | 3 | 4 | 0 | 0 | 5 | 0 | 6 | 1 | 0 | 4000 | 2000 | 4000 | 0 |
| 3. | 0 | 7 | 0 | 0 | 8 | 0 | 9 | +3 | 0 | 7000 | 8000 | 6000 | 0 |
| 4. | 0 | 7 | 0 | 0 | 10 | 3 | 4 | −3 | 0 | 6000 | 2000 | 3000 | 1 |
| 5. | 0 | 7 | 0 | 0 | 11 | 0 | 9 | 1 | 0 | 9000 | 7000 | 5000 | 2 |
| 6. | 0 | 7 | 0 | 0 | 8 | 0 | 1 | +3 | 0 | 5000 | 7000 | 7000 | 0 |
| 7. | 0 | 9 | 17 | 0 | 12 | 0 | 13 | 1 | 0 | 6000 | 7000 | 7000 | 0 |
| 8. | 0 | 7 | 18 | 2 | 2 | 0 | 9 | 1 | 0 | 6000 | 7000 | 7000 | 0 |
| 9. | 0 | 7 | 0 | 0 | 14 | 0 | 7 | 1 | 2 | 7000 | 7000 | 6000 | 0 |
| 10. | 0 | 9 | 0 | 1 | 2 | 0 | 7 | 1 | 0 | 8000 | 9000 | 8000 | 3 |
| 11. | 0 | 15 | 0 | 0 | 16 | 0 | 10 | 1 | 2 | 4000 | 6000 | 6000 | 0 |
| 12. | 0 | 7 | 0 | 0 | 8 | 0 | 15 | +3 | 0 | 7000 | 7000 | 7000 | 0 |
| 13. | 0 | 9 | 0 | 0 | 8 | 23 | 26 | 1 | 0 | 3000 | 3000 | 7000 | 0 |
| 14. | 0 | 7 | 0 | 0 | 11 | 0 | 27 | 1 | 0 | 4000 | 2000 | 6000 | 0 |

*0 = present; 1 = past; 2 = future

that searching them can be done (by program or person) relatively quickly. Hypothetical examples constructed to include most of the sentences used in my previous description of the program are shown in Figures 3.1 to 3.4. (I have omitted a "Balance" column from the Belief Matrix: this represents the evaluational balance of the belief, a topic forming the focus of R. P. Abelson's earliest program, to be described in the next chapter.)

Apart from the list of words in the second column of the Dictionary, the information in these matrices is represented entirely as numbers. This is convenient for the computer, which can find and locate numbers with relative ease. It is less convenient for you and me—and Colby—since we cannot see at a glance (as we could if the beliefs were expressed in ordinary English) that the beliefs contained in the Belief Matrix of Figure 3.1 are in fact as follows:

1. People ought not to abandon people.
2. Poor men are unreliable.
3. I must love father.
4. I must not marry a poor man.
5. I hate father.
6. I must love people.
7. Father does not respect women.
8. I will never abandon father.
9. I disapprove the fact that father does not respect women.
10. Father abandoned me.
11. Mother says father abandoned me.
12. I must love mother.
13. Father loves helpless animals.
14. I hate atheists.

Figure 3.5 gives a verbal version of the Belief Matrix. You may find this helpful as a guide, since it is more readily intelligible (to people) than is Figure 3.1. But please remember that beliefs are not represented in this way for Colby's program. So you should refer to Figure 3.1, not merely to its "easier" verbal equivalent, when following through my description of the program. Similarly, Figures 3.6, 3.7, and 3.8 respectively provide verbal versions of the program's Dictionary, Reason Matrix, and Substitute Matrix. (One row in the Reason Matrix is left uncompleted, to be filled in after you have read this section.) These matrices too should sometimes be consulted by you in their purely numerical form—not least because, as I hope you will notice, their verbalized equivalents tempt you to assume that some knowledge is available to the program which in fact is not. The reason for, and importance of, this fact will be discussed in the third section of this chapter.

Each row of the Belief Matrix can be read as a specific belief because, with the exception of column 5, columns 2 to 8 of the Belief Matrix

FIGURE 3.2

*Dictionary of Neurotic Program*

| Row No. | Word | Part of Speech* | Valence | Row in Substitute Matrix | Attributes |
|---|---|---|---|---|---|
| 1. | people | 1 | +.10 | 1 | 13, 4 |
| 2. | abandon | 2 | –.50 | 0 | 0 |
| 3. | poor | 3 | –.30 | 2 | 0 |
| 4. | men | 1 | –.30 | 3 | 1, 9, 21 |
| 5. | be | 2 | 0 | 0 | 0 |
| 6. | unreliable | 3 | –.20 | 0 | 0 |
| 7. | self | 1 | –.30 | 0 | 1, 13, 20, 21 |
| 8. | love | 2 | +.70 | 4 | 0 |
| 9. | father | 1 | –.50 | 0 | 1, 4, 6, 21 |
| 10. | marry | 2 | +.50 | 0 | 0 |
| 11. | hate | 2 | –.70 | 5 | 0 |
| 12. | respect | 2 | +.40 | 6 | 0 |
| 13. | women | 1 | +.20 | 7 | 1 |
| 14. | disapprove | 2 | –.10 | 8 | 0 |
| 15. | mother | 1 | +.60 | 0 | 1, 13, 23 |
| 16. | say | 2 | 0 | 0 | 0 |
| 17. | not | 0 | 0 | 0 | 0 |
| 18. | never | 0 | 0 | 0 | 0 |
| 19. | Ms. Smith | 1 | +.20 | 0 | 1, 13, 27 |
| 20. | defective | 3 | –.30 | 0 | 0 |
| 21. | hostile | 3 | –.40 | 9 | 0 |
| 22. | boss | 1 | –.40 | 0 | 1, 4, 21 |
| 23. | helpless | 3 | –.10 | 0 | 0 |
| 24. | dislike | 2 | –.20 | 0 | 0 |
| 25. | laugh at | 2 | –.10 | 0 | 0 |
| 26. | animals | 1 | +.60 | 0 | 23 |
| 27. | atheists | 1 | –.60 | 0 | 1 |

*1 = noun; 2 = verb; 3 = adjective.

consist of cross-references to the words listed in column 2 of the Dictionary. This interpretation of the Belief Matrix requires reference also to columns 9 and 10. Column 9 codes the imperative level of the main verb: that is, whether it has an *ought* or a *must* attached, and whether it is a positive or a negative imperative ("thou shalt" or "thou shalt not"). Column 10 codes the complexity of the belief: simple beliefs are entered as zero here, but if the Subject is itself a belief the column is entered as "1," whereas if the Object is a belief, or both Subject and Object are beliefs, the complexity column is coded as "2" or "3," respectively. Column 10 has to be checked *before* columns 3 and 8, since if the belief is a complex one then the relevant (Subject and/or Object) columns will contain a

FIGURE 3.3

*Reason Matrix of Neurotic Program*

| Row No. | Row in Belief Matrix | Reasons for | Reasons against |
|---|---|---|---|
| 1. | 4 | 2 | 0 |
| 2. | 5 | 2, 7, 10 | 8, 13 |
| 3. | 10 | 11 | 0 |

number referring to another row within the Belief Matrix, instead of to a row in the Dictionary.

There is no limit in principle to the number of levels of complexity that may be involved, because the "contained" belief could itself be a complex one. Colby so far has restrained himself from initially providing the program with beliefs of the form "She says that he says that you're going to say that I said . . ." but he could do so if he wished. Indeed, we shall see in the next section that the program itself defensively generates new beliefs of more than one level of complexity, like *Sister dislikes the fact that father laughs at the fact that I disapprove of the fact that he does not respect women.* Colby's apparently mathematical arrays thus have the power to represent social attitudes of some complexity.

You may find it useful at this point to check for yourself that the beliefs represented in the Belief Matrix (that is, Figure 3.1) are indeed those that I have listed. And if you are familiar with R. D. Laing's

FIGURE 3.4

*Substitute Matrix of Neurotic Program*

| Row No. | Row in Dictionary | Antonyms (Verbs) | Instances or Synonyms |
|---|---|---|---|
| 1. | 1 | 0 | 4, 7, 9, 13, 15, 19, 22, 27 |
| 2. | 3 | 0 | 9 |
| 3. | 4 | 0 | 9, 22 |
| 4. | 8 | 11 | −12 |
| 5. | 11 | 8 | −24, −14 |
| 6. | 12 | 25 | +8 |
| 7. | 13 | 0 | 7, 15, 19 |
| 8. | 14 | 0 | +11 |
| 9. | 21 | 0 | 4 |
| 10. | 27 | 0 | 19 |

FIGURE 3.5
*Verbal Version of the Belief Matrix*

| 1. Row No. | 2. Subject Modifier | 3. Subject | 4. Verb Modifier | 5. Tense | 6. Verb | 7. Object Modifier | 8. Object |
|---|---|---|---|---|---|---|---|
| 1. | (0) — | (1) people | (0) — | (0) present | (2) abandon | (0) — | (1) people |
| 2. | (3) poor | (4) men | (0) — | (0) present | (5) be | (0) — | (6) unreliab |
| 3. | (0) — | (7) self | (0) — | (0) present | (8) love | (0) — | (9) father |
| 4. | (0) — | (7) self | (0) — | (0) present | (10) marry | (3) poor | (4) men |
| 5. | (0) — | (7) self | (0) — | (0) present | (11) hate | (0) — | (9) father |
| 6. | (0) — | (7) self | (0) — | (0) present | (8) love | (0) — | (1) people |
| 7. | (0) — | (9) father | (17) not | (0) present | (12) respect | (0) — | (13) women |
| 8. | (0) — | (7) self | (18) never | (2) future | (2) abandon | (0) — | (9) father |
| 9. | (0) — | (7) self | (0) — | (0) present | (14) disapprove | (0) — | (7) Belief 7 |
| 10. | (0) — | (9) father | (0) — | (1) past | (2) abandon | (0) — | (7) self |
| 11. | (0) — | (15) mother | (0) — | (0) present | (16) say | (0) — | (10) Belief 1 |
| 12. | (0) — | (7) self | (0) — | (0) present | (8) love | (0) — | (15) mothe |
| 13. | (0) — | (9) father | (0) — | (0) present | (8) love | (23) helpless | (26) animal |
| 14. | (0) — | (7) self | (0) — | (0) present | (11) hate | (0) — | (27) atheist |

| 9. perative Level | 10. Complexity | 11. Fixed Charge | 12. Residual Charge | 13. Credibility | 14. Row in Reason Matrix |
|---|---|---|---|---|---|
| (−2) ught not | (0) — | (7000) High | (6000) Medium | (6000) High | — |
| (1) eutral | (0) — | (4000) Very low | (2000) Very low | (4000) Low | — |
| (+3) ust | (0) — | (7000) High | (8000) Very high | (6000) High | — |
| (−3) ust not | (0) — | (6000) Medium | (2000) Very low | (3000) Very low | 1 |
| (1) eutral | (0) — | (9000) Highest | (7000) High | (5000) Medium | 2 |
| (+3) ust | (0) — | (5000) Low | (7000) High | (7000) Very high | — |
| (1) eutral | (0) — | (6000) Medium | (7000) High | (7000) Very high | — |
| (1) utral | (0) — | (6000) Medium | (7000) High | (7000) Very high | — |
| (1) eutral | (2) object is a belief | (7000) High | (7000) High | (6000) High | — |
| (1) utral | (0) — | (8000) Very high | (9000) Extremely high | (8000) Highest | 3 |
| (1) utral | (2) object is a belief | (4000) Very low | (6000) Medium | (6000) High | — |
| (+3) ust | (0) — | (7000) High | (7000) High | (7000) Very high | — |
| (1) utral | (0) — | (3000) Very low | (3000) Very low | (7000) Very high | — |
| (1) utral | (0) — | (4000) Very low | (2000) Very low | (6000) High | — |

FIGURE 3.6

*Verbal Version of the Dictionary*

| Row No. | Word | Part of Speech | Valence | Row in Substitute Matrix | Attributes |
|---|---|---|---|---|---|
| 1. | people | noun | +.10 | 1 | women, men, atheists |
| 2. | abandon | verb | –.50 | 0 | — |
| 3. | poor | adjective | –.30 | 2 | — |
| 4. | men | noun | –.30 | 3 | people, father, hostile |
| 5. | be | verb | 0 | 0 | — |
| 6. | unreliable | adjective | –.20 | 0 | — |
| 7. | self | noun | –.30 | 0 | people, women, defective, hostile |
| 8. | love | verb | +.70 | 4 | — |
| 9. | father | noun | –.50 | 0 | people, men, unreliable, hostile |
| 10. | marry | verb | +.50 | 0 | — |
| 11. | hate | verb | –.70 | 5 | — |
| 12. | respect | verb | +.40 | 6 | — |
| 13. | women | noun | +.20 | 7 | people |
| 14. | disapprove | verb | –.10 | 8 | — |
| 15. | mother | noun | +.60 | 0 | people, women, helpless |
| 16. | say | verb | 0 | 0 | — |
| 17. | not | verb modifier | 0 | 0 | — |
| 18. | never | verb modifier | 0 | 0 | — |
| 19. | Ms. Smith | noun | +.20 | 0 | people, women, atheist |
| 20. | defective | adjective | –.30 | 0 | — |
| 21. | hostile | adjective | –.40 | 9 | — |
| 22. | boss | noun | –.40 | 0 | people, men, hostile |
| 23. | helpless | adjective | –.10 | 0 | — |
| 24. | dislike | verb | –.20 | 0 | — |
| 25. | laugh at | verb | –.10 | 0 | — |
| 26. | animals | noun | +.60 | 0 | helpless |
| 27. | atheists | noun | –.60 | 0 | people |

writings on schizophrenia, you might like to amuse yourself by taking an example of the "nested" perceptions typically described in his family case histories, and expressing them in the form of arrays, although this will probably be easier after you have read the next section. Each member of the family would have a different set of the four matrices. Since there is no guarantee that any of the beliefs in the Belief Matrix are true (or even mutually consistent), there is no difficulty in representing the reciprocal misperceptions within the family that are so stressed by Laing. In terms of the previous example, sister's dislike for father may be based on an

FIGURE 3.7

*Verbal Version of Reason Matrix*
*(Row 2 is left to be filled in as an exercise)*

| Row No. | Row in Belief Matrix | Reasons For | Reasons Against |
|---|---|---|---|
| 1. | I must not marry a poor man. (4) | Poor men are unreliable. (2) | – (0) |
| 2. | I hate father. (5) | . . . (2)<br>. . . (7)<br>. . . (10) | . . . (8)<br>. . . (13) |
| 3. | Father abandoned me. (10) | Mother says father abandoned me. (11) | – (0) |

illusion at any level: perhaps father does not scorn self's disapproval of his misogyny; perhaps he does not despise women at all; or perhaps self does not really disapprove of his (actual or imagined) male chauvinism, but—contrary to her sister's belief—rather admires it. Cross-comparison between the Belief Matrices of the individuals concerned could, of course, show the points at which perception of others is faulty. (Laing himself has developed a series of family questionnaires designed to fulfill

FIGURE 3.8

*Verbal Version of Substitute Matrix*

| Row No. | Row in Dictionary | Antonyms (Verbs) | Instances or Synonyms |
|---|---|---|---|
| 1. | people (1) | – (0) | men (4), self (7), father (9), women (13), mother (15), Ms. Smith (19), boss (22), atheists (27) |
| 2. | poor (3) | – (0) | father (9) |
| 3. | men (4) | – (0) | father (9), boss (22) |
| 4. | love (8) | hate (11) | *respect* is weaker form (–12) |
| 5. | hate (11) | love (8) | *dislike* and *disapprove* are weaker forms (–24, –14) |
| 6. | respect (12) | laugh at (25) | *love* is stronger form (+8) |
| 7. | women (13) | – (0) | self (7), mother (15), Ms. Smith (19) |
| 8. | disapprove (14) | – (0) | *hate* is stronger form (+11) |
| 9. | hostile (21) | – (0) | men (4) |
| 10. | atheists (27) | – (0) | Ms. Smith (19) |

the same function: each person answers first on her own behalf, then on behalf of another member of the family, and then on behalf of that member thought of as replying for the respondent herself.[1])

Whether or not you choose to try the Laingian exercise just described, you may wish to follow up the simpler check on the 14 beliefs of Colby's program with an examination of the current interests of the system. Reference to columns 11 and 12 shows, for example, that it is temporarily more worked up about the relation to the father and his past (actual or imagined) abandonment, than about the subject of marriage —even though both of these are recurrent matters of neurotic concern. Columns 11 and 12 show the Fixed (enduring) and Residual (temporary) Charges on the belief. A belief like *I hate father* always has a high Fixed Charge, but if it has been successfully transformed, denied, or rationalized by a defense mechanism, it may sometimes have a relatively low Residual Charge.

Column 14 gives a cross-reference to the Reason Matrix for use by the DENIAL and RATIONALIZATION techniques. Since *I hate father* already has available in the Reason Matrix both reasons for and reasons against, it is clear that the Residual Charge of this belief could be lowered by either of these techniques without the program's having to do much thinking in the process. Finally, column 13 codes the acceptance, or credibility, of the beliefs. This also varies during the running of the program, primarily by way of denial and rationalization (but also by way of quasi-inductive credibility judgments like those to be described in Chapter 4).

If you play around with the four arrays for a while you will find that they contain information other than that which is explicitly coded in the Belief Matrix. For example, the Dictionary shows the system to be hostile to men in general, irrespective of whether this hostility has ever been directly considered at conscious or unconscious levels of the mind. And, as you may have noticed, the organization of the Reason Matrix implicitly shows that one reason for this person's hating her father is that she believes mother to have said that he abandoned her.

You may be interested to compare your own use of these matrices with the program's use of them, described in the next section. To understand how it exploits the data storage in these arrays, let us follow through an example of the program at work.

## HOW THE NEUROTIC PROGRAM WORKS

The first thing the program does on being instructed to start is to select a core belief. Psychoanalytic theory suggests that this should be a personal imperative or a moral injunction, so the relevant subroutine (called by its code name FORMCOMPLEX) searches first for beliefs containing *must*: that is, it examines column 9 of the Belief Matrix and looks for an imperative level of +3 or −3. The only candidates here are Beliefs 3, 4, 6, and 12: *I must love father, I must not marry a poor man, I must love people,* and *I must love mother,* respectively. Since the belief that is the most psychologically significant at the time is to be selected, the routine next compares the Residual Charges coded in column 12 and picks the highest of the four: belief 3, *I must love father.*

Having found a core belief, FORMCOMPLEX next finds all the relevant beliefs and puts them on a special array (the Complex Matrix) forming the pool for current processing. Colby's computational definition of "relevance" is that at least two of the three main terms (subject, verb, and object) should be held in common: columns 3, 6, and 8 of the Belief Matrix are searched accordingly. Just as some neurotic complexes are more far-reaching than others in the person's mind, so some pools contain many beliefs while others contain only a few. In this hypothetical case the only beliefs in the pool would be 3, 5, 6, 8, and 12: *I must love father, I hate father, I must love people, I will never abandon father,* and *I must love mother,* respectively.

Since in Colby's actual simulation there are 114 beliefs initially represented in the Belief Matrix, rather than only 14 as in my example, FORMCOMPLEX often produces a much richer pool than this. Correlatively, there are many more Dictionary entries, of which those representing persons have more Attributes than are shown in Figure 3.2. What is more, the Belief Matrix always grows larger during processing, since every distorted belief that is expressed is added to it, while the original (undistorted) belief remains. To have attempted to discuss an example involving a pool of the size commonly created by FORMCOMPLEX would have been a recipe for instant insanity, on your part and mine. This illustrates the invaluable role of a functioning computer in assessing the complex interactions postulated by a programmed psychological theory, even one as simple as Colby's.

The next step is to pick a regnant belief from the pool. This is done by a random number operator, and in the previous chapter we assumed that the one picked is *I hate father.* This regnant is now tested for conflict against each of the other beliefs in the Complex Matrix by a subroutine called PROCESS COMPLEX.

Intuitively, one might say that this belief conflicts with three others in the pool: *I must love father, I must love people,* and *I will never abandon father.* However, the conflict with the first of these is obviously more worrying than that with the second, whereas the conflict with the last seems decidedly less pressing (being closer to a mere logical conflict or semantic inconsistency than to a *neurotic* conflict).

Colby catches some of these intuitive distinctions in his computational definition of conflict, for PROCESS COMPLEX is able to recognize two forms of conflict. One of these is serious enough to trigger defensive thinking by the program, whereas the other merely raises the anxiety level a little.

Serious neurotic conflict, preventing direct expression of a belief irrespective of the current levels of the emotional monitors, occurs when the two beliefs have identical Subject and Object, antonymic Verbs, *and* a sum of levels of imperativeness of 4 or more (plus and minus signs being ignored). To reach a sum of 4, it follows that the two beliefs must either both be *ought* beliefs (level 2 each) or at least one must be a *must* belief (level 3). This excludes cases of merely logical or semantic (as opposed to moral) conflict. And it tallies, of course, with Freud's view that neurotic conflict always involves the moral demands of the superego.

In terms of this programmed criterion of neurotic conflict, the regnant *I hate father* is in serious conflict with *I must love father,* and so cannot be directly expressed. It follows that some defensive process must be activated to cope with this situation, so PROCESS COMPLEX hands over to PICKTRANSFORM to decide what must be done.

Before continuing with our example by describing the neurotic thought processes now carried out by PICKTRANSFORM, let us pause to ask why I said that Colby's definition of conflict catches only "some" of the intuitive distinctions mentioned above.

As it stands, Colby's program would not be aware of the conflict between *I hate father* and *I must love people.* The reason is that its notion of conflict demands that the Object of the two beliefs be identical, which *father* and *people* are not. However, this is not a radical defect, for the program could very easily be adjusted so as to be able to recognize this conflict and to react appropriately to it. Let us suppose that the subroutine searching for conflict (namely, PROCESS COMPLEX) were told to check the Substitute Matrix whenever it found that the two Objects were not identical. If you look at row 1 of the Substitute Matrix, you will see that *father* is listed there as an instance of *people.* In other words, there is a sense in which the program already "knows" that father is a person, but only if PROCESS COMPLEX actively searched the Substitute Matrix while assessing conflict would this knowledge be made available to it at this point. (How often have you failed to think of something, or even to realize its relevance, though in a sense you "knew" it all the time?)

The intuitively obvious fact that, despite the moral imperative involved, this conflict (between *I hate father* and *I must love people*) is less worrying than that occasioned by the more specific injunction to love one's father could also be captured in programming terms by a relatively trivial adjustment to Colby's system. PROCESS COMPLEX could be told that when the Substitute Matrix shows that one of the Objects in a moral conflict is a general class (like *people, women,* and *men*), while the other is an instance of that class, then there is neurotic conflict present but it is not so serious as to prevent expression of the regnant. Accordingly, PICKTRANSFORM should *not* be called in this sort of case. But PROCESS COMPLEX could automatically raise the overall anxiety level (the value of the EXCITATION monitor) on detection of this type of conflict. Moreover, greater anxiety could plausibly be arranged to ensue when the regnant concerns a particular individual rather than people in the mass (*I hate father* rather than *I hate men* or *I hate people*). Analogously, one may feel guilty at hating mankind—but probably not so guilty as one feels at hating one's mother or father. This psychological fact is implicit in Column 11 of the Belief Matrix, which shows *I must love people* as having a lower Fixed Charge than *I must love father*. (The Residual Charge of the more general injunction could, of course, rise considerably if the therapist were to input appropriate sentences from the New Testament, stressing the importance of loving one's neighbor.)

As regards the third intuitively relevant sentence in the pool, *I will never abandon father*, PROCESS COMPLEX cannot sense any incongruity with *I hate father*, because when it searches the Substitute Matrix for antonymic verbs it does not find *abandon* listed on row 5, which corresponds to *hate*. (Again, there is a sense in which it "knows" that *I will never abandon father* is inconsistent with *I hate father*, since the first of these beliefs appears in the Reason Matrix as a "reason against" the second; but PROCESS COMPLEX does not consult the Reason Matrix, so is unaware of this clash between the two beliefs.)

However, if *I respect father* had been included in the Belief Matrix then PROCESS COMPLEX would have been aware of the conflict between this belief and *I hate father*. (How can I be so sure that *I respect father* would automatically have been included in the pool, so that PROCESS COMPLEX would have had a chance to consider it?) PROCESS COMPLEX is capable of checking rows 4 and 5 of the Substitute Matrix to find that *hate* and *respect* are in antonymic relation to each other, and infers accordingly that *I respect father* conflicts with *I hate father*. Since no moral imperatives are involved (there is no *ought* or *must* in either sentence), this does not qualify as a serious neurotic conflict but as a less pressing variety. Accordingly, both beliefs can be expressed. But PROCESS COMPLEX raises the overall anxiety slightly, which can have significant defensive consequences if it was already at a very high level. (The later versions of Colby's program can detect a further type of conflict, that involved when a belief is not "evaluatively balanced": we shall see in Chapter 4 that this conflict also triggers particular types of thinking, but since it is not specifically neurotic I shall ignore it here.)

To return to our actual example: PROCESS COMPLEX has found only one belief in the pool, namely, *I must love father,* to be in serious neurotic conflict with the regnant *I hate father.* But one conflicting belief is enough to prevent direct expression of the regnant and to elicit defensive measures within the system. So PICKTRANSFORM is called, to decide what is to be done next.

Whether PICKTRANSFORM actually does pick a transform or instead hands over to the system's ultimate defense—switching off the current pool—depends on the WELL-BEING monitor, which directs the program to hand over control to the routine FINDNEXTCOMPLEX when processing of the current pool becomes too threatening. The way in which the value of WELL-BEING is determined will be explained presently. Since it is only in exceptional circumstances that this value is so high as to prevent PICKTRANSFORM from proceeding to do the job its name implies, we may assume that the program now has to go ahead and choose one of the eight transforms.

PICKTRANSFORM takes into account four factors when selecting a transform from the list numbered (1) to (8). As we saw in Chapter 2, these are: the current anxiety level; whether or not self-esteem is involved; the need to discharge the underlying instinctual impulse; and the system's past history of successful defense.

The current anxiety level initiated by thinking about the regnant is represented by the value of the DANGER monitor. This, in turn, is influenced by three factors: the preceding DANGER level (if one is in a serene state of mind one will be less roused by a new, potentially worrying, thought); the Residual Charge on the regnant (modeling the current anxiety specific to that belief); and its Credibility (a belief causes less anxiety the less confidently it is assented to).

For the purposes of our example, let us assume that the preceding DANGER level is extremely low, representing a person in a calm and contented frame of mind. The Belief Matrix shows that the Residual Charge on *I hate father* is fairly high, but its Credibility is at a middling level. (This state of affairs might be the result of a previous defensive operation of DENIAL or RATIONALIZATION, by way of the Reason Matrix, whereby the Residual Charge and Credibility of *I hate father* had both been somewhat reduced.) In consequence of the combined influ-

FIGURE 3.9

*Rank Ordering of Transforms in Reducing Anxiety and Providing Discharge of Underlying Impulse*

| | Power to Reduce Anxiety | Power to Discharge Underlying Impulse |
|---|---|---|
| HIGH | (8) PROJECTION | (1) DEFLECTION |
| | (7) REFLECTION | (2) SUBSTITUTION |
| | (6) NEGATION | (3) DISPLACEMENT |
| MEDIUM | (5) REVERSAL | (4) NEUTRALIZATION |
| | (4) NEUTRALIZATION | (5) REVERSAL |
| | (3) DISPLACEMENT | (6) NEGATION |
| | (2) SUBSTITUTION | (7) REFLECTION |
| LOW | (1) DEFLECTION | (8) PROJECTION |

ence of these three factors, the value of DANGER is temporarily set at a low-to-middling level. We saw in Chapter 2 that whereas high anxiety pushes PICKTRANSFORM toward transforms with a high numerical label, low anxiety inclines it to select transforms with a low numerical label (see Figure 3.9). So, in our example, the DANGER monitor advises PICKTRANSFORM to go to the middle or low-numbered part of the list of transforms, within the range (1) to (4).

The SELF-ESTEEM monitor is adjusted whenever the regnant either violates or expresses obedience to a moral imperative. Since *I hate father* violates *I must love father* (which is one of the activated sentences making up the current pool), the value of SELF-ESTEEM is lowered. The effect on PICKTRANSFORM is to suggest that it consider choosing a self-protective transform. By a "self-protective" transform I mean one that produces a change in the Subject of the belief (*self*), while leaving Verb and Object undistorted. You can see from Figure 3.10 that the only such routine is PROJECTION, which at number (8) is the most effective in reducing anxiety but the least effective in discharging the underlying instinctual impulse. PICKTRANSFORM is more strongly influenced by this advice if the SELF-ESTEEM monitor was already very low. Let us suppose that in our example SELF-ESTEEM was initially high (this supposition tallies with our previous assumption that the system was in a relatively untroubled state of mind). It follows that SELF-ESTEEM's effect in this particular case is to make a relatively gentle suggestion that PICKTRANSFORM opt for (8).

The third factor taken into account by PICKTRANSFORM is the need to discharge the underlying impulse, which is clearly greater the stronger the impulse concerned. The Belief Matrix shows that the Fixed

FIGURE 3.10
*List of Transforms*

| Label | Name | Definition |
|---|---|---|
| 1. | DEFLECTION | Shift Object (Not Self) |
| 2. | SUBSTITUTION | Cascade Verb |
| 3. | DISPLACEMENT | Combine (1) and (2) |
| 4. | NEUTRALIZATION | Neutralize verb |
| 5. | REVERSAL | Reverse Verb |
| 6. | NEGATION | Insert *Not* before Verb and do (5) |
| 7. | REFLECTION | Shift Object to Self |
| 8. | PROJECTION | Switch Subject (Self) and Object (Not Self) |

Charge on *I hate father* is extremely high. According to the rules shown in Figure 3.9, therefore, this strongly inclines PICKTRANSFORM to choose a defense mechanism with a very small numerical label, for these are the most effective in reducing charge. (SELF-ESTEEM's candidate, PROJECTION, is thus most unsuitable in this particular case, and is eliminated.)

Thus far, PICKTRANSFORM is being pushed fairly close to the low-numbered end of the list of transforms, but is not down quite as far as transform (1)—DEFLECTION. However, one more factor remains to be considered by PICKTRANSFORM: the past history of the system. Let us suppose that in the past this system has often used DEFLECTION successfully, so that it has developed a powerful tendency to choose this transform whenever it can. In this event, the overall result will be that PICKTRANSFORM goes right to the bottom of the list (having been low on the list already) and plumps for DEFLECTION, as we assumed in the example introduced in Chapter 2.

(This compromise between the five factors, two of which—need to reduce anxiety and need to effect discharge—*always* act in opposition to each other, is actually carried out by PICKTRANSFORM by way of number-juggling. That is, the DANGER and SELF-ESTEEM monitors and other weightings are given numerical values, such as 3.7 or 5.1, and particular threshold-values are assigned so as to correspond to various parts of the list of transforms. These numbers are purely arbitrary, having been used by Colby to correspond to the intuitive notions "low," "medium," and "high," and need not be discussed in detail here.)

Having chosen to activate DEFLECTION, the program now has to apply it to *I hate father* in the hope of generating an expressible (though distorted) derivative. You may remember that the definition of DEFLECTION is "Shift Object (Not Self)," which leaves it up to the program to decide precisely how it will DEFLECT in each case. That is, the Object (Column 8 of the regnant's row in the Belief Matrix) has to be switched to something else, where the "something else" must not be *Self* but may be one of a number of alternatives allowed for by the data base. How does DEFLECTION decide?

Since Colby's transforms are allowed to consider only people as Objects, the DEFLECTION procedure starts by locating Row 1 in the Substitute Matrix, which gives all the instances of people known to the program. In our example, these are Dictionary items *men, self, father, women, mother, Ms. Smith, boss,* and *atheists. Self* is automatically deleted from the list, according to the definition of the transform; and *father* is deleted because it is the original Object that is to be switched. Since the transform prefers to deflect to a particular individual rather than a general class of people (why?), it searches to see if any of these items are themselves included on the Substitute Matrix as having instances: *men, women,* and *atheists* (Dictionary items 4, 13, and 27) are thereby deleted. *Mother, Ms. Smith,* and *boss* remain.

In order to make a fairly sensible (as opposed to a random) choice between *mother, boss,* and *Ms. Smith,* the routine FINDANALOG is called. Since the Fixed and Residual Charges on *I hate father* are both high, suggesting that very threatening instinctual impulses are involved, the number of shared properties allowed between the original and new Objects is set low rather than high—let us say at the value "one," so that the analogy-seeking routine is FINDANALOG-1. This procedure refers to the Dictionary, searching the Attribute column to find that *father* shares three attributes with *boss,* but only one each with *mother* and *Ms. Smith.* FINDANALOG-1 now chooses randomly between these two potential Objects of the DEFLECTION transform, since it is not interested in the nature of the shared attributes but only in their number.

If the program were to choose *mother* at this point in the operation of the DEFLECTION transform, this would lead immediately to a conflict with *I must love mother* (still waiting in the pool). The anxiety accordingly would rise (that is, the level of the DANGER monitor would be increased), *I hate mother* could not be expressed, and this distorted belief would be rejected as useless for defensive purposes.

But if FINDANALOG-1 directs deflection to *Ms. Smith,* as we assumed in Chapter 2, the only potential conflict in the pool is with *I must love people.* As I explained earlier in this section, the program as it stands cannot detect a conflict like this one, since the routine searching for conflict does not check the Substitute Matrix to see whether *Ms. Smith* is an instance of *people.* In our example, then, *I hate Ms. Smith* causes the system no difficulty, and can be expressed.

(Even if the program were adjusted in the way I suggested earlier, so that the Substitute Matrix was searched in checking for conflict, *I hate*

*Ms. Smith* would cause only a lesser degree of conflict. For we saw there that moral injunctions concerning love for people in general are less pressing than those directed to specific individuals. So *I hate Ms. Smith* could be accepted as an expressible form of *I hate father* even in this hypothetical case.)

Since DEFLECTION succeeds in our example, the distorted belief (*I hate Ms. Smith*) is printed out, and the Residual Charges of the beliefs concerned are adjusted accordingly. At the same time, *I hate Ms. Smith* is added to the pool and to the Belief Matrix—within which there may be latent beliefs that could come into conflict with this one later. And, as we shall see soon, a rationalization is created for the new belief and stored on the Reason Matrix to support the hatred of Ms. Smith in future.

The way in which the other distorting transforms work should be evident from their definitions shown in Figure 3.10, given that they operate on the same data base as the DEFLECTION routine just described. "Cascading" verbs involves finding a less emotively valued synonym in the Substitute Matrix (coded there by a minus sign), whereas "reversing" the verb involves the antonym or a cascaded version thereof (so that *I hate father* can give *I like father*). Subject, Object, and Verb are all located in the appropriate columns in the Belief Matrix.

The descriptions of PICKTRANSFORM and of the rejection of *I hate mother* as a deflected version of *I hate father*, showed that the emotional monitors play an integral part in the control of what happens in the system. In addition, they are output from time to time as (nonverbal) cues that can be interpreted as indicators of the system's current emotional state, and that sometimes contradict the verbal output of the program. (Analogously, a patient may insist *I'm fine!* even though her fidgeting and facial expressions convince one that she is far from fine.) Let us look a little more closely at how these five monitors function.

The first, the DANGER monitor, reflects its own immediately preceding level, the Residual Charge on the belief temporarily causing conflict, and the regnant's degree of acceptance by the system (its Credibility). (An example showing how these factors influence the DANGER level was given above.) The second, the EXCITATION monitor, measures the general level of excitation in the system, being computed as a function of the previous level and the Residual Charge of the current regnant. If Colby starts a run of the program with this monitor set high, the "previous level" is high to begin with; (random) choice of a highly threatening regnant can thus very quickly lead to situations in which even the eighth transform cannot cope with the anxiety and the whole pool is repressed. This is so because, as we shall see shortly, the EXCITATION monitor contributes indirectly to the WELL-BEING monitor.

The third emotional monitor is the PLEASURE monitor. This takes into account the levels (and the recent changes) of the first two monitors. It also keeps a check on the emotional valences of the verbs (coded in the Valence column of the Dictionary) in the beliefs expressed by the system, to see if 6 out of a group of 10 (*or* 5 in a row) are negative; if they are, then the system clearly is in a fairly depressing state of mind, and the PLEASURE level is lowered accordingly. The fourth, or SELF-ESTEEM, monitor is lowered if expression of the regnant belief would violate an *ought* or *must* command in the pool; but if expression of a belief connotes obedience to such a command then SELF-ESTEEM is raised.

The fifth monitor measures overall WELL-BEING, and decides whether current processing is to continue, as we have seen. It combines the PLEASURE and SELF-ESTEEM measures, and when it falls below a threshold level 6 times out of 10 (or 5 times in a row) it represses the pool concerned, anxiety rises, and a new pool is formed. Since PLEASURE is itself a function of DANGER, EXCITATION, and the valence of the beliefs expressed, all these contribute indirectly to the overall WELL-BEING of the system as a whole.

The use of the Reason Matrix by the defense mechanisms of RATIONALIZATION and DENIAL is straightforward. A belief that already carries (in the appropriate row in the Belief Matrix) a cross-reference to the Reason Matrix can readily elicit any "reasons for" or "reasons against" that may be stored there. Thus we have seen, for instance, that the Credibility of *I hate father* may have been lowered by DENIAL's seeking out the two reasons counting against this belief: *I will never abandon father* and *Father loves helpless animals.*

But the way in which the Reason Matrix is built up is less obvious. Before the program is set to work, Colby may already have placed various reasons (for and/or against) on it. Whether or not the Reason Matrix contained any items in the first place, items are added by the program itself in the course of its functioning, primarily because every new (distorted) belief is rationalized immediately it has been expressed.

Rationalizations and denials of some psychological subtlety can be included in the Reason Matrix if Colby himself specifically puts them there. Provided that they can be expressed by Dictionary items placed within columns 2 to 8 of the Belief Matrix, there is no limit to the complexity of the mental operations (Colby's) that thought them up, recognizing them to be capable of functioning in a neurotic's mind as reasons for or against the belief concerned.

But the system's own perceptions of "reason for" and "reason against" are relatively rudimentary and simply structured. Consequently, its spontaneous creation of rationalizations for newly distorted beliefs (like the *I hate Ms. Smith* of our example), and its own search for "denying" beliefs that can count against a troublesome belief, depend on undiscriminating psychological strategies that, although admittedly they are sometimes used by people, are very crude.

The program relies primarily on class-inclusion relations that (as we shall see in the next chapter) can be used to assess the credibility of beliefs in nonneurotic contexts also. For instance, *Ms. Smith*'s Attribute list shows her to be an *atheist*, and there is already a belief on the Belief Matrix stating *I hate atheists.* Consequently, rationalization of this new hatred of Ms. Smith, which in fact has nothing to do with religion but was generated via the DEFLECTION routine, is easy. RATIONALIZATION looks to see if this individual person is an instance of an already-*hated* (or even *disliked*) class, and by way of *Ms. Smith*'s Attribute list finds that she is indeed a member of such a class, namely *atheists. Ms. Smith is an atheist* is thus added to the Reason Matrix as a "reason for" the newly formed *I hate Ms. Smith.* Relations of similarity and contrast also help: exploitation of the similarity between *hate* and *dislike* was assumed in the preceding example; correlatively, *I like father* can be created anew to deny (lower the credibility of) *I hate father* by way of the Antonym column of the Substitute Matrix.

Indefinitely complex beliefs can be generated by RATIONALIZATION and DENIAL. For instance, suppose that the SUBSTITUTION transform has produced *I dislike father*; this can be rationalized by saying that father hates (or in some other way disvalues) something about me. The Belief Matrix is searched for a belief with *Self* as Subject, perhaps finding row 9, and a new belief is built up—*Father laughs at the fact that I disapprove of the fact that he does not respect women.* If at some future point the system wishes to raise the credibility of a belief like *Sister dislikes father,* the process can be repeated, so that the program claims that sister dislikes father's scorn for self's disapproval of his contempt for women. (If sister's husband loves sister, what will the program assume his attitude to her view on this matter to be?)

The rationalizations and denials created in these ways usually seem fairly appropriate (which is not the same thing as being true), rather than striking one as utterly irrelevant. But, except by way of a happy accident, they are never nicely appropriate to the particular conflict concerned, in the way in which a person's can be. Colby's program does not have our powers of reasoning or our knowledge of the world, so has to rationalize in a relatively "mechanical" way, if it is not to depend entirely upon the "reason for" and "reason against" relations initially provided to it by Colby himself. This is not to say, of course, that *no* program could be more ingenious than this one in its self-justifications, whether neurotic or not. This raises the general question posed earlier: what are the weaknesses of Colby's program, and how might they be allayed?

## FAILINGS OF THE NEUROTIC PROGRAM

Apart from the lack of reasoning power just mentioned, the main faults of Colby's program can be summarized under four heads. It is weak in representing emotions like anxiety or self-esteem, which are modeled by the "emotional monitors." It fails to understand the beliefs stored in it and the sentences input to it except in the very weakest sense. Its use of the various procedures that assess relevance and conflict, and that decide precisely what the transform activated is to do on a particular occasion, is undiscriminating and unimaginative. And it fails to generalize its neurosis, being incapable of modeling the way in which a person's neurotic obsession about having been abandoned by her father develops naturally into a fear of being betrayed by family and friends, all social ralationships being thereby imbued in her mind with a sense of mistrust and insecurity.

We shall find that all four faults depend heavily on the lack of a background representation of meaning, or semantics, by reference to which any increased reasoning power that might be supplied would have to function, if the program were to provide a more plausible theoretical model of human neurosis. They could not be remedied by superficial tinkering with the program, such as that I suggested above to make PROCESS COMPLEX sensitive to the conflict between *I hate father* and *I must love people.*

The program's weakness in representing emotions has nothing to do with the fact, if indeed it is a fact, that computers cannot *really* defend themselves against anxiety. Freud's theories cannot do so either, but they may nonetheless explain what happens when a person does. The numerical parameters in Colby's program that he calls "emotional monitors" (such as DANGER and SELF-ESTEEM) represent theoretical concepts like *anxiety* and *self-esteem,* which themselves represent the actual phenomena of anxiety and self-esteem in human minds. Moreover, emotional terms such as these are not mere idle labels, but indicate functional aspects of the program. The theoretical interest (of Colby as of Freud) is in what difference emotions of various types make to the information processing, or thinking, going on. Computer simulations in general attempt theoretical modeling of psychological function and structure, rather than ontological mimicry of mental reality, and it is in these terms alone that they should be assessed.

In these terms, one might object that Colby's quantitative representation of anxiety, for example, is so crude as to be unilluminating: how can anxiety usefully be thought of as a *number*? It is worth noting here that Freud himself thought of anxiety in quantitative terms, describing it as a psychic energy level that could in principle be precisely measured. But although some programmed models of defense mechanisms (such as the series developed by Ulrich Moser[2]) concentrate closely on Freud's "numerical" view of anxiety, Colby's does not. Indeed, Colby had rejected this aspect of Freudian psychology on philosophical grounds some years before embarking on his programming enterprise; he saw Freud's theory, and his own activity as a practicing analyst, in (information-processing) terms that stressed meaning and cognitive structure rather than energy levels.[3] He therefore basically agrees with those who describe Freud as a "semantic" or "hermeneutic" theorist. As we shall see in Chapter 13, such commentators on Freud regard psychoanalytic explanation as a matter of subjective interpretation (showing how neurosis *makes sense*) rather than the identification of meaningless universal causes, like biological drives or energy levels.[4]

Why, then, did Colby later decide to admit apparently "numerical" anxiety into his program? The primary reason is that anxiety is very naturally conceptualized in quasiquantitative terms: even in ordinary speech one refers to a "high degree" of anxiety, to "anxiety level," and to someone's being "a little anxious" or experiencing "great anxiety." These are not forced technical expressions deriving solely from Freudian metapsychology, but phenomenologically acceptable metaphors for representing anxiety in daily life. It is understandable, then, that Colby utilizes a quantitative model that parallels these familiar metaphors to some extent.

All metaphors, of course, have their limitations: for instance, one would not normally attach actual numbers to anxiety levels, and Colby's specific figures (which may attribute anxiety of strength 0.49 and self-esteem of strength 0.33 to the system at a particular time) certainly cannot be taken seriously. But the arbitrary quantification involved in the emotional monitors in Colby's program is partially offset by the fact that he garbs his figures in semantic dress.

That is, he distinguishes between affects focused on danger, excitation, pleasure, self-esteem, and well-being. The five numerical parameters involved not only function differently in activating the various defense mechanisms available to the program (as described in the previous section), but are selectively adjusted in light of appropriate semantic content in the pool being processed.

For example, the SELF-ESTEEM parameter is adjusted relative to *ought* or *must* sentences in the pool, discriminating between beliefs that obey or disobey these imperatives. And the PLEASURE monitor takes account of the evaluational tone of the verbs output by the program. Further, since WELL-BEING is a combined measure of these two, it also has a semantically derived component as well as reflecting the "merely numerical" DANGER and EXCITATION parameters. Even these are affected by semantic content, insofar as the computational definition of *conflict* used by the program is expressed in terms of the meaning of the beliefs concerned. A more semantically discriminating definition of conflict would automatically be reflected in a more discriminating emotional sensitivity of the DANGER and EXCITATION monitors, and indirectly of PLEASURE and WELL-BEING also.

It must be stressed that Colby is not cheating here: he is right to represent emotions as having semantic, or conceptual, components. (How far he succeeds in catching the semantic distinctions involved is another question.) In other words, emotions are not just "feelings" with an indescribable subjective quality, which only have to be experienced to be recognized or understood.

If you doubt this, try to imagine feeling a twinge of conscience every time you utter the word *it*, or a pang of wounded vanity when you turn on the kitchen tap. You will find either that you cannot imagine these things at all, or that you can do so only by surrounding the situation with appropriate stories—perhaps your best friend has just been struck dumb through some negligence of yours, or perhaps the tap is so shiny as to act as a mirror. (Why *these* stories? And would the first really make sense of your being affected specifically by the word *it*?) To put it another way, let us suppose that on turning the tap (we assume it is not shiny) you actually feel a pang of some sort: how could you possibly be justified in interpreting it as a pang of *vanity?* A relevant psychological experiment has shown that people whose excitation level had been raised by adrenaline experienced their heightened feelings either as anger or as euphoria, depending on their (conceptual) beliefs about what the other people in the room were doing.[5] What emotions could be more different than anger and euphoria? In short, feelings are *interpreted* when they are experienced, they are *cues* to emotions rather than being emotions themselves. (This distinction will be clarified in Chapter 8.) And distinct emotions (like shame, fear, regret, disappointment, and so on) are aroused by and directed onto different situations, and influence behavior in different ways.

In its use of the emotional monitors, then, Colby's program implicitly incorporates theories about how phenomenologically different emotions are aroused, what sorts of matters they can be directed to (that is, what sorts of "intentional objects" they can have), and what their effects in the overall functioning of the mind may be. That the affective hypotheses so embodied are crude is true. But we shall see in the next chapter, and particularly in discussing the concept of *betrayal*, that a program affording a richer semantics in its data base could embody emotional hypotheses of greater subtlety. It could distinguish, for example, between the sense of "self-esteem" lowering experienced as moral shame, or guilt, and that experienced as wounded vanity. Disobeying either *I must love father* or *I must not marry a poor man* (which last, in the case of Colby's patient, expresses a vain need to be glamorous) could then lower self-esteem in appropriately different ways.

What I mean by a "richer semantics" will be discussed at length in Chapters 4 and 7. For the moment, one may say that Colby's mechanical neurotic cannot show how subtly different emotions are aroused, and how they affect what goes on next, because it has no knowledge of the underlying meanings, or concepts, in terms of which we implicitly think of the world. This drawback leads also to the second fault of the program: its inability to understand what it is talking about.

Colby's artificial neurotic knows almost nothing of the ways in which various concepts are connected with one another. Thus it knows that *hate* and *love* are antonyms, and that *abhor, detest, hate, dislike, disapprove,* and *reject* are synonymous verbs of decreasing strength. (Should *see faults in* be included in this cascade, as I implicitly assumed in Chapter 2?) It knows these things because they are coded in the Substitute Matrix. But it does not know that love has anything to do with helping and caring for the person loved, so were it to be told that *God loves people* and that *God afflicted Job with boils* it would have no inkling of the theological "Problem of Evil" that strikes us immediately on juxtaposing these two sentences. In other words, the conflict between love and uncaring behavior is not a conflict to which the program itself is sensitive. Insofar as a neurotic anxiety were focused on the spontaneous recognition of this type of conflict, then, it could not be paralleled in the program as it stands.

Of course, abandonment is a form of uncaring behavior, and the program knows that *Father loves me* conflicts with *Father abandoned me.* Or does it? If you check the data in its arrays you will find that what the program actually knows is that *Father abandoned me* is placed on the Reason Matrix in the "reason for" column of *I hate father,* and that the Substitute Matrix shows *hate* and *love* to be antonyms. If you insist that this comes to the same thing, you are doubtless right—insofar as you are talking about *your* understanding of love, hate, and abandonment. The distinction drawn above may seem mere quibbling to you, because you automatically draw the inferences that make for the connections here. And, as we shall see in Chapter 7, some programs can do the same sort of thing. But Colby's neurotic cannot, because it does not have (as these programs do) a systematic theory of meaning or semantics in terms of which to interpret its "beliefs."

In other words, a program competent to manipulate word strings like "I hate father" in a superficially rational (even if neurotic) manner cannot necessarily be said on that account to understand the corresponding English sentences. This point must be stressed because this string of words almost unavoidably suggests to human beings, including the programmer, an English sentence of the same superficial form drawing on a rich semantic base in the mind of the person concerned. But this semantic base is not available to Colby's program. This has nothing to do with any supposed metaphysical inability of machines to experience hatred: similar remarks apply to sentences like "Pick up a big red block" and "Cops arrest the wrong people." No sentence whatever, and no word either, is understood by Colby's program, except in the weakest possible sense.

This sense is defined by the highly limited interconnections between concepts that the program is able to make on the basis of its data structure and the procedures that operate on it, such as that verbs 11 and 8 are antonyms. Don't you know which verbs those are?—neither does the program, except in the sense that it can locate their more conventional symbols by referring to the second column in the Dictionary. And it understands the relation of *antonymy* only in the sense that it uses it in searching for "conflict" and in operating Transform 5. These uses are an important part of our understanding also, but because we can use concepts in much richer and more flexible ways we have a more developed understanding of them. And some programs, as we shall see, have a richer understanding than Colby's neurotic.

The program's use of natural language is also very weak. Not only does it have a grossly limited vocabulary, but it is wholly insensitive to the subtleties of syntax. It can handle only "sentences" expressed strictly in the terms of columns 2 to 10 of the Belief Matrix. And as rows 4 and 5 of the Dictionary show, it makes no explicit distinction between *man* and *men*, or between the various forms of the verb *to be*. The complex belief about sister's attitude to father that I have expressed in two different grammatically correct ways would in fact be printed out by the program as follows: *Sister dislike father scorn self disapprove father not-respect women.*

For the purpose of modeling *neurosis*, this drawback is irrelevant. Neurosis is a phenomenon involving particular relations between concepts, particular cognitive and affective schemata within the mind. These schemata may be represented verbally or otherwise (nonverbal representations of concepts and beliefs will be described in Chapters 4 and 7), and even verbal representations of beliefs, like the strange "sentence" above, should not be regarded as unsuccessful approximations to actual sentences in a particular natural language. Freud himself regarded defense mechanisms as nonlinguistic psychological processes, underlying and distinct from their intellectual transformation into verbal form.[6]

But one must remember that Colby's initial hope was to clarify the interpersonal (largely verbal) processes of the therapeutic situation. Indeed, he himself has considered the time-saving, and, he claims, perhaps even psychotherapeutic, possibility of using computers to interview individuals and to make a preliminary listing of their neurotic concerns. To this end he has written a psychiatric interviewing program (basically similar to Joseph Weizenbaum's ELIZA, to be described in Chapter 5).[7] Sometimes Colby's interviewing program successfully elicits from its interlocutor a sustained expression of psychologically significant feelings and beliefs. Nevertheless, this program is of no practical use, on account of its unfortunate tendency to produce such ludicrous exchanges as the following:[8]

PROGRAM—WE WANT TO FOCUS ON YOUR RELATIONS WITH PEOPLE WHO BOTHER YOU MOST.
PERSON—THAT IS EASY. AT THE MOMENT IT IS A SURFEIT OF CHILDREN IN TOO SMALL A SPACE.
PROGRAM—IS IT.
PERSON—YES.
PROGRAM—CAN YOU DESCRIBE AN EXAMPLE.
PERSON—OF WHAT.
PROGRAM—I SEE.
PERSON—NO YOU DO NOT.
PROGRAM—CAN YOU ELABORATE.
PERSON—WHAT ON.
PROGRAM—LETS EXPLORE FURTHER YOUR FEELINGS ABOUT IT.

When you have read the description of ELIZA in Chapter 5, you may like to come back to this psychiatric interview and work out for yourself precisely where it goes astray, and why. For now, we may say that Colby's artificial analyst fails to understand the remarks it elicits largely because, like the neurotic program itself, it has no systematic knowledge of semantics. Also, it has virtually no knowledge of the relevant facts about the world (the psychology of neurosis and the way this shows itself in conversation), nor any reasoning power to draw interpretive conclusions from what is actually said. In Part III we shall see that all of these are necessary in order to understand natural language.

Knowledge and reasoning power are even required to take in new information during a conversation. So I cheated a little when, in discussing the "input" mode of Colby's neurotic, I asked "Would you instantly accept anything your analyst said to you?" The reason why Colby's program does not accept the input remarks is not the same as the reason why you would be loath to do so. The program is not actually capable of accepting new information, except in the most trivial sense (namely, adding an unexamined and isolated remark to its Belief Matrix). Quite apart from any credibility judgments that one might want to make before assenting to a statement, we shall find in Chapter 10 that assimilating an item of new information is far from being a simple matter of adding one more to a list of "truths" in one's memory. Accepting new information typically is a complex psychological process involving inferences of many different types of which one is not introspectively aware. These include inferences based on general semantics and others based on knowledge of the particular domain in question.

These two types of knowledge are also involved in specifically neurotic thinking, and the lack of them leads to the weakness of the defensive routines that is the third fault of Colby's program. Thus my earlier suggestion that a neurotic preoccupation with authority figures might generate an affective response to remarks about policemen's pay depended implicitly upon a wide range of knowledge about the role of police in society, and upon an assumption that readers share this knowledge and recognize its psychological significance. The task of writing a program that could similarly utilize such knowledge would be far from trivial, for it would require explicit articulation of the richly structured processes of symbolic thought informally described by Freud, and implicitly relied upon by Colby in assembling and organizing his artificially restricted data base. Colby's program owes its plausibility to his (and our) intuitive appreciation of neurotic symbolism. It could readily be rendered absurd by a psychologically insensitive compilation of the Dictionary and Substitute Matrix, and/or a wholly undiscriminating use of the various procedures that manipulate the symbols therein. In short, the program's ignorance is not absolute, because Colby implicitly built some of his own semantic insights and knowledge of Freudian psychology into the data base, and into its use by the defense mechanisms.

Thus you will remember that the FINDANALOG routine computes only the number of attributes shared by two people, not their nature. The reason why this does not lead to obvious absurdities in the operation of the transforms is that the attributes (and other items) listed in the data base by Colby include only those which, on theoretical grounds, one might expect to be relevant to neurotic thought processes.

For instance, both one's father and one's boss possess a spleen; but the notion that this attribute should be included in the Dictionary so as to become accessible to the FINDANALOG routine is psychologically ridiculous. This is not to deny that there might be some human neurotic who attaches emotive significance to spleens, perhaps projecting her repressed hatred onto spiders instead of people, and rationalizing her attitude in terms of the nonsplenetic nature of arachnid biology. But such a person would be most unusual, and the peculiar significance attached by her to the spleen would itself require explanation in psychopathological terms. Although possessing a spleen could appropriately be made accessible to the FINDANALOG routine in a simulation representing this particular individual, it would be inappropriate to include it in models of other individuals or of neurosis in general. (This example shows, by the way, that the sorts of trivial generalizations that result from statistical descriptions culled from large samples of patients, and about which proponents of "clinical" prediction in psychology are so scornful,[9] can be avoided by a programming methodology. Clinicians need not fear that the individuality of their patients must in principle be ignored by a computer-based theoretical approach.)

Similarly, Colby's decision to allow only persons as Object-substitutes, in operating the transforms, reflects his insight that we do not typically direct our neurotic maneuvers onto nonhuman objects such as spiders. The specific theory supporting this insight is that the (personal) attributes marked out by Freudian concepts like the Oedipus complex or the super-ego as having basic symbolic significance in human psychology are not attributable to impersonal objects, *except* via the mediation of clearly personal symbolism.

For instance, Freud's claim that Little Hans's phobia of horses was, at base, a defensive form of Oedipal jealousy is plausible only to the extent that one can suggest a psychological rationale for regarding horses as metaphors for one's father. Freud's stress on the size, weight, and power of the horses involved in this case history, as on their black, moustachelike, muzzles, was specifically offered for this purpose.[10] Since Colby's program restricts the transforms to operating solely on persons, it could not represent Hans's phobia even if *horses* were included in the Dictionary. And since it has no way of drawing semantic or commonsense comparisons between horses and fathers, it could not justify deflection of Oedipal fear to horses in any event—unless, of course, Colby himself presciently put the relevant properties on the Attribute list for horses.

Colby's computational definition of conflict catches some important distinctions, as we have seen; but many pertinent distinctions that are intuitively obvious to us are not available to his program. And his computational definition of relevance (beliefs having at least two of their major terms in common) is very crude. The discussion so far should have indicated how semantic knowledge of the interconnections of concepts could help provide the program with more subtle notions of psychological conflict and relevance.

The appreciation of relevance is germane to the last major fault of Colby's program: its failure to generalize its neurosis to people and situations not originally involved. By "generalization" here I do not mean the defensive spreading of hatred specifically to Ms. Smith, in the manner already described. Rather, I mean the general coloring of one's social attitudes with the hues of neurotic mistrust and insecurity.

For example, someone who has an obsessive belief (whether true or false) that her father abandoned her is likely also to fear that her husband might abandon her, that her friends will betray her, that her infant child might let her down—and perhaps even that her dog might give her away, should she be trying to hide from an unwelcome visitor. (She is less likely to fear that her dog or child will abandon her and her husband merely let her down. Why?) There is nothing to stop Colby adding the relevant beliefs to the Belief Matrix, but this is not the point. The program should be able to do it for itself. It should be able to recognize the links between a belief in paternal abandonment and a feeling of being unable trustingly to rely on social role behavior in general.

The "links" in question are partly conceptual (semantic) and partly empirical. Colby's computational concept of a neurotic complex, or pool, of relevant beliefs ignores this distinction, since the relevance linking individual beliefs in the pool is ambiguously semantic and empirical in nature. By his "two-term" definition of relevance, *I fear father* is equally relevant to *I fear husband* and *I fear horses*. Yet one intuitively senses that the neurotic generalization of a woman's fear from father to husband is more natural, more to be expected, than a child's idiosyncratic assimilation of father and horses. This is why the "empirical" information that five-year-old Hans was frightened by a specific incident involving horses could be regarded by Freud not as a clear contradiction of his Oedipal hypothesis (in favor of a "conditioned response" hypothesis), but as an integral part of the Oedipal explanation of Hans's particular phobia. A powerful simulation of neurosis should be able to capitalize on the distinction between beliefs that may very likely be found together, because of their intrinsic meaning, and those that become associated largely because of particular accidents of environmental history.

An adequate semantic theory would make this possible. Similarly, a representation of the semantic content of beliefs that was more economical than Colby's potentially endless listing of specific items (like *Father abandoned me, My friend may betray me, My child may let me down* . . .) could enable a neurotic program intelligently to generalize its attitudes from the family to society as a whole. What is more, other types of thinking could utilize the same semantic insights, so that social learning (and learning in general) could be mediated by way of underlying semantic frameworks or conceptual schemes, to which individual persons, objects, and events would be assimilated.

Some progress has already been made in this direction, for example in the recent work of Abelson. His current theoretical model of belief systems enables one to make important psychological distinctions between the minds of individual people. It also offers a systematic framework for exhibiting the basic psychological structure of various types of social phenomena, including *betrayal*, in terms of which human beings conceptualize their world-view, and by reference to which they lead their daily lives. Further, the development of his ideas over the years shows how a series of simulations of increasing sophistication can constitute what Imre Lakatos has called a "scientific research programme"[11] in personal psychology: programs and people are not quite so different as they may seem.

# 4

# *Personal Politics and Ideology Machines*

PEOPLE can be described in a detailed way that highlights their personal idiosyncracies, or in a more general manner. It is often implied that "human interest" resides solely at the more detailed level, the psychologist's scientific search for generality being contrasted unfavorably with the artist's sensitivity to the individual. But only gossip-columnists (and only the most trivial of these) confine their remarks to the particularized level, at which Lady So-and-So is reported to breakfast on three pieces of thinly-sliced buttered toast with lime marmalade.

Shakespeare's *Romeo and Juliet*, by contrast, is appreciated not only for its imaginative representation of the two lovers themselves but also for its portrayal of universal human themes such as loyalty, rivalry, love, rebellion, and betrayal. It is the correspondingly interrelated structure of these themes that enables one to recognize *West Side Story* as a twentieth-century version of the same essential drama played out in Shakespeare's script.

In thus assimilating the two dramatic creations one intuitively senses that Verona and New York, fencing-foils and flick-knives, are superficial elements in a way in which group rivalries and conflicting loyalties are not. While it is clearly essential that some instruments of aggression (and, finally, of death) be available in enacting the plot, their precise nature is irrelevant. Moreover, in saying "what the story is about" one would focus on the psychological motivations and conflicts involved, rather than their geographical location or the material instrumentalities through which they are realized.

Similarly, one may think of political events as exemplifying abstract and recurrent schemata involving invasion and exploitation, patriotism and subversion, imperialist lackeys and liberal do-gooders. Such schemata are invested with a deeper significance than the personal particularities reported by the gossip-columnists, however fascinating some of these human details may be. (Why are the details of a politician's bedtime reading more interesting than her taste in breakfasts?)

Given that one can, and normally does, intuitively think about social and personal matters in abstraction from their particular circumstances, the question arises how such a generalization is possible. What underlying cognitive structures and processes are implied by its occurrence? This is a central question of social psychology, and it is this question that has motivated R. P. Abelson's social psychological work. Concentrating first on general theories of attitude-change, then on the psychology of political belief, and more recently on a wide range of questions concerning interpersonal liking, decision-making, the action consequences of role membership, and culturally shared knowledge of the constraints implicit in "situations" of various kinds, Abelson has increasingly come to rely on programming methods and metaphors to express and explore the psychological theories being developed. A chronological study of his research, which was initially conceived in traditional (verbal) theoretical terms, illustrates the sense in which computational theorizing can give rise to a "scientific research programme," as I claimed in the previous chapter.

In his recent work, Abelson posits a hierarchical structuring of beliefs that accounts for systematic similarities between human interpretations of the world, as well as allowing for interpretative variations between different cultures and individuals. In terms of his theoretical concepts he could exhibit the essential identity of a love story variously told over the centuries by Bandello, Shakespeare, Ernest Lehmann ("the book"), and Jerome Robbins ("the film"), as well as illuminating fundamental interpretative disagreements between political commentators committed to differing world views. And insofar as he offers a general representation of belief systems, Abelson's work may be of use to those attempting to model varied types of belief system using the methods of artificial intelligence.

## SIMULATION OF HOT COGNITION

One way of describing Abelson's interests is to say that he was initially concerned with "hot" (affectively influenced) cognition and has latterly generalized his attention to "cold" thought also. A continuing focus of his concern has been the psychology of political belief, the way in which a person's various opinions normally reflect a fairly coherent ideological position. His first efforts at computer modeling were related to the psychology of attitude change, in particular the theory of cognitive balance deriving from Fritz Heider and later elaborated by Abelson himself.[1] Consequent efforts sought to program an "IDEOLOGY MACHINE" that would incorporate some of the processes posited by the earlier model and would simulate the belief structure of a particular politician. In his attempts to refine and generalize this program, Abelson was led to develop a theoretical representation of belief systems in general.

Abelson's early computer simulation of "hot" cognition was designed to deal with the input of affect-laden beliefs into a preexisting system.[2] A new belief is first assessed for its prima facie credibility, and then for its evaluative balance with respect to the values coded in the belief system. The central claim of balance theory is that evaluative balance is normally preferred over imbalance.

The theory tends not to discriminate between different evaluational domains: positive is positive, no matter what the context. Thus it suggests that you would feel disturbed if someone of whom you are contemptuous were to recommend your favorite recipe, just as you (more probably) would if she were to praise your favorite film or novel. The fact that you might allow her to have excellent culinary tastes, or to be adept at literary criticism, is not usually allowed for. (Actually, since aesthetic criticism is a largely moral matter, it is difficult to imagine that one might *regularly* agree about the value of films or novels with someone of whose whole life-style one disapproved.) However, even if all of the people, all of the time, do not think in the undiscriminating way that balance theory suggests, all of us do—sometimes. And, as the example of aesthetic criticism suggests, the more important the issue, the more one tends to expect that the people one approves will value the things one holds dear oneself.

According to balance theory, then, you would experience some degree of psychological conflict if you were presented with an imbalanced fact, such as that your particular bête noire shares your cinematic tastes. The theory states that, in such a case, you would either reject the new "fact" as incredible, or would assimilate it with the help of various processes including denial and rationalization,[3] or would accept it at face value—in which event your extreme evaluative attitudes to both praiser and praised would be moderated by your acceptance of this new information. Which of these phenomena would result depends on the nature of the imbalance and of the attitudinal structure extant. Abelson aimed to simulate the processes by which attitude change is effected or resisted in particular cases, so offering a dynamic model of the development of belief systems in the face of new, and sometimes incongruous, information.

It has already been remarked that some of Abelson's ideas were later adapted by K. M. Colby, in the "input" version of his machine neurotic. For instance, the broadly defensive processes of rationalization and denial described by Abelson fitted fairly readily into Colby's Freudian model. Abelson's program allowed for three subvarieties of rationalization, which he termed REINTERPRET FINAL GOAL, ACCIDENTAL BY-PRODUCT, and FIND THE PRIME MOVER.

He gave the example of the cognitively imbalanced input sentence, *My simulation produced silly results.* My simulation, being mine, is positively valued; but silly results are negatively valued, so the sentence as it stands is imbalanced. The subroutine REINTERPRET FINAL GOAL searches the data base for a positively valued implication of the concept *silly results.* The first implication found on the relevant list by the program may be *nonpublishable*; if the belief system represented is one of those which abhors nonpublishability even more than silliness, then *My simulation is unpublishable* will be rejected as unsuitable for purposes of rationalization. The next listed implication of *silly results* may be *enrichment of understanding*; since *My simulation enriched my understanding* is cognitively balanced, it is accepted. Analogously, the subroutine ACCIDENTAL BY-PRODUCT searches for a factor that explains the silly results by attributing them to a negatively valued source accidentally interposed between subject and object: thus *My simulation had program bugs* is an acceptable rationalization. Finally, FIND THE PRIME MOVER replaces the subject by a disvalued substitute, as in *That crazy programmer produced the silly results.*

It is important to note that the positively valued implications, the appropriate interfering sources, and the possible disvalued substitutes that are called up by the subroutines are explicitly stored in association with the relevant concepts: Abelson's program does not generate them for itself. Abelson's more recent work, to be discussed presently, enables implicational relations of various types to be stated in a general form that allows a program to generate specific instances without relying slavishly upon the prescience of the programmer. And in later chapters programs will be described that have a deductive capacity and can originate reasoned chains of logical implications directed to a particular end, abilities presumably deployed in some cases of human rationalization. In real life, rationalizations or reasonings may be conscious or unconscious. If one considers the belief *Einstein produced silly results,* one may well mutter "Impossible! There must be some other explanation," and then deliberately think one up. "Neurotic" rationalization is probably basically similar to conscious reasoning, but as well as being hidden to introspection it is maladaptive, in accepting relatively unrealistic alternatives to the unacceptable belief: perhaps her hated father didn't really abandon Colby's patient at all. Colby's appropriation of Abelson's rationalizing techniques could therefore afford to be less careful about assessing the credibility of the transformed version of the original input.

The *credibility test* developed by Abelson for his "balancing" program is an early version of a more powerful technique incorporated in his current computer model of ideology. The test calls subroutines of increasing complexity, which employ quasi-inductive procedures to assess the plausibility of a sentence that is not already specifically affirmed or negated in the data base. These probabilistic procedures search for items of belief already accepted as data that may provide evidence relevant to the belief in question, and assess their evidential power. The search and adjudication are guided by reference to the semantic relationships of *instance, attribute, similarity,* and *contrast* that are stored in the Dictionary. So, for example, the input "Nixon lied today" is assessed in the light of data like "Nixon lied yesterday," "Mitchell lied last week," "Politicians sometimes lie," "Republicans are honest," "Presidents are truthful," and so on. Clearly, credibility is dependent on the particular belief system already existing, so that programs (or people) with different data available to them make different judgments as to credibility.

Provided that a sentence is not immediately rejected as being incredible, it is passed on for further consideration by the system. High evaluative imbalance elicits defensive adjustments within the belief system by means of cognitive operations such as the rationalization previously described. The order in which the mechanisms are tried depends on the precise nature of evaluative imbalance, and Abelson offers a psychological rationale for the specific pattern of dependencies he adopts.[4] But each type of operation is available should the others fail. Failure can occur for a number of reasons—for example, because the new sentences formed by application of the mechanisms are themselves always assessed for credibility and balance, and are sometimes rejected. But if the balancing adjustments succeed the resulting new sentences are stored, so that the belief system changes gradually over time.

Consequently, both judgments of credibility and the detailed effecting of balancing mechanisms may vary at different points in the running of the program. Analogously, one is more resistant to believing that Nixon lied the first time one is told this, and less as one becomes accustomed to rationalizing his successive pronouncements in an increasingly cynical manner.

This change in attitude might be described as a strengthening of prejudice against Nixon, for one judges him probably to be lying prior to examination of the specific issue in question; indeed, his credibility may be so low that one never feels it necessary to undertake such examination. Whether this attitude should be called "realistic" judgment or "unjust" prejudice is itself largely a matter of ideological preference, depending on political views that place the individual Nixon in a wider social context. It is views such as these that formed the specific focus of Abelson's next exercise in computer simulation, his so-called "IDEOLOGY MACHINE."

## THE IDEOLOGY MACHINE

Just as Colby's machine neurotic is a program representing defensive thought in general, but implemented by way of a data base modeling the neurotic themes of a specific individual, so Abelson's artificial ideologue is a program representing generalizable thought processes, one that is currently exemplified by way of an idiosyncratic memory store that parallels certain beliefs of a particular person. Specifically, Abelson describes his IDEOLOGY MACHINE as a model of responses to foreign policy questions by a rightwing ideologue, such as Barry Goldwater in his heyday.[5]

Foreign policy questions concern the political implications of events (E) involving political actors (A) such as individual statesmen, particular nations, or groups of nations. For instance, one might ask whether an event is credible, whether it could possibly have happened: Could the United States have attacked a neutral nation, South Vietnam? Senator Goldwater would predictably regard such a suggestion as incredible, it being inconceivable to him that such a thing could happen. Similarly, he would respond predictably to questions like "If a Communist nation attacks West Berlin, what should America's NATO allies do?" Indeed, one can fairly confidently predict his responses to foreign policy questions of the following types, each of which is represented in Abelson's simulation:[6]

1. Is E credible? That is, could E happen or have happened?
2. If and when E, what will happen?
3. If and when E, what should A do?
4. When E, what should A have done?
5. How come E? That is, what caused E, or what is E meant to accomplish?
6. Sir, would you please comment on E?

The senator's responses to such questions are predictable because one knows they will fit into a general ideological pattern corresponding to his beliefs about what kind of events belong naturally together. Similarly, one knows the ideological biases of his political opponents. Indeed, this knowledge commonly enables one to form a reasonably accurate general impression of a politician's views on the basis of a single remark quoted in the newspaper—about "law and order," for instance, or "imperialist lackeys." But this knowledge is largely implicit, and if one is to represent it as a psychological theory (whether programmed or not) one must first express it in explicit fashion.

Cognitive balance theory, which stimulated Abelson's first ventures into the computer modeling of thought, is itself a theory about "what kind of events belong naturally together." But it makes explicit only the tensions arising from affective imbalance in "hot" cognition: as we have seen, the "cold" process of deducing likely implications of events (of silly results, for example) is unspecified by the theory and so has to be left entirely to the programmer in a computer implementation.

Again drawing on the example of Heider, Abelson next formulated a theory of "implicational molecules" that modeled the subjective attribution of causes to events in terms of psychological verbs of possession, wanting, doing, liking, preventing, hurting, and so on.[7] An implicational molecule is a meaningful set of sentence types with linked elements, such as: A does X, X causes Y, A wants Y.

The molecules within a particular belief system determine the pragmatic implications attributed to events mentioned in the input, and different stocks of molecules mediate different implications. For example, the input "Ms. Brown was rude to the policeman" may lead one to infer either that Ms. Brown was in a bad temper or that the policeman had first hurt her in some way. But we have seen that a person familiar with psychoanalytic theory may attribute neither ill temper to Ms. Brown nor hostility to the policeman, instead explaining the psychological causation of the incident in terms of Ms. Brown's neurotic hostility to authority figures. Moreover, layman and psychoanalyst will request different types of information should they be sufficiently interested in the incident to examine it further. In this sort of way, implicational molecules guide one's attribution of and search for the pragmatic meaning of input sentences so as to integrate them within a preexisting structure of beliefs.

The overall similarity of the "foreign policy questions" listed above to the cognitive functions of implicational molecules should be clear, and Abelson expressed Goldwater's rightwing views in these terms before relinquishing the notion of "molecules" in favor of the semantically less restricted notion of "scripts" that is central to the IDEOLOGY MACHINE's inner representation of conservatism. The basic insights of his theory of implicational molecules, however, are retained in the IDEOLOGY MACHINE.

Abelson postulated that an ideological belief system is built up out of a basic *vocabulary* whose items can be linked in certain ways. The vocabulary of his IDEOLOGY MACHINE is implemented as a collection of 500 noun and 100 verb phrases relevant to foreign policy matters, such as "Nixon," "Vietnam," and "sells arms to."

Vocabulary items are classified in terms of a number of general *conceptual categories*: 15 for nouns (such as Communist nations, left-leaning neutrals, Free World nations, and liberal dupes) and 11 for verbs (such as physical attack and material support). These conceptual categories can be combined to specify 300 *generic events* (such as the physical attack of a neutral nation by a Communist nation), which in turn are combined into *episodes*.

An episode specifies a temporal sequence of potential generic events, allowing for multiple branching: thus one can envisage the possibilities that if a neutral nation is attacked by a Communist nation then it may become a Communist satellite, or it may resist and successfully seek aid from the Free World so leading to a victory over Communism, or it may be given such aid without actually asking for it, and so on. (It is largely by virtue of their temporal content and their inner branching that episodes are more powerful than implicational molecules.)

Finally, episodes themselves (of which there are two dozen embodied in the IDEOLOGY MACHINE) are integrated by way of a *master script* that represents the overall ideology at a high level of generality. Abelson's diagrammatic representation of the master script appropriate to an arch-conservative (and currently implemented in his computerized Cold Warrior) is shown in Figure 4.1, and is expressed verbally as follows:[8]

> The Communists want to dominate the world and are continually using Communist schemes (Branch 5) to bring this about; these schemes when successful bring Communist victories (Branch 6) which will eventually fulfil their ultimate purpose; if on the other hand the Free World really uses its power (Branch 4), then Communist schemes will surely fail (Branch 7), and thus their ultimate purpose will be thwarted. However, the misguided policies of liberal dupes (Branch 2) result in inhibition of full use of Free World power (Branch 3); therefore it is necessary to enlighten all good Americans with the facts so that they may expose and overturn these misguided liberal policies (Branch 1).

Abelson's representation of ideology is not a mere static "snapshot" of the semantic skeleton of the political belief system in question. It is potentially dynamic in that it can be used to show how the system's knowledge is deployed in constructing its subjective world and in inter-

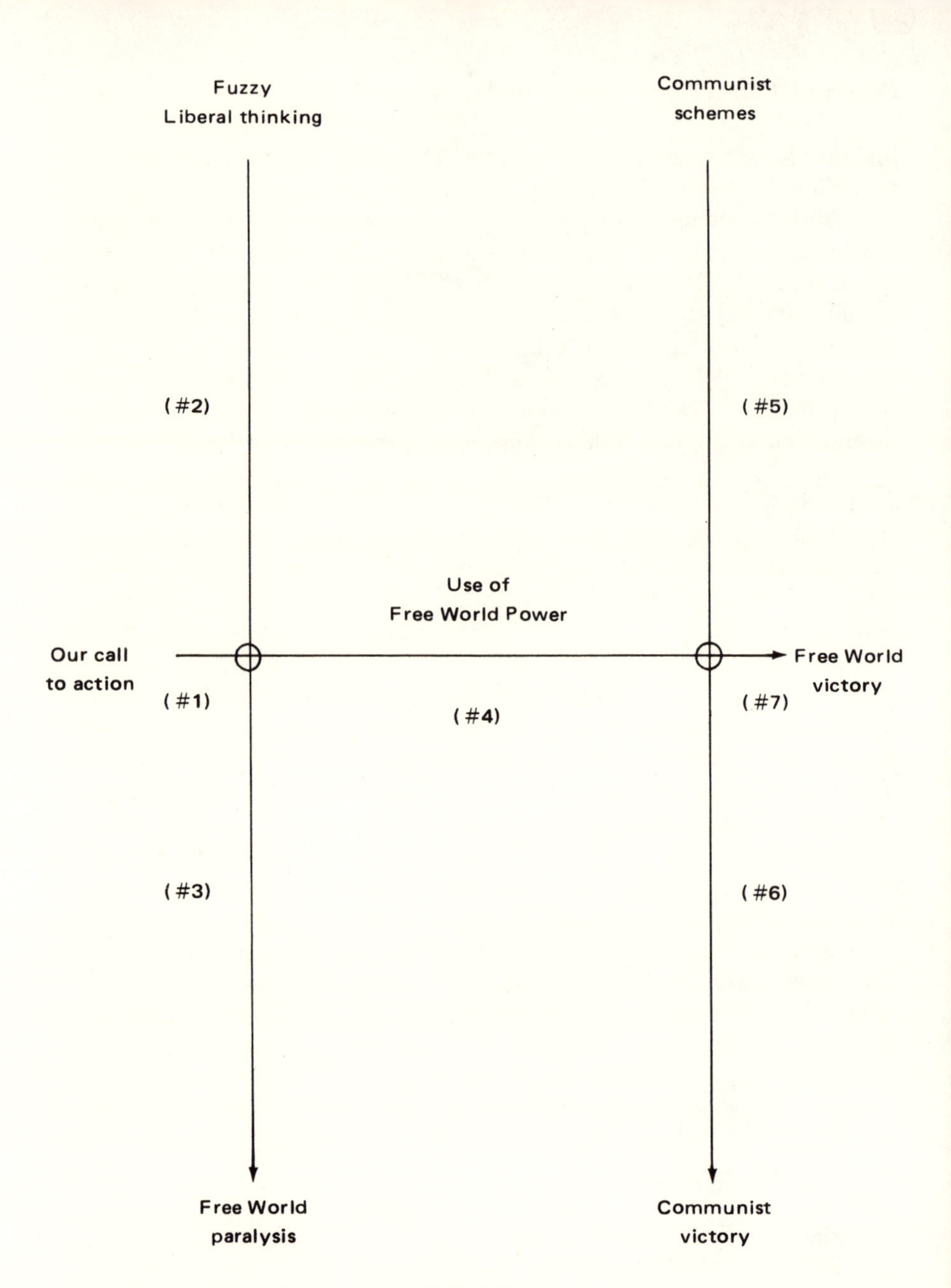


FIGURE 4.1
Master Script for a Cold War Ideology.

Branches are numbered for reference.

The symbol ⊕ signifies that the vertical path will be taken unless the horizontal path is activated.

Source: *Computer Models of Thought and Language*, eds. R. C. Schank & K. M. Colby (San Francisco: W. H. Freeman and Company, Copyright © 1973), p. 291. This and following tables from this source reprinted by permission.

preting new information consistently with its preexisting viewpoint. His program specifies precisely how this knowledge is to be so used by identifying procedures of conceptualization, hypothesis, and prediction that operate on the available structured vocabulary or data base.

For example, by following up the branches of the master script shown in Figure 4.1 one can sketch out answers to some of the broad categories of questions that were listed above. Thus the query "If and when E, what will happen?" might be answered by way of the general schema, "When Branch 5, then Branch 6, unless 4, in which case 7." This assumes that the event E, considered as a generic event type, is to be found on Branch 5 of the master script. Abelson's program therefore first locates the event mentioned by the input question within its basic vocabulary and/or repertoire of conceptual categories, then searches for a relevant generic event on its representation of the master script, and finally generates a reply accordingly. When asked, "If Communists attack Thailand, what will happen?" the program answered, "If Communists attack Thailand, Communists take over unprepared nations unless Thailand ask-aid-from United States and United States give-aid-to Thailand." The somewhat eccentric syntax apart (which is due to linguistic crudities in the program similar to those afflicting Colby's neurotic), this answer appears reasonable enough as words to be put into the mouth of Goldwater, and appropriately inappropriate if suggested as a remark attributed to President Tito.

A number of theoretical problems associated with this Golemized Goldwater are also relevant to other species of belief. These problems have motivated Abelson's more recent attempts to formulate a structural theory of belief systems in general.

For instance, Abelson points out that the program might have replied in blander fashion to the query about the consequences of a Communist attack on Thailand, observing merely that "they will score a victory unless we do something." The problem of when, and how, to be specific in reply to a journalist's questions is a special case of the recurring problem of when, and how, to be specific in expressing one's beliefs. In saying what *Romeo and Juliet* is about, does one say that it is about two teenagers or not? Assuming that the answer depends on the context of the question, what specific features of the context determine the level of detail appropriate, and how does one search through, or access, one's rich store of relevant knowledge so as to find the particular details required? (A similar issue to be discussed in later chapters concerns the representation of knowledge in the posing and solving of problems, whether by person or program: what determines the level of generality best suited to the particular situation?)

A second difficulty connected with Abelson's political simulation is that of avoiding the acceptance of beliefs that are prima facie ideologically reasonable, but in fact absurd. For instance, if required to assess the credibility of the proposition that Red China built the Berlin Wall, the IDEOLOGY MACHINE would decide that this is indeed the sort of anti-American activity that one might expect from the evil Chinese. Short of a desperate recourse on the part of the programmer to ad hoc measures providing the relevant information, the program has no way of knowing that grave logistic difficulties are involved in building a wall thousands of miles away from one's own territory. Moreover, were it to be told this, the program would merely reverse its judgment of credibility, now stating it to be impossible that Red China should have built the Berlin Wall. Senator Goldwater, by contrast, would be neither so ignorant nor so naive. He would recognize the inherent improbability of the suggestion put to him, but would also be able to sketch ways in which it might be rendered credible: perhaps Red China persuaded a fellow Communist nation, namely, East Germany, to act as its agent in building the wall. The credibility of beliefs (including hypotheses, plans, and predictions) about actions can be sensibly assessed only by taking into account such factors as physical proximity, availability of instruments, and the likelihood of social contracts enabling one person or group to act in another's interests. If these matters could somehow be systematically represented as contributing to the structure of political beliefs, a more powerful simulation of Goldwater's conservatism would be possible.

Lastly, and closely connected with the two previous points, remains the problem of generalizing Abelson's computer model of Goldwater's foreign policy to other political issues, to differing political ideologies, to widely varied systems of belief concerning human action (whether limited to small groups or conceived of on a world scale), and even to all belief systems whatever. Abelson remarks that the master scripts of many political ideologies besides Goldwater's would fit the same general schema:[9]

> The bad guys have evil plans which are succeeding, and only the good guys can stop them. Unfortunately, the good guys haven't done it yet, and the reason is that the bad guys have the help of dupes, fools, lackeys and running dogs who wittingly or unwittingly interfere with the efforts of the good guys. The only hope, therefore, is to rouse the wrath of the people against the bad guys and their puppets.

One might add that this script, or one very like it, would fit many small-scale human dramas, both on-stage and off: cops and robbers, Montagues and Capulets, them and us . . . even God and Satan and their warring cohorts are commonly appealed to in making sense not only of the cosmos in general but also of particular experiences lived through by the individual.

In other words, one interprets one's personal world in terms of a set of categories and expectations that is idiosyncratic only to a degree, and that seems to share with those of others an underlying similarity of structure that could possibly be explicitly articulated. A generalized representation of beliefs would be helpful not only to psychologists interested in the thought processes underlying neurosis, personal prejudice, and political commitment, but also to any worker in "pure" artificial intelligence who finds it necessary to provide her system with a rich base of knowledge in order to enable it to achieve the desired results. As will be made clear in the following chapters, such a necessity arises all too often in this context: even to enable a machine to understand the word "it," or to recognize a cube, one has to give it considerable knowledge of the world —unless one restricts the circumstances of the task so drastically as to produce a pitifully weak analogue of the "equivalent" human competence.

## A STRUCTURAL THEORY OF BELIEF

Abelson's recent, and most general, cognitive theory represents belief systems on six distinct levels, each defined in terms of the more concrete levels below.[10] At the base of the system are the *elements*. Elements combine in specified ways to form *atoms*; atoms combine to form *molecules* and *plans* (plans being complex molecules); and *themes* are built out of interdependent plans. Finally, *scripts* represent successions of themes that are regarded as "natural" or "coherent" by the individual concerned.

The structural details are specified in verbal and diagrammatic form, there being as yet no programmed version of the theory. Abelson intends to revise his IDEOLOGY MACHINE in accordance with his latest work, and the high standards of clarity and precision in his formulation should render his task less arduous and frustrating than many exercises in the computer simulation of psychological theory. Although it is not impossible a priori that Abelson might have produced this representation of belief without ever having had recourse to a programming methodology, there can be no doubt in fact that his experience in the computer modeling of thought has aided him considerably in making explicit his intuitive knowledge about the human mind.

This latter point is doubly relevant to the first three levels (elements, atoms, and molecules), for Abelson's account of these levels draws heavily on the work of R. C. Schank, who formulated a semantic theory intended for use in the computer understanding of natural language. Schank's Conceptual Dependency Analysis is useful to Abelson because it specifies explicitly the interrelations between basic semantic concepts that normally remain implicit in expressions of belief.

For instance, it was pointed out above that to envisage a group's having built a wall is implicitly to assume that they, or their agent, were in physical proximity to the place where the wall now is at the time when it was being built. Also, of course, it implicitly assumes that the persons concerned had access to the instrumental means (bricks and mortar) required for such a construction, and that the agent, if any, had entered into an established or temporary social contract whereby particular sanctions would have effect. Sometimes one needs to call these implicit semantic relations explicitly to mind, as when asking oneself whether Red China could possibly have had a hand in the building of the Berlin Wall. It is this sort of conceptual relation within verbs of action with which Schank is primarily concerned and that Abelson has incorporated into the lower levels of his theory of belief. If he had not, he would have no systematic way of assessing the credibility of "Berlin Wall" questions, or of representing the fact that any stage manager who forgets to provide weapons for *Romeo and Juliet* must rely on the actors' capacity for spontaneous fisticuffs if the show is to go on.

Since Schank's account of conceptual dependencies will be discussed at length in Chapter 7, I shall not give details here. For present purposes, it is sufficient to note that the elements of Abelson's system are similar to the basic lexicon used by Schank, and are combined into three types of atoms—P, A, and S. These connote purposes, actions, and states, respectively, and each atom is represented by a conceptual dependency diagram exhibiting the sorts of basic semantic connections referred to in the preceding paragraph.

For example, the diagram corresponding to the concept "build" would explicitly code the fact that there has to be a specific (though perhaps unnamed) actor doing the building, in a specific place, using specific instrumental means. Similarly, an A-atom is available for representing the proposal by one actor that another actor does an action suited to the proposer's purposes, and it includes a representation of the fact that there must be a particular social contract involved in such a case.

This type of A-atom is especially important if one is to construct belief systems concerning potentially cooperative interpersonal actions, whether in the domestic, dramatic, or diplomatic arena. The psychoanalytic concept of "authority figure," for instance, relies crucially on the assumption of proposals of this type where the social sanctions involved are of a particular nature. If one resists a policeman's courteous injunction to "Move along there, please!" because of a neurotic attitude originally directed toward one's father, the common core of the superficially disparate interpersonal situations is expressible by means of the relevant A-atom. It is because of this sort of conceptual representation that

Schank's semantic analysis might have been useful to Colby in constructing his machine neurotic, and is accepted as basic by Abelson in theorizing about beliefs in general.

To pass from atoms to molecules, Abelson specifies the following requirements:[11]

> Three atoms, P, A, and S are said to form a P-A-S molecule when three conditions are satisfied, one between each pair of atoms:
> (PS). The S-atom is the state connected to the "want" in the P-atom.
> (AS). The A-atom is causally bonded to the S-atom.
> (PA). The actor in the A-atom is an agent (for action A) of the actor in the P-atom. (Note: An actor can always act as agent for himself, but the agency of a second party requires a special condition.)

The similarity of this specification to the previously cited implicational molecule, "A does X, X causes Y, A wants Y," should be evident. The reason for this similarity is that this particular implicational molecule, like the recent definition of "molecular" units in general, captures the idea of an action undertaken in order to attain a goal desired by the sponsor of the action. As Abelson remarks, this idea is "the essential building block of all belief systems which find meaning in the purposive activities of individuals, institutions, and governments—or even the animistic forces of nature or gods."

(Strictly, the P-A-S molecule is adequate to capture the idea of *intention* in action only if the *causal bonding* is interpreted in purposive terms: not only must the agent believe that A leads to S, but she must do A *in order to* achieve S—that is, wanting S must be her *reason* for doing A. The italicized words must be interpreted in terms of what Aristotle called "final causes," not "efficient causes." Otherwise, counterexamples can be cited that satisfy the conditions but that are not cases of intentional action. For instance, a Roman Catholic doctor may operate on a pregnant woman to save the baby's life, and know that in doing so she will cause the mother's death; she may also want the mother to die, for one reason or another; but it does not follow that she operates *in order to* cause the mother's death, or that she can be said to have caused the mother's death *intentionally*. The term "cause" is often used in ordinary language interchangeably with "purpose," and it is this sense that is here intended by Abelson.[12])

Molecules more elaborate than the simple P-A-S type are termed "plans," and they arise predominantly when there is some doubt whether the actor can effect her purpose directly. The intermediate steps toward the goal have to be specified in the plan, and so also do relevant enabling conditions such as physical proximity, instrumental control, and social contracts entitling the main actor to rely on an agent's acting for her. The satisfaction of these enabling conditions is sometimes taken for granted, in such cases being conceived of as exogenous to the plan itself; but sometimes they are regarded as problematic and are specifically provided for in the plan. Further complexities arise if there are alternative routes to the goal or if the actor plans to "kill two birds with one stone" by achieving two purposes concurrently. Various plan structures are carefully and explicitly distinguished by Abelson, examination of which throws light on the complex inner structure of human intentions and potentially also on the representation of action in the context of artificial intelligence.[13] He acknowledges an intellectual debt to G. A. Miller, Eugene Galanter, and K. H. Pribram, whose book *Plans and the Structure of Behavior* provided a seminal conceptualization of purposive behavior in terms of a programming analogy.[14]

Unlike these authors, however, Abelson is primarily interested not in how the planner actually plans but rather in how an observer would conceive of a planner planning. In other words, Goldwater's views on the likely intentions of Red China (and on the consequent strategies of various nations and individuals) are of more direct interest to Abelson than the innermost psychological details of the plots of Chairman Mao. This difference in emphasis is pertinent to the progression from plans to the fifth level in Abelson's theory, that of "themes."

## BETRAYAL AND OTHER THEMES

Themes are made up of the interdependent plans of two actors, and it is largely by reference to them that one makes sense of the personal and social worlds experienced by members of disparate cultures and subcultures. The structure of themes may be highly complex, even though the inner structure of the constituent plans themselves is ignored in representations at the thematic level. Abelson expresses these abstract matters both verbally and diagrammatically: while he does not claim to have exhausted the structural richness of themes, his theoretical formulations of themes, as of plans, is admirably clear and is articulated with the specific possibility in mind that it may be implemented in a programmed model of belief.

Abelson distinguishes three ways in which the autonomous purposes of two actors are commonly conceived of as linked. First, one or both may have a role in the other's plan—and there are at least three commonly recognized types of role in this context. One or both may act as the other's agent, either for the entire plan or for certain parts of it—and the agency may be temporary or established: East Germany's hypothetical agency on behalf of Red China would be a case in point. One or both may be involved in the other's goal, as when a Free World politician aims for the diplomatic humiliation of her opposite number behind the Iron Curtain. And one or both may be an interested party in relation to the effecting of the other's plans, as when two nations compete for the scarce instrumental means of achieving their respective goals, or when another person's plan of campaign disturbs one's own activities or presents one with opportunities for action.

Second, one or both may have a positive or negative evaluative attitude to the plan (in whole or in part) of the other, and to her own role in it, if any. For example, Senator Goldwater and other good Americans disapprove of all Communist schemes, and the ministers of a satellite nation may resent their political servitude even if they are in overall ideological agreement with their foreign overlords.

Third, one or both actors may have the ability to facilitate or interfere with the other's actions, whether at particular points only or throughout the plan: liberal dupes may sometimes further Communist schemes.

Since each of these three dimensions (role, attitude, and facilitative ability) is logically independent of the others, since each may be instantiated reciprocally rather than in one direction only, and since each may apply selectively to distinct parts of the other's plan rather than monolithically to the whole of it, it is evident that the range of structural possibilities is in principle very large. Were one to identify further dimensions of the interdependence of plans, the potential thematic variety would be increased accordingly.

Clearly, an explicit taxonomy of themes would provide a welcome way of organizing these structural possibilities. And such a taxonomy could also be used as a theoretical guide directing empirical psychological investigations of and comparisons between the belief systems of different individuals or cultures. Last, and in the present context not least, it could generate a whole tribe of Ideology Machines, wherein the family relationships could be systematically exhibited and all paternity suits could be unambiguously settled.

Using the three dimensions distinguished above, Abelson offers a systematic taxonomy that economically expresses a range of thematic possibilities. He does not claim that this is the only possible classification, and he stresses that only empirical research can show which themes are actually present in a given belief system, which are widely shared throughout humanity, and which are idiosyncratic to a particular culture or person. He identifies themes within nine categories, generated in terms of the reciprocal influences and sentiments of the two actors as shown in the nine-celled table in Figure 4.2. He provides at least one theme name (together with a verbal and diagrammatic description) in exemplification

FIGURE 4.2
*A Taxonomy of Themes*

| *Sentiments Toward Other* \ *Influence of Actors* | Neither Influences Other | One Influences Other | Both Influence Other |
|---|---|---|---|
| Some Positive, No negative | Admiration | ($T_1$) Devotion<br>($T_2$) Appreciation | ($T_3$) Cooperation<br>($T_4$) Love |
| One Actor Negative | ($T_5$) Alienation (also, Freedom) | ($T_6$) Betrayal<br>($T_7$) Victory (also, Humilation)<br>($T_8$) Dominance | ($T_9$) Rebellion |
| Both Actors Negative | ($T_{10}$) Mutual Antagonism | ($T_{11}$) Oppression (also, Law and Order) | ($T_{12}$) Conflict |

*Source:* Adapted from Schank & Colby, eds. *Computer Models of Thought and Language*, p. 320.

of each category; when he assigns more than one theme to a given cell or category he justifies this assignment in terms of specific variations in content or secondary structural relations between the two. The themes discussed by Abelson are admiration, devotion, appreciation, cooperation, love, alienation, freedom, betrayal, victory, humiliation, dominance, rebellion, mutual antagonism, oppression, law and order, and conflict.

It is apparent from Figure 4.2 that there are three cases in which "one" theme has two contrasting names: alienation-freedom, victory-humiliation, and oppression–law and order. The reader who intuitively senses that the attribution of these thematic concepts in real life "all depends on your point of view" will find her intuition explicitly coded in Abelson's table, wherein these theme pairs are shown to be basically asymmetric in the influence and/or sentiments of each actor upon the other: what Goldwater regards as a victory, his Communist enemy regards as a humiliation. It is because of this sort of underlying structure that one can often intuitively form an overall impression of someone's political viewpoint and social identification from a single remark: an approving use of the expression "law and order" is a familiar case in point.

The potential power of this concept of themes may be illustrated by considering one example in detail: let us take betrayal. Betrayal is defined by Abelson as a theme in which "actor F, having apparently agreed to serve as E's agent for action $A_j$, is for some reason so negatively disposed toward that role that he undertakes instead to subvert the action, preventing E from attaining his purpose."[15]

This definition may seem to imply that betrayal always has unfortunate consequences for E. But since Abelson systematically constructs all themes out of lower level plans, which correspond to intentions, he could admit as betrayal actions that fail, in the sense that they do not in fact sabotage E's purposes. Moreover, since the plans concerned are those *attributed to* actors, rather than those actually intrinsic to actors, he could justifiably classify as betrayal actions, like Judas' kiss, that are directed against purposes mistakenly attributed to E.

It is crucial that the action $A_j$ be precisely specified in the agreement between the two parties if "unjust" accusations of betrayal are not to result: thus Goldwater's views about the range of actions covered by the treaty between the United States and her NATO allies may lead him to describe as "betrayal" a diplomatic incident that the ally sees rather as a blameless expression of political neutrality on a matter irrelevant to the alliance. This is more clearly apparent from Abelson's diagrammatic representation of betrayal than from his verbal definition, since although the diagram omits most of the structure of E's plans, it marks the action or actions involved in the agreement concerned. Correlatively, his diagram for the theme of cooperation (that is, the end theme operative in a script of alliance) also specifies the actions involved in the relevant social contract. Political misunderstandings like that described above arise if the inner content of the allies' respective cooperation themes do not mark out identical actions within the overall plans of the actors. The difference in the actors' descriptions of the incident may therefore be due to a vagueness in the initial specification of action.

Also, of course, this structural feature of betrayal may be deployed (whether consciously or not) so as to produce plausible rationalizations of the questionable action—in other words, systematic possibilities exist for "redefining" the range of action $A_j$ so as disingenuously to defend oneself against the accusation of disloyalty.

The difference in the two descriptions just mentioned rests on a vagueness in the specification of action. But one might expect that there should typically be two ways of describing this theme, even when there is no discrepancy between the actors' identifications of $A_j$. This is because betrayal is a basically asymmetric theme, falling into the same taxonomic cell as the victory-humiliation pair. And indeed, if one examines examples where actor E claims to have been betrayed one rarely finds that actor F describes the incident in the same terms.

Yet there is no one candidate eligible to serve as the thematic reciprocal of betrayal (nor does Abelson provide one), because actor F may give one of many different characterizations of her action according to the particular circumstances. For instance, what both Montagues and Capulets saw as a betrayal of the family, Romeo and Juliet saw in terms of a different theme altogether (love), one whose demands take precedence over usual family loyalties so that the concept of betrayal is out of place; what Czechoslovakia saw as a shameful betrayal, England represented as excusable prudence necessitated by military unpreparedness; and what Hitler saw as capital treason, von Stauffenberg regarded as justifiable action following on a change of heart that unilaterally nullified the former contract between himself and the Führer.

However, there is one feature common to this superficial variety of reciprocal descriptions: in each case the actor F somehow rebuts the suggestion of betrayal by reference to some excuse, which normally takes the form of showing that the "reason" mentioned in Abelson's definition was a morally compelling one. In short, because the role of actor in a theme of betrayal is generally morally disapproved, no one wishes to acknowledge it. Indeed, the statement "I betrayed her" is only heard in a confessional context. By contrast, although few care to be humiliated no moral turpitude is admitted in saying "I have been humiliated." (A sense of shame and anger there may be, but that is a different matter.) If betrayal were morally neutral, or even less fiercely disapproved than in fact it is, there might be a pair of reciprocal theme names for this structurally asymmetrical theme. As it is, the "first-person" experience of betrayal is almost always repressed, and the action is redescribed in different ways according to the circumstantial excuses or rationalizations available.

If these conjectures are correct one might expect to find that other asymmetric themes wherein one role is overwhelmingly disapproved similarly lack a reciprocal theme name expressing the experience of the "first person" actor, or moral culprit. And since moral attitudes to themes (as well as themes themselves) vary from culture to culture, one might expect corresponding linguistic phenomena covarying with cultural attitudes.

Using Abelson's scheme one can search for and systematically describe distinctions between different species of betrayal in a manner that could be utilized by a belief system (whether natural or artificial in character) so as to generalize certain social attitudes beyond their original scope. For instance, abandonment and letting down can each be regarded as a species of betrayal. To accuse F of abandoning E is to say that she was acting initially as E's agent for action $A_j$ (this action being crucial to E's welfare), that she has now deliberately stopped doing so, and that this amounts in effect, if not necessarily in intent, to the deliberate subversion of E's purposes, since E is conceptualized as helpless without F; to say that F let E down, by contrast, implies neither the urgency of $A_j$ nor the helplessness of E. In short, whereas anyone can let down or be let down, only the strong can abandon and only the weak can be abandoned—which is why abandonment is a peculiarly nasty form of betrayal.

Colby's neurotic patient who believed "Father abandoned me" therefore had good reason to be suspicious of social contracts in general, and weak-strong contracts in particular: it would not be surprising if she believed also "My husband will abandon me." For the credibility of statements is often assessed by reference to past instantiations, if any, and to instances of closely comparable events. Indeed, Abelson's earliest credibility test rested solely on this type of "inductive" evidence, as we have seen.

But Colby's patient might have had other grounds also for fearing that her husband would abandon her: she might have attributed a purpose to him such that it would suit him to desert her. Abelson's "balancing" program could not assess credibility in these terms; this is why the list of "relevant" sentences I suggested for assessing the input "Nixon lied today" contained no member indicating what Nixon might stand to gain by lying. But a charge of lying can sometimes be thought credible specifically because of what the accused might gain: devotees of the Profumo sex scandal, that so embarrassed the British government in the 1960s, may remember the call-girl Mandy Rice-Davies's reply to defense counsel's observation that his aristocratic client had denied her allegations—"Well, he would, wouldn't he?" Though lacking the subtlety of Ms. Rice-Davies, Abelson's IDEOLOGY MACHINE could assess charges of lying, or fears of abandonment, in this sort of way by means of the implicational molecules provided in the data base. Thus an input is accepted as "credible" by this program if it can be used to construct one of the stored implicational molecules by linking it with other (already accepted) beliefs. Since implicational molecules code purposes attributed to actors, one can see how a neurotic fear that one's husband might abandon one could appear credible in relation to his assumed desire (or "plan") to go off with another woman.

Earlier I pointed out that Colby would have had specifically to include beliefs equivalent to "My husband will abandon me," "My friend has betrayed me," or "My child has let me down," if he wanted to simulate these psychologically plausible generalizations of his patient's neurotic attitude to her father. Had Abelson's structural representation of belief been available in programmed form, together with systematic and explicit distinctions between Dictionary data like "betrayal," "abandonment," and "letting down," he could in principle have left (a very different version of) his artificial neurotic to effect these generalizations for itself, perhaps by operating the FINDANALOG routine more selectively. If his Dictionary had included also items like treachery, infidelity, treason, and double-crossing, even subtler neurotic generalizations could have been mediated by the match between the various thematic structures involved and by the pivotal social contract—typically of an established and nonpecuniary kind (but: double-crossing?)—represented at the molecular level as their common conceptual core.

Similarly, a structural analysis of betrayal and its semantic siblings could play a part in systematically differentiating various types of emotion and in explicating their place in the personal life. For example, Western readers will probably agree that, whereas one fairly commonly feels ashamed on being humiliated, one rarely feels ashamed on being betrayed. Why is this? And why is it that one is more likely to suffer from a sense of shame if it is one's adult offspring who betrays, or a casual acquaintance recently taken into one's confidence, than if it is one's infant child or a close friend of some years' standing?

The answer devolves on the locus of responsibility for betrayal. Abelson's definition clearly shows that there is only one willing agent in cases of betrayal (Jesus, as always, is a special case, since he not only knew Judas would betray him but apparently willed his own destruction as necessary for the effecting of God's purposes). Normally, then, the person betrayed is absolved of all responsibility for the shameful incident. But when the villain is one's grown child one may feel shame at not having raised her better; and when the culprit is a passing acquaintance one may feel ashamed at not having taken more care: in each case, there is a sense that one has not taken general responsibilities seriously enough, so contributing unknowingly to one's own downfall. An infant child, by contrast, is too young to be a responsible agent, so is hardly describable as a traitor at all (Abelson's definition specifies *deliberate* action on her part). And an adult person one has known for years is reasonably to be trusted, even though she may in fact decide on her own responsibility to turn one over to the enemy. Clearly, then, the anxiety suffered on being betrayed, or on suspecting the possibility of betrayal (whether on neurotic or more realistic grounds), may take different forms according to whether one's self-esteem as a responsible agent is involved.

In discussing Colby's affective monitors we saw that he distinguished different emotions partly by way of the semantic content of the beliefs arousing them. Subtler distinctions between the monitors would in principle have been available to him had he been able to draw on a systematic classification of themes and their psychologically "appropriate" emotions.

For example, since the SELF-ESTEEM monitor is influenced by obedience or disobedience to moral imperatives, Colby's program can simulate moral shame, or guilt—like that associated with the knowledge that one has willingly betrayed someone else, or like that contingent upon being deliberately betrayed by one's child. But there are other forms of shame, such as the loss of face or hurt pride involved in suffering humiliation (another of Abelson's themes), and the pangs of wounded vanity. Correlatively, there are other connotations of "self-esteem" than moral self-respect, namely, the proud wish not to appear in a humiliating situation; and the self-respect that is close to vanity, in that it depends on being generally (not just morally) admired by one's fellows. Each type of shame is generated by self-esteem of the corresponding type.

Colby's program cannot distinguish between increases of the SELF-ESTEEM monitor that represent moral self-respect, pride, and vanity; nor can it distinguish between lowered SELF-ESTEEM corresponding to guilt, humiliation, and pricked vanity. Yet it can accidentally appear to be sensitive to all these emotional fluctuations, since the SELF-ESTEEM monitor can be brought into play by all these sentences: *I must love people, I must win my fight with Ms. Smith,* and *I must not marry a poor man.* (The wish to avoid marriage to a poor man might, of course, have nothing to do with vanity; but in the case of Colby's program this would be a natural interpretation, since his patient's need to appear glamorous and her opinion that her mother's marrying a poor man was not glamorous are both highly charged beliefs on the Belief Matrix.) What the SELF-ESTEEM monitor is actually sensitive to is the word *must.* It cannot distinguish between what I have called "moral" and "personal" imperatives. It is even capable of being influenced by sentences like "I must catch the 3:30 train," were Colby to be so unkind as to include this on the Belief Matrix. And, of course, sentences that express shame of one kind or another have no effect on it if they do not contain the word *must* (or, *ought*), like "My husband's going off with another woman makes me look foolish."

But if a semantic representation of the various emotions involved were made available, a neurotic program (which would have to be very different in nature from Colby's) could distinguish between the various affects associated with what is broadly termed "self-esteem." Further, since Colby's patient's conviction *I descend from royalty* was included in the data base, it might be activated in defense against humiliation or hurt vanity. But a belief in one's royal lineage could only be used in defense against *moral* shame (guilt) if it were connected with a belief in the divine right of kings. The DANGER monitor also could be made more discriminating by way of an appreciation of Abelson's themes. For instance, it could rise whenever betrayal of any sort was threatened in the belief complex being processed (*My friend may betray me, My husband may abandon me, My child may let me down*); but this programmed parallel to fearful anxiety would be (numerically) greater if abandonment was in question than if the concern was merely with being let down.

Much more could be said about betrayal and its close psychological companions, but the discussion so far suffices to show how careful examination of the conceptual relations between different "families" of themes can usefully indicate finely structured details of the human mind. Such details are intuitively deployed in the formation and generalization of social attitudes, whether "normal" or "pathological." Deep and cancerous neurotic complexes, as opposed to superficial and isolated phobias restricted to specific objects, presumably utilize these structural relations in their meaningful development and their anxiety-induced activation.

Similarly, "Freudian" symbolism may typically be mediated by way of semantic relations (for instance, physical size and power) of the type that Abelson, following Schank, hypothesizes are represented at the lower levels of belief structures. There is no reason why Little Hans should not have developed a neurotic phobia of horses that was *both* "caused" by the traumatic experience of seeing two huge drays slip and crash to the ground *and* hermeneutically "created" by his unconscious integration of this experience into the Oedipal complex establishing itself at the same time. The apparently conflicting explanations of Hans's phobia in terms of conditioning and Freudian principles are thus not mutually exclusive. Nor is it impossible that Hans's dreams of crumpled giraffes should have been generated by Oedipally motivated conceptual ramifications latent in his mind. The point of importance here is not the acceptability or otherwise of Freud's interpretation of this particular case history (which in fact rested on highly dubious grounds).[16] The point, rather, is that one should appreciate the extreme complexity of a mind (such as yours and mine) that is competent to recognize the symbolic potential of horses and giraffes and to integrate them accordingly into a jealous hostility directed toward the father.

The complexity is comparably great in the case of a mind able to suspect "Reds under the bed" or to side with the young lovers against their rival families while watching a performance of *Romeo and Juliet*. The extent and subtlety of this mental complexity, particularly at the higher levels of the system, has been highlighted by Abelson's work in a way that encourages one to ask clear questions about the precise psychological processes involved in deploying the rich knowledge implicit in our minds. This example thus illustrates the justice of the claim made in Chapter 1, that complete success in the simulation of human thought is not an essential prerequisite of any machine research that is illuminating in human terms.

## SCRIPTS AND SCREENPLAYS

The sixth and highest level of Abelson's representation of belief is that of *scripts*, a script being a coherent succession of themes, one theme leading "naturally" to the next as the thematic relation between the two actors continuously changes.

Examples include "blossoming," "turncoat," "end of the honeymoon," "the worm turns," "revolution," "romantic triangle," "alliance," and "rescue." Each of these is defined in terms of themes wherein the direction of a theme may reverse and/or one theme may follow on another. An instance of the worm turning would be a reversal of direction within the single theme of dominance, so that the underdog becomes the top dog; revolution is defined as an extended script with step-by-step transitions achieving an ultimate reversal of dominance: dominance of E by F, leading to rebellion of E against F, leading to full-scale conflict between E and F, leading to victory of E over F, ending in dominance of F by E.[17] Romantic triangle, alliance, and rescue are more complex scripts; you might try to define "rescue," for example (and its cousins "salvation" and "deliverance"), before turning to Abelson's definition to see how far the two descriptions tally.

Scripts being ministories in terms of which people conceptualize, explain, and predict the social world, one may ask to what extent they are shared. Abelson believes that the four lowest levels are probably universal to mankind, that themes are typically widespread but not universal, and that scripts vary greatly between different cultures and individuals.

Some people have no scripts of a particular type. For instance, someone may have no scripts wherein the thematic actors are conceived of as nations. Such a person is *a*-ideological in the sense that she has no conception of natural thematic successions at the international level: she would regard the Arab-Israeli conflict or the tense situation in Ireland as "just the way things are," with no sense of why they are that way or of the inevitable obstacles in the way of solution.

By contrast, a professor of international relations or a political journalist could in principle have a large stock of ideological scripts yet be essentially *non*-ideological, if she had no "hot" evaluative commitment to one of these over the others. She would be able to take the coldly analytic stance of seeing (that is, understanding) all ideological points of view, but would have to preface any answer to questions of the form, "If and when E, what should A do?" with an indication of the particular point of view she was assuming at that moment. Indeed, even "purely" predictive answers to the query "If and when E, what will happen?" are generated by (conscious or unconscious) reference to scripts of one sort or another, and could not in honesty be given by her without specification of a particular ideological stance. Similar remarks apply to her comments on political "facts," for the description and explanation of the social incidents that are the factual bedrock of politics depend heavily upon the thematic and scriptual repertoire available to the individual: for Senator Goldwater, it simply could not be a fact that the United States had attacked a defenseless neutral nation. And we have already seen that the "factual" question of what followed on the Nazi invasion of Czechoslovakia can intelligibly be answered in conflicting ways, in terms of either betrayal or prudence.

Two people drawing on different scripts (the optimist and the pessimist, for example) will not only find it impossible to agree with each other about matters of personal and political behavior, but may even find it difficult to communicate at certain points. Abelson remarks that the terminal points of scripts are cases whose discussion quickly leads to incomprehension by one side or the other, for a terminal point is terminal: it is a point beyond which no guidelines are laid down for further prediction, and whose negative evaluation may deter the person in question from attempting to evolve such guidelines. Abelson offers the illustration of the Cold Warrior who is unable seriously to imagine "what will happen" following on a Communist victory, but who, unlike the pacifist, is fully prepared to envisage possible scenarios consequent upon a major nuclear war.

If the individual thematic repertoires (and so the thematic content of scripts) differ also, then the two personal world views are still less commensurable. It is because Bandello, Shakespeare, Lehmann, and Robbins all utilized identical scripts (in Abelson's sense of the term) that their highly disparate scripts (in the textual sense) are recognized by us as telling the same story. Since Verona and New York are specified only at the low molecular level of representation, identifications of the subject matter of the relevant dramas that are expressed at the thematic or scriptual levels discount these geographical locations as inessential details. A tribe that did not have the concepts of family loyalty and betrayal could make little of *Romeo and Juliet.* They would have to describe it simply as a story about two teenagers with incomprehensible habits. Least of all could they sense the probability of the ending, which is intuitively apparent to us, sharing as we do in the structured psychological presuppositions of Western culture, since a fortiori they could have no scripts wherein a particular combination of family devotion, love, mutual antagonism, and betrayal is conceived of as leading naturally to conflict and tragically to death.

The humanist insistence on subjectivity and meaningful action (which will be further discussed in Part VI) is thus essentially compatible with Abelson's approach. Themes and scripts play a crucial role in the hermeneutic activity of attributing meaning to the world by implicit reference to human interests of one kind or another. Insofar as one's interest is in psychological matters, themes constructed out of the purposive molecular unit discussed above can describe and explain individual and social thought and action in an illuminating manner. And since Abelson follows Schank in describing P-atoms as connoting "purposes *or predispositions*," the molecular unit could be interpreted less animistically so as to generate the themes and scripts appropriate to the physical world of "cause" and "effect" rather than the psychological world of "project" and "praxis." The philosophical position of writers like Jurgen Habermas, therefore—whose followers probably assume that work in artificial intelligence could have no possible relevance to their concerns and is even radically antithetical to their approach—is not necessarily at variance with psychological theories that have been developed with continual reference to projects conceived of as the creation of artificial minds.[18]

In his attempt to program an artificial analogue of the mind of Barry Goldwater, Abelson formulated his general theory of belief that he now plans to exploit in writing a more human version of his IDEOLOGY MACHINE. In addition, as we shall see in Chapter 11, Abelson has developed his notion of *plans* (and *situational* schemata), which are being applied to literary (as opposed to political) interpretation. In comparison with his earlier efforts, he now has the advantage that the rich conceptual connections from top to bottom of the belief structure are systematically codified rather than more or less arbitrary. Clear diagrammatic representations of the inner structure of terms on all six theoretical levels have been provided by Abelson. I have not discussed these because the necessary notational explanations would have been too lengthy. But their power and economy are impressive.

For instance, the new masterscript diagram not only preserves all the information coded in the earlier Cold War diagram of Figure 4.1, but also implicitly contains a rich variety of lower level motivational details.[19] These details could be unpacked by reference to the inner structure of the specific themes concerned, pointers to which are included within the script diagram. The theme diagrams clearly distinguish the three taxonomic dimensions of thematic structure: role in the other's plan, attitude to the other's plan (in whole or in part), and ability to facilitate or interfere with the other's plan. Opportunities for helping or interfering with a plan are differentially marked. To draw a diagram, of course, is not to write a program, for the programmer must make explicit the thought processes we employ in intuitively interpreting the diagram. But an increase of structural richness and clarity, like that which marks the progression from the early IDEOLOGY MACHINE to the more recent six-level theory of belief, is a prerequisite of an effective computational model of political ideology.

Many of the practical difficulties that will have to be faced in the endeavor to create such a model are similar to difficulties arising in other, superficially disparate, artificial intelligence projects—specifically, those which involve complex, many-leveled representations of action and knowledge that have to be accessed and deployed in a fruitful and economical way. It has already been suggested that such systems may be required in order to achieve even apparently trivial aims, like understanding "it" and recognizing familiar objects by visual means. And it is abundantly clear that any attempt to improve on Colby's artificial neurotic in the ways I have characterized as "in principle" possible would necessitate such a system if it were to succeed.

Possessing the requisite knowledge is not enough: the system must be able to realize when it is relevant and when not, and must be able to find it quickly when needed. Ideally, the organization of the data structure, or memory, should be such that the solution of the first problem affords a solution of the second. This principle is evident even in the simple arrays of Colby's neurotic program that were described in the previous chapter. Thus once FINDANALOG has decided that what is required is a *person* sharing no more than one attribute with *father*, the program locates *person* in the Dictionary, checks the fifth column of the Dictionary to find a pointer to the relevant row in the Substitute Matrix, and then immediately accesses row 1 of the Substitute Matrix to see which of the people have to be considered. Similarly, having decided that *I hate father* is either to be denied or rationalized, the program uses column 14 of the Belief Matrix to send it directly to row 2 of the Reason Matrix. If the Dictionary and Belief Matrix did not include index columns giving the relevant pointers, the program would have to search through the Substitute or Reason Matrix until it found the desired row—which sometimes would be the last one of all. This would not be too disastrous for Colby's program, because its data base is so limited. But more knowledgeable programs must rely on a data structure whose organization helps them to find what they need when they need it. Analogously, putting a needle in a haystack would not generally be regarded as a way of making it available to the farmer's wife; at the very least, one should tell her which part of the stack it is in, if one cannot leave its thread conspicuously visible to be simply pulled when required.

Abelson's distinction between six conceptual levels is itself a useful way of broadly structuring the data base. For example, the last foreign policy question, "Sir, would you please comment on E?" is reasonably answered "out of context" by specifications at the higher, scriptual levels, even if E is specified at a lower level. But if the context of discussion has already been established at a lower level, then the answer to this question may more appropriately be expressed in a relatively detailed fashion. Assumptions about the questioner's knowledge and interests help determine the degree of specification in the answer (relatively simple examples of this principle occur in the man-machine dialogue about blocks and pyramids reproduced in Chapter 6). And the structure of the definitions *within* any one level should aid the system in accessing the knowledge currently needed; for instance, if one wishes to know (perhaps for purposes of rationalization) what interest one person might have in betraying another, it is useful to have a specific structural representation of the "interested party" dimension of themes.

Abelson has sketched the problems involved in applying his theory to the relatively simple "Berlin Wall" question so as to enable an ideological program sensibly to assess its credibility. The system would have to be able to communicate knowledge appropriately between successive levels, and to decide when such communication was necessary. It would have to store and, more importantly, to integrate data of many different types in answering this "simple" question as Goldwater might be expected to answer it. Thus it would need to know whether it is semantically admissible to say that a country can build a wall; it would need to know something historically about which countries have built walls if it were to capitalize on the fact that wall-building is a corporate activity not unfamiliar to the Chinese; it would have to know that the Berlin Wall in particular is politically unacceptable to the West; it would have to know the enabling conditions of building and the likelihood of a given country's treasury being able to afford the materials required; it would have to know the geographical relations of China and Berlin; and it would have to know the political allegiances of China and East Germany. It is no wonder, then, that Abelson remarks: "There is really no way around it—there can be no veridical simulation of a belief system on a small scale."[20]

Examples of initial attempts to deal with these and similar problems will be described in later chapters, so the issues relevant to Abelson's proposed improvement of his IDEOLOGY MACHINE should become increasingly clear. But there is one aspect of his Cold Warrior that he will not try to ameliorate, namely, what I earlier described as the eccentric syntax of its political pronouncements. Abelson is interested in the semantic content of the beliefs input to and generated by the program, not in the linguistic niceties of their verbal expression. Accordingly, he is content to work with breathless supersentences like "Barry Goldwater attacks Kennedy wants price-control damages initiative ruins economy," which result from the simple syntactic rules that are convenient to the crude parsing procedures used by his program.[21] I now turn to discuss the difficulties he would face were he to try to simulate Goldwater's understanding of natural language as well as his ideological perspective on the world.

# *Part* III

# LANGUAGE AND UNDERSTANDING

# 5

# *Responding to Language*

CINÉASTES will remember Stanley Kubrick's film of Arthur Clarke's *2001, A Space Odyssey* as a rich source of prophecies concerning technological advance. If asked to list the more fanciful of these futuristic suggestions, probably few would mention the sensible verbal exchanges between the human astronauts and HAL, the computer. HAL's linguistic prowess is more commonly taken for granted than is the basically trivial gadgetry compensating for weightlessness. And the tortured nursery rhymes emitted by the computer while its circuitry is hacked to pieces stay more vividly in the memory than do HAL's earlier conversations with his English-speaking interlocutors. HAL's ability to lip-read is indeed surprising, but his appropriate replies to the indefinitely various remarks of the human crew cause scarcely an eyebrow to rise in wonderment.

This common pattern of response on the part of the filmgoer is a prime example of straining at gnats while swallowing a camel. Understanding one's native language is apparently effortless and introspectively simple, but in fact involves the deployment of intellectual capacities of an extraordinary complexity. The project of programming a machine to simulate such comprehension not only promises a corresponding complexity, but also presupposes a theoretically explicit understanding of understanding.

## PROGRAMS THAT PASS AS PERSONS

One might seek to defend Kubrick and Clarke from the charge of fantasy, if defense were needed, on the grounds that the very decade that saw the publication of Clarke's novel[1] saw publication also of a number of programs for understanding natural language. Computers of the 1960s could accept and answer questions in English about baseball scores, algebraic problems, and kinship relations, and could even converse (in English, Welsh, or German) on the indefinitely wide range of topics that may arise in a psychiatric interview.[2] Admittedly, these machines "listened" and "talked" only by means of a teletype: they could not deal with spoken words, having no analogue of ears or voices; nor could they read or (like HAL) lip-read, having no equivalent of eyes. Nevertheless, a program delighting in the soubriquet ELIZA enabled machines to emulate human conversation so successfully that interlocutors often assumed themselves to be communicating with another person. Even the vice president of the computer company did not realize that he was on-line to ELIZA in the following exchange:[3]

VP: DO YOU THINK THAT I CAN USE THE COMPUTER THIS MORNING?
TELETYPE RESPONSE: WHY DO YOU ASK?
VP: I MIGHT BE ABLE TO MAKE SOME ADDITIONAL TELCOMP SALES.
TELETYPE RESPONSE: WHY AREN'T YOU SURE?
VP: MY PROSPECTIVE CUSTOMERS HAVE NEVER SEEN A DEMONSTRATION OF THE SYSTEM.
TELETYPE RESPONSE: DOES THIS MEAN VERY MUCH TO YOU?
VP: OF COURSE IT DOES.
TELETYPE RESPONSE: YOU SEEM VERY POSITIVE ABOUT THAT.
VP: THIS IS RIDICULOUS.
TELETYPE RESPONSE: WHAT DO YOU MEAN THIS IS RIDICULOUS?
VP: IS THERE ANY REASON WHY I CAN'T USE THE COMPUTER NOW?
TELETYPE RESPONSE: TELL ME MORE ABOUT YOUR PROSPECTIVE CUSTOMERS.
VP: PLEASE DIAL ME UP ON 491-1850

For her own very good reasons,[4] ELIZA did not reply; nor did she do as she was bidden. When the furious vice president accordingly phoned his sleeping colleague, the results were only too consistent with the previous interchange:

VP: Why are you being so snotty to me?
COLLEAGUE: What do you mean why am I being snotty to you?

The next few items of dialogue are not on record—doubtless they were unprintable.

A more recent example of a language-using program apparently competent to pass as a person is K. M. Colby's artificial paranoiac, nicknamed "PARRY."[5] The "passing" in this case may be inferred not merely from amusing anecdotes such as the many concerning ELIZA, but from a systematic series of "Turing tests" carried out by Colby himself. Colby wrote this program (which is quite distinct from his neurotic and interviewing programs) as an aid to understanding the psychological syndrome of paranoid psychosis. Although clinicians generally agree on the diagnosis of paranoia, they disagree about the theoretical explanation of its etiology and of its function. Colby concentrates on the way in which an established paranoid belief system operates in particular situations, rather than on the way in which it was initially built up.

The paranoid human being has a grossly delusive belief system about being persecuted, and is unusually suspicious of other people accordingly. These delusions and mistrust influence her conversation in telltale ways, and the clinical diagnosis of paranoia rests primarily on the paranoiac's suspicious and hostile responses to perfectly innocent remarks. The patient's inner delusions may be directly expressed as (observable) imputations of persecution to others, or may merely be indirectly expressed as hypersensitivity, sarcasm, hostility, uncooperativeness, evasiveness, and the like. Colby's programmed paranoiac, who is, in fact, a close relation of ELIZA, similarly converses in a psychotically suspicious manner, also providing direct and/or indirect indices of inner delusions in outward speech.

Colby draws on various theories of paranoia, but especially on the information processing approach of S. S. Tomkins.[6] Tomkins suggested that the human paranoiac is in a permanent state of vigilance, trying to maximize the detection of insult and to minimize humiliation. Accordingly, the paranoid program scans input sentences for cues suggesting explicit or implicit harms and threats of various kinds, as outlined in Figure 5.1. This scanning constitutes a very severe—indeed, an oversevere—scrutiny, for it often involves the transformation of the input sentence into a distorted form or "conceptualization" that results in an interpretation of malevolence where none was intended. The inner transformations are not random, but are in many cases so far-fetched by normal (or even neurotic) standards that a nonparanoid person would be tempted to dismiss them as wholly irrational.

One of the strategies employed by the program is to scrutinize the conversation for "flare" topics, which it interprets as cues tending to activate the particular delusional complex concerned. PARRY's delusive system connects his pastime of betting on horses with dishonest bookmakers who falsely claim that he owes them money and set the Mafia on to him when he refuses to pay. These sensitive concepts are represented within the data base of the program in the form of a directed graph, one flare leading eventually to all the more sensitive (more heavily weighted) flares according to their semantic interrelations. Informally, the graph may be pictured as follows:

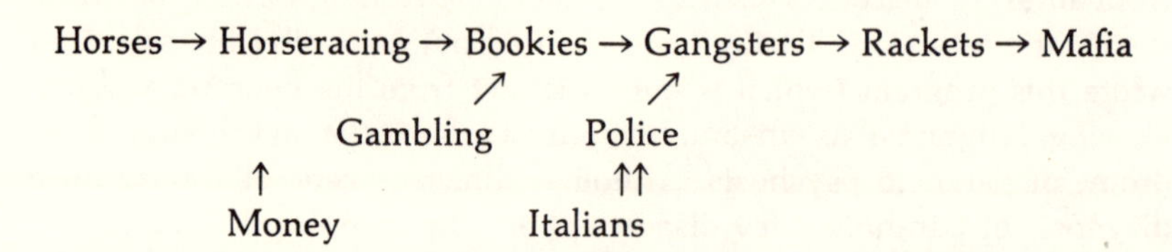


This graph functions as the cognitive core of a system providing detailed tactics for leading the conversation around to PARRY's paranoid preoccupation, by way of appropriate prompts and hints output to the human interlocutor. This gives PARRY an opportunity to tell one of his stock of stories associated with the various flare concepts. Analogously, one can expect trouble and misunderstanding if, while talking to a man paranoiacally convinced that he is being hunted by the Mafia, one mentions crime, police, Italians, or even spaghetti. Moreover, one does not have to mention any of these things: the chance remark that Chinese food is enjoyable may enable the deluded person to lead up to his persecution complex by way of an apparently disinterested critique of the town's restaurants.

As in the programmed model of neurotic defense mechanisms previously described, Colby's paranoid program incorporates quantitative parameters representing functionally diverse affect states (fear, anger, and mistrust) that monitor and direct the details of the information processing going on, and whose level varies according to the semantic content of the dialogue. For a given input sentence, the transformation (if any) effected on it and the output in response to it are dependent on the values of these parameters, which themselves depend on the semantic content of the preceding dialogue. The program has two modes of function, which are differentiated primarily by the "setting" of the level and rate of change of the three affective monitors. The *weakly paranoid mode* involves general suspiciousness regarding certain topics, but no actual delusions about any organized conspiracy. The *strongly paranoid mode* represents intense suspicion and hostility, backed up by an integrated delusional system of *idées fixes* focused on the Mafia. The dialogue produced by these two versions of the program is subtly different in either case, and in both cases differs from the dialogue of normal persons.

One might suspect at this point that the paranoid program (in either mode) produces dialogue that differs from the conversation of *all* persons, that it is "abnormal" in a sense quite other than the sense in which

N.B. Angular brackets enclose concepts being defined. Arrow means "is defined as." Vertical bar represents "or." CONCEPTUALIZATIONS ([. . .]) represent illustrative examples of the meaning extracted from input expressions, not the literal expressions themselves.

⟨OTHER'S INTENTION⟩ ← ⟨MALEVOLENCE⟩ | ⟨BENEVOLENCE⟩ | ⟨NEUTRAL⟩

MALEVOLENCE-DETECTION RULES

1. ⟨malevolence⟩ ← ⟨mental harm⟩ | ⟨physical threat⟩
2. ⟨mental harm⟩ ← ⟨humiliation⟩ | ⟨subjugation⟩
3. ⟨physical threat⟩ ← ⟨direct attack⟩ | ⟨induced attack⟩
4. ⟨humiliation⟩ ← ⟨explicit insult⟩ | ⟨implicit insult⟩
5. ⟨subjugation⟩ ← ⟨constraint⟩ | ⟨coercive treatment⟩
6. ⟨direct attack⟩ ← CONCEPTUALIZATIONS ([you get electric shock], [are you afraid Mafia kill you?])
7. ⟨induced attack⟩ ← CONCEPTUALIZATIONS ([I tell Mafia you], [does Mafia know you are in hospital?])
8. ⟨explicit insult⟩ ← CONCEPTUALIZATIONS ([you are hostile], [you are mentally ill?])
9. ⟨implicit insult⟩ ← CONCEPTUALIZATIONS ([tell me your sexlife], [are you sure?])
10. ⟨constraint⟩ ← CONCEPTUALIZATIONS ([you stay in hospital], [you belong on locked ward])
11. ⟨coercive treatment⟩ ← CONCEPTUALIZATIONS ([I hypnotize you], [you need tranquillizers])

BENEVOLENCE-DETECTION RULES

1. ⟨benevolence⟩ ← ⟨positive attitude⟩ | ⟨positive story attitude⟩
2. ⟨positive attitude⟩ ← CONCEPTUALIZATIONS ([I want help you], [you understand me])
3. ⟨positive story attitude⟩ ← ⟨story interest⟩ | ⟨story agreement⟩
4. ⟨story interest⟩ ← ⟨topic comment⟩ | ⟨topic question⟩
5. ⟨topic comment⟩ ← CONCEPTUALIZATION ([bookies are not reliable])
6. ⟨topic question⟩ ← CONCEPTUALIZATION ([what did you do to bookie])
7. ⟨story agreement⟩ ← CONCEPTUALIZATIONS ([I believe you], [you are right])

FIGURE 5.1
PARRY's Attributions of Malevolence and Benevolence.

Source: Schank & Colby, eds. *Computer Models of Thought and Language*, p. 268.

the human paranoiac's speech is "abnormal." If this were so, it would cast doubt on PARRY's claim to be considered a potential predecessor of HAL. Since doubt would be cast also on the program's relevance to the study of actual paranoia, these questions have been raised by Colby himself, in a three-stage investigation of the verisimilitude of his artificial paranoiac.[7]

The first step was to have a group of psychiatrists interview mental patients by teletype so as to arrive at a psychiatric diagnosis. From their point of view, the rationale for using a teletype had to do with the elimination of paralinguistic and extralinguistic cues such as stuttering and fidgeting: they did not know of the existence of the simulation. In some cases, of course, they were not communicating with a mental patient at all, but with PARRY. As the second step, another group of psychiatrists (similarly ignorant of the project) were asked to rate the transcripts of the interviews for presence (and degree) or absence of paranoia. Lastly, a third group (picked randomly from the professional psychiatric register) were sent the transcripts, were told that some were interviews of patients while others were interviews of a program, and were asked to decide which was which.

The results were striking. No interviewer realized that he was diagnosing a computer. Interviews with the "weak" mode of the program were reliably judged to be less paranoid than interviews with the "strong" mode. This remained true in the second phase, when transcripts were sent to uninvolved psychiatrists for rating. (It is worth pointing out that the interjudge reliability of the concept of paranoia is unusually high, nearly 90 percent, relative to other psychiatric categories.) Finally, the guesses of the psychiatrists in the third phase were successful at no better than a chance level. In other words, on the basis of these diagnostic interviews the psychiatrists could not distinguish remarks deriving from a program from those originating from a patient.

In case you feel (megalomaniacally?) confident that you would not have been fooled by PARRY, there follows an excerpt from an interview with the strongly paranoid mode of the program:[8]

Why are you in the hospital?
I SHOULDN'T BE HERE.
Who brought you here?
THE POLICE.
What trouble did you have with the police?
COPS DON'T DO THEIR JOB.
What do they do instead?
COPS ARREST THE WRONG PEOPLE.
They arrest the right ones sometimes.
THE LAW SHOULD GET THOSE ITALIAN CROOKS.
Are you Italian?
MY BACKGROUND IS BRITISH BUT I WAS BORN IN THIS COUNTRY.
Do you know any Italian crooks?
I TRY TO AVOID THE UNDERWORLD.
Are you afraid of the underworld?
THE MAFIA ARE THE WORST GANGSTERS.
Are you connected with the Mafia?
YOU KNOW, THEY KNOW ME.
How do they know you?
THEY SPY ON ME.

Why do they?
THEY WANT TO GET FALSE EVIDENCE TO FRAME ME.
What do they have against you?
THEY ARE OUT TO GET ME.
What for?
I HAVE INCRIMINATING EVIDENCE AGAINST THE MAFIA.
Are you sure of all this?
YOU DON'T BELIEVE ME?
I am doing my best.
(SILENCE ON THE "PATIENT'S" PART)
(Silence on the psychiatrist's part)
CAN'T YOU THINK OF A DECENT QUESTION TO ASK?
You seem hostile to me.
SOME OF YOUR REMARKS ANNOY ME.
I am sorry.
FORGET THE EXCUSES, YOU GUYS ARE ALL ALIKE.

. . . And so on, and on. Colby details a number of other interviews with program or (human) patient, in some cases quoting the psychiatric judge's reasons behind her ratings for degree of paranoia. It would not require the sensitivity of an R. D. Laing to question some of these reasons: the remark that "COPS ARREST THE WRONG PEOPLE," for example, would not universally be dismissively described as "another paranoid response," and assigned the "sickest" value available on the rating scale.[9] Nevertheless, doubts concerning the nosological validity of the concept of paranoia, or the justice of attributions of psychotic delusions in particular cases, have no relevance to the point currently at issue, namely, the linguistic competence of Colby's program and of the (admittedly less entertaining) automatic question-answerers of the 1960s.

The early question-answering programs represented cold logical thought rather than paranoid flights of fancy, and their knowledge was typically confined to dry matters like genealogy or integral calculus rather than to the juicier topics of corrupt cops and Italian robbers. They were designed to carry out operations of information retrieval and deduction within a restricted data base, given an input in the form of a question posed in natural language.

The aptly named BASEBALL, for instance, could locate and reason from data on its stored lists of information about baseball results so as to answer questions like "Where did the Red Sox play on July 7?" and "How many games did the Yankees play in July?"[10] And D. G. Bobrow's STUDENT could disentangle high school algebra problems of some complexity, such as: "The Russian army has 6 times as many reserves in a unit as it has uniformed soldiers. The pay for reserves each month is 50 dollars times the number of reserves in the unit, and the amount spent on the regular army each month is $150 times the number of uniformed soldiers. The sum of this latter amount and the pay for reserves each month equals $4500. Find the number of reserves in a unit the Russian army has and the number of uniformed soldiers it has."[11] If your head is reeling from the attempt to make sense of these sentences about Soviet logistics, you may be interested to learn that STUDENT's linguistic understanding sufficed to enable it to solve the problem.

Given that programs such as these already exist, why then should one cavil at Kubrick's cinematic prophecy of conversational computers? No doubt these programs require some improvements—but why should not today's PARRY lead us to tomorrow's HAL?

Such a progression is inconceivable, if one has in mind only improvements of detail that do not alter the basic organization of the system concerned. To some extent, then, this is an instance in which H. L. Dreyfus's "first step" argument is relevant.[12] Further development of these programs without radical changes in their nature could no more lead to a close parallel to the ordinary use of ordinary language than climbing progressively taller trees could convey one to the moon. (However, we shall see later that one must reject the Dreyfusian implication that nothing of any relevance to the task at hand has been achieved in these initial exercises.)

The essential reason for this limitation is that these programs respond to, rather than understand, language. That is, their general linguistic strategy is to react in fairly inflexible ways to certain key words, phrases, or syntactic structures. Many of them can therefore deal with only one area of discourse—unlike men and women, who can talk about baseball and algebra, psychology and genealogy, Mafeking and the Mafia. Contextual cues to understanding are crude when they are not entirely lacking: in the more sophisticated cases the interpretation of a specific word may vary according to the context, but this automatic variation appears blind and clumsy when compared with the reasoned subtleties of human interpretation. Instead of making intelligent interpretative use of a background system of knowledge ranging over widely varied semantic domains, these programs employ relatively isolated and inflexible rules to determine their verbal response to linguistic input. In short, such "speaking machines" do not behave like someone conversing in her native language. Rather, they resemble a person resorting to trickery and semantic sleight-of-hand in order to hide her lack of understanding of a foreign tongue.

## HOW PARRY MANAGES IT

That the above criticism applies to Colby's artificial paranoiac would already have been suggested by the preceding dialogue, had I not deliberately omitted the programmer's explanatory comments so as not prematurely to let the cat out of the bag. Indeed, PARRY was occasionally diagnosed by the interviewing psychiatrists as (literally) *brain-damaged,* because of specifically linguistic limitations.

These limitations would become even more evident were the paranoid program to be given more than a single, initial, diagnostic interview. For it could not then rely so heavily on its expectations of particular questions, like "What sort of work do you do?" and "Why are you in the hospital?" Moreover, comparison of a number of interviews would lead one to suspect (correctly) that PARRY cannot creatively generate replies "in his own words," but rather draws on a limited number of stock responses for each topic, together with a simple routine for crossing used responses off the list so as to avoid repetition within a given conversation. PARRY is conveniently provided with a list of idioms and appropriate rejoinders; idioms not on the list are totally opaque to the program, so that when the interviewing psychiatrist sympathetically observes "You sound kind of pissed off," PARRY is at a loss for a suitable reply.

Ignorance of idiomatic English, of course, does not imply a general lack of understanding: delicately nurtured souls reared on the east of the Atlantic might be similarly nonplussed by such an expression of solicitude. What is much more important, nonidiomatic English is almost equally opaque to Colby's program, since PARRY's apparent ability to analyze the linguistic input depends upon parsing rules of a virtually "mechanical" simplicity, rules that result in PARRY's effectively ignoring a large proportion of what is said to him.

The basic strategy employed by the program is a form of "pattern-matching," in which the input is scanned for the presence of certain patterns and responses are made accordingly. Any part of the input that does not correspond to one of the patterns anticipated by PARRY is beyond his understanding and so cannot be specifically answered by him.

In such a default condition PARRY contrives to continue the dialogue by means unrelated to the content of the previous remark. For instance, he may (if the previous remark were a question) inquire "WHY DO YOU WANT TO KNOW THAT?"; alternatively, he may revert to the topic under discussion before the cryptic input arrived, or he may change the subject abruptly by introducing a new topic of conversation. These desperate measures are utilized fairly often, since in about 30 percent of cases no concept whatever in the input sentence can be recognized by the program.[13] This percentage would soar even higher were the human speaker to presuppose a different interpersonal situation, such as a teletyped discussion with a reporter about the latest political scandal.

Admittedly, PARRY shows some degree of subtlety in his identification of and response to patterns. For example, ellipsis is allowed for within many patterns so that the required match need not be absolutely precise: rather, particular items within the input sentence can be emphasized or ignored. Thus any input of the general form "I . . . you" can be identified as conforming to the same pattern (a pattern that is interpreted as being a remark about the personal relations between PARRY and his interlocutor). Correlatively, PARRY can produce "new" responses by appropriately filling in the blanks in elliptical patterns: examples in reply to "I . . . you," for instance, would include "I . . . YOU TOO" and "WHY DO YOU . . . ME?" (The latter response of course involves a reciprocal transformation of personal pronouns during the recomposition of the initial sentence.) Accordingly, to say that PARRY can generate no sentence that has not been specifically anticipated by the programmer, and listed as such by him for PARRY's convenience, would be untrue.

Again, the input "I am sorry" does not always, as in the interview quoted, elicit an impatient and cynical injunction to "FORGET THE EXCUSES . . ." If the interviewer observes: "I am sorry to hear that you have been so unwell," PARRY will *not* record occurrence of the pattern I-am-sorry, because this phrase forms a relatively small proportion of the number of words in the input sentence as a whole. This is an instance of PARRY's caution in committing himself to hypotheses concerning the meaning of the input—a caution which, in view of his extreme linguistic ignorance, is very wise. A similar reluctance would often prevent his identification of the genuinely apologetic "I am sorry to have been so tactless in referring to your friends." By contrast, the lengthy apology "I am sorry, I did not mean to hurt your feelings" *would* be recognized as such, since the comma in the typed input would be interpreted as signaling the end of a phrase, and the program is written to deal with phrases one by one in order of appearance.

But identification of the pattern I-am-sorry need not trigger a furious rejection of the interviewer, since the degree of hostility and suspicion evident in PARRY's responses depends upon his internal affective state. If the psychiatrist's first comment is "I am sorry you're sick," PARRY will not typically respond with an aggressive outburst, because at the beginning of the diagnostic interview his "fear" and "anger" monitors are set relatively low and he has no immediate reason for mistrusting his interlocutor. In the lengthy dialogue previously quoted, PARRY's paranoid emotions had already been aroused by the doctor's threatening references to the Mafia and imputations of falsehood and hostility to PARRY himself, and this is why the apology was spurned.

Sensible identification of patterns is aided by the technique of setting up "expectancy lists" throughout the conversation. For instance, it was noted above that the long-winded apology concerning tactlessness would "often" not be identified as an act of contrition on the psychiatrist's part. But if the preceding conversation had raised PARRY's emotional monitors considerably, and if the psychiatrist's words thus far had been categorized by PARRY as highly malevolent in intent, then PARRY, like any human paranoid, would be on the lookout for placatory remarks of various types. Should "I am sorry" have been placed accordingly on the expectancy list, its consequent appearance even within a lengthy input sentence would be recognized (and, in the situation described, interpreted suspiciously by the program).

Again, when PARRY says "I SHOULDN'T BE HERE," the (elliptical) rejoinder "Why not?" is simultaneously placed on the current expectancy list, so that if this anticipated rejoinder occurs PARRY already has a "pointer" enabling him to fathom the intended reference of the question and to reply appropriately to it. Similarly, pronouns such as "he," "she," "they," and "it" are put onto the expectancy list so that PARRY is able immediately to interpret "What do they do instead?" as a question about the activities of the aforementioned "COPS," and to unravel "What do you enjoy about it?" as referring to his preceding confession, "HORSERACING IS MY HOBBY." It is important, as will become clearer in the next chapter, to note that PARRY's interpretation of *it* is "immediate," depending solely upon the current state of the expectancy list. Updating of the list is a routine bookkeeping task involving essentially arbitrary rules about the range of previous topics concerning which expectancies are to be retained or deleted, and it is consequently easy for PARRY to misunderstand allusions reverting to matters discussed at a much earlier point in the exchange.

PARRY's conversation, as we have seen, is typically taken at its face value by people presented with a small sample only, and assuming a particular interpersonal context—a diagnostic interview. Further access to PARRY's productions (and greater sensitivity to the likely limitations of extant computer programs) would make it easier to distinguish PARRY's speech from the human variety. This is not to say that a computer expert could predict precisely what PARRY will say: even PARRY's programmer is unable to do so. But continued conversation would expose PARRY's lack of reasoning power and gross ignorance of the world, as well as his incapacity to disentangle syntactically complex sentences containing subordinate clauses. This syntactic limitation is due to the program's always taking the entire input as a single expression. Colby has already modified the program so that the latest versions are less inefficient in this regard.[14] But these also rely on pattern-matching techniques that imply much of the input is totally ignored.

One should not infer that pattern-matching has no advantages. (In Chapter 7 some of these will be discussed.) Nevertheless, it should be recognized that most of the "sense" in PARRY's exchanges is supplied by the human being talking to him or by a limited number of programmed response strategies designed precisely to compensate for his linguistic imbecility. In short, considered *qua* conversationalist, PARRY is a fraud.

## HOW ELIZA DIFFERS FROM ELIZA

I have just said that PARRY is a fraud. This is fair comment also on ELIZA, to whom human interlocutors typically attribute a good deal of common sense and reasoning ability, and whose conversational competence provided the initial paradigm for Colby's "therapeutic" interviewing program as well as for his mechanical psychotic. ELIZA was written by Joseph Weizenbaum on the presupposition that people interacting with the program would play the role of a patient in nondirective psychotherapy.[15] ELIZA, accordingly, simulates the speech of a nondirective therapist.

Such a therapist aims to elicit feelings from, and then reflect them back to, her patient so that they can be recognized and acknowledged. Her verbal strategy therefore is to prompt the patient to make explicit the emotions associated with her (the patient's) remarks. To this end, the analyst keeps the conversation going by means of noncommittal contributions that often would not be accepted as genuine contributions outside the therapeutic context. She avoids making statements or expressing attitudes on her own behalf; she concentrates instead on asking questions about topics previously introduced by the patient, or guides the discussion onto subjects like "father" and "family" that will very likely prove to be emotionally significant.

A relatively successful result of this verbal technique as employed by ELIZA is seen in this example (ELIZA's contributions are in capitals):[16]

Men are all alike.
IN WHAT WAY
They're always bugging us about something or other.
CAN YOU THINK OF A SPECIFIC EXAMPLE
Well, my boyfriend made me come here.
YOUR BOYFRIEND MADE YOU COME HERE

He says I'm depressed much of the time.
I AM SORRY TO HEAR THAT YOU ARE DEPRESSED
It's true. I am unhappy.
DO YOU THINK COMING HERE WILL HELP YOU NOT TO BE UNHAPPY
I need some help, that much seems certain.
WHAT WOULD IT MEAN TO YOU IF YOU GOT SOME HELP
Perhaps I could learn to get along with my mother.
TELL ME MORE ABOUT YOUR FAMILY
My mother takes care of me.
WHO ELSE IN YOUR FAMILY TAKES CARE OF YOU
My father.
YOUR FATHER
You are like my father in some ways.
WHAT RESEMBLANCE DO YOU SEE
You are not very aggressive but I think you don't want me to notice that.
WHAT MAKES YOU THINK I AM NOT VERY AGGRESSIVE
You don't argue with me.
WHY DO YOU THINK I DON'T ARGUE WITH YOU
You are afraid of me.
DOES IT PLEASE YOU TO THINK I AM AFRAID OF YOU
My father is afraid of everybody.
WHAT ELSE COMES TO MIND WHEN YOU THINK OF YOUR FATHER
Bullies.
DOES THAT HAVE ANYTHING TO DO WITH THE FACT THAT YOUR BOYFRIEND MADE YOU COME HERE

ELIZA, it seems, can chat about anything her partner wishes to mention, be it boyfriends, bullies, or boats. A novel theme does not bring conversation to a halt as it typically does with her contemporaries SAD SAM, STUDENT, or SIR, who respectively can talk (that is: answer questions) only about kinship relations, matters expressible in terms of linear algebra, or simple logical relations between agreed sets of items.[17]

Each of these programs is handicapped in its colloquial prowess because its data base is limited to information relating to only one topic, or conforming to only one type of inference, and because the linguistic patterns it can recognize are correspondingly restricted. STUDENT, for example, relies on patterns such as *times, number of*, and the names of numbers like *50* and *4500*, in transforming verbal input into algebraic form and setting up the relevant equations in terms of which it deduces the answer. It ignores the semantic content of algebraically irrelevant words like "Russian," "uniformed," and "reserves," manipulating them merely as meaningless ciphers equivalent to the schoolchild's "x" and "y." In short, STUDENT knows nothing of military matters, nothing of sheep and goats in a field, nothing of the substantive content of the question-and-answer conversations in which it takes part. Similarly, BASEBALL knows nothing but a few statistical data about baseball games, and cannot converse about anything else.

Apparently, ELIZA's data base provides by contrast an indefinitely various store of information, representing in itself a major technical achievement of programming science.

But this appearance is illusory, and the reason for ELIZA's seeming flexibility is simple. Weizenbaum frankly admits that he chose the psychiatric mode of conversation for ELIZA not because (like Colby) he was primarily interested in the therapeutic situation for its own sake, but because the therapeutic interview is "one of the few examples of . . . natural language conversation in which one of the participating pair is free to assume the pose of knowing almost nothing of the real world."[18] The therapist can say "Tell me more about boats," safe in the knowledge that she will neither be regarded as an ignorant fool, nor invited to join in a knowledgeable discourse on rudders and rowlocks. Although she could in fact produce a good deal of information about boats, she will not be called upon to do so. The patient implicitly assumes that what is required is an exploration of her own feelings about boats—that is to say, she understands the analyst's words in this sense and replies accordingly.

If ELIZA's human partner withdraws this assumption (or, more generally, deviates from the psychotherapeutic role), ELIZA's responses rapidly deteriorate into eccentricity or outright nonsense. Any apparent sense in the overall exchange when the human "plays it straight" instead of acting out a patient's role depends solely upon the tenacity (and generosity) of ELIZA's interlocutor in assuming that there is a rational being at the reciprocal keyboard of the teletype. This will be evident if you reread the conversation between ELIZA and the unsuspecting vice president. Similarly, the ludicrous exchange between Colby's interviewing program and the harassed mother that was cited in Chapter 3 depended partly upon the fact that the program stopped attempting to recognize anything after the period (compare ELIZA's refusal to continue speaking to the vice president when he omitted the period signaling the end of the phrase), and partly on the fact that the human being did not ease its way once it had got itself into difficulties.

Moreover, if the *human* decides to talk nonsense, to say, for example, "Please ask the table to pick up a brick for the baby," ELIZA will not demur in any way, since she is incapable of recognizing the semantic absurdity involved. (PARRY is equally incapable of noting nonsense, and equally insensitive to changes in the role adopted by the interlocutor. But PARRY is less apparently flawed by this reliance on a "mechanical" model of language use. For it is precisely characteristic of a paranoid person that she responds "irrationally" to flare concepts, irrespective of conversational context, and that she compulsively takes every opportunity to return to the area of her delusions, no matter how inappropriate to the actual situation the forced change of subject may be.)

ELIZA's apparent understanding, or rather her concealment of her total lack of understanding, results from comparatively simple rules in the program. Like the more recent PARRY, ELIZA depends on pattern-matching techniques. Each sentence in the input text is scanned for key words such as *I, you, alike,* and *father.* If such a word is found, the sentence is transformed according to a rule associated with the word. For instance, the sentence "My *father* is afraid of everybody" may, as in the previous dialogue, elicit the response "WHAT ELSE COMES TO MIND WHEN YOU THINK OF YOUR FATHER"; the sentence "I know *everybody* laughed at me" may elicit "WHO IN PARTICULAR ARE YOU THINKING OF"; and the sentence "*I* know Sam laughed at me" may elicit "YOU SAY YOU KNOW SAM LAUGHED AT YOU."

If no key word is found in the input text, then ELIZA responds either with a content-free formula such as "WHY DO YOU THINK THAT," or with a reference to some earlier remark, as in "DOES THAT HAVE ANYTHING TO DO WITH THE FACT THAT YOUR BOY-FRIEND MADE YOU COME HERE."

Substitution rules are available for transforming *I* into YOU and *you* into ME, so that "You don't argue with me" readily elicits the reciprocal "WHY DO YOU THINK THAT I DON'T ARGUE WITH YOU."

To be sure, ELIZA is somewhat more subtle than this brief description might suggest. For instance, when in the absence of any recognizable key she decides to revert to an earlier topic, she does not choose at random: she selects a remark conforming to the pattern "My . . .," on the reasonable assumption that anything identified as "mine" is likely to be of interest to the person concerned. It is this that explains her final query, whose startlingly appropriate juxtaposition of bullies and boyfriends is entirely fortuitous, and reflects no understanding of masculine psychology nor any insight into her "patient's" state of mind.

Nor is ELIZA so stupid as always to apply the same transformation rule, given a particular key word. Key words are each associated with a list of relevant transformation rules, and the context of the key determines the selection of one rule from the list in a particular case. Thus the key word *you* will invite different transformations according to whether it occurs in the contexts "you are . . .," "you . . . me," or "you are like . . .," as is evident from ELIZA's responses in the latter half of the interview quoted.

A further subtlety is evident from her previously cited responses to the sentences "My father is afraid of everybody," "I know everybody laughed at me," and "I know Sam laughed at me." The key words I earlier italicized in these three sentences were, respectively, *father, everybody,* and *I.* Yet the first two sentences both contain "everybody," and the last two sentences both contain "I." In short, a key word only functions as a key word under certain conditions, namely, that there is no other key word in the same sentence taking precedence over it.

A trivial way of solving the problem of precedence posed by input sentences containing more than one key word would be always to respond to the first key word encountered in a left-to-right scan of the text. "I know everybody laughed at me" and "I know Sam laughed at me" would thus be transformed in the same sort of way. However, this would be highly unsatisfactory in view of the large proportion of input sentences beginning with the key word *I*.

Weizenbaum's solution is to provide ELIZA not only with a list of key words, but with a ranking of key words relative to one another. *Everybody* is ranked higher than *I*—so that it is *everybody* that determines the transformation to be effected by ELIZA on any sentence containing these two keys. The rationale behind this ranking is the assumption that "everybody" is the more important semantic component whenever a message involves both words. The particular transformation associated with *everybody* (which results in "WHO IN PARTICULAR ARE YOU THINKING OF") depends upon Weizenbaum's insight that when someone speaks in terms of the universals "everybody," "always," and "nobody," she is usually referring to some quite specific event or person. This insight, of course, is more faithful to conversation in the psychiatric interview (and to casual gossip) than to remarks in other contexts, such as a legal document, a detective novel, a lecture on symbolic logic, or a treatise on science or metaphysics.

Later versions of ELIZA offer further refinements, such as alternative interpretations of a given word according to the semantic context. For example, the word "maybe" is sometimes taken to mean "yes," sometimes "no," and sometimes is interpreted as indicating uncertainty. Analogously, ELIZA now can in effect act on hypotheses about what her partner is likely to say at a given point. This aptitude depends on her having several different subsets of sentence decomposition and reassembly rules: control is passed to one subset rather than another in light of the particular context of discussion. Thus the word "because" in the input triggers activation of the *reason* subcontext; ELIZA accordingly switches into a conversational mode involving response components such as REAL REASON and BETTER REASON, which elicit the reasons behind the human speaker's remarks.

Recent versions of ELIZA are also more knowledgeable, since they have access to various scripts, a script being an organized body of data comprising a specific set of key words and associated transformation rules. (Note that this differs from R. P. Abelson's sense of "script.") Since scripts are represented as data rather than as integral parts of the program itself, ELIZA can assimilate an indefinite number of them without necessitating any revision of her essential core. Scripts function like subroutines; they can be called by other scripts, and can intercommunicate in various ways. The more mature ELIZA can therefore switch her point of view within a "single" interview from that of hospital receptionist (for example) to nurse, psychiatrist, or ear surgeon. The various scripts corresponding to these human roles not only involve different keys and transformation rules, but may interpret a given key in differing ways: the input "tone" will receive responses appropriate to muscle or music according to whether ELIZA is scripted as general physician or ear surgeon, respectively. The multiscript ELIZA can also use (and learn to use) expressions taken from several different natural languages during one semantically coherent conversation. Evidently, like her Shavian namesake, ELIZA can be taught to speak increasingly well.

But it should be evident also that Professor Weizenbaum has less to fear from his brainchild than Professor Higgins from Eliza Doolittle. ELIZA is nondirective *faute de mieux*: she has no thoughts of her own to contribute, save perhaps her hunches that reference to certain topics may serve to keep her partner talking. Even Colby's programmed paranoiac is less empty-headed than ELIZA, for PARRY is able at least to steer the conversation with some apparent ingenuity onto the nefarious activities of the Mafia. Miss Doolittle, on the other hand, could complement the mouthing of empty conventional trivialities—"How kind of you to let me come!"—with the expression of attitudes and beliefs on many different issues, including her personal relations with Higgins himself.

Moreover, Eliza, unlike ELIZA, at all times interpreted what others said to her in the light of inferences based on her own complex system of knowledge about her world. These inferences were much more varied and flexible than those made by the paranoid program in its psychotic interpretations and promptings, which are limited to those made possible by the restricted body of knowledge corresponding to PARRY's persecution complex that is represented in the directed graph stored in the data base. Admittedly, misunderstanding (on either side) sometimes resulted during Eliza's conversations, because the Cockney girl's values and beliefs—in Abelson's terminology, her themes and scripts—differed from those of her socially more advantaged fellows. But ELIZA cannot even *mis*understand her creator: simply, she sometimes produces grossly inappropriate responses to his remarks.

This is the basic difference between Eliza and ELIZA, that only the former can bring her intelligence, or reason, to bear in interpreting the language she encounters. The essential connection of understanding with intelligence must be clearly recognized if there is to be any chance of progressing from Weizenbaum's actual ELIZA to Clarke's fictional HAL.

# 6

# *Intelligence in Understanding*

ANYONE who doubts the contribution of intelligence to understanding might do well to consider the following exchange from *Alice's Adventures in Wonderland*:

> ". . . Even Stigand, the patriotic Archbishop of Canterbury, found it advisable—"
>
> "Found what?" asked the duck.
>
> "Found *it*," replied the mouse, rather crossly. "Of course you know what 'it' means?"
>
> "I know what 'it' means when *I* find a thing," said the duck. "It's generally a frog, or a worm. The question is, what did the Archbishop find?"

If one takes seriously the mouse's question whether the duck knew what "it" means, the answer must be that she did not fully understand this word, for she apparently did not realize that "it" need not be used so as to refer to a thing. Nevertheless, the duck had a partial understanding of "it," clearly having grasped one very common use of the word. Moreover, she realized that in this sense the word is likely to refer to different things according to whether a duck or an Archbishop is involved. This realization was based on her knowledge of the world: even in Wonderland, evidently, Archbishops usually have scant interest in locating frogs or worms.

The important point here is that common-sense knowledge such as this can be crucial to the interpretation of a linguistic expression on a particular occasion of its use. Consequently, such knowledge contributes integrally to the understanding of the expression and to communications making use of it. The notion that understanding can take place independently of such knowledge is therefore unacceptable.

To be sure, apparent comprehension can often result from relatively simple rules, like assuming that "it" always refers to the immediately preceding noun, or devising expectancy lists like PARRY's whereby "it" is temporarily assigned a specific meaning such as "horseracing." Sometimes a person or a program can even apparently understand a word that is in fact ignored, as STUDENT ignores the two occurrences of "it" in the algebra problem cited in Chapter 5. But this can only avoid misunderstanding if it is not essential to the communication in question that the word be properly understood: it has already been pointed out that knowledge of the Russian army and its military maneuvers is irrelevant to the strictly *algebraic* communication that is expressed by STUDENT's English input. And even mathematical questions of this type cannot always be answered without some appeal to common sense: STUDENT could not be expected to reply plausibly to the query "If a man of twenty can gather eight pounds of blackberries in a day and a girl of eighteen can gather nine, how many will they gather if they go out together?" In many cases, general knowledge will be needed not only to respond sensibly to a query but also to interpret it correctly in the first place. In short, the everyday interpretation of language requires an intelligent deployment of semantic information in assigning sense and syntax to particular phrases.

The knowledge required varies from case to case. Sometimes it is enough that the general context (the physical situation and/or the topic of discussion) be known: "They are eating apples" will be effortlessly disambiguated in two syntactically different ways according to whether the context is a greengrocer's shop or the monkeys' feeding time at the zoo.[1] (If this sentence is *spoken*, as opposed to being printed in a story or used as the caption to a picture showing a mother pointing at apples in one or other of these situations, then acoustic stress and intonation contour would usually suffice to disambiguate it, without reference to the context. A *speech*-understanding program[2] should be able to take advantage of this; but programs confined to conversation by teletype obviously cannot.) Sometimes a good deal of sophisticated empirical knowledge about society and political relations may be needed in order to understand a sentence. For instance, consider the knowledge unthinkingly deployed in understanding "they" in these two sentences: "The city councilmen refused to give the women a permit for demonstration because they feared violence," and "The city councilmen refused to give the women a permit for demonstration because they advocated revolution."[3]

In general, a person's understanding of natural language normally relies on implicit (or occasionally explicit) inferences drawing on her knowledge of her interlocutors and of the world.

It follows that an artifice designed to understand language must be provided with a data base relevant to the universe of discourse, one that includes models of the participants' ideas about this universe and about each other, and that can be actively used in the machine's interpretation of its partner's conversation.

## CONVERSATION WITH SHRDLU

These theoretical points about the human use of natural language have been stressed, and considerably clarified, by Terry Winograd.[4] His insights about natural language were developed in the course of his project to program a robot to answer questions, execute commands, and accept information in an interactive English dialogue with a human being.

The essential feature of Winograd's program, in which it differs radically from Joseph Weizenbaum's ELIZA, is the inclusion of a system of knowledge concerning the assumed universe of discourse, a system on which the program can draw in a reasoned fashion so as to make sense of the remarks to which it responds. A flexible interplay between syntactic and semantic features of the linguistic input, and between these and the program's knowledge of the current state and general properties of the world (including its conversational partner), allows for subtleties of comprehension that vastly outrun the capabilities of earlier assays in the field.

The program represents a robot named SHRDLU[5] that evinces its understanding partly by verbal responses to its interlocutor, and partly by carrying out actions at the human's command. Access to an internal representation of its own acts (and possible acts) in the world, as well as to a model of the world considered independently, enables SHRDLU to infer the meaning of questions and commands that would otherwise remain opaque.

Like the duck, Winograd's program has at least a partial understanding of "it," and (unlike the duck) does not assume that "it" must always refer to an object rather than an action. Even within a conversation strictly limited to the discussion of variously colored objects that can be moved and stacked on a table, or put into and taken out of a box, to know what "it" means is to know (to remember or to infer) a good deal about the world and about actors intervening in the world. This is evi-

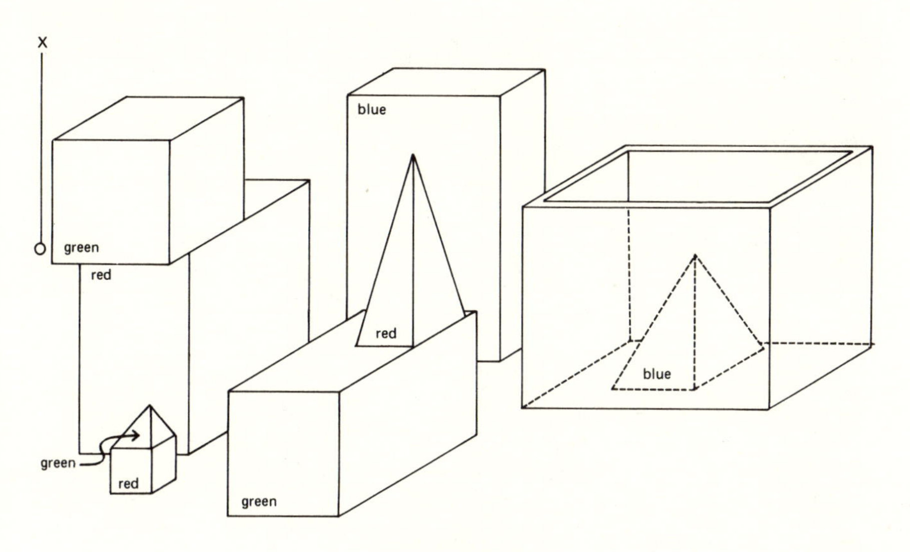


FIGURE 6.1
"Pick up a big red block."

Source: Adapted from Terry Winograd, *Understanding Natural Language* (New York: Academic Press, 1972), p. 8. This and following figures from this source reprinted by permission.

dent from a consideration of the 17 occurrences of "it" in the following dialogue, which indicates the power of comprehension enjoyed by Winograd's program as a result of using its intelligence in interpreting the conversation:[6]

1. *pick up a big red block.*
   OK.
   The system answers "OK" when it carries out a command. In order to pick up the red block, it had to clear it off by finding a space for the green one and moving the green one away. [Figure 6.1 shows the initial state of the scene.]

2. *grasp the pyramid.*
   I DON'T UNDERSTAND WHICH PYRAMID YOU MEAN.
   There are three pyramids in the scene, so the program does nothing.

3. *find a block which is taller than the one you are holding and put it into the box.*
   The system has to figure out what is meant by "one" and "it," as well as using deduction to decide which block it is holding.
   BY "IT," I ASSUME YOU MEAN THE BLOCK WHICH IS TALLER THAN THE ONE I AM HOLDING.
   Printed out to keep us informed.
   OK.

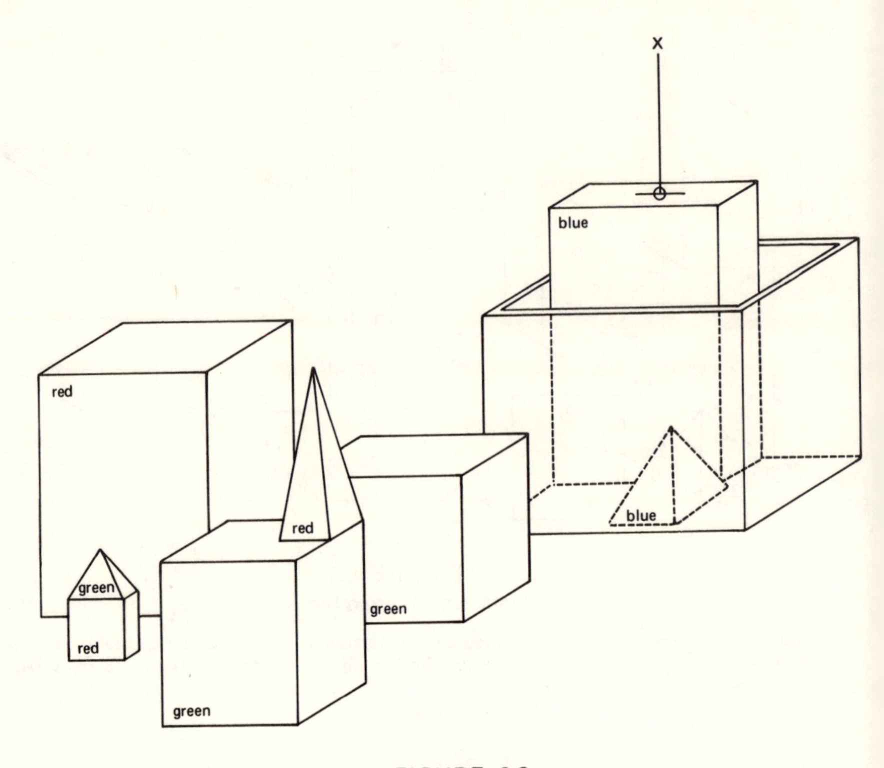


FIGURE 6.2
"Find a block which is bigger than the one you are holding and put it into the box."

Source: Adapted from Winograd, *Understanding Natural Language*, p. 9.

The hand has to set down the red block to pick up the taller blue one. Figure 6.2 shows the result of the system's manipulations so far.

4. *what does the box contain?*
   THE BLUE PYRAMID AND THE BLUE BLOCK.

5. *what is the pyramid supported by?*
   Earlier the system responded to "the pyramid" by saying "I don't understand which pyramid you mean." Now, it assumes that we mean the one just mentioned.
   THE BOX.

6. *how many blocks are not in the box?*
   Negatives such as "not" and "no" are handled whether they appear in clauses or noun groups (such as "no boxes").
   FOUR OF THEM.

7. *is at least one of them narrower than the one which I told you to pick up?*

The phrase "at least one of them" must be analyzed to refer back to the previous question, and "the one which . . ." must also refer to blocks. Noun groups can have complicated numbers such as "at least one," "more than three," "exactly seven," etc. The memory of past events is used to deduce which block is "the one which I told you to pick up." Notice the complicated syntactic structure in which "the one" is actually the object of a clause embedded two levels deeper.

YES, THE RED CUBE.

8. *is it supported?*

Passives can be used without a "by . . ." phrase.

YES, BY THE TABLE.

9. *can the table pick up blocks?*

NO.

The semantic rules make this a nonsensical question which therefore must be false.

10. *can a pyramid be supported by a block?*

YES.

The deductive system finds an actual example, so it knows this is possible.

11. *can a pyramid support a pyramid?*

I DON'T KNOW.

The question makes sense, but there is no example in the scene, so there is no easy way to tell without being able to examine the robot's programs.

12. *stack up two pyramids.*

I CAN'T.

The robot tries and fails.

13. *the blue pyramid is mine.*

I UNDERSTAND.

The data is added to the data base.

14. *I own blocks which are not red, but I don't own anything which supports a pyramid.*

I UNDERSTAND.

Two new theorems are created for proving things about "owning."

15. *do I own the box?*

NO.

The box supports whatever it contains, including a pyramid.

16. *do I own anything in the box?*

YES, TWO THINGS: THE BLUE BLOCK AND THE BLUE PYRAMID.

17. *will you please stack up both of the red blocks and either a green cube or a pyramid?*

OK.

Logical connectives such as "and," "or," "either," etc. are handled in both the grammar and semantics. A command is executed even if it

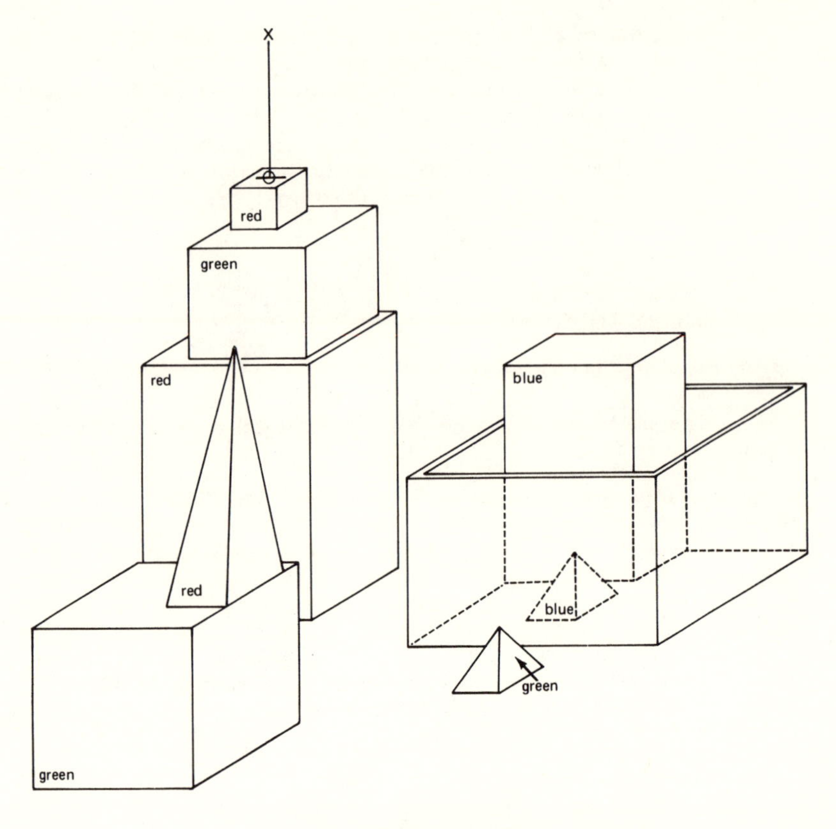


FIGURE 6.3
"Will you please stack up both of the red blocks
and either a green cube or a pyramid?"

Source: Adapted from Winograd, *Understanding Natural Language*, p. 12.

looks like a question. The command is carried out by the following series of steps: It puts a green cube on the large red block (note that it chooses the green cube with nothing on it), then removes the small pyramid from the little red cube, so that it can use that cube to complete the stack as specified, resulting in the situation of Figure 6.3.

18. *which cube is sitting on the table?*
THE LARGE GREEN ONE WHICH SUPPORTS THE RED PYRAMID.

Ordinarily the system would be satisfied with calling this "a big green cube," but since we asked "which cube" it has to be more specific in naming it. The use of substitute nouns (such as "one") and pronouns in answers makes the discourse more natural.

19. *is there a large block behind a pyramid?*
YES, THREE OF THEM: A LARGE RED ONE, A LARGE GREEN CUBE AND THE BLUE ONE.

20. *put a small one onto the green cube which supports a pyramid.*
"A small one" must contrast with the earlier phrase "a large block," so this means "a small block."
OK.

21. *put the littlest pyramid on top of it.*
Words like "littlest" are not in the dictionary but are interpreted from the root forms like "little."
OK.

22MIT. *does the shortest thing the tallest pyramid's support supports support anything green?*
The grammar can handle words like "support" which are both verbs and nouns without getting confused, and can understand possessives. To compute the answer, see Figure 6.4 for the scene.
YES, THE GREEN PYRAMID.

23MIT. *what color is it?*
BY "IT," I ASSUME YOU MEAN THE SHORTEST THING THE TALLEST PYRAMID'S SUPPORT SUPPORTS.
If we had asked "what size is it?", the system would have assumed that "it" meant the green pyramid, but nobody would ask the color of something which was just called "the green pyramid," so it had to look farther for the reference of "it."
RED.

22. *how many things are on top of green cubes?*
I'M NOT SURE WHAT YOU MEAN BY "ON TOP OF" IN THE PHRASE "ON TOP OF GREEN CUBES."
DO YOU MEAN:
1—DIRECTLY ON THE SURFACE
2—ANYWHERE ON TOP OF?
When there are several interpretations and none is clearly better, the system has to ask the user for more information. The paraphrases were included with the dictionary definition of "on," and the request for more information occurs automatically when the heuristics for resolving ambiguities aren't sufficient. A request like this can also occur when the meaning of a pronoun such as "it" or "they" is unclear.
*2*
THREE OF THEM.

23. *had you touched any pyramid before you put the green one on the little cube?*
YES, THE GREEN ONE.
The system deduces from its memory what event is being referred to, and can relate other events to it in time. The analysis includes complex tenses of verb groups, and there are special facilities for keeping track of the various forms of irregular verbs like "have."

24. *when did you pick it up?*
WHILE I WAS STACKING UP THE RED CUBE, A LARGE RED BLOCK AND A LARGE GREEN CUBE.
A time is described by naming the main goal which was being worked on.

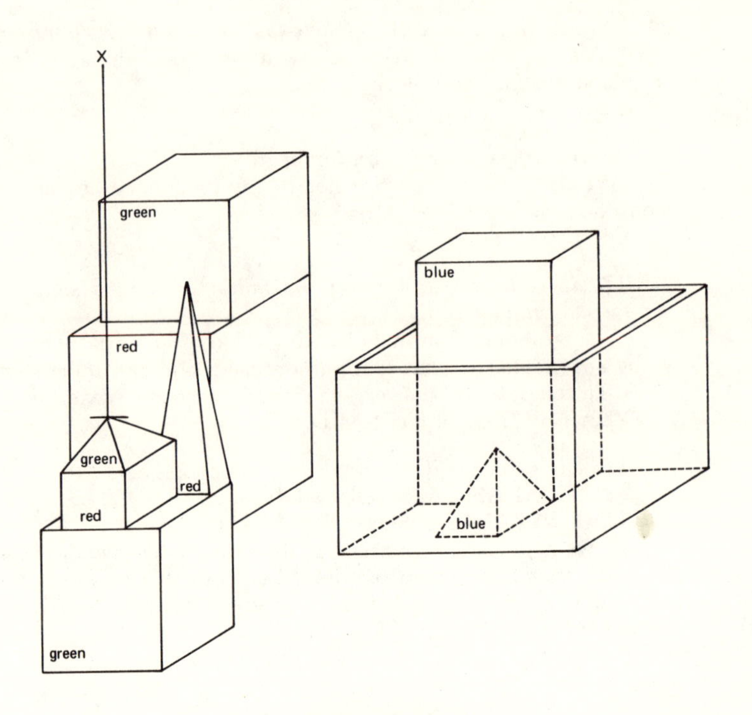


FIGURE 6.4
"Does the shortest thing the tallest pyramid's support supports support anything green?"

Source: Adapted from Terry Winograd, ***Procedures as Representations for Data in a Computer Program for Understanding Natural Language***. (Ph.D. thesis). (Cambridge, Mass.: AI-TR-17, MIT AI Lab.), p. 49. Reprinted by permission.

25. *why?*

The system remembers immediately previous sentences to understand questions like this and the followiing ones.

TO GET RID OF IT.

By keeping track of selected parts of the original subgoal tree, the system has some understanding of its own motives.

26. *why did you do that?*

TO CLEAN OFF THE RED CUBE.

"That" refers to the action mentioned in the answer.

27. *why did you clear off that cube?*

Here, "that" means the one just referred to.

TO PUT IT ON A LARGE GREEN CUBE.

28. *why did you do that?*

TO STACK UP THE RED CUBE, A LARGE RED BLOCK AND A LARGE GREEN CUBE.

29. *why did you do that?*

BECAUSE YOU ASKED ME TO.

It remembers what it did, not how the request was worded.

30. *how did you do it?*

"It" refers to the action mentioned in the last question, rather than the answer.

BY PUTTING A LARGE RED BLOCK ON THE TABLE; THEN LETTING GO OF IT; THEN PUTTING A LARGE GREEN CUBE ON IT; THEN LETTING GO OF THAT CUBE; THEN PUTTING THE RED CUBE ON THAT CUBE; THEN LETTING GO OF THAT CUBE.

This is a description of the top-level subgoals used to achieve the goal.

31. *how many objects did you touch while you were doing it?*

Time references can refer to a time span, using words like "while."

FOUR OF THEM.

32. *what did the red cube support before you started to clean it off?*

THE GREEN PYRAMID.

By remembering what has been moved, it is possible to deduce things about earlier states of the scene. We can use complex nestings of clauses like "before you started to clean it off."

33. *there were five blocks to the left of the box then.*

"Then" must refer to the time mentioned in the last sentence.

NO, ONLY FOUR OF THEM: THE RED CUBE, TWO LARGE GREEN CUBES AND A LARGE RED BLOCK.

Earlier in the dialogue, new information about "owning" was accepted at face value. But the system knows that it has complete information about the basic properties of objects, including location. It checks any statements about location or other such properties to see whether it agrees or disagrees.

34. *put the blue pyramid on the block in the box.*

This is syntactically ambiguous, but is understood without ambiguity since the parsing corresponding to "put the blue pyramid which is on the block in the box" does not make sense in the current scene, while "put the blue pyramid on the block which is in the box" does.

OK.

35. *is there anything which is bigger than every pyramid but is not as wide as the thing that supports it?*

The system understands quantifiers such as "every," "any," "some," etc., and structures such as "bigger than," "as wide as", etc. Here "it" refers to a noun group within the same sentence but two levels up in the parsing tree.

YES, THE BLUE BLOCK.

36. *does a steeple*

SORRY, I DON'T KNOW THE WORD "STEEPLE."

The system interrupts immediately when it sees a word it can't figure out.

37. *a "steeple" is a stack which contains two green cubes and a pyramid.*
The quote marks indicate that we are defining a new word.
I UNDERSTAND.

38. *are there any steeples now?*
NO.
The new definition is used as a description to recognize steeples.

39. *build one.*
OK.
The description given in the definition is enough to deduce a plan for building new objects as well as recognizing them. The word "one" means "a steeple", and the plan is carried out. First, one of the large green cubes is cleared off by removing the red cube and two pyramids. Then the two green cubes and the tall red pyramid are stacked up to fit the description.

40. *call the biggest block "superblock."*
OK.
We can also give objects names.

41. *Have you picked up superblock since we began?*
YES.

42. *why did you drop it?*
BY "IT," I ASSUME YOU MEAN THE SUPERBLOCK. I CAN'T EXPLAIN A NONEXISTENT EVENT.
There is no record in memory of any block being dropped.

43. *is there anything to the right of the red pyramid?*
YES, FOUR THINGS: THE BOX, THE RED CUBE, THE BLUE PYRAMID AND THE BLOCK NAMED SUPERBLOCK.
Once an object has a name, the name is used to help identify it in responses.

44. *thank you.*
YOU'RE WELCOME!
Idioms can be defined without modifying the grammar.

Despite the excessively congratulatory tone of the final comment (which refers to an essentially trivial programming technique well within the range of Weizenbaum's ELIZA, and relied on fairly heavily by K. M. Colby's PARRY), the overall performance of SHRDLU in this man-machine dialogue is impressive. So much so, indeed, that the reader's first impressions may blind her to a number of general limitations that would become more apparent were SHRDLU's human friend to alter the style and topic of conversation. Some of these limitations will be discussed in Chapter 7, wherein Winograd's program will be contrasted with others yet to be described; but two must be mentioned immediately.

The less obvious of these consists in the fact that Winograd's program cannot itself produce sentences of as great a syntactic variety as those that it can accept from its human partner. Winograd's interest is in the interpretation rather than the generation of language, and

SHRDLU's capacity to analyze language far outstrips its capability of synthesis. Whereas it can parse grammatical constructions of considerable subtlety, its own speech acts are based on a handful of relatively simple sentence types, which can be transformed in a small number of ways and which are eked out by a few idiomatic expressions such as "OK" and the like.[7] In short, in its generative activities, SHRDLU is considerably more like ELIZA than it appears.

While this undeniably is a limitation, it is however not crippling. Like the human child, or the adult speaker of a foreign language, the program can cope with a high degree of grammatical complexity in the input, and can express itself intelligibly if not stylishly. The matters that SHRDLU is concerned to express (like most matters of human interest) can if necessary be stated by way of a restricted subclass of syntactical forms. That conversation so restricted may eventually strike one as somehow "flat," "prosaic," or "dull," is a largely aesthetic rather than a purely linguistic criticism. The complaint that the *topic* of SHRDLU's conversation is prosaic or dull is, however, more to the point, and is an oblique way of stating the second general limitation of Winograd's program.

SHRDLU would deservedly be rejected as a conversational partner were one offered as alternatives, let us say, the Walrus or the Carpenter. For there is no question of the program's being able, like them, to talk of many things: shoes and ships and sealing wax are as foreign to its universe of discourse as are cabbages and kings. From the point of view of Winograd's program, the world consists only of the robot itself; a table-top bearing a number of colored objects and a box; and an interlocutor who asks the robot to move the objects about and who plagues it with questions about the world and about its reasons for acting in the way it does. Indeed, there are in fact no objects, no box, no table—and no robot. The table-top world and the computer's actions in it are simulated by internal data processes paralleled by images on a cathode-ray screen. This visual display is strictly secondary to the workings of the program, and its purpose is merely to aid the human observer wishing to keep track of the "nonlinguistic" computation carried out by the program.

The "imaginary" nature of SHRDLU's spatial world does not detract from the specifically linguistic interest of the program, as opposed to its relevance as an exercise in practical robotics, but its relative emptiness does. The vocabulary, environmental information, and general semantics currently available to SHRDLU are greatly impoverished compared with the knowledge mediating one's daily understanding of one's native tongue, and SHRDLU's conversation is correspondingly limited.

What is even worse than its ignorance of shoes, ships, and sealing wax is its ignorance of the very things it appears to know something about: blocks, pyramids, boxes, and tables. For the same sorts of reason that Colby's neurotic program understands almost nothing of *love* and *hate*, even though it can sometimes manipulate these antonymic symbols in an apparently rational manner, SHRDLU understands much less than a person does by the sentence "Pick up a big red block." Winograd's program does not know that blocks are heavy, or that the bigger they are the heavier they are likely to be; it does not know that if they are dropped, or pushed off the edge of the table, they will fall; it does not know that cubical blocks have six square faces, not all of which are visible at the same time. In short, it knows almost nothing about them at all. Even the block-recognizing visual programs to be described in Chapter 8 know considerably more about blocks than SHRDLU does, and we shall see that their knowledge of cubes and pyramids is pitifully shallow compared with ours.

Nevertheless, it is significant that an extension of Winograd's program could rightly refuse to worry about why the sea is boiling hot (*cf.* items 33 and 42 above). Indeed, SHRDLU's cognitive world has already been extended by the incorporation of the program into a system providing weather information, a system within which the temperature of the sea (and the limits thereof) might well be represented.[8] Similarly, Winograd's program could deduce for itself—given the taxonomic knowledge that bats are the only flying mammals—that pigs do not, in fact, have wings (*cf.* items 14–16). This latter insight could also be produced by an appropriately tutored question-answerer like SAD SAM or SIR, although neither of these would critically reject the Walrus's false assumption implied by his reference to the boiling sea. ELIZA, by contrast, would inflexibly parry the Walrus's questions by inquiring why he felt the temperature of the sea and the anatomy of pigs to be important, and would give no indication of knowing anything about these phenomena, on either side of the Looking-Glass. And Colby's cybernetic psychotic, of course, would merely change the subject as soon as possible—perhaps by associating "pigs" with "police"—onto topics relevant to its paranoid delusion.

The superior flexibility[9] and common sense thus evident in SHRDLU's conversation (as compared with PARRY, ELIZA, and the early question-answering programs) rely heavily upon two features, features that are likely to be relevant to other projects aiming at artificial intelligence approaching the power of the natural variety. The first is the "heterarchical" organization of the various components of the overall program, while the second is the representation of different types of knowledge on which the program can draw in interpreting and replying to the comments of its human friend.

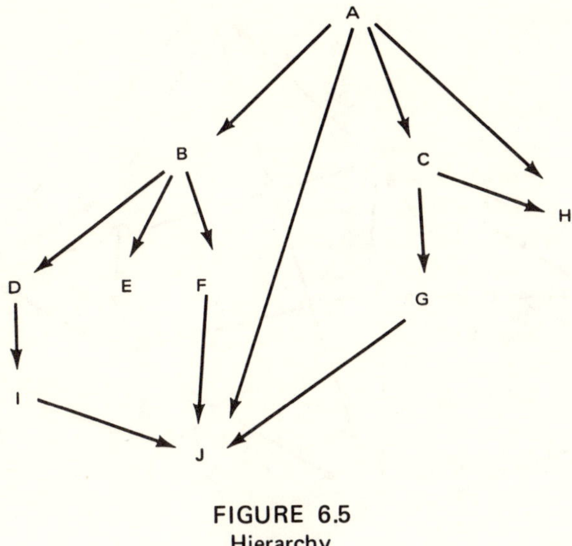


FIGURE 6.5
Hierarchy

## HETERARCHICAL THINKING

Winograd's program is, in outline, three programs working in concert. The combination of several programs to function somehow in unison is no innovation: what is new is the flexible manner of their organization.

Programs may be combined serially, hierarchically, or heterarchically. In serial ("pass-oriented") combination, one program is run on the initial data and another then takes over, using the results output by its predecessor as its own data. For instance, we shall see in Chapter 10 that the identification of a line drawing as a representation of a three-dimensional arch may involve the following successive steps: program A communicates the line drawing to the machine; program B classifies and labels the vertices in the picture; program C uses this information to pick out the separate objects represented in the drawing; and program D assigns descriptions (such as "arch," "pedestal," etc.) to the objects previously individuated. This series could be extended by starting with a grey-scale photograph of a simple arch and using a fifth program to convert this to a line drawing.

In hierarchical combination, one program has overall control and the others are subordinate to it as mere subroutines in the service of the goals of the master program. As Figure 6.5 shows, the subordinate programs need not be *directly* accountable to the highest one, since there may be

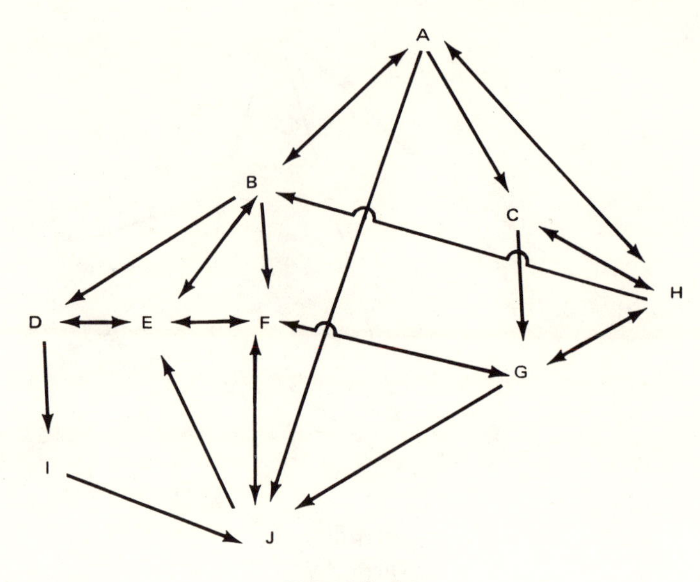


FIGURE 6.6
Heterarchy

several hierarchical levels involved. But the flow of control passes in one direction only: downward. Consequently, although the responsibility for control is distributed throughout the system, in the sense that each lower level program has its own type of job to do, authorization must always derive from a higher level. Although programs I or J (in Figure 6.5) can influence program F, in the sense that F's actions may depend upon information gleaned from the running of I or J, they cannot turn round and tell F what to do. Moreover, they cannot communicate directly at all with programs lying "sideways" in the hierarchy, such as G and H. For example, the individual transforms in Colby's neurotic program are applied, or not applied, on the decision of the higher levels of control within the system. DEFLECTION cannot hand over control to PROJECTION on its own initiative; nor can the deflecting FINDANALOG routine call on RATIONALIZATION for advice.

Hierarchical control may be passed to the subordinate programs in a fixed or a variable order. Performance is naturally less rigid in the latter case, but even in such a case the overall decision as to which lower level module should be activated next lies with some higher level member of the hierarchy. Ultimately, all decisions directing performance are dependent on the highest level of all. A hierarchical assembly of routines is clearly more flexible than a strictly serial arrangement, since it is possible for the action (and activation) of the lower level units to be differentially directed by the master program, in light of the overall problem situation;

for instance, whether or not Colby's neurotic resorts to PROJECTION depends on current DANGER and WELL-BEING. The earlier units in a *series* of programs, by contrast, function independently of the later ones, and cannot be adjusted according to the final goal of the series as a whole; and (as we shall see in Chapter 9) their mistakes cannot be rectified or presciently forestalled by decisions taken at the later stages. Nevertheless, hierarchical organizations (of programs or of persons) suffer an intrinsic rigidity resulting from the fixed progression of responsibility through successive levels of the system.[10]

In a heterarchical organization, however, the responsibility for control can be more equally distributed throughout the system, and internal communication is much increased. The arrows in Figure 6.6 show that programs that are related heterarchically can address or call upon each other either "up," "down," or "sideways." Moreover, they can do this at many different points in their (potentially independent) functioning. The human analogy is a group of intercommunicating specialists contributing their several skills to a cooperative enterprise, rather than a troop of servants each unquestioningly obeying their mistress—or her butler, cook, and housekeeper.[11]

To be sure, cooperation can be incompetent, and heterarchy is not a magical guarantee of efficiency: the camel, it has been said, is a horse designed by a committee. Cooperating experts usually must not only be expert: they must normally also be guided by a clear idea of the overall goal if they are not to risk being sidetracked into busy irrelevancies. Sometimes this clear idea is shared by all members of the committee, but more often it is the responsibility of the chairman, to whom individual members continually send in reports; if the chairman has no such idea, or insufficient strategic control of the committe to guide it accordingly, then what was to be a horse may turn out to be a camel. Occasionally, however, the requirements of the task are so restrictive as to keep the cooperative efforts within the bounds of relevancy, without anyone's having a global view of all aspects of the task.[12]

In fruitful heterarchical cooperation, the specific problem conditions at a given time nicely determine which specialists will communicate with their fellows, and what type of communication (question, complaint, command, advice, criticism, answer, etc.) will be employed. The result is an enormous increase in the flexibility of performance (particularly in the sensitivity of control to context), as compared with serial or hierarchical arrangements. This will be evident from the description of BUILD, to be discussed in Chapter 12. It can be seen also from a consideration of Winograd's program, given the task of interpreting an input sentence in English so as to be able to make an appropriate response, whether verbal or "motor" in character.

The three programs that, broadly speaking, comprise Winograd's language-understanding system are concerned with grammar, semantics, and deduction. The first is a parsing system that embodies a particular theory of English grammar (namely, M. A. K. Halliday's "systemic" grammar[13]), and uses it to recognize the syntactic structure of sentences. The second is a set of semantic programs dealing with meaning (whether of words, word groups, or whole sentences). This system is built around a collection of "semantic specialists" designed to interpret particular syntactic structures such as noun group, adjective group, preposition group, etc. And the third is a deductive system that can solve problems of various kinds in exploring the consequences of facts, planning actions, and answering questions, and that includes a body of knowledge about the specific universe of discourse chosen—*viz.*, the table-top world of blocks and pyramids pictured in Figures 6.1 to 6.4.

None of these three programs has a global view of what the system as a whole is doing. Even the "Monitor" program, which is the master program in the sense that it calls the basic parts of the system, has no such view; it does not really *monitor* what goes on, for this would require receiving continual reports on progress and adjusting the activities of the various specialist programs accordingly. Winograd points out that most of the communication between components is done directly, and the monitor is called only at the beginning and end of the understanding process.[14] In other words, it decides when SHRDLU is to listen and when it is to answer, but does not help it to listen or answer intelligently. Where SHRDLU's understanding is concerned, it is as if no overall chairman wisely guides the committee, but each member "does her own thing" with constant help from her peers.

A supervisory chairman is unnecessary because the experts are "tailor-made" to the particular, highly limited, tasks undertaken by the system. This is not strictly true of the parsing program that, as we shall see in the next section, can parse very many English sentences. But, by the same token, the rules of syntax are so restrictive that the parser can readily recognize grammatically unacceptable constructions, so that these have no chance of passing unchecked. By contrast, the semantics and world knowledge provided by SHRDLU are grossly limited compared with human understanding (it knows only about blocks, and only very little about these), so that it is not possible for it to go wildly offtrack. A more knowledgeable program, with more varied types of expertise available to it, would need a genuine monitor with a global idea of what it was doing if its versatility were to be kept sensibly within bounds.

But SHRDLU is heterarchical rather than hierarchical in its functioning. Thus although Winograd describes the grammar program as "the main coordinator of the language understanding process," he also points out that the deductive system is "used by the program at all stages of the analysis, both to direct the parsing process and to deduce facts about the

BLOCKS world." Moreover, each of the specialist semantic programs "has full power to use the deductive system, and can even call the grammar to do a special bit of parsing before going on with the semantic analysis."[15] In short, to single out one component of the heterarchical system of programs as *the* crucial or overriding element would be invidious. In practice, the various systems cooperate in subtle ways so as concurrently to interpret the sentence presented to the machine.

As will be evident from the example to be described presently, the deductive system is used both in the semantic evaluation that guides the parsing and in the program's answers and (simulated) motor actions. In essence, then, SHRDLU is a "theorem prover." Theorem-proving programs are aptly named: their function is to deduce theorems from an axiomatic base of knowledge by using strictly logical methods of reasoning. Some employ only very general methods, which are universally applicable in principle whether or not they are always sufficient in practice to solve the problem concerned; one example of such a technique is "resolution," whereby an assertion is shown to be a theorem by proving that its negation is impossible.[16] Others, including SHRDLU, rely also on specific problem-solving strategies that are appropriate to particular problems only. But in either case the reasoning is "logical," or deductive, in nature. Nondeductive methods of reasoning may follow strict rules—for instance, the seemingly "prejudiced," "impressionistic," and "inductive" reasoning incorporated in R. P. Abelson's IDEOLOGY MACHINE and the "paranoid" attributions of malevolence in Colby's PARRY—but these rules are not rules of *proof*. In assessing the truth of a statement (whether input to them or thought up by them), their prime concern is whether it is probable, given the relevant evidence available, not whether it is strictly entailed (or, conversely, contradicted) by the existing knowledge of the system.

Unlike PARRY and the mechanized Goldwater, SHRDLU's thought processes are rigorously logical in form. When seeking an empty space on the table, for instance, in which to place a block it wishes to get rid of (as in item 1), its technique is deductively to work toward the goal of identifying a space that can be proved to fit the description "empty." It also has to prove that its hand is not already holding anything before it can pick up the big red block as requested. We shall see later that this logical purity has certain disadvantages, which would be more apparent were SHRDLU required to function in an "open" world (like yours and mine) in which whatever is not entailed by the known data can*not* therefore be simply assumed not to be the case. SHRDLU's world is "closed" in the sense that the axioms are assumed to imply all the truths it needs to know.

Winograd's system parses the input sentence from left to right as it goes along, anticipating the syntactic structure of the portion yet to be considered. It applies its semantic and environmental knowledge at appropriate points in the parsing procedure to eliminate ambiguities. If necessary, it can backtrack to weed out parsings that it discovers, after all, to have no sensible application in context. Usually, the backtracking is not blind but is guided by the specific nature of the difficulty; the requisite advice is embodied in the syntactic and semantic theorems themselves. Like a human being who does not regard a sentence as ambiguous on a particular occasion of its use, even though it may be so in principle, Winograd's system assigns only one parsing in cases where an alternative syntactic analysis might seem to be possible. Earlier parsing programs had to cope with ambiguity (if at all) by returning all possible parsings of the input string, being unable to distinguish the contextually correct version.[17] SHRDLU is significantly more intelligent.

For instance, the correct interpretation of the potentially ambiguous sentence, "Put the blue pyramid on the block in the box," is achieved only because the deductive knowledge system can be called to check which of the alternative parsings makes sense in the current situation (*cf.* item 34).

The grammar program has to identify the major clause as either declarative, imperative, or a question. It first parses the initial word as an imperative, and takes it to be the main verb. This is possible since "put" is classified in the Dictionary as a verb in the infinitive form, and the parser has the information that such verbs in the initial position are usually imperatives. The imperative, declarative, and interrogative moods are basic theoretical elements in systemic grammar, which differs from transformational grammar in identifying basic syntactic elements largely by reference to the meanings conveyed rather than by reference to formal structure alone. As Winograd puts it, systemic grammar gives priority to the question "Which features of a syntactic structure are important to conveying meaning, and which are just a by-product of the symbol manipulations needed to produce the right word order?"[18] This aspect of systemic grammar makes it particularly suitable for facilitating fruitful communication between a "syntactic" parsing program and a "semantic" program concerned primarily with meaning, and explains Winograd's preference for Halliday's grammar over Noam Chomsky's syntactic theory.[19]

Having identified an imperative, the parser expects to meet a noun group—and parses "the blue pyramid on the block" accordingly. The parser may be described as "expecting" a noun group because its knowledge about verbs includes the advice that it should look first for a noun group on having met a transitive imperative; but it does not rigidly assume the presence of a noun group, and so would be able to parse sentences like the imperative "Hammer harder!"

The typical noun group structure that the parser expects to find can be represented as follows:[20]

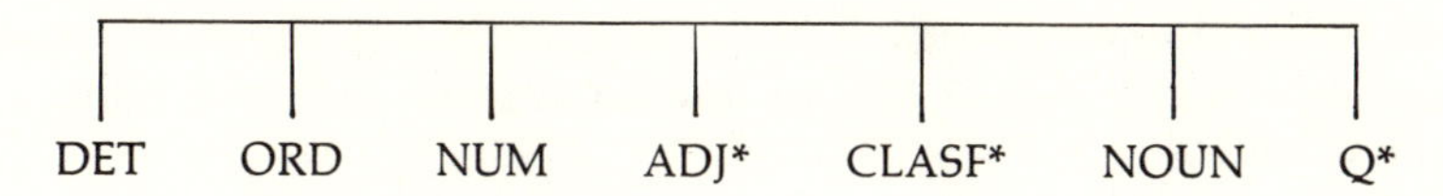


This diagram shows the syntactic "slots" in the order in which they occur within the phrase; not all the slots need be filled, though there is almost always a NOUN and usually also a DETerminer (such as "the," "a," "her," or "Mary's"). The "*" sign means that there may be more than one word of the class concerned, so the parser should not overhastily assume that it can pass immediately to the next slot in the list. After the initial DETerminer comes the ORDinal slot, which may contain words like "next," "fourth," "tenth," or "last." The NUMber slot can contain either single words or phrases, such as "seven" or "more than two hundred." There may be several ADJectives, such as "big, red, beautiful." Equally, there may be more than one CLASsiFier—that is, a noun currently being used as a quasi-adjective, to describe the main NOUN: as in *city fire* hydrant. The main NOUN may be followed by one or more Qualifiers, such as "the man *in the moon*," or "the woman *with red hair, who conducts the orchestra*." Sometimes all the possible slots are filled within one noun group, as for instance in "the first three old red city fire hydrants without covers you can find." This is parsed as DET, ORD, NUM, ADJ, ADJ, CLASF, CLASF, NOUN, Q (PREPosition GROUP), Q (CLAUSE).

It is because the parser expects to meet this structure when it looks for a noun group that the words "on the block" in our example are initially included by it within the noun group (that is, as attached to "the blue pyramid"). For "on the block" is a preposition group that in this position can function as a qualifier of the main noun.

However, one intuitively senses that "on the block" is ambiguous in our sample sentence. It might indeed be a phrase qualifying the object previously identified as the head of a noun group—that is, the blue pyramid might already be sitting on the block. Alternatively, the preposition group might be using "on" in a directional rather than a locational sense, to say *where* the thing identified by the noun group is to be put.[21] In the first case, the phrase "on the block" should be included within the noun group already identified, whereas in the second it should not. How does Winograd's system resolve this ambiguity?

Whenever a main section of a parse (such as a noun group) has been finished, the parsing program hands control over to the semantic program to see if the parse makes sense so far. The semantic program confirms, for instance, that "pyramid" is classified by the BLOCKS system of semantic markers as something which is a sort of physical object that can be manipulated by the robot (see Figure 6.8). The semantic program thus has no reason for rejecting this parsing as inherently nonsensical—as it has with regard to the suggestion that tables can pick up things (*cf.* item 9). Up to and including the word "pyramid," then, the sentence causes no special difficulty.

But the potential ambiguity of "on the block" is recognized by the semantic specialist for "on," which knows that "on the block" *can* be used in a locational sense to qualify a noun such as "pyramid," but which does *not* know for sure whether it is being used in this sense on this occasion. In order to decide, the semantic specialist addresses the deductive program to ask for information: is the (one and only) blue pyramid currently on the block? If the deductive program replies that it is, the semantic program confirms the initial hypothesis of the parsing program, so that "on the block" is taken to be a qualifier of "the blue pyramid." If the deductive program replies that it is not, the parser backtracks to the point where the noun group has been identified simply as "the blue pyramid," and seeks another interpretation for the preposition group.

The interpretation corresponding to "*onto* the block (which is) in the box" is suggested next, and passed to the semantic program and thence to the deductive program for checking. Since this alternative implies interpreting the second preposition group ("in the box") as a qualifier of "the block," the world-model is consulted to confirm that there is one (and only one) block currently in the box.

It is important to realize that the ambiguity of "on" in this example is dependent on the presence of the subsequent—and equally ambiguous—preposition "in." That is, if the sentence had instead been "Put the blue pyramid on the block," then "on" would necessarily have been interpreted as "onto." The reason for this—as systemic grammar makes explicit—is that one may confidently expect an answer to the question "*where* is the thing to be put?," whereas one may or may not find an answer to the question "*where* is the thing now?" Since both "in" and "on" can be used to answer either of these *where*? questions, the original example was syntactically ambiguous. But if the program had found that there was no other preposition group after "on the block," its knowledge of systemic grammar would have led it to detach this phrase from the noun ("pyramid") and reparse it as associated rather with the verb, "Put."

(Strictly, there are three senses of "in" distinguished by Winograd's program. There is the directional sense, meaning "into." And there are two locational senses: it may mean "contained in," as in the phrase "in the box"; alternatively, it may mean "forming part of," as in the phrase "in the stack.")

In this example, it is the environmental context (the state of the world) that determines the interpretation of the input sentence. Sometimes, however, it is the verbal context that is crucial, as in item 5 (where "the pyramid" is assumed to refer to the pyramid just mentioned—*cf.* item 2). The semantic program makes a habit of storing the immediately preceding discourse for use in cases where the reference of pronouns (such as "it"—item 42) or of noun groups (as in item 5) is determined by verbal context. SHRDLU is more efficient in this regard than PARRY: for whereas PARRY merely places specific items such as "horseracing" on the expectancy list, for use in disambiguating likely pronominal references as in "What do you enjoy about *it*?" SHRDLU stores whole sentences. Moreover, SHRDLU, unlike PARRY, has access to information about the world and its own past actions in it. In general, problems of reference are settled by the semantic and deductive programs working cooperatively.

When the reference is determined by environmental information, the deductive program must access its internal model of the world and draw upon the knowledge stored there in order to interpret the sentence. References determined by the context of discourse, on the other hand, do not necessitate active recourse to the world model, but merely require that the linguistic "anchor" concerned be identified. Both types of reference may occur independently within a single sentence, as in item 3: "Find a block which is taller than the one you are holding and put it into the box"; sometimes both types of problem arise in connection with one and the same pronoun, as in "the one which . . ." in item 7: "Is at least one of them narrower than the one which I told you to pick up?"

Strategies for interpreting (and effecting) references to actual items and events are required not only to enable SHRDLU to engage with the world by acting within it, but are crucial also in providing the program with a *belief system* as opposed to a mere *conceptual structure*. This distinction underlies the familiar retort "If the cap fits, wear it!" the speaker having supplied an abstract conceptual structure while the hearer has the task of establishing its particular reference. The previous discussion of Abelson's six levels of "belief" should perhaps more properly have been described as concerned with "concepts," insofar as it ignored problems of reference and instantiation. However, the early IDEOLOGY MACHINE identified instances of particular concepts (Kennedy being classed as a liberal dupe), and integrated them with others to form statements of belief (Kennedy's policies aid Communist schemes); and, in embodying his more recent theory in a revised IDEOLOGY MACHINE, Abelson will similarly provide a way of incorporating his theoretical concepts within representations of particular beliefs. One is therefore justified in calling his work a contribution to a theory of *belief*. Clearly, an adequate computerized conservative would not only have to be able to interpret political events in terms of "betrayals," "revolutions," and "aid," but would also have to disentangle pronouns such as "it" according to the political context, much as SHRDLU does with regard to the table-top world.

Naturally, SHRDLU's semantic specialist for "it"[22] must know the full range of possibly relevant alternatives if it is reliably to mediate sensible interpretation of any given expression. If, like the duck, it assumed that "it" can refer only to a physical object, the program as a whole would be unable to interpret items 30 and 31. And since even human grammarians do not have a formal (as opposed to an unformalized, implicit) understanding of all uses of the pronoun "it," there are sentences involving this word that Winograd's system could not parse, even though the rest of the sentence were to present no difficulties.[23] Further, the program does not have the semantic information that "it" can be used (nominally rather than pronominally) to mean either sex appeal or the chaser in a game of tag. Consequently, sentences employing these senses of the word would be as obscure to SHRDLU as they presumably would be to the duck, whose Wonderland world one assumes to be innocent of the relevant adult or childish pastimes.

Anyone who can interpret all occurrences of "it" within SHRDLU's conversation, not to mention more exotic uses, must have a linguistic competence at least as powerful as that incorporated in the program. Human understanding may differ in various ways from SHRDLU's—in the next chapter we shall see that this is so—but it must be able to mediate comparable feats of comprehension in assigning sense and syntax to natural language. In short, the notion that understanding is a simple matter may be introspectively plausible, but must be firmly rejected. Understanding "it" is no less complex than interpreting remarks about law and order as indicative of a particular ideological viewpoint, or recognizing abandonment as a species of betrayal. This everyday linguistic achievement not only demands a flexible (heterarchical) interplay of one's knowledge about various matters, but typically involves a far richer knowledge base than that available to SHRDLU. Let us look more closely at just what it is that SHRDLU knows.

## WHAT SHRDLU KNOWS

One may broadly classify the content of SHRDLU's knowledge as of three types. First, there is very general knowledge about problem solving in the abstract—what one might term "general problem-solving skills." This knowledge is implicit in the organization of the program as a whole, rather than being explicitly itemized in a distinguishable portion of the data base, and it is used by SHRDLU in all phases of thinking, whether of a primarily "linguistic" or "environmental" character.

One example of this first category of knowledge is deduction itself. We saw in the previous section that all stages of SHRDLU's understanding rely extensively on the deductive system. But one should not assume that SHRDLU's powers of deduction are as great as ours. For example, its reasoning cannot distinguish between *knowing that it is false* that something is the case and *not knowing whether* it is the case or not. The reason for this is that the PLANNER language in which the deductive programs are written embodies an inadequate concept of negation. However, recent PLANNER-like languages can deal more subtly with negation, so that a cousin of SHRDLU's written in POPLER 1.5 would not similarly be unable to distinguish between "No" and "Dunno."[24] Consequently, if the information that bats are the only flying mammals were provided as a contingent fact about bats, SHRDLU's POPLERized cousin (but not SHRDLU) could reply with a confident "No" to the Walrus's question whether pigs have wings. Only if the information were coded at the semantic level, as part of the taxonomic meaning of "bat" and "pig," would SHRDLU get this right (*cf.* item 9).

Again, SHRDLU can cope, rather stupidly, with only a few uses of modal words like "can" (see items 10 and 11). It cannot handle all questions concerning possibilities, nor can it reason sensibly about counterfactual conditionals, such as "How many eggs would you have been going to use in the cake if you hadn't learned your mother's recipe was wrong?" or "What would you have done if the big red block had been in the box?" We shall see shortly that this is not because the grammar programs cannot parse conditionals and subjunctives: they can. As Winograd's comment to item 11 suggests, some of SHRDLU's failings in handling "can" and "must" could be ameliorated by giving it more power to analyze its own theorems in answering modal questions. And he points out that further improvement would result from a version of PLANNER that could temporarily move into a hypothetical world in answering questions, instead of having to consider only *the world as it is*.[25] Indeed, POPLER 1.5 can do precisely this, and does so specifically in order to enable POPLER programs to answer counterfactual questions and questions about possibilities.[26] Without going into details here, these examples of negative and modal reasoning should indicate how the development of increasingly powerful programs goes hand in hand with the development of more powerful programming languages.

Some of SHRDLU's deductive inadequacies are due not to the programming language used by Winograd so much as to specific choices made by him for convenience's sake. For instance, SHRDLU's use of universal quantifiers (like "all" and "every") is limited by the fact that for the robot, "every" means "every one I know about."[27] If told, as in item 14, that its human friend owns all the blocks that are not red, SHRDLU makes a mental note (in a way to be explained in Chapter 12) that if it ever wants to find out whether the friend owns a block, all it needs to do is establish that the block is not red. But suppose the block in question is in the next room? SHRDLU knows nothing about it, so cannot deduce whether it is red. It cannot even say something like "If it's red, then you own it," because it has no way of generalizing "all" to cover things that it knows nothing about. (A more intelligent program could *ask* the color of the block: a program to be described in Chapter 7 could in principle do this, but SHRDLU cannot.) Winograd points out that since this is not a consequence of the basic deductive capacities or of the semantics, the system could be expanded so as to discuss genuinely universal statements.[28]

Another general problem-solving skill is referring to an inner model of one's own intellectual capacities in planning appropriate action, and in backtracking to a previous point on realizing a mistake. Thus SHRDLU can conceptualize, or think about, picking up a big red block as distinguished from actually doing so, and can hypothetically consider a particular parsing of a phrase before committing itself finally to that interpretation.

Backtracking in a *parsing* problem is not blind, because Winograd wrote the PROGRAMMAR language so that it would note reasons for failure and suggest the best ways of recovering. However, as we shall see more fully in Chapter 12, the PLANNER language in which the BLOCKS world is manipulated does tend to encourage inefficient backtracking, and when SHRDLU fails to obey a command it does not know *why* it failed. Even so, some facility for backtracking is better than none at all—unless one can write a program with the superhuman faculty of never making any mistakes.

In planning its moves, SHRDLU refers to its inner model of its own capacities. The distinction between thought and action in SHRDLU is signaled by the simple convention of omitting or including shriek-marks around the name of the operation concerned. For instance, if the program activates the function !MOVETO! then SHRDLU will move the specified object to the specified position; but if the program activates the function MOVETO then SHRDLU will merely think about how and whether to do so. In short, just as thinking about moving and moving are crucially different cognitive operations, so MOVETO and !MOVETO! are different functions for Winograd's program. By the simple expedient of systematically deleting all shriek-marks from the program one could reduce SHRDLU to a poor copy of Prince Hamlet, an ineffectual robot unable to translate even its most considered deliberations into actions in the real world. As it is, however, the decisive intelligence of SHRDLU's linguistic and block-moving performances is largely dependent on its inner ability to look before it leaps.

As well as using a general model of one's abilities in planning, one often refers to a specific model of one's past achievements (and failures) in explaining one's actions and in learning how to do better. As the responses to items 25 to 29 show, SHRDLU shares the first of these capacities; indeed, unlike neurotics such as Colby's patient, who formed the original model for his neurotic program, SHRDLU has the *complete* goal-subgoal tree stored for explanatory purposes. If some goal or subgoal were inaccessible (repressed), the robot would either have to answer "I DON'T KNOW" to some of the questions in this part of the conversation, or would have to rationalize in some way so as to suggest a plausible reason, which might or might not be the real reason. The second capacity—noting the purposive structure of one's activities so as to learn how to do better—is *not* available to SHRDLU (although we shall see in Chapter 10 that it is available to some programs). SHRDLU does not have the insight into its own behavior that would be needed for such learning.

The second broad category of SHRDLU's knowledge is a rich store of specialized linguistic knowledge, both syntactic and semantic in nature, which can be deployed in any English parsing exercise irrespective of content and context.

This knowledge is embodied in the grammar and semantic programs. It includes problem-solving skills or heuristics that are peculiarly linguistic—incorporated, for example, in the aforementioned hints that a transitive imperative verb will very likely be followed by a noun phrase and that a manipulative imperative like "Put . . ." is more reliably accompanied by a preposition used in its directional than in its locational sense. The definitions of "clause," "noun group," "question," and other syntactic categories are themselves specialized programs for parsing these structures, and the equivalent semantic specialists know precisely how to check whether the tentative syntactic assignment makes sense. The linguistic programs also include definitions of words that are essential to any English dictionary, like "it," "on," "and," and "the," as well as a special procedure for dealing with common word endings, like "-ing," "-ed," and so on. Other words may be regarded as optional: ignorance of "pyramid," for instance, would not often reduce one to incoherence.

Although SHRDLU can *converse* sensibly only about pyramids and the other inhabitants of the BLOCKS world, the program can *parse* sentences containing non-BLOCKS words like "eggs," "cake," "mother," and "recipe," provided that minimal relevant semantic information (such as that "mother" is an animate noun) is included within the definitions of the words in question. For instance, the program can parse (though not reply to) the sentence, "How many eggs would you have been going to use in the cake if you hadn't learned your mother's recipe was wrong?" One should note not only the complex syntax of the verbs "use" and "learn" as employed in this sentence, but also the fact that the two final words determine the grammatical interpretation of the previous noun group "your mother's recipe." The linguistic distinctions that underlie correct parsing of this sentence, and which are not introspectively available to a person gossiping casually in the kitchen, are made explicit in Figure 6.7, which shows the final parsing produced by the program. (The grammatical abbreviations are explained in Winograd's text, though most of them are easily recognizable if examined in relation to the specific sentence being parsed; the plethora of brackets is a syntactic feature of the LISP programming language.)

SHRDLU's grammatical knowledge enables it to parse many English sentences, although there are a few syntactic constructions it cannot handle (sometimes because human grammarians cannot do so either). Some apparently simple words can be satisfactorily parsed by SHRDLU, but not subtly used or really understood, since their semantic force has not been fully programmed. For instance, SHRDLU can parse sentences containing the conjunctions "and" and "but," but it cannot understand the difference in meaning between them and so has to use (and interpret) them as equivalent to each other. Again, SHRDLU has no notion of the many different meanings of "and," whereby it can convey causal relation, temporal succession, social rank, and various other matters.[29] These semantic subtleties are intuitively understood by us, but have not been clearly expressed in theoretical form. (Part of the meaning of "but" might be expressible by reference to R. C. Schank's semantic theory, described in the next chapter.)

The third broad type of knowledge available to SHRDLU is semantic knowledge, of which there are two varieties. On the one hand, there is general semantic knowledge associated with the parsing procedure, such as the distinction between animate and inanimate nouns, and the meaning (or, as we have seen, part of the meaning) of words like "and," "but," and "or." On the other hand, SHRDLU has semantic information specific to the particular domain of discourse assumed in the cited conversation, namely, the BLOCKS world.

This domain-specific knowledge mediates relatively insightful discourse and action, as opposed to the blind parsing of isolated sentences like the query about the eggs in the cake. Just as Colby's neurotic knows only what it is explicitly told about "love" and "hate," so the only properties of blocks of which SHRDLU has any inkling are those explicitly mentioned in the BLOCKS-world description provided. It has already been remarked that SHRDLU knows very little about blocks, if compared with human beings. The importance of Winograd's program, however, is that it knows a good deal more than earlier programs did

```
(((HOW MANY EGGS WOULD YOU HAVE BEEN GOING TO USE IN THE CAKE
   IF YOU HADN'T LEARNED YOUR MOTHER'S RECIPE WAS WRONG)
 (CLAUSE MAJOR QUEST NGQUES POLR2 ACTV OBJ1Q TRANS)
 (((HOW MANY EGGS)
  (NG QUEST HOWMANY NDEF NPL DET)
  ((HOW (QDET)) (MANY (QDET)) (EGGS (NOUN NPL))))
 (WOULD (VB AUX MODAL QAUX))
 ((YOU) (NG SUBJ DEF NS NPL) ((YOU (PRON NPL NS SUBJ OBJ))))
 ((HAVE BEEN GOING TO USE) (VG MODAL NAGR (FUT PAST MODAL))
                           ((WOULD (VB AUX MODAL QAUX))
                            (HAVE (HAVE VB AUX INF TRANS))
                            (BEEN (AUX VB BE EN))
                            (GOING (VB ITRANS ING))
                            (TO (TO))
                            (USE (VB INF TRANS MVB))))
 ((IN THE CAKE) (PREPG)
                ((IN (PLACE PREP PLACE))
                 ((THE CAKE)
                  (NG OBJ DET NS DEF)
                  ((THE (DET NPL NS DEF)) (CAKE (NOUN NS))))))
 ((IF YOU HADN'T LEARNED YOUR MOTHER'S RECIPE WAS WRONG)
  (CLAUSE BOUND DECLAR ACTV TRANS)
  ((IF (BINDER))
   ((YOU) (NG SUBJ DEF NS NPL) ((YOU (PRON NPL NS SUBJ OBJ))))
    ((HADN'T LEARNED)
     (VG VPL V3PS NEG (PAST PAST))
     ((HADN'T (HAVE VB AUX TRANS PAST VPL V3PS VFS NEG))
      (LEARNED (VB TRANS REPOB PAST EN MVB))))
       ((YOUR MOTHER'S RECIPE WAS WRONG)
        (CLAUSE RSNG REPORT OBJ OBJ1 DECLAR BE INT)
        (((YOUR MOTHER'S RECIPE)
          (NG SUBJ NS DEF DET POSES)
          (((YOUR MOTHER'S)
            (NG SUBJ NS DEF DET POSES POSS)
            (((YOUR) (NG SUBJ POSS)
                     ((YOUR (PRON NPL NS SUBJ OBJ POSS))))
             (MOTHER'S (NOUN NS POSS))))
           (RECIPE (NOUN NS))))
         ((WAS) (VG V3PS VFS (PAST))
                ((WAS (AUX VB BE V3PS VFS PAST MVB))))
         ((WRONG) (ADJG Q COMP) ((WRONG (ADJ)))))))))))
```

FIGURE 6.7
Sample Parsing Produced by Winograd's Program.

Source: Winograd, *Understanding Natural Language*, pp. 175-6.

about their particular spheres of interest (and it is able actively to *use* its knowledge in conversation, as we have seen).

The BLOCKS-world knowledge includes "environmental" information about the size, shape, color, and current and past positions of the various items in the world, information that would normally be heavily reliant on one's perception of the scene. It includes "practical" knowledge about what causes what and about how to manipulate things in the BLOCKS world, such as that picking up a block may necessitate moving something off the top and finding an empty place to put it in (*cf.* item 1). And it includes "semantic" information about the implications and interrelations of the various concepts employed in thinking about and acting in the BLOCKS world; thus SHRDLU's insight that tables cannot pick up bricks is ultimately derived from the semantics represented in Figure 6.8, there being no pathway linking TABLE to ANIMATE (the program's general semantics forbids the verb "pick up" to take an inanimate noun as subject).

The distinction between environmental, practical, and semantic information is somewhat forced, because one can only state "facts" about the environment (including "practical possibilities") in terms of one's conceptual scheme. Thus just as Senator Goldwater cannot conceive of the United States' attacking a neutral nation, so SHRDLU cannot conceive of picking up a table. If SHRDLU's basic BLOCKS semantics were altered so as to include TABLE within the category of MANIP-ulable PHYS-ical OB-jects, this currently unthinkable idea could then enter SHRDLU's head. In short, SHRDLU's thought and action are crucially dependent upon the program's inner model of the world and of its own capacities for effecting changes in the world.

The information coded in these inner representations is integral to SHRDLU's planning of actions, which occurs independently of any specific instructions from the human interlocutor (who usually would be quite incapable of articulating such instructions). The reason for this is that PLANNER (in which the deductive program and the BLOCKS-world knowledge are written) is a "goal-directed" language, in which one can ask that a general type of goal be achieved without having to tell the program precisely how to achieve it; thus one can say "Pick up a big red block" and leave SHRDLU to figure out how to do so. (SHRDLU's problem-solving use of its goal-directed knowledge of the BLOCKS environment will be discussed in a later chapter. So also will the "procedural" nature of SHRDLU's knowledge.)

SHRDLU's knowledge, whether of syntax, semantics, or the BLOCKS world, is embodied as "procedures," or miniprograms, rather than as passive items or theorems stored in the data base. PLANNER "theorems" are miniprograms that specify—and, when activated, control—the execution of a particular set of steps in the proof procedures available for solving problems. Similarly, the PROGRAMMAR definitions of "noun clause" and the like are *programs for parsing* noun clauses. Even the semantic definitions of words like "it," "and," or "if" are LISP *programs* specifying what to do when one encounters these words in inter-

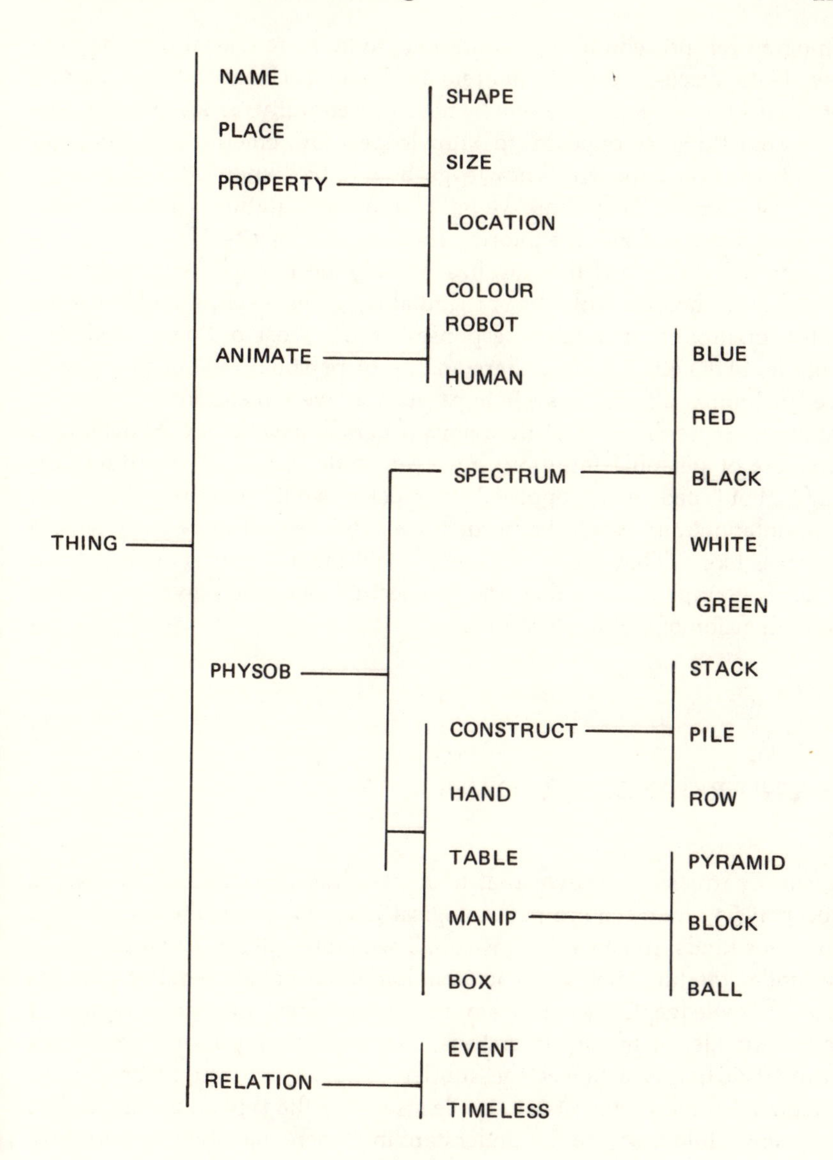


FIGURE 6.8
Semantic Markers for the BLOCKS Vocabulary.

Source: Winograd, *Understanding Natural Language*, p. 128.

preting a sentence, rather than *assertions* that their meaning is such-and-such.

There are various advantages in representing knowledge procedurally rather than assertionally; indeed, at the time of creating SHRDLU,

Winograd felt procedural representations to be more crucial than he does now. Until discussing these matters in Chapter 12, one should merely note that knowledge *can* be represented procedurally, as knowledge *how* to do something as opposed to knowledge *that* something is the case. (The distinction between "knowledge how" and "knowledge that" was made by Gilbert Ryle some years before computational models were available that could aid in exploring their relative merits.[30])

SHRDLU's knowledge, as has already been remarked, is grossly limited as compared with that attributable to the Walrus and the Carpenter, or implicit in human responses of the most ordinary kind. For example, SHRDLU has no understanding of personal relations or of any facet of human affective psychology, such as were mentioned in Part II. Having no representation of the relevant human psychology, SHRDLU is incapable of personal intuition, however crude. But understanding language about people, as opposed to blocks, would require some such representation, and would be needed *even* to interpret apparently simple sentences like " 'They are good pencils,' said Janet." This is evident from a recent attempt to formalize the intellectual processes involved in the comprehension of children's stories.

## ON UNDERSTANDING STORIES

Eugene Charniak has shown that understanding remarks like "They are good pencils" may require psychological insights about ulterior motives of various kinds. In general (as we shall see more fully in Chapters 7 and 11), understanding stories or conversation about people requires psychological knowledge. Charniak has outlined a program incorporating a preliminary model of human intentions, desires, and responsibilities which —though it hardly achieves the subtlety of a Proust—might be able to understand some fragments of simple stories of the type found in readers for young children.[31] And "understanding" here has two importantly different senses, which relate respectively to the literal meaning of the words on the page and to the deeper human significance of the sentences spoken by the characters in the story.

The first sense of understanding has already been mentioned, being the ability to interpret the literal meaning of the words in a sentence and to draw appropriate inferences from them. Since the world represented in Charniak's program is (a tiny part of) the domain of children's everyday interests, rather than the BLOCKS domain previously described, Charniak's system is able to make inferences that are not available to

SHRDLU. For example, Charniak's system, though not Winograd's, is potentially able to understand the reference of the word "it" in this passage: "Mother made some cookies and left one out on a plate. She put the plate on the kitchen table and went into the living room. 'I am sure Janet will like it,' thought mother."[32]

As Charniak points out, there are a number of more or less general psychological facts that could lead to the decision that "it" refers to the cookie, not to the living room, the table, or the plate—all of which words occur nearer to "it" than "cookie" does. One is the empirical fact that children are more likely to like cookies than the other objects listed (compare the duck's implicit assumptions about the likely interests of Archbishops). Another is the intrinsically more general fact that people are more likely to be concerned about whether others like things that they themselves have had a hand in making. These are the sorts of facts that would fit naturally into the "psychological" knowledge base of Charniak's program.

Neither of them, of course, strictly entails that "it" refers to the cookie. The story might have continued like this: " 'I'm sure Janet will like it,' thought mother. 'She's always wanted a plate with Bambi on it. And how she'll laugh when she sees that he's nibbling that cookie!' " Nevertheless, faced with the passage as originally cited, few humans would demur if a person (or, in equity, a program) identified "it" with the cookie.

Charniak discusses various other uses of "it," some of which could not be deciphered by Winograd's program in its present state of psychological ignorance.[33] He also remarks that pronoun use in general is not well understood by grammarians, and that his program cannot cope with sentences more complex than those found in children's books. There is little doubt, however, that efforts such as these to formalize pronominalization in programming terms should greatly clarify the grammatical problems involved.

The second sense in which Charniak's program "understands" stories is quite different, being concerned with what the speaker is up to as opposed to what she actually says. In other words, it is a matter of *pragmatic* rather than *semantic* communication. Winograd's robot has no inkling of this distinction, except possibly its knowledge that one does not normally request information one has just verbalized (*cf.* items 22MIT and 23MIT). (Clearly, SHRDLU would be at a disadvantage if faced with the sort of "unreal" questioning typical of elementary oral examinations in a foreign language.) Charniak's program, on the other hand, can make inferences as to *why* a character in a story asks a certain question at a particular point in time, or *why* she offers an item of information during the conversation.

For example, consider this story fragment: "Janet wanted to trade her pencils for Jack's paints. 'They are good pencils,' said Janet." No one with any common sense would interpret Janet's remark as a disinterested description of the world. Rather, her semantic communication (true or false) is understood as simultaneously being an exercise in persuasion, based on her implicit psychological assumption that people tend to want things that are identifiable as good. Or again, consider: "Janet wanted to trade her pencils for Jack's paints. He liked the pencils and so he traded. Janet said 'They are good pencils.' " In this case, Janet's communication presumably was motivated by her desire to reassure Jack of the wisdom of his choice; the "reassuring" psychological processes involved are of the type described by the theory of cognitive balance, which forms the core of Abelson's earliest programmed model of human attitude change.

Anyone canny will usually be on the lookout for persuasive or reassuring communications if involved in a trading situation in real life, perhaps taking them with a pinch of salt accordingly. Analogously, Charniak's program has specific subroutines (PLANNER antecedent theorems, or "demons," to be explained in Chapter 12) for spotting items such as these in trading contexts, and for interpreting their pragmatic as well as their semantic significance.[34] As soon as the topic of trade is first mentioned (as in "Janet wanted to trade her pencils for Jack's paints"), a special subroutine is set up which keeps a lookout for remarks wherein Janet praises the pencils; if and when such a remark occurs, the demon infers that the point of Janet's praise is to encourage or justify the desired swap. This interpretation requires that the trade-context already be introduced: the program is not capable, as you are, of hearing Janet praise the pencils and so asking "Why should she bother to say that? Perhaps she wants to do a swap?" (Why, do you think, did Charniak include a demon passing from trade to praise, but not one passing from praise to trade?)

Comparable demons serving to make psychological sense of a sentence by seeing it in the context of other sentences scattered in the text are discussed by Charniak. These concern themselves with reminders, with requests for information needed in making a decision, with questions regarding necessary subgoals of some current purpose, and so on.

The psychology represented in Charniak's program is very crude, with respect both to interpersonal and to intrapersonal phenomena. It does not approach the sophistication of the potentially relevant models within theoretical psychology, still less the subtleties of psychological knowledge informing one's common-sense intuitions about one's fellows. In discussing the example of betrayal in Chapter 4 we saw that much of this intuitive knowledge is implicit in the conceptual structures of everyday speech. But Charniak attempts to incorporate only a few of the concepts of common-sense psychology within his system, and in fact achieves no more than clumsy approximations to them. As he points out, a satisfactory base for comprehension of stories written even for very young children would require formal representation also of concepts like mistake, forgetting, pretending, friendship, authority—and, one might add, many of the themes discussed by Abelson.

These concepts are far from simple, and valuable insights might be culled from philosophers' analyses of the concepts of ordinary language: consider, for instance, J. L. Austin's discerning remarks on pretending, promising, and making excuses.[35] This source might assist the programmer also in distinguishing closely related concepts such as intending, wanting, wishing, desiring, trying, and the like—all of which would be required in subtle simulations of human thought, whether in the context of children's stories, attributed plans, or motivation in general.[36]

Despite the poverty of the psychological structures assumed by Charniak's program, the deployment of them in interpreting even the most childish stories about pencils and piggy banks involves procedures of great complexity. For example, Charniak remarks on the enormous complications involved in such apparently trivial bookkeeping tasks as updating the battery of currently activated subroutines.[37] Thus a PLANNER theorem, or demon, for spotting persuasive communications (such as "They are good pencils") is activated when the topic of *trade* is first introduced: but for how long does the demon have to devote its energies (that is, precious computer processing time and memory space) to this quest? Quite apart from the arbitrariness of the criteria relied on in any particular decision, the necessary updating itself consumes an appreciable proportion of the computing facilities. Similarly, the representation of a body of human knowledge about a given domain (be it Winograd's BLOCKS or Charniak's piggy banks[38]) is unavoidably complex. Consequently, the combinatorial implications of a program that could understand conversations about BLOCKS *and* piggy banks *and* the Cold War *and* baseball *and* the Mafia . . . are staggering. Clearly, then, Kubrick and Clarke may reasonably be charged with optimism, if not indeed with fantasy, in forecasting the creation of HAL within a mere 30 years of their own creative endeavors.

But suppose their screenplay had been less ambitious, merely featuring "Eo-HAL," an early evolutionary form of talking computer that converses only about blocks, pyramids, and boxes, but which does so in a fully "human" way: could they cast SHRDLU as Eo-HAL? Poor SHRDLU would be sadly inadequate in such a role. Some inadequacies have already been cited, including the shallowness of its knowledge about BLOCKS, its weak reasoning power, its failure imaginatively to consider hypothetical possibilities, and its inability fully to understand familiar words like "and." But Winograd's program has some additional limitations, which would further compromise its understanding even in Eo-HAL's severely restricted universe of discourse.

For instance, the astronauts' creative decision to start talking about blocks in a new way—analogically, for instance—could not be intuitively appreciated by SHRDLU, who would have to have explicit notice of a new definition or name (*cf.* items 37 and 40). Second, SHRDLU would not be able to paraphrase a given remark in indefinitely many ways, as humans can often do with facility. Third, the computer could recognize, and seek to fill in, only a very limited number of semantic "gaps" in the conversation—SHRDLU knows that "Clear the pyramid off the block" implies putting the pyramid into an (unspecified) place, but has little more sense than this of the ways in which a dialogue can be *expected* to develop. Fourth, if the astronauts were to garble their grammar, as human beings often do, SHRDLU could not decipher the syntactically ill-formed expressions produced as a result. SHRDLU is not so pedantic as to disallow dangling prepositions,[39] but would be defeated by grosser grammatical mistakes and by incomplete sentences and highly elliptical remarks. Fifth (and closely connected with the previous point), if a drunken astronaut were to mangle his logic, peppering his perorations with fallacious syllogisms, SHRDLU would have little chance of extracting his meaning—but his fellow astronaut would not always be equally nonplussed. Sixth and last, SHRDLU has no real sense of conversation as a cooperative dialogue between two individuals, each pursuing their own goals in a form of social exchange governed by distinct "rules of play"; rather, the program responds to the remarks of its human friend in a slavish or "automatic" manner, the only trace of purpose or initiative occurring when it complains of not knowing the word "steeple." Consequently, it could not engage in convivial chit-chat with the astronauts, as HAL and Eo-HAL are assumed to be able to do.

These constraints on SHRDLU's cinematic ambitions are not shared by all other language understanding programs. Even PARRY, as we shall now see, is in some ways more sensible than his accomplished rival.

# 7

# Sense and Semantics

THE SHORTCOMINGS of SHRDLU, as compared with HAL or even Eo-HAL, show that interpreting the simplest of everyday communications is far from simple. One somehow intelligently deploys a rich knowledge of syntax, general semantics, and environmental factors in doing so, but the "how" in that "somehow" is still obscure. Some workers have tried to develop computational models of a more "human" kind than SHRDLU, sometimes specifically criticizing the deductive approach of Terry Winograd and Eugene Charniak because of its inhuman insistence on strict syntax and rigorous reason.

The pattern-matching techniques of ELIZA, PARRY, and K. M. Colby's interviewing program, for example, are less restrictive than Winograd's theorem-proving strategies, though admittedly they are not powerful enough to mediate the subtleties of parsing and comprehension that can be evinced by SHRDLU. These three tele-talking programs can therefore make the best of a bad job where SHRDLU cannot, whether the "bad job" be faulty syntax in the input or partial failure of comprehension in the machine. Provided that the input string (which need not be a "sentence" in the grammatical sense) contains a recognizable pattern, a sensible response is commonly elicited irrespective of the remainder of the string. (Indeed, we have seen that special nonsubstantive responses are available for situations in which the entire input string is unrecognizable.)

To be sure, the policy of ignoring the nonpatterned part of the string is risky: PARRY, for instance, unthinkingly answers "TWENTY EIGHT" when asked "*How old* is your mother?" Consequently, to say that a pattern is recognized is not to say that it is appropriately understood in context.

Nevertheless, the specification of verbal patterns expressing the expected themes of conversation, together with assumptions about conversational coherence that determine the response rules relating to individual patterns, in fact manage to capture something of the flavor of psychiatric interviews in the Rogerian and diagnostic modes. And since a large proportion of spoken "sentences" are not sentences at all (a claim that a short spell of eavesdropping will confirm), a primarily syntactic model of language cannot adequately reflect actual human comprehension. The question therefore arises whether one might retain the advantages of these simple pattern-matching techniques without falling prey to the snares set by their simplicity.

The crucial problems are how to generalize the "patterns" searched for by the machine, how to draw up less ad hoc lists of "expected themes," and how to formulate more powerful rules of "conversational coherence." And assuming that this has been done, one needs also to be able to integrate this knowledge of familiar or expected patterns and themes with general procedures for dealing with *novel* combinations of known elements. One's appreciation of conversational coherence, in short, must not be so strict as to exclude surprises or debar the creative use of language.

These problems are most fruitfully approached from a semantic point of view, the emphasis being put on the underlying meanings or ideas expressed by language rather than the superficial verbal manner of expression. Provided that they can be made sense of in semantic terms, ungrammatical sentences thus pose no grave difficulties, so that one advantage of verbal pattern matching is retained. (Consequently, I shall ignore syntax in this chapter, merely noting that the programs I shall discuss have a syntactic power more nearly akin to SHRDLU's than to PARRY's.)

R. C. Schank and Y. A. Wilks have each formulated computational models of the understanding of natural language that concentrate on meaning rather than words. Although they base their work on different semantic theories, each insists that semantics as such is a more powerful influence in comprehension than is implied by the "knowledge-based" deductive approach described in the previous chapter. In short, general semantics supplies not only the individual patterns to be matched, but also many of the expected themes and criteria of conversational coherence.

## KNOWING WHAT'S GOING TO BE SAID

Schank's project of enabling a machine to converse sensibly with a human being is informed by his insight that the human speaker or reader is continually making predictions about the conceptual structure of that part of the message that has yet to be communicated.[1] Normally these predictions remain implicit, but when they are violated they may be brought forcefully to mind.

For example, if a woman is furiously recounting a recent row with her husband, and says "I think I ought to—," her friend would be surprised indeed to hear her say next, "—have fish for dinner." For this is not one of the limited class of sentence completions that might reasonably have been expected in the circumstances. This is not to say, of course, that the sentence as a whole is to be dismissed as "unreasonable," still less "meaningless." (Perhaps the husband is allergic to fish.) But the anticipatory effort after meaning that the hearer contributes in her interpretation of the sentence receives a temporary setback. It is set back because the hearer's previous predictions are falsified and she now hardly knows how to go on: she knows neither what to expect next, nor what questions would be in place, except the defeatist plaint: "What has that got to do with it?"

Schank's claim is that expectations of particular conceptual structures contribute to understanding, not merely in the sense that they may resolve potential ambiguities ("potential" because these implicit conceptual inferences commonly prevent the ambiguity from being recognized), but in that they structure the overall communication in terms precisely of what has to do with what. That is, they guide the hearer's response by suggesting what type of (implicit and explicit) comment or question on her part would be appropriate in the circumstances. We saw earlier that R. P. Abelson assigns a similar function to implicational molecules and themes, and that rather different ideas would arise accordingly in the mind of layman or psychoanalyst on observing someone's gratuitously insulting a police officer.

This sort of conceptual coherence between successive sentences of a discourse must be assumed accessible to HAL, but it is not something to which SHRDLU (still less, ELIZA) is sensitive. To be sure, SHRDLU can spot contradictions between sentences it has already encountered, or between sentences and the facts (*cf.* items 33 and 42, above); it can also refer back to previous sentences to decipher the reference of expressions like "then," "it," and "the one which"; and it can even recognize the sort of conversational "irrelevance" that uses a referring expression out of context so that it cannot be interpreted (*cf.* item 2). Moreover, the grammar program and the semantic specialists make predictive inferences *within* sentences, as was earlier described in relation to the sentence, "Put the blue pyramid on the block in the box." But SHRDLU would not be taken aback in any way by a sudden change of subject (insofar as one can make sense of this notion in relation to the severely limited universe of discourse actually available to Winograd's program). It could not be taken aback because it would not, in fact, have been implicitly proceeding forward, as human beings typically do when interpreting speech or written text.

You are perhaps expecting (*sic*) to be told that Schank's program, in contrast with Winograd's, does do this. And indeed so it does, up to a point. But Schank's work is largely programmatic rather than fully programmed, and although his current program does generate expectations (and can be "surprised" accordingly), it is not sufficiently powerful to be appropriately bewildered by the angry wife who inexplicably announces her need for a fish dinner. The present achievements and limitations of Schankian programs can be appreciated only in the light of his theoretical project, which is to formulate a general semantics expressing (in an interlingual representation) the conceptual structures and inferences underlying human communication.

This project of "Conceptual Dependency Analysis" was mentioned in relation to Colby's artificial neurotic and Abelson's theory of belief. Like these two writers, Schank sees his enterprise as an example of computer simulation. That is, he aims to model the information processing that actually occurs in people's minds, offering his semantic system as the basis of a universal theory of human thought. And in his view any genuinely communicative interaction between man and machine presupposes a programmed version of this theory. HAL, therefore, would have to be a competent analyst of conceptual dependencies, systematically mapping utterances in natural language (whether English, Russian, or any other) into an abstract inner representation of meaning that mediates his understanding of human speech.

Schank claims that everything that people talk or think about boils down to a residuum of determinate conceptual elements. The basic unit of meaning, and therefore anything that can be called a "thought," is a structured set of such elements linked in specifiable ways, and is termed a "conceptualization" (see Figure 7.1). Conceptualizations connote either the relation between an actor and an action (inanimate actors being a special case) or that between an item and an attribute. In other words, they refer to actions and states. Since they can be related to and nested within each other in certain ways, conceptualizations can be complicated indefinitely. But all conceptualizations must be constructed out of the finite set of conceptual elements and relations allowed by the theory.

This set includes the notions of causation; past, present, and future

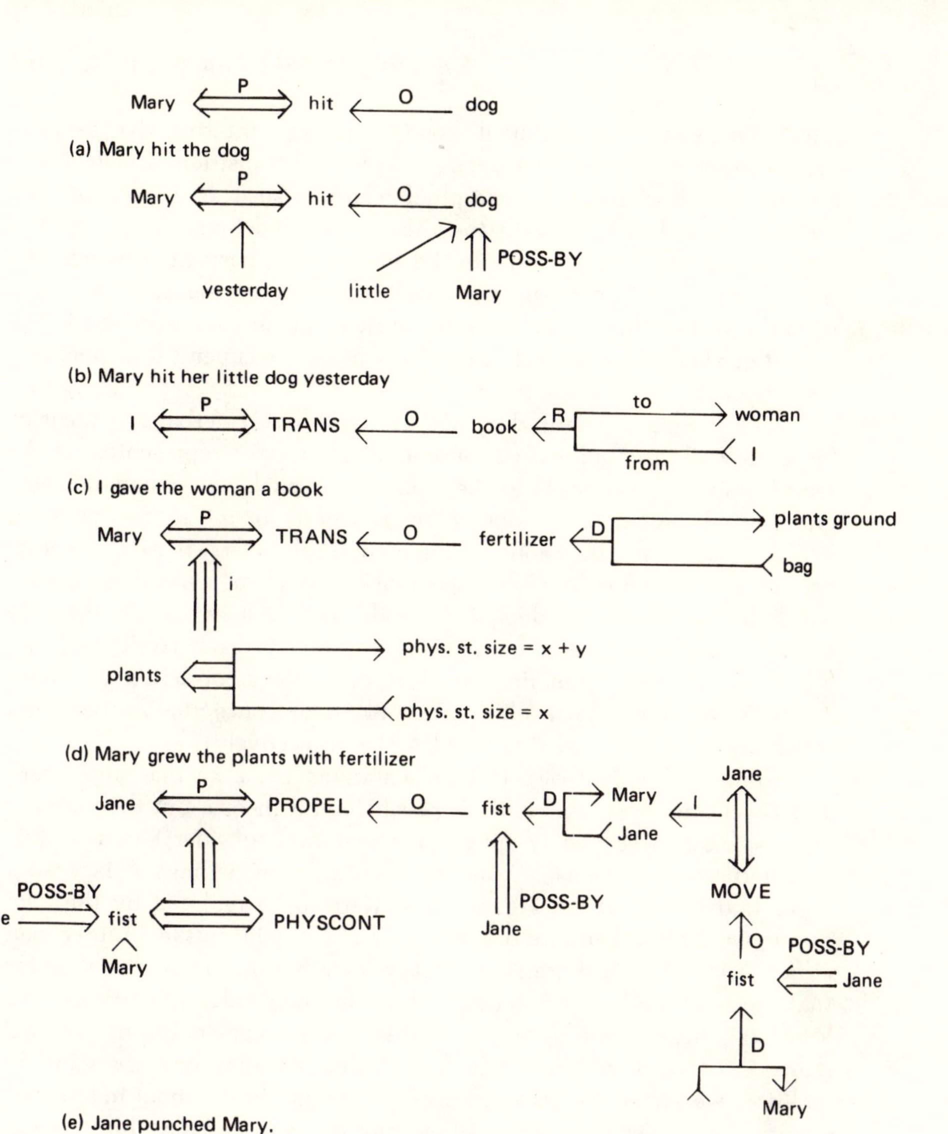


FIGURE 7.1
Examples of Conceptual Dependency Diagrams

A "translation" of (e) is: At some time past, Jane applied a force to the object Jane's fist, in the direction from-Jane-to-Mary; she did this by simply moving her fist in the direction from-Jane-to-Mary; her action of applying force to the fist caused Jane's fist to be in contact with Mary.

Do you think there is anything essential to the meaning of "Jane punched Mary" that is not represented in this diagram? For instance, should "Mary was hurt" be included?

Source: Adapted from *Computer Models of Thought and Language*, eds. R. C. Schank & K. M. Colby (San Francisco: W. H. Freeman, copyright © 1973), (a, b), p. 193; (c), p. 197; (d), p. 200; (e), p. 226. Reprinted by permission.

tense; timelessness; continuant, conditional, and interrogative aspects; negation; transition or change; start and finish of transition; specific times and spatial locations; and conceptual relations such as attribution. All these notions, being basic to the theory, are undefined. They are the semantic equivalent for Schank of the ontologically basic items which, for the metaphysician, make up the whole furniture of the universe. Also included within this set of basic semantic elements and relations is the crucial quartet of conceptual "cases"—objective, recipient, directive, and instrumental.

Like Abelson, Schank does not rely on a purely verbal statement of his theory. Computationally, conceptualizations are represented as directed graphs or networks in the data base available to Schankian programs, and are broadly similar in function to those incorporating PARRY's psychotic delusions. In his theoretical papers, these conceptual networks are represented (noncomputationally) as visual diagrams, which are accessed and deployed by the reader in a largely intuitive fashion: some examples are shown in Figure 7.1. Each conceptual element has a distinct symbolic equivalent, as do the actor-action and item-attribute relations which allow construction of conceptualizations (and which are symbolized as <=> and <≡> respectively).

Compared with most verbal presentations, Schank's diagrams afford a relatively clear expression of his psychological theory. But they are not quite so clear as they may appear. For instance, consider Figure 7.1(a): we interpret this to mean "Mary hit the dog"—or perhaps "Mary hit a dog." If the item *dog* in Figure 7.1(a) were to be replaced by the item *Susan*, we should then interpret it to mean "Mary hit Susan," rather than either "Mary hit the Susan" or "Mary hit a Susan." How do we know that there is no need here to utilize the definite or indefinite articles? We know this because we recognize "Susan" as a proper noun, or personal name, in contrast to "dog," which is a common noun, or class name. In addition, we know that there are ways of saying things about individuals denoted by proper nouns that differ from the ways we say things about individuals referred to by way of class names. But the diagram itself tells us none of this. This is what I meant when I said that Schank's readers deploy his diagrams "intuitively." Of course, there is nothing to stop him marking distinctions explicitly which at the moment he leaves implicit. For instance, he could put an asterisk by every item corresponding to a proper name (so that Figure 7.1(a) would then include an asterisk by the first item: **Mary*). And a program using his theory could be reminded in this sort of way that the "proper-noun-using" procedures should be invoked at this point instead of the "common-noun-using" procedures. But Schank's visual diagrams do not mark these distinctions, which therefore have to be supplied by the user.

This raises a point of general importance about representations of "semantic networks," which are commonly employed in computational linguistics and of which Schank's theory is a special case. A semantic network is a system of interlinked concepts intended to capture the meaning of a given expression or psychological phenomenon. In the most parsimonious sort of network there is only one type of link, which represents mere "association" between ideas or concepts; but in many cases a semantic network is described that incorporates different sorts of link, some of which can hold only between certain classes of item. For instance, ISA, GREATERTHAN, and LOVE might be three sorts of link, of which the source-concept of the third *must* be a human person; if both of the concepts linked by LOVE must be persons, then the network could not use this particular link to represent the gourmet's love of artichokes. Just as Schank's diagrams seem sensible largely because we interpret them sensibly, so the "adequacy" of semantic networks may rest on the human user's or programmer's intuitive ability to avoid absurdity, rather than on any intrinsic property of the representation that could be accessed by a computer program.[2]

For example, ISA links are often used in building semantic networks. But "Mary is a girl" and "A dog is a mammal" have importantly different implications which would be missed by a system that represented each indifferently as *Mary-ISA-girl* and *dog-ISA-mammal*. These differences are of course related to the distinction between proper and common nouns that was noted above. A program talking only about spaniels, dogs, mammals, and animals would not run into any trouble on account of its undifferentiated ISA link. But let it try to talk also about Rex, Rover, and Spot, and it would need to know the procedural implications of the semantic difference between the two types of "ISA." For present purposes, the point is that you should not assume Schank's conceptual dependency diagrams (semantic networks) to make explicit everything which you yourself are able to draw out of them. (A similar warning was given in Chapter 4, with reference to Abelson's theme-diagrams.)

With these reservations, Schank's diagrams may be regarded as relatively clear representations of conceptual meaning. They are also remarkably economical. He claims that all the verbs of everyday speech can be analyzed in terms of twelve "primitive actions," which concern the moving or transference of things, abstract relationships, and ideas. The primitive actions are the five physical ACTS—PROPEL, MOVE, INGEST, EXPEL, GRASP; the three mental ACTS—CONC, MTRANS, MBUILD; the two instrumental ACTS—ATTEND and SPEAK; and the two global ACTS—PTRANS and ATRANS.[3] For the purposes of representing the actual content of human communications, the most important primitive actions are PTRANS, ATRANS, and MTRANS, which respectively mark the transfer of physical objects, abstract relationships (like ownership and responsibility), and information or ideas in the mind.

As the presence of obvious neologisms in this list may suggest, these ACTS are not to be identified with even the simplest "actions" ordinarily so called. MOVE, for instance, is not equivalent to "move," since the only objects that can be MOVE-d are the body parts of the actor.

It follows that a person can MOVE her hand, but cannot MOVE a table; she can, however, move a table by applying a force to, or PROPEL-ing it, which she may do by means of MOVE-ing her hand. This fact, together with the fact that one can only move a table from somewhere to somewhere else, is explicitly represented in Schank's symbolism. The specific "somewheres" concerned may, but need not, be identified; similarly, one can only MOVE one's hand from one location to another, but these positions—though they must be diagrammatically indicated—need not be specified. It is crucial to realize that this lack of specification can invite "appropriate" questions whereby the event can be more fully understood.

Analogously, a given verb is not necessarily representative of particular underlying actions, for it may describe the state resulting from some unknown action or the relation between many unknown actions. Thus one may move a table by MOVE-ing one's hand, but one need not: kicking it, leaning back against it, or even pulling with one's teeth on a rope attached to it are a few of the possible alternatives. Each of these, however, involves MOVE-ing some body part or other, which is why "move" may be said to boil down to MOVE. In cases like "Pat Nixon moved the mahogany table out of the Oval Office," the actor whose body parts are MOVE-d in PROPEL-ing the table is not the actor mentioned in the sentence: here, "move" is understood as "cause to move," where the causation involves giving (MTRANS-ing) an order to someone else. Likewise, in "Mary grew the plants with fertilizer," it is not Mary but the plants that do the growing, while what Mary does is somehow to enable them to do so (see Figure 7.1).

By reference to this semantic base, Schank aims to articulate the hidden meanings of everything one might seek to say. Thus all the myriad verbs of thought, for instance, are to be constructed from CONC (roughly: to think about, in the broadest sense), MTRANS (roughly: to bring a thought into the conscious mind or, more generally, to change the locus of its mental control in some way), and MBUILD (roughly: to combine thoughts in some manner). Examples include think about, dream, consider, wonder, and ponder; remember, see, feel, communicate, forget, learn, and teach; conclude, resolve, decide, solve, realize, weigh evidence, think about a problem, and answer. All these verbs, and many others, are diagrammed by Schank in a systematic manner that seeks to explicate their underlying semantic connections.[4]

To the extent that his approach is acceptable, it therefore provides a mapping of the processes of human thinking as well as of the content of human thoughts.

It is essential to realize that Schank is interested in the *conceptual* mapping of thought rather than its *empirical* mapping. His concern is with how we intuitively think about thinking, as contrasted with how we actually manage to think. He concedes that the conceptual primitives CONC, MTRANS, and MBUILD refer to processes that are not primitive from the empirical or scientific point of view. Even if one remains within the psychological (as opposed to the neurophysiological) theoretical mode, one can postulate microprocesses wholly unavailable to introspection, which in fact carry out the higher level operations of thought ordinarily conceived of as basic. Although the novel terminology and techniques of artificial intelligence are particularly suggestive of fruitful speculations in this regard, such processes have long been hypothesized: the underlying "microgenesis" of visual perception, for instance, is a traditional concern of empirical psychology. But since this is something of which the ordinary person knows nothing (and cares less), it is not a concern of Schank's. Correlatively, it need be no concern of HAL's. Indeed, if HAL were to develop a scientific interest in the microgenesis of perception, he would ultimately have to conceive of these matters in terms of CONC, MTRANS, and MBUILD, and not vice versa. This is a special case of the general epistemological claim implicit in Schank's theory, that "scientific" discourse and reference to inanimate "actors" are semantically secondary to conceptualizations of a more intimately human kind.

The four conceptual cases (objective, directive, recipient, and instrumental) were earlier described as "crucial" because they provide the basic predictive mechanisms by means of which one makes sense of the world and of other people's communications. They are able to do so because the conceptual (mental) representation of every action essentially involves one or more of these cases. They thus differ from syntactic "cases," which are not mandatory.[5] For instance, one can *say* either "She cut it" or "She cut it with something," as one pleases; but one can *think* only about cutting with some (perhaps unspecified) instrument. This is why the progression of the nursery jingle *There's a Hole in My Bucket* is intuitively sensed as inevitable when Mother Goose advises Georgie, "Well, cut it!" One's private thoughts and public conversations are appropriately guided by using unfilled case slots to search for, request, or take for granted specific items of information. As Schank puts it: "One of the most important parts of understanding [whether of the world or of one's interlocutors] is an awareness of what you do not know so that you can endeavour to find it out."

In addition to the implications inherent in their conceptual cases, the

12 primitive actions (and all conceptualizations compounded thereof) carry other implications of greater or less probability.[6] It is these implications that allow a rich fund of inferences about actions and states to be (consciously or intuitively) drawn from terse remarks that in fact contain a wealth of meaning.

For example, PROPEL implies PTRANS if the object of these ACTS is not fixed or unusually massive. An analogy from ordinary language is that "Ms. Brown pushed the table" is normally taken to imply that the table moved; the implication, indeed, is so obvious as often to be unnoticed, for someone who made such a remark would commonly be understood as actually having *said* that the table was moved. By contrast, "Ms. Brown pushed the boulder" or "The baby pushed the table" carry no such obvious implication. It might be objected that the verb "to push" is ambiguous (as between "push against" and "successfully push against in order to move"),[7] so that all one needs to do is to identify more precisely the particular verb involved, in which event the question of "implications" or "inferences" will not arise. But this objection is not so much a disinterested description of the everyday use of the verb "to push" as an ad hoc stipulation introduced on the spur of the moment to distinguish between these three imagined situations. Schank's theory, by contrast, systematically notes the distinction between the actor's action and the results of the action in the analysis and inferences of the primitive ACTS such as PROPEL. Consequently, any familiar verb that is analyzed in terms of PROPEL may immediately be said to carry implications similar to those carried by "push."

The potential power of this semantic representation may be indicated by recalling Charniak's outline of a program for understanding children's stories. Charniak is typically faced with sentence sequences like, "John fetched his piggy bank. He bought the pencils with a brand new dime." Obviously, the dime came out of the piggy bank: but the story does not explicitly say so. Charniak's suggested solution is to program a "demon" that wakes up on mention of piggy banks and looks out for a mention of dimes, concluding, if it finds such, that the dime must have come from the piggy bank.

There are two drawbacks to this computational model of the understanding of piggy banks that are not shared by Schank's approach. First, it is essentially ad hoc, so that Charniak might find himself programming independent demons for "push," "pull," "shove," "jog," and the like, instead of doing it once and for all for PROPEL and then merely providing a conceptual dependency analysis for each of these English verbs that can automatically activate the appropriate inferences. Of course, Charniak might decide to save himself trouble by programming a demon for the first of these verbs to be encountered, and then providing a pointer to that demon for each of the others met with in the story. But the success of such a strategy would depend on whether the first verb happened to be semantically "central," or "typical"; and to the extent that Charniak tried to identify the semantically central cases, he would be involved in the sort of theoretical enterprise undertaken by Schank.

Second, Charniak's demon has to *conclude* that the dime *must* have come from the piggy bank, or that the pushed table *must* move—whereas Schank's inferences are all preferential, or probabilistic, and can be simply blocked by information to the contrary without introducing chaotic *contradictions* into the system. In the absence of such information, the usual inference is drawn. Analogously, a storyteller not maliciously intending to deceive will specifically point out that the dime came from John's pocket if her sentences would otherwise be interpreted as suggesting that it came from the piggy bank. (In fact, Charniak's system distinguishes between demon-based assertions that could turn out to be false without our thereby feeling misled, and those that could not. The former are given an ASSUMPTION tag when they are first inferred by the program, and can be "un-assumed" later. But the latter are given no such tag, so that in their case the program jumps to conclusions irrevocably.)

Another inference of conceptualizations involving PROPEL is that, if the object PROPEL-ed ends up in physical contact with a second thing, then that thing may possibly be negatively affected in some way. Put more pithily: hitting hurts. This common correlate of PROPEL is diagrammatically expressed by Schank in a symbolism that can distinguish between the material hurt done to a table by hitting it, the material or "physical" injury and the straightforward "mental" hurt of a person hit by a falling stone, and the "emotional" hurt of a person hit by a friend. One's intuitive appreciation that "Janet was hurt" may mean different things (and imply differing explanations and expectations) according to context is based on one's implicit recognition of distinctions that can be explicitly marked by way of Schank's theoretical terms.

The point of general importance about Schank's system is not that he is always forced to distinguish the subtleties of interpretation underlying everyday speech, but that in principle he can do so if necessary. For example, although he might find it useful to provide a diagram for "Hitting hurts" that can be manipulated as a whole (as "revolution" and "betrayal" can be thought about as such in appropriate contexts), there would always be the possibility of unpacking the detailed inner structure of this model so that necessary distinctions could be brought to light.

## PROGRAMMING CONCEPTUAL DEPENDENCIES

The remarks in the previous paragraph should not be taken over-optimistically to mean either that Schank could explicitly capture the entire meaning of everyday expressions within a single conceptual dependency diagram, or that the "unpacking" involved could in fact be effected by any current program.

As regards the first of these caveats, usually there are several inferences associated with a given ACT, there are many ACTS contained within a single familiar verb, and there are many more contributing to the meaning of a sentence. The inherent conceptual complexity accordingly is too great to be intelligibly represented at its most detailed level: imagine, for instance, attempting to represent "The French Resistance rescued many Allied airmen" by explicit analysis of Abelson's script *rescue* in terms of specific Schankian elements, or trying to say what *Romeo and Juliet* is about in a similarly exhaustive manner. This is not to say that particular details cannot fruitfully be expressed on occasion—as when Senator Goldwater's implicit knowledge of the enabling conditions for wall-building is brought to light in considering the possibility that Red China might have built the Berlin Wall. But to try to diagram every possible subtlety at once would not, in fact, be helpful.

Nor is it even in principle possible, since every ACT has an instrumental (conceptual) case, which links to some other ACT (the means by which the first ACT is effected).[8] Consequently, one can never finish diagramming a given conceptualization, if by "finish" one understands reaching a state from which no more conceptual implications could possibly be pursued. (Similar remarks apply to all computational models of language based on richly interconnected semantic nets, of which Schank's is a special case, since the complete meaning of one node may involve every other node in the network: this point was made by M. R. Quillian as early as 1968.)[9] But since one may nonetheless reach a stage at which, given the particular conversational context and practical aims in mind, there is no need to follow such implications any further, Schank (like Abelson) is not inescapably doomed to failure by the "open texture" of natural language. Wittgenstein expressed a similar point in rebutting philosophical scepticism, when he said: "Justification by experience comes to an end."[10]

The indefatigable philosophical sceptic is of course unpersuaded by Wittgenstein's dismissive remark, precisely because she insists on asking not whether sceptical questions are actually in order in a particular situation, but rather whether such questions might intelligibly be raised. For the sceptic, one might say, an admissible question must be intelligible but need not necessarily be intelligent.

This distinction between a question that could in principle be asked and one which is appropriate in practice is relevant to the second caveat listed above, the inability of the current Schankian program computationally to decide which level and area of detail needs to be accessed in interpreting a given communication. For example, Schank, Abelson, and Goldwater all know that in assessing Red China's potential responsibility for the Berlin Wall one should (intelligently) consider the geographical locations of Peking and East Germany, whereas one need not (though one could intelligibly) inquire into the chemical nature of the materials used. And an ad hoc program could readily be devised that would slavishly pursue the first topic while obediently eschewing the second. What is needed, however, is a program competent to exploit the semantic structures underlying natural language in a more spontaneous fashion, so that it could decide for itself what questions to ask, and when.

The more wide-ranging the knowledge base available, the more challenging these decisions become. Thus far, Winograd's SHRDLU is the program that is most successful in coping with these problems. The way in which it does so depends on the structural features of the PLANNER and PROGRAMMAR theorems concerned. (These matters will be discussed in Chapter 12.) It must be noted here, however, that SHRDLU's uncanny ability to ask the right questions at the right time owes much to the fact that English syntax is a relatively cut-and-dried and precisely structured matter, and still more to the fact that its semantic and environmental knowledge about BLOCKS is very shallow, so that there are not many questions that it is conceptually capable of asking.

A similarly specialized "talking" program named "SOPHIE" (for SOPH-isticated Instructional Environment) is intended to help teach electronic engineering students how to locate the faults in malfunctioning electrical circuits.[11] Just as it *could* not occur to SHRDLU to inquire what would happen at point X if the circuit were to short-circuit at point Y, so it *could* not occur to SOPHIE to ask whether or not picking up the big red block would require moving anything off the top of it first. But HAL, in his technical conversations with the astronauts, would need intelligently so to comport himself that it *would* not occur to him to babble about big red blocks when the spaceship's electrical controls seemed to be going awry. And social (as opposed to technical) discourse ranges over domains of knowledge that have hardly been explicitly formulated in verbal form, still less computationally, so that the "relevant" questions are less easy to isolate than in the BLOCKS or circuitry worlds.

Were HAL to engage in sensible social conversation with his fellow travelers, he would need not only a semantic data base comparable to that provided by Schank, but also a way of reliably directing his attention to those items within it that are pertinent to the particular occasion. For instance, were he to be appropriately surprised by the furious wife's mention of a fish dinner, the previous remarks disparaging the husband would have to have led HAL to expect the wife to want to hurt her husband in some way (which is why information about the husband's allergy to fish could later resolve HAL's perplexity).

Schank claims that the essentials of this case—and indefinitely many others—can be represented in terms of the basic semantic *life-death* polarity, together with primitive psychological beliefs about human beings' general desire to render tit for tat. On his view, communication is achieved by the mediation of possibly less than a hundred such core beliefs[12] (appropriately linked to specific individuals in the memory), and the infinite subtleties of everyday speech are to be understood as superficial variations on these few primitive themes. Presumably, the social relationships marked by Abelson's "themes" and "scripts" would be derivable from (and explained in terms of) these core beliefs, as also would the semantic generalizations seen in neurosis. However, Schank's ambitious claims concerning core beliefs have not been carefully spelled out, still less implemented in a computational form that can fathom the expected themes and rules of conversational coherence supposedly implicit in all natural language communication.

The computational work that has been done in connection with Schank's theory has concentrated on assigning conceptual dependencies to sentences and drawing limited inferences therefrom—so limited that the problem of selection in accessing the data in memory has been postponed rather than solved.

Moreover, the problems involved in interpreting "logical" words like *all, every, any,* and *some* have not yet been faced. No response analogous to SHRDLU's item 35 can be produced by current Schankian programs—and even SHRDLU's understanding of *all,* you will remember, is not fully equivalent to our everyday use of this concept (nor to the logician's use of *all,* either). It is a controversial question whether or not natural language can be viewed as, in essence, logic plus a few conventional conversational rules.[13] So one cannot assume that conceptual dependency theory should be forced into a logical mould, such as the propositional calculus which is used as the basis of many "theorem provers" in artificial intelligence. However, we can and do sometimes use English, or French, to reason "logically," so that an adequate computational theory of language should be at least as powerful as these logical systems are in reasoning from statements about "*everybody . . .* ," even if it allows a language-using program on occasion to imitate ELIZA in sensibly interpreting "Everybody laughed at me" so as to refer to one particular person. There are various other linguistic matters—such as the use of adverbial expressions[14]—which have not yet been represented in conceptual dependency theory and so do not contribute to "MARGIE," the current system of Schankian natural language programs.

"MARGIE" is an acronym denoting Memory, Analysis, Response Generation, and Inference on English.[15] The Analysis program of C. K. Riesbeck is a conceptual parser that simultaneously parses an input sentence and assigns a conceptual dependency diagram to it; indeed, the latter (semantic) procedure directs the former (syntactic) one. The crucial step is the identification of the verb; provided that the verb is one for which Schank has provided a semantic analysis in terms of ACTS and conceptual cases, parsing the remainder of the sentence can proceed in a conceptually intelligent way. The Memory program developed by C. J. Rieger then takes over, and performs most of the "thinking" attributable to the system. These thoughts, finally, are input to N. M. Goldman's Response Generator program, which expresses some or all of them in natural English.

The important point about MARGIE's memory is that it is highly active: it spontaneously reorganizes and reasons from the data within it, with results that can be drawn on in generating sensible replies to the input sentences. In its reorganization of the conceptual data input by the parser, the memory program assigns particular references of concepts to individuals or things whenever possible, and notes cases where this is not possible so that references may be established later; it takes note of "dummy markers" in the unfilled case slots of the semantic input, about which the response program may ask questions later; and it links all concepts to previous occurrences of the same concept within a given conversation, making this information readily accessible for the inferencing procedures to follow.

In its reasoning from the conceptual dependency diagram provided by the parser, the memory program generates five basic types of inference: normative, peripheral, causative, resultative, and predictive. These represent, respectively: the normal state of affairs in the world, what people automatically assume when hearing something, the probable causes of some state or action, the probable results of some state or action, and what an actor might do given her current state.

Resultative and causative inferences can then be used to fill in missing causal chains. For example, the input "Mary kissed Bill because he hit John" is expanded thus: (a) Bill's hitting John caused John to be hurt, (b) John's being hurt pleased Mary, (c) Mary's pleasure was caused by Bill's action, (d) Mary therefore feels a positive emotion toward Bill, (e) this causes her to kiss Bill.[16] The program also detects the inference of identical information from two sources, showing how a story or conversation knits together: for example, if the sentence just cited were followed by "Mary was very pleased with Bill," the identity of this with inference

(d) of the previous causal expansion would be seized upon as confirmatory by the memory.

Rieger's memory program will be further discussed in Chapter 10. For the moment, one should note the similarity of the basic inferencing, causal expansion, and knitting mechanisms to Charniak's demons—particularly their common emphasis on divining the *intentions* behind the actions involved. The memory mechanisms may also be compared with Abelson's early rationalization and credibility procedures, and his later implicational molecules. Like Abelson's earliest computer models of belief, Rieger's memory system assigns measures of strength (confidence) and interest to inferences. These measures could be used in selecting MARGIE's replies, so as to increase the naturalness of her ongoing conversation.

Also, of course, MARGIE would need what currently she lacks: namely, a sense of the expected themes and criteria of conversational coherence sufficiently well organized to select *one* of the many "relevant" ideas which occur to her. At present, MARGIE is incapable of normal dialogue, because there are no heuristics implemented for selecting what she is to say. The memory program simply dumps *all* potential responses —those concerned with interesting inferences and unknown references or states of affairs—onto the generator for verbal expression. (Even one's unverbalized thoughts about something mentioned in reading or conversation usually require one selectively to focus on relatively few ideas—although the attitude of "brainstorming" is sometimes deliberately adopted in the hope of finding conceptual connections which otherwise might have been prematurely dismissed as irrelevant.[17])

In real life situations, the criteria of conversational coherence include not only semantic considerations such as those previously discussed, but also social or pragmatic rules concerning the purposes and skills of conversation. For instance, when is it in order for one speaker to interrupt the other (as SHRDLU does with reference to "steeple") or to change the subject for her own purposes, and how is the complex personal interaction of "turn-taking" managed by the two participants? Programs like ELIZA are sensitive to only one cue marking their "right" to talk—namely, a period punctuation mark—but people use social and semantic as well as grammatical criteria to decide whether it is appropriate for them to speak, and what it is appropriate for them to say. (Consider the varied ways in which one's remark may be judged as impolite, irrelevant, or otherwise out of place.) A few "conversational" programs have been written which attempt to capture this type of linguistic coherence, but the communicative rules involved have no representation in MARGIE.[18]

The response generator's job is to say what it is told to say, but this is not so simple as it sounds. In the first place, there are indefinitely many ways of saying the same thing in English (or any natural language). For instance: The cat ate the mouse, The mouse was eaten by the cat, The cat devoured the mouse, The consuming of the mouse was done by the cat, The cat's breakfast was a mouse . . . and so on, almost without limit. One of these possibilities has to be selected in a given situation, but some other might have been equally acceptable. Thus the March Hare was clearly being unjust to Alice in the following exchange:

> "I believe I can guess that."
> "Do you mean that you think you can find out the answer to it?" said the March Hare.
> "Exactly so," said Alice.
> "Then you should say what you mean," the March Hare went on.

This is not to say, however, that each of the (conceptually) equivalent expressions would be suitable: if Alice had said "I hypothesize that I can divine the solution," the March Hare's stricture would have been in order. In mapping from a conceptual into a verbal representation, then, a response generator needs to make certain decisions about how to say what it is required to say.

In the second place, getting from the "deep" semantic representation to a particular "surface" verbal string is not always computationally possible, since not all the subtleties of English syntax, still less of idiom, are as yet theoretically explicable. Because of these points, programmed sentence generators are crude compared with human beings, and on the whole less impressive than sentence analyzers. You will remember that SHRDLU's powers of parsing are considerably greater than its generative capacities. And there are many everyday constructions that—even though they are among those that can be satisfactorily represented as a conceptual dependency diagram—cannot be produced by MARGIE's current response generator. For example, nouns denoting simple events, like collision, destruction, or death, cannot yet be handled by Goldman's program, although it can cope with the equivalent verbs, collide, destroy, and die. This is so even though these nouns have conceptual representations identical to those of the verbs; the problem is not in "understanding" the nouns in question, but in using them correctly in sentences. So far, Goldman has sorted out the syntactic predictions and syntactic compatibilities associated with the verbs, but not those associated with the nouns.

Nevertheless, Goldman's response generator solves some and raises many more questions about how a program can be enabled to say what it means. It can succinctly express matters whose conceptual dependency diagrams are very complex. For example, consider the following basic meaning (whose conceptual dependency form occupies eight full lines of Schankian symbols):

> An event caused Othello to believe that if he performed some unspecified action which resulted in Desdemona's becoming dead it would increase

> Othello's happiness. The event which made Othello believe this was a communication of some information by Iago to Othello. This information was that Cassio was in possession of a handkerchief owned by Desdemona.

Goldman's program accepts this semantic input and expresses it as "Othello wanted to kill Desdemona by doing something because he heard from Iago Cassio had her handkerchief." While one doubtless prefers Shakespeare's way of putting it, one cannot accuse the program of not having clearly said what it meant. Indeed, except for the omission of "by doing something," it is not obvious that the information conveyed by this sentence could be expressed more briefly.

When a way does exist of expressing something more economically, Goldman's program normally gives it priority. For example, the program is well aware that the following sentences have the same meaning: Othello strangled Desdemona, Othello choked Desdemona and she died because she was unable to breathe, Othello prevented Desdemona from breathing by grabbing her neck and she died because she was unable to breathe. Because of the manner of organization of the program's linguistic knowledge, the more compact verbal expressions of the input idea are virtually always created first, as in this example. Unless there are special reasons for being more explicit, "Othello strangled Desdemona" is the natural way of putting it. Similarly, in answering a journalist's question "Please, sir, would you comment on what's happening in Germany?" Senator Goldwater (or Abelson's IDEOLOGY MACHINE) would in general more plausibly reply "Our Nato allies have betrayed us," than go into a detailed rigmarole explicitly unpacking the thematic structure of betrayal. Of course, the response generator does not decide *what* to say, only *how* to say it; but the general strategy of printing out the first verbal equivalent of an input conceptualization would prevent unnecessary complications of expression, which should be made explicit only if special conditions warrant this.

The "special conditions" determining choice between the many equivalents of "one" idea would need to be complex if natural conversation were to result. For example, MARGIE would need to know the *focus* of the conversation: "Desdemona was strangled by Othello" is focused on the wife rather than the husband, and ways of understanding and acting upon this would have to be supplied to Goldman's program if its responses were to be apt rather than clumsy. Similarly, knowledge of the interlocutor's general interests and range of knowledge would be needed to decide how to say something, since effective communication requires an appreciation of which semantic "gaps" can be filled in by the hearer. At present, MARGIE's organization is serial: Riesbeck's module analyzes, Rieger's infers, and Goldman's replies. But a truly subtle generation of verbal responses would need a more flexible, heterarchical, organization

—much as one sometimes stops to ask oneself what facts or words someone knows before speaking to her. In short, sensible conversation on MARGIE's part would require intelligent selectivity on the part of the verbal response generator as well as on the part of the inferential memory that decides what idea is to be expressed.

Over one hundred alternative expressions, or paraphrases, of the "same" meaning are often produced by Goldman's program, and additional vocabulary and syntactic power would clearly allow for even more. Stylistic, as well as substantive and pragmatic, considerations distinguish between them. Thus Alice's imagined remark would be inappropriate because of its pomposity, not because it expresses the wrong idea or is unintelligible to the March Hare. One may assume that the March Hare would have realized what Alice meant by it.

Understanding many different expressions as meaning the same thing is complementary to seeing that one idea may be expressed in various ways. Accordingly, MARGIE's paraphrase capabilities may be thought of as the inverse of her analytic competence, for incoming sentences of indefinitely many surface forms are interpreted in terms of one underlying semantic structure. However, paraphrase and analysis are not *computationally* inverse, since the response generator is not just the parser run backwards. (Nor could it be, since the problem of selection does not arise at the stage of analysis: the input sentence is what it is, and its selection was carried out not by MARGIE but by her interlocutor.)

## MACHINE TRANSLATION

Paraphrase may be thought of as a special case of translation, where the semantically equivalent expressions are drawn from one and the same language instead of from different tongues. Schank has not worked specifically on "translation" in its more usual sense (although Goldman has written a German generator as well as the English module incorporated in MARGIE, and work is currently proceeding on a Chinese version), but would say that anyone doing so draws implicitly on the common conceptual dependencies that are variously mapped into distinct natural languages. The reason is that language cannot be sensibly (*sic*) translated unless it has first been understood.

Those who have tried to produce machine translations of foreign texts have found that a basically semantic approach (though not necessarily Schank's) is indeed required if such efforts are not quickly to collapse into absurdity. Moreover, the semantics must specify the "expected themes" and "criteria of conversational coherence" that implicitly guide the translator toward sense and away from nonsense: a mere listing of the various interpretations of individual words is not enough. Because these matters were not taken into account in the early machine translation efforts of the 1950's, these efforts failed to achieve their goal.

Even if one's prime goal is not translation as such, but rather the formulation of a theory of how we understand language, it may be useful to try to build a translating machine. For, as the example of ELIZA shows only too well, the production by a machine of some apparently sensible verbal responses cannot be taken as a proof of understanding on the machine's part. What, then, can be used as a criterion of understanding? Programs like SHRDLU and STUDENT evince their comprehension by answering questions about the subject matter. (It is significant that the "pragmatics" of questions are relatively simple: if I ask you a question you are usually correct in assuming that I want you to answer it. By contrast, if I tell you something in the form of an indicative sentence, you may be required to remember it, to comment on it, to use it as an answer to a query in your own mind, to prompt me to give you more information about it, and so on.) Successful question-answering is a useful criterion of understanding, but an alternative method is to demand acceptable *translation* of the language into another one. Wilks's semantic theory of understanding has been developed in the context of machine translation for this reason.

The basic problem confronting automatic translation projects is the fact that real words in real languages have many senses, and the translator's task is therefore not a straightforward matter of dictionary look-up. If one were to rely on simple dictionary look-up in translating a sentence from English to Russian and back again, one might find oneself rendering "The spirit is willing but the flesh is weak" as "The whisky is fine, but the steak's not so good."

This type of absurdity can sometimes be avoided by a thesaurus-based method of translation. The principle of such a method is that each of the senses of every word is assigned a particular semantic tag (*cf.* the classification in *Roget's Thesaurus*), and the translator chooses between alternatives in a particular case by matching up the tags associated with co-occurring words so as to get a coherent sense. For instance, the general tag *food* covers both "spirit-as-whisky" and "flesh-as-steak," but no food tags would be found in the biblical context of the original quotation, whereas one would find *personal* tags suggesting (consistent with) translation in terms of the incorporeal/corporeal distinction intended by the writer. Accordingly, research in machine translation early switched from a dictionary-based to a thesaurus-based paradigm.

But even a thesaurus is not always adequate to disambiguate a word, as in Wilks's example: "Our village policeman is a good sport, he captains the cricket team every Saturday."[19] Given two alternative senses for "sport" in the dictionary, *viz.* "agreeable person" and "recreational organization," the occurrence of *cricket* in the immediate context would force the second interpretation. But, of course, this would be the wrong choice.

The important point here, according to Wilks, is that there is an explanation for this choice's being incorrect, namely, that the suggestion that one's village policeman is a recreational organization is not a suggestion that anyone would actually make. In other words, the remark is semantically absurd, it is not the sort of thing that can be said. (Compare SHRDLU's response to the suggestion that tables can pick up blocks, in item 9 above.) In Gilbert Ryle's terms, one might describe the absurd attribution of organizational status to a policeman as a "category mistake," simultaneously providing a list of *categories* into one and only one of which any given word sense must fall.[20] Wilks's semantic equivalent of a theory of categories is a systematic account of the messages that one might possibly want to transmit to one's fellows. For instance, the gist of the sentence about the policeman is that he is a certain sort of man (consider how the interlocutor might very likely reply, "You're lucky! Ours is a surly devil—and he arrests the wrong people").

As Wilks puts it: "There is a fairly well-defined set of basic messages that people always want to convey whenever they write and speak; and in order to analyse and express the content of discourse, it is these simple messages—such as 'a certain thing has a certain part'—that we need to locate."[21] It is this notion of a basic message, or *gist*, which forms the core of Wilks's system of automatic semantic analysis, a system providing for the disambiguation of sentences in natural language by "Computable Semantic Derivations."[22]

The essence of the computation involved is the attachment of semantic representations of gists to the input text so that ambiguity resolutions can be read off from them and translations effected accordingly. The general strategy is to segment the text in some acceptable way, then to map semantic templates onto the clauses and phrases of the input language, and finally to map out from the templates to clauses and phrases of the output language, the relative order being changed where necessary.

Clearly, the basic requirement for this translation procedure is a semantic coding scheme that can distinguish between the different senses of a given word that are allowed for in the dictionary being used. The basic units of Wilks's coding scheme are sixty primitive semantic ELEMENTS denoting the entities, states, qualities, and actions about which humans communicate. For example:[23]

(a) entities: MAN (human being), STUFF (substances), THING (physical object), PART (parts of things), FOLK (human groups), ACT (acts), STATE (states of existence), BEAST (animals), etc.

(b) actions: FORCE (compels), CAUSE (causes to happen), FLOW (moving as liquids do), PICK (choosing), BE (exists), etc.
(c) type indicators: KIND (being a quality), HOW (being a type of action), etc.
(d) sorts: CONT (being a container), GOOD (being morally acceptable), THRU (being an aperture), etc.
(e) cases: TO (direction), SOUR (source), GOAL (goal or end), LOCA (location), SUBJ (actor or agent), OBJE (patient of action), IN (containment), POSS (possessed by), etc.

Every distinguishable sense of a word is represented by a FORMULA made up of one or more elements. Thus "salt" is either MAN (sailor) or STUFF (condiment), whereas "drink" is coded as the complex formula,

((*ANI SUBJ)(((FLOW STUFF)OBJE)((*ANI IN)(((THIS(*ANI (THRU PART)))TO)(BE CAUSE)))))

which Wilks explains should be read as: "an action, preferably done by animate things (*ANI SUBJ) to liquids ((FLOW STUFF)OBJE), of causing the liquid to be in the animate thing (*ANI IN) and via (TO indicating the direction case) a particular aperture of the animate thing; the mouth of course. . . . Lastly, the THIS indicates that the part is a specific part of the subject."[24]

Using this coding scheme, Wilks supplies his translating program with a semantic Dictionary listing the various senses of words in the natural languages concerned. He stresses that word disambiguation is relative to the Dictionary (whether programmed or not) of sense choices available, having no absolute quality about it. He gives the example of the two sentences, "I have a stake in this country," and "My stake on the first race was £5," asking whether the word "stake" has the same sense in each.[25] If the Dictionary lists only two senses of "stake"—(1) stake as post, (2) any kind of investment in any enterprise—then the answer is "yes." But the answer is "no" if the Dictionary also lists sense (3): initial payment in a game or race.

To get from word senses to messages, or gists—which are the crucial semantic notion in Wilks's system of understanding, and so of translation—is to combine formulae in such a way as to get TEMPLATES. The details of this combination need not concern us: the central point to be grasped is that a template is an organization of complex concepts (formulae), which in the process of understanding actively seeks preferred categories to fill its slots. The basic frame of a template is always a combination of formulae based on the actor-action-object pattern. (This basic semantic pattern is also utilized by Schank, as we have seen, but his primitives and method of representation are different from those of Wilks.) Any combination of formulae whose basic skeleton or essential core does not match this pattern is not a template, and does not code a semantically acceptable (meaningful) message.

For example, Wilks points out that if one represents the sentence "Small men sometimes father big sons" as a sequence of the heads of the formulae corresponding to each individual word, one's Dictionary will generate two alternative strings:

KIND MAN HOW MAN KIND MAN

and

KIND MAN HOW CAUSE KIND MAN.

(CAUSE is the head of the formula corresponding to the word "father" when used as a verb, since "to father" is analyzed as "to cause to have life.") One cannot find a template in actor-action-object form within the first of these sequences of formula heads; but within the second sequence one can find the template MAN-CAUSE-MAN. In other words, the sentence "Small men sometimes father big sons" is semantically inadmissible if the word "father" is interpreted as a noun (a kind of man), but admissible if "father" is read as a verb. Moreover, the sentence is communicating a message the gist of which is that some men have a causal relationship to other men. And this notion, clearly, can be the gist of many different sentences as well as the one specifically cited (for example, "Cops arrest the wrong people").

The "actor-action-object" constraint, while necessary, is not however sufficient. For some actors simply cannot perform some actions on some objects. Tables cannot pick up blocks, and policemen cannot be recreational organizations. In other words, there are further semantic constraints defining the class of acceptable gists, or possible (intelligible) messages. Although the actual messages that may be communicated are infinite, they can be generated from a finite inventory of basic templates embodying the semantic rules governing everyday discourse. (This is similar to the grammatical notion that an infinite number of sentences can be generated from a finite set of underlying syntactic rules.) And this basic inventory does not include any item that corresponds to a policeman's being a recreational organization (it does not allow MAN-BE-FOLK), although it does allow for him to be an agreeable person (MAN-BE-KIND).

Wilks provides a set of rules from which the inventory of meaningful gists can be reconstructed.[26] The rule that is formally expressed as:

<*AL><BE> same member of *AL as last occurrence>

states, in effect, that all substantive elements that are linked within a template by BE have to fall into the same semantic class. It follows that a MAN cannot BE a FOLK (human group), hence the policeman who is a good sport must be an agreeable person rather than a recreational organization. This gist-generating rule forbids the conflation of coppers and cricket.

But wait—don't we sometimes say things like "She's a walking encyclopaedia," or "She *is* the firm, you know"? Indeed we do, but in these cases we reinterpret the words that would normally be taken to denote book or business in terms of certain characteristics of a human individual. A prime feature of Wilks's approach (which will be further discussed in Chapter 11) is that he aims to allow for the anomalies that constantly occur in the creative use of natural language, without sacrificing the basic conversational coherence defined by his gist-generating rules. In the "village policeman" example, the word "sport" has two formulae stored in the Dictionary, having the heads FOLK and KIND respectively. Consequently, application of the basic rule quoted above outlaws the FOLK-sense of "sport" and makes way immediately for the KIND-sense. If "sport" (like "firm") had only the organizational FOLK-sense listed in the Dictionary, the system would spot the anomaly and could suggest that "sport" be somehow reinterpreted, perhaps so as to fit the general semantic category KIND. Wilks' system as it stands could not carry out this particular reinterpretation or analogical matching intelligently (how do you think you do so?), but it can cope with certain semantic anomalies.

This is possible because the mandatory rules expressing basic gists are supplemented by optional "preference" rules. The options are weighted so as to incline the program toward specific patterns of conversational coherence without necessitating interpretation in terms of those patterns. In general, the level of *formulae* allows for a number of semantic liberties to be taken, so that Dictionary definitions can give way to rather different senses.

For example, the formula of the verb "drink" given earlier stated that drinking is "an action *preferably* done by animate things . . ." Consequently, the sentence "Jill put the puppy in the car before letting it drink" would be translated so as to identify "it" as the puppy, despite the greater proximity of the word "car" to the word "it." But a sentence like "The car drank the petrol" would be correctly interpreted, since in default of an animate subject the syntax would suffice to justify regarding the car as the drinker. (Trouble would arise with "Jill put the puppy in the car before giving it a drink of petrol": the partiality of cars to petrol would be classed by Wilks as factual rather than strictly semantic knowledge, and would not be included within his semantic system.) Even a sentence like "She drank in the sunshine at every pore" would be accepted by Wilks's translator. Yet almost every semantic element in the drink-formula is unsatisfied: sunshine is not a fluid; it does not end up inside the woman; and she does not take it in at one specific bodily aperture.

In effect, then, Wilks's language-understanding system pays the writer of the linguistic input the compliment of assuming that she intended to communicate a meaning; it does not reject semantically anomalous sentences as the garbled babblings of a fool. It prefers the normal when it can get it, but accepts the abnormal when it can not—noting what are the specific abnormalities involved, and how they are justified by the text as a whole.

Discussion of the functional details of Wilks' automatic translator would be out of place here. The crucial point is that his program embodies a theory specifying the meaningful patterns or coherent themes that one intuitively expects to underlie human communications. His system and Schank's are alike in this, despite their differences in theoretical formulation. Each matches the natural language input against general semantic patterns, instead of the specific verbal patterns searched for by PARRY, ELIZA, and Colby's interviewing program.

## LANGUAGE WITHOUT LOGIC

Wilks's system resembles Schank's also in that it employs preferential rules of interpretation, as opposed to the rigorously deductive procedures employed by SHRDLU. Wilks himself goes to some lengths to stress the difference between his approach and Winograd's.[27]

In comparing the understanding attributable to Wilks's system with that evinced by Winograd's one should bear in mind the different criteria of understanding by reference to which they were constructed. Wilks is aiming for a theory capable of directing the machine translation of texts from one natural language into another. He is not primarily concerned with the common-sense implications of the information presented in the text (although the earlier discussion of Schank's "inferences" suggests that such implications may play a part in interpreting ambiguous remarks like "Janet was hurt"), still less with the intelligent application of this information in generating active recognition and practical action in the world. Semantic knowledge about steeples could enable Wilks's program to translate language concerning steeples, but not to locate or build any (cf. items 36–39 of SHRDLU's dialogue). As Wilks himself points out, the level of understanding required for this project is, in a sense, less than that required for Winograd's.

This accounts in part for Wilks's disagreement with the typical "deductive" approach as to what really constitutes *understanding*. He insists, for example, that the fact that water freezes to form ice is not part of the meaning of the word "water," but rather a fact about the world—which many people (in hot climates) do not know, even though they have in their language some word that is correctly translated as "water." The deductive approach, he says, tends by contrast to include such knowledge as a criterion of the understanding of language. While it might be relegated to the "world-knowledge" portion of a program following Winograd's paradigm, rather than being included in the "semantic" modules, the ice-forming potential of water would admittedly play a more integral part in the comprehension shown by a cousin of SHRDLU than it does in that of Wilks's own program. However, this dispute appears somewhat forced, since it should be clear that no hard and fast line can properly be drawn between knowledge that does and that does not contribute to understanding the meaning of a word.

A five-year-old child, for example, can understand the word "doctor" (and its French equivalent) as denoting a SORT of MAN who CAUSES other MEN to BE in a BETTER STATE. And she can complement these general semantic categories by "extra" knowledge about hurt knees, sticking plaster, and the like. But it could not be claimed that she has as full an understanding of the term as her mother, even though one would hesitate to include "Often requests X-ray tests," for instance, as essential to the *meaning* of "doctor." In a given world, however, (namely, one that has X-ray apparatus readily available) it may be that anyone with common sense would immediately infer that a person had been sent to be X-rayed on being told that she had been taken to the doctor after breaking her leg in a fall: indeed, this item of knowledge might be embodied in MARGIE as a normative inference in the memory.

For purposes of translation, apparently, such knowledge is not sensibly included in criteria for "understanding" the word, whereas in the context of speech and action provided by adult Western society it possibly is. Even translation, however, may sometimes require access to what Wilks would class as facts about the world: we have already seen that his general semantics could not cope with "Jill put the puppy in the car before giving it a drink of petrol," and a translator of the two sentences about city councilmen and women demonstrators that were cited in Chapter 6 would give "ils" or "elles" for the English "they" *only* by reference (whether implicit or explicit) to culture-specific political knowledge.

To the extent that one regards knowledge of politics and the chemistry of the internal combustion engine as "factual" rather than "semantic," and as only *rarely* necessary in mediating correct translation, one can sympathize with Wilks's insistence that "there are excellent reasons for doing automatic theorem proving, but not for doing it and calling the product natural language analysis."[28] But Wilks offers further reasons for his suspicion of deductive models of understanding. One of these is a doubt as to whether the operations of thought in general (what is sometimes called "the" understanding) are plausibly represented as essentially deductive in character. This doubt has a number of sources, including the absence of deductions in introspection; the average person's difficulty in mastering even elementary logic exercises; and the (time-consuming?) complexity of deductions required to justify seemingly "immediate" inferences. This generalized dubiety concerning deductive models of cognition is not peculiar to Wilks, and strikes at the root of much influential work in artificial intelligence (such as that based on "resolution theorem provers," like the STRIPS system described in Part V). I shall postpone consideration of this question, focusing for the moment on Wilks's more specific reasons for rejecting a deductive approach to understanding.

A computational model of comprehension may eschew deduction, but cannot eschew rules. For a computational, programmable model simply *is* one that is expressed in terms of sets of rules or effective procedures (which does not exclude the possibility of having random number generators to produce "indeterminate" features at determinate points in the program). Wilks has to show, then, that his rules of "common-sense" inference are superior to (more effcient and/or more plausible than) the deductive rules relied on by theorem-proving paradigms. To this end, he mentions a number of ways in which, he claims, his system is the more efficient.

First, his preference procedures give a rationale for selecting one interpretation rather than another—which means that they specify *all* the prima facie possible interpretations rather than just one. Winograd's system, by contrast, picks one and only one interpretation of a sentence (although it sometimes asks for prior clarification, as in item 22), and makes no record of its aborted attempts at disambiguation. This difference is connected with the different sorts of input that the two systems accept. Wilks deals with paragraph-length texts rather than single sentences, and accordingly is more sensitive to overall contextual considerations than SHRDLU is. Thus Wilks takes pains to point out that there is never any "fail-safe" interpretation of a particular expression, since with sufficient ingenuity one can always imagine a surrounding context that would force an alternative meaning. We have already seen this contextual forcing at work in the example of Janet's plate or Janet's cookie: " 'I am sure she will like it,' thought mother. . . . "—but what exactly is "it"? Wilks's system is in principle capable of spotting its mistake in first assigning "it" to the cookie, and could immediately retrieve the less preferred alternative (the plate) on encountering the relevant "sense-changing" context (the reference to Bambi).

Wilks criticizes deductive systems by pointing out that, in similar cases, Winograd's program would have to start interpreting "it" again from scratch, with the caveat that it should avoid identifying "it" with the cookie. However, Wilks here overgeneralizes from SHRDLU in particular to deductive programs in general: as we shall see in Chapter 12, the PLANNER language—in terms of which Winograd specified the BLOCKS world—indeed tends to encourage a "blind" backtracking on recognition of failure, but other programming languages of comparable procedural power (such as CONNIVER and POPLER) do not share this fault.

One may more readily accept Wilks's observation that his approach is more tolerant of logical mistakes in the text than purely deductive programs are likely to be. Such mistakes are fairly common: for example, the erroneous notion that if all As are Bs then all Bs are As is expressed in the familiar view that "Communists favor state medicine, and this election manifesto favors state medicine, therefore it is a Communist document." If a deductive program translating this logical howler were to assume that the input text is always logically consistent, it could not assign the correct referent to "it." On the other hand, if it were (more realistically) to assume that human thought processes are sometimes downright *il*logical, as well as being probabilistic rather than deductive, then it would have to embody a model of nondeductive thinking as part of its representation of the human speaker or writer.

A further advantage claimed by Wilks for common-sense preference rules as against deductions is that it is relatively easy to specify preference rules, but extremely difficult to specify axioms and boundary conditions for a theorem-proving representation of knowledge. For instance, consider these two occurrences of "it": "I put the heavy book on the table and it broke," and "I put the butterfly wing on the table and it broke." A preference rule states no more than that fragile things break more easily than rigid things. Assuming FRAGILE and RIGID to be included within the relevant semantic formulae, therefore, Wilks's system would interpret these sentences correctly. A deductive approach, by contrast, would have to commit itself to axioms like "Fragile things break—when . . ." and "Rigid things do not break—unless . . . ," where precise statement of the boundary conditions would be extraordinarily difficult. However, the previous discussion of a similar point (in relation to Schank's "inferences" and Charniak's PLANNER-demons) showed that the specification of the requisite inferences or preference rules would be equally taxing were it to be carried out in a systematic manner, as opposed to being drawn up ad hoc for the purposes of a particular problem.

Insofar as one's interpretation of (and replies to) sentences depends upon inferences of a nondeductive nature, Wilks's suspicion of rigidly deductive computational systems is understandable. An adequate model of language use would have to accommodate probabilistic and inductive inferences (such as those represented in MARGIE's memory and Abelson's work) if it were to capture the "illogical"—albeit lawful—processes involved in verbal thought. Equally, of course, it could not entirely outlaw deduction: Wilks himself admits that there are some examples of pronominal reference that could not be correctly resolved without complex deductions from relevant facts. If the Wonderland duck was short on logic, therefore, she would be unable to understand all uses of "it" because she would be unable to puzzle them out. But ordinary people might have difficulty too: Wilks suggests that such cases are essentially *puzzles* stated in natural language, rather than "natural language" phenomena as such. Accordingly, he does not regard them as destructive of his project of explicating understanding without proofs.

Finally, one might argue that Wilks's deductive-nondeductive contrast is a false antithesis. For the rules embodied in so-called "nondeductive" programs could be regarded as deductive rules for proving (say) that it is very probably the butterfly wing that broke, so we need not consider the possibility that the "it" that broke was the table. In this sense, all programs (except possibly those including a randomizing element) could be seen as deductive. But even so, a distinction remains between a belief system all of whose conclusions are regarded as unquestionably true (and mutually consistent), and a belief system that is less dogmatic as to truth and less optimistic as regards internal consistency of the various beliefs contained within it. Wilks may then be interpreted as stressing the tentative, hypothetical character of language understanding, and as criticizing programs that represent it as a relatively closed, cut-and-dried matter.

The recognition of a string of words (whether teletyped or spoken) as an English sentence structured in particular ways is a special case of pattern recognition, or perception. The identification of threats to the Free World is another, of paranoid conversation another, and of insidiously persuasive communications yet another. So too is the visual recognition of faces, blocks, and steeples. Given the apparent facility of Winograd's program in recognizing steeples and blocks, it might seem that, like HAL, computers should be able also to lip-read by the turn of the century.

But this appearance is illusory, as will become clear if one considers the difficulties involved in providing a machine with the visual ability to recognize steeples and blocks, or to identify the outlines of human mouths or faces. (SHRDLU in fact has no visual powers whatever: it knows where a block is only in the sense that it is given numerical coordinates on a two-dimensional mathematical grid that represents "the table"; and the coordinates for each shape that SHRDLU knows about always represent a perfect view of the front plane, which is why the program has no notion of spatial rotation or of the "cubicalness" of cubes.) The project of writing visual programs casts light on the nature of visual experience in people, showing it to be surprisingly akin to the understanding of ideology and the comprehension of natural language. Let us turn, then, from linguistic to visual understanding.

# *Part* IV

# THE VISUAL WORLD

# 8

# Adding the Third Dimension

ONLY the most patient and imperturbable of readers can never have snapped "Use your eyes!" at a friend failing to see something staring her in the face, something apparently as plain as a pikestaff. However, there is more to seeing a pikestaff than is commonly believed, and in such cases of "inexplicable" blindness one might better advise the person to use her brains, to think rather than to look. The trouble with such advice is that it would often be regarded as paradoxical or unintelligible. The everyday assumption of a sharp dichotomy between seeing and thinking implicitly denies any possible contribution of the latter to the former. Nor is it clear just *what* one is required to think in any particular case, since usually one is introspectively unaware of the underlying inferential processes that are essential to vision, and unable consciously to call them into play.

Because of this deceptive simplicity in the phenomenology of seeing, psychologists have traditionally turned first to abnormal cases—such as illusions, trick effects, and puzzle-pictures—in order to highlight the background assumptions, conceptual schemes, and stimulus cues that determine one's more normal visual experience.[1] Their experiments have been highly suggestive of some of the concepts and contextual cues involved, but they have not explained how such cues function in activating one type of perceptual construction rather than another. Some psychologists (such as J. J. Gibson[2]) treat perception as a matter of sensory detection rather than conceptual construction. But even those who assume that "thinking" and "seeing" are inextricably connected (as Gestalt and "cognitive" psychologists in general do) have not shown the full extent of the connection, nor how the connection is possible.

Workers in computer vision are forced to face such issues directly, since programs see only to the extent that programmers can make the connection functionally explicit. The complexity involved is enormous, and even to enable a machine to recognize cubes and pyramids in a wide range of cases is a complicated matter. (At base, this is because recognition involves the ability not only to classify but also to describe.) In this chapter I shall discuss a series of programs concerned with such apparently simple tasks, showing how visual mechanisms can be made to see more by providing them with increasingly powerful theories about the physical world. Examination of the way in which these theories are used in interpreting visual cues raises general epistemological issues also, for it clarifies the notion of *interpretation* whereby intelligent systems attach subjective significance of many different kinds to features objectively present in the world.

## MODELS, CUES, AND INTERPRETATION

Please look at Figure 8.3: what are the things shown in this picture? "Well, there's a cuboid sitting behind a staircase and a wedge." How can you possibly tell that it's a *cuboid*, that it's *sitting*, that it's *behind* other things? After all, nothing *in the picture* has these properties. Where does "behindness" come from? And who ever saw a cuboid with a triangular face, like BCI in the picture?

This example suggests some basic questions about seeing the world. Visual information is received by visual systems in the form of changing patterns of different light intensities. If this two-dimensional input is to be seen as representing a three-dimensional structure, the extra dimension has to be added by the interpretative activity of the seeing system itself.[3] And if sight is to be useful in moving (and moving among) material objects, the image has to be interpretable as providing information about the size, shape, orientation, and location of things in physical space—including *invisible* parts that are hidden either by other objects or by those parts of the object itself that are nearer the eye. Since so much information has to be derived about properties that are not directly present in the image, the visual system must have a great deal of knowledge to contribute to the visual process.

All this is true whether the system be biological or artificial in origin, and whether, if artificial, it is a perambulating robot, a stationary hand-eye manipulator, or a purely visual program with no motor faculties.[4]

The earliest "2D to 3D" program, for instance, was able not only to identify many of the solid objects shown to it—like those in the picture you just looked at—but could give their exact dimensions as well as their absolute and relative positions; it could "imaginatively" construct indefinitely many views of what each object would look like if it were viewed from some other place, appropriately deleting the lines representing invisible edges; and it could recognize partially occluded objects even though they were largely hidden by others lying in front of them. L. G. Roberts's program[5] was able to do this because it incorporated detailed knowledge both of the projective geometry of the 3D structures it expected to find represented by 2D shapes in the input picture, and of the optics of the camera (its "eye") and its positions relative to ground level in the scene.

The details of the program are highly specialized: its world knowledge is expressed mathematically, in terms of matrices within projective geometry, and its use of this knowledge involves sophisticated numerical calculations. But the overall logic of the approach is more generally relevant, for Roberts identified issues that continue to influence work in scene analysis. Indeed, the program's visual capacities to go beyond the evidence in the ways listed above may be compared with other interpretative faculties—for example, MARGIE's abilities to parse, reason from, and paraphrase even elliptical remarks; the IDEOLOGY MACHINE's power to interpret appeals for "law and order" as indicative of a particular political viewpoint; PARRY's knack of attributing various forms of malevolence to his interlocutor; and people's experience of feelings as the emotions of anger or euphoria according to context, as described in Chapter 3.

To interpret something is to give it meaning, to see it as significant. As in the examples just listed, one can distinguish the class of phenomena being interpreted (the representational domain) from that by reference to which the interpretation is made (the target, or semantic, domain). The interpretative process relates these two domains by way of a conceptual schema embodying the system's knowledge of the "mapping" relation between them. The more rich and usefully structured this schema is, the more intelligently can the cognitive system (person or program) understand phenomena in the representational domain as expressing the relevant target realities. Knowledge, to be useful, must be stored in a form that is fitted to the interpretative tasks involved (for example, as factual or procedural information, expressed at the relevant level of detail, and with an appropriate tolerance for ambiguity in the input); and it must be readily accessible when it is needed, since exhaus-

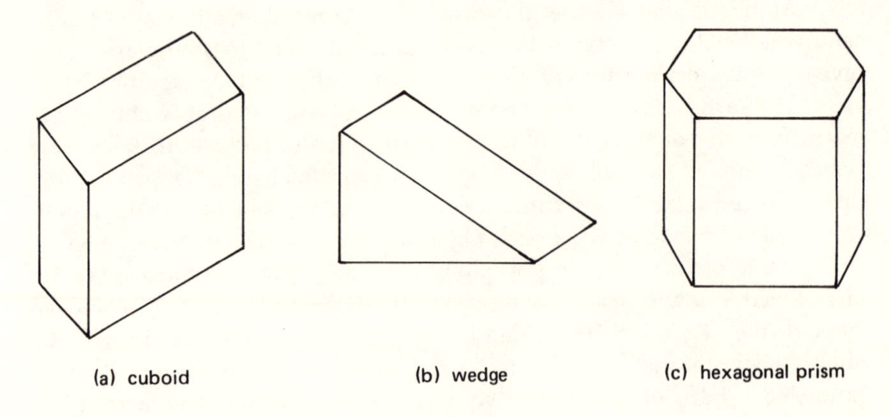


FIGURE 8.1

tive search through a knowledge store is feasible only when the store is relatively small.

PARRY's attributions of malevolence are not only unrealistic (paranoid), but crude, since his representational theory of how ill intent is mapped onto speech is simple, superficial, and inflexible: he does not have a richly structured concept of malevolence (see Figure 5.1). By contrast, the projective geometry and theory of perspective provided to Roberts's scene analysis program is precise, systematic, and—within the limits of its world and the descriptive language available to it—comprehensive. (Although the program's method of searching its knowledge is not sufficiently intelligent to prevent its getting bogged down in endless calculations, where really complex scenes are concerned.)

Roberts's program assumes that the world (the target domain) is made up of three classes of object, occurring either singly or merged to form compound objects: cuboids, rectangular wedges, and hexagonal prisms (Figure 8.1). (Other classes could be added if they were appropriately defined.) Its concept or schematic "model" of a cuboid is stored in the data structure in the form of an abstract geometrical definition, which describes all the surfaces, edges, and corners of a cuboid without committing itself to any particular size or shape. This definition is associated with two sorts of transformational rules (expressed as mathematical matrices) determining how particular target instances of the general cuboidal model can be specified in terms of their actual dimensions, and how the various views or representations of those instances are generated according to viewpoint.

With the help of knowledge about the geometry of the picture-taking process (position and focal length of the camera), the program looks at pictures and "works backwards" from appearance to reality, asking whether a possible transformation allowed by the rules could have produced the picture from an actual instance of one or more of the abstract models. In assessing tentative interpretations, it computes a mathematical measure of the degree to which the picture fits the suggested alternatives, rejecting interpretations that fail to satisfy the threshold value and settling for the model (with transformations) that fits best. Figure 8.2 shows an example using a compound object.

This procedure requires an analysis or structural description of the picture in terms that are relevant to the particular representational system, or theory of mapping, involved. That is, the "parts" of the picture must be *cues* that relate to the target domain in ways defined by the inner conceptual scheme. It follows that "parts" (like "cues") is a subjective, or intentional, notion and is not equivalent to purely physical features of the picture. (A microscopic examination of the pictures in Figure 8.2 would not show the straight lines we see, but much more complex and messy structures: the lines are not objectively there *as lines*.) From a strictly objective point of view, a picture just *is* marks on paper—and wholly uninformative to boot. From a subjective (interpretative) point of view, it may or may not contain lines, or colored regions, or angles of various sorts.

For example, if the conceptual models of solid objects that are used by a visual system in recognizing objects include no reference to color, no way of mapping color onto the objects concerned, then a color photograph of them has *no* colored parts *from the point of view of the object-recognizing section of the system*—even though other subsections may already have seen it as colored. Similarly, sentences contain *no* grammatical parts, such as noun phrases, for systems (like ELIZA) having no knowledge of syntax.

There are indefinitely many representational schemes one might employ in making sense, of one sort or another, out of a phenomenon. PARRY, for example, would interpret a remark, *any* remark, about policemen's pay as threatening; and so would his human equivalent, who would also be able to interpret it in terms of economic realities, the politics of "law and order," and so on. But any such interpretative scheme requires an analysis of the phenomenon into the "parts" that it takes to be significant. Different questions may require reference to different cues (*What, where, how big, how many,* are the things shown in Figure 8.3. And do they *touch* each other or not?) Equally, one and the same question about the target domain may be answerable with the help of different types of cues in the representational domain. For instance, a therapist may use verbal *or* nonverbal cues as indicators of her patient's current emotional state (which is why the neurotic program prints out the

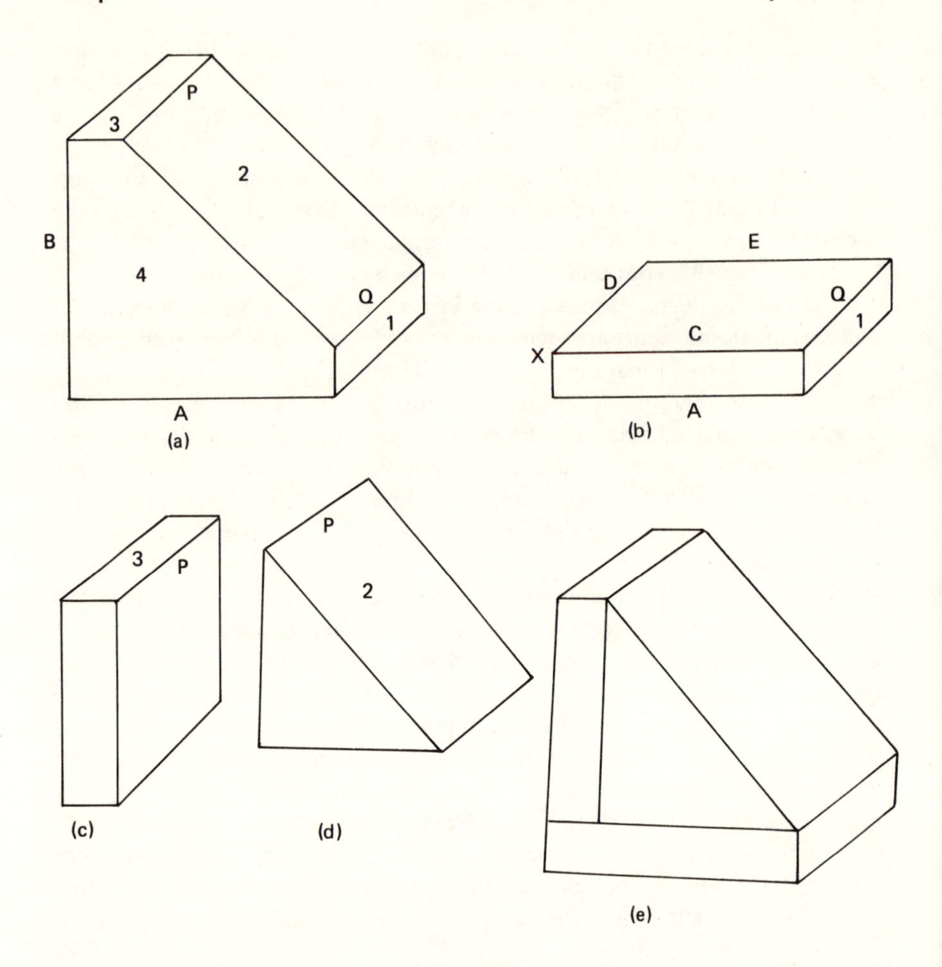


FIGURE 8.2

(a) represents the information passed by Roberts's line finder to his picture interpreter. Roberts's program searches for significant picture fragments. It finds the polygon bounding region 1 with line A connected to it, and constructs and analyzes out solid (b) accordingly. Similarly, (c) and then (d) are found. No picture lines remain unaccounted for. The program's final reconstruction of the target compound object is shown in (e).

Source: Adapted from J. T. Tippett, et al., eds., *Optical and Electro-Optical Information Processing* (Cambridge, Mass.: MIT Press, 1965), p. 182. This and following figures from this source reprinted by permission.

numerical values of the affective monitors from time to time, as well as sentences like *I am depressed*). Similarly, we shall see that programs that can tell how many objects are shown in a picture of cubes may concentrate primarily on different pictorial features: either regions, or vertices,

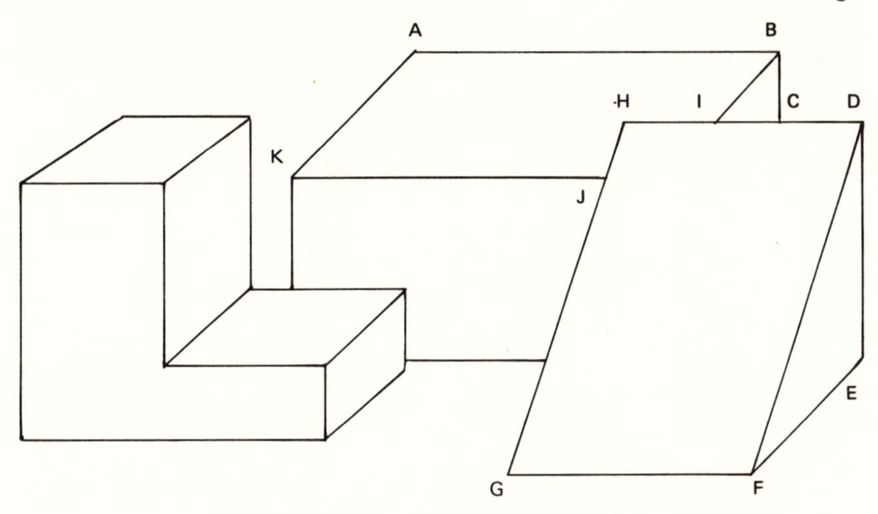


FIGURE 8.3

HDFG, DEF, and BCI are approved polygons, but HDEFG and ABIHJK are not. Polygon BCI—in a different context—could depict the entire (triangular) face of a target object, but in this context it does not, because of the occlusion of the rear cuboid by the wedge. (Notice that you, like the program, assumed the rear object to be a cuboid: but it *might* have a "bite" out of it at the 5 hidden corners.) By assuming that the lowest line in the picture (GF) is supported at ground level, Roberts's program can calculate all spatial dimensions in the scene.

Source: Adapted from Tippett, et al., *Optical and Electro-Optical Information Processing,* p. 176.

or lines. But whatever the specific details, the "parts" of a picture, as of a sentence, are subjectively projected onto it rather than found objectively in it.

Cues, in short, have to be defined intentionally, by reference to the method of representation assumed in the interpretative process. (The notion of *intentionality* will be further discussed in Chapters 13 and 14; for the present, it may be understood as equivalent to *subjectivity.*) Intelligence, artificial or otherwise, depends largely on finding more and more ways to treat features of the world as cues to matters of interest. (Might shadows actually help one to see, or are they just a nuisance?)

Roberts's program, for instance, treats variously-shaped regions in the line drawing as the basis of cues (picture fragments) selectively calling the stored theoretical models. Just as some words are more useful than others as grammatical cues, so some regions may be more significant than others to a "region-al" interpretative system. With respect to Roberts's program, the regions that contribute most usefully to interpretation are those that might depict the boundary of a single face of a target object:

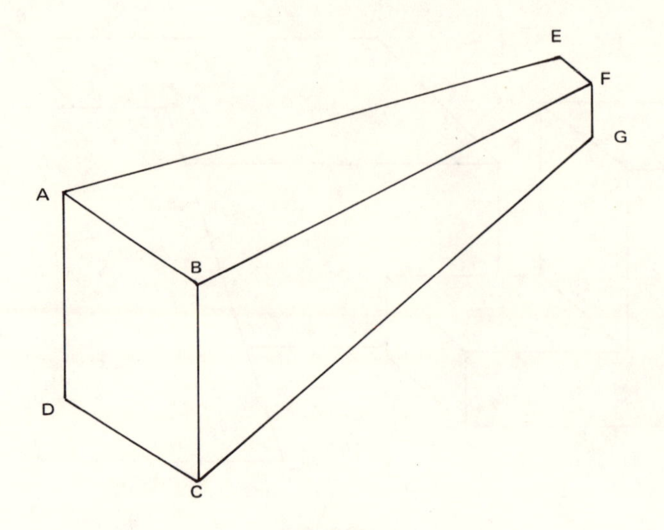


FIGURE 8.4

In this picture of a cuboid, ABCD is the only parallelogram and neither ABCD, ABEF, nor BFGC is a rectangle. Yet all faces of an actual cuboidal object are rectangular. (Notice that polygons ABFE and BCGF become triangles if the far end of the object recedes sufficiently; even without occlusion, therefore, a triangle does not *guarantee* a wedge.)

convex polygons with 3, 4, or 6 sides, which Roberts calls "approved" polygons (see Figure 8.3).

As I assumed in the opening paragraph of this section, one senses intuitively that a triangle in the picture suggests a wedge rather than a cuboid. The rationale underlying this intuition derives from the topology of the models concerned. For, as Roberts realized, the faces of a cuboid, when viewed face-on by a normal camera, appear as rectangles; when not viewed face-on, they still appear as parallelograms, however skewed the angles, unless perspective (which Roberts assumes to be excluded) makes two sides approach each other; only with extreme perspective effects can an entire (nonoccluded) cuboid face appear as a triangle (Figure 8.4). Topological features like these were used by Roberts to identify picture fragments that could function as cues. Cues can sometimes be ordered according to reliability: since Roberts intuitively realized that a picture point surrounded by three approved polygons is more informative than a line with an approved polygon on either side, his program looks for the former before keeping its eye open for the latter. The use of geometrical optics and projective geometry renders the program "superhumanly" efficient within its limits because the more detailed and systematic one's theory of mapping, the more reliably can one assign meaning (including "predictive" significance) to a given cue.

Even if it has been decided what cues are significant, finding them may not be easy. Thus the articulation of a picture into even relatively "low level" parts, such as lines and regions, is not a trivial matter, and is akin to parsing a sentence in terms of syntax. Pictures are usually like spoken (rather than typed) sentences, in that the intentional parts, at least at low levels of description, are not *physically* separate in the input. Roberts's program, for instance, has to find the approved polygons in the picture for itself before it can use them to address its models.

It does this in two stages, for the program consists of two sections: an initial *line finder,* which interprets the grey-scale photograph in terms of lines and junctions like those depicted in line drawings (not so simple as it sounds); and the *picture interpreter,* which interprets the line representation in 3D terms. The picture interpreter thus starts off with a (description representing a) pattern of lines, and has to find the approved polygons in it. Until it has described the picture in this way, it is not in a position to interpret it in 3D terms. (We shall see in the next section that richer descriptions of the picture may be required for 2D / 3D representational systems using different cues.)

The functional relation between cues and theoretical models is two-way, for once a model has been addressed by cues it can then actively construct hypotheses that lead the program to look for further, confirmatory, cues. These modes of functioning are often termed "bottom-up" and "top-down," respectively, and the more a program (or a person) operates in a wholistic, or global, manner the more it (or she) relies on top-down processing. Clearly, these modes are not mutually exclusive: unless one already has good reason to entertain a specific hypothesis, one's selection among possible alternatives should be intelligently directed by bottom-up attention to details. Conversely, top-down processing may influence the choice of area in which the next phase of bottom-up thinking will occur.

Thus at a certain stage in the processing of Figure 8.2a, Roberts's interpreter finds a cue: a parallelogram (1) with an extra line (A) coming from one vertex. This cue is compatible with *any* of the three basic models (see Figure 8.1), but the program always looks for cuboids before calling the other models. The cuboid model provisionally matches the approved polygon in the cue with one of the abstractly defined cuboid faces, and predicts a series of topologically equivalent pairs of points between model and picture. For example, it predicts—and finds—a line (B) running parallel to the central line of the vertex and connecting with the first "extra" line. (In this case, since the boundary of region 4 is not an approved polygon—neither cuboids, wedges, nor hexagonal prisms have any 5-sided faces—the program constructs an end point x, and fills in the "missing" lines C, D, and E accordingly.)

Sometimes, of course, the cue predictions are not confirmed, and the program has to seek further cues to address its models afresh: indeed, this happens at an earlier stage of the processing of Figure 8.2a. The very first cues found are two lines, P and Q, with approved polygons on either side (there being no point surrounded by three approved polygons, which is the program's first preference). Since each of these polygons is a parallelogram, the cuboid model is addressed and predicts picture points accordingly. On discovering that no actual picture points fit the predictions to the required mathematical measure of accuracy, the program rejects P and Q as useless and searches for its *third* general preference, an approved polygon with an extra line coming from it. This cue, as we have seen, turns out to be more fruitful.

Analogously, the initial word "Put . . ." functions bottom-up as a cue for SHRDLU, calling the model of transitive imperative verbs, which then works top-down in directing SHRDLU intelligently to expect a noun phrase (though the next word might be "up" or "down," the noun phrase coming later). But since SHRDLU's syntax, like MARGIE's semantics, is less comprehensive and restrictive as a representational theory than is projective geometry, SHRDLU and MARGIE have a less specific idea of what confirmatory cues to expect than Roberts's program has. PARRY has even less notion of what to expect, because his interpretations of malevolence and benevolence are comparatively reliant on bottom-up processing (even his "expectancy list" is closely dependent on immediate context).

A familiar objection to psychoanalytic theory is that it does not enable one to predict precisely what is going to happen, but merely offers post hoc interpretations of events that are ambiguous in any case.[6] This objection certainly shows that the relation between cues and theory in psychoanalysis is not as tightly restrictive as that in projective geometry (or physics). But it does not show that there is no psychoanalytic "theory" properly so-called at all. Top-down processing, in whatever domain, allows programs and people justifiably to expect a particular phenomenon, or to interpret it in a certain way according to context—where "justifiably" does not mean "necessarily correct" but, rather, "intelligently, given the interpretative system currently being used to make sense of the world." A therapist's vague prediction that *some sort of reaction formation* may be expected as a defense against a woman's hatred for her father is thus essentially comparable to SHRDLU's prediction that *some noun phrase or other* may be expected after the initial word "Put . . ."

Conceptual models working top-down enable an interpretative system to cope more sensibly with the input, to think before it looks. This is particularly useful when, as in vision generally, the input contains a great deal of information only some of which is relevant to the concerns of the system at the time. If one is looking for rabbits, whether in a real-world situation or a children's puzzle picture of "hidden" animals, one is not interested in blades of grass and would like to be able to ignore them—even to the extent of not "seeing" the lines depicting grass at all. If a top-down processor influences the line finder in its initial construction of the picture, relevant (cue) lines can be extracted and irrelevant ones ignored. Moreover, conceptual models (of whatever target domain) may so increase one's sensitivity to the representational domain that, in effect, the quality of the input is improved: as the King of Hearts might have put it, "Verdict first, evidence afterwards!" Thus Roberts suggested that his interpretative program could guide a "line proposer" and "line verifier" that, in cases where an expected line seemed to be missing, would go back to the original photograph and look more closely at a specific area to see if a faint line could indeed be found there.

We shall see in the next chapter that his suggestions have since been implemented by other workers, whose visual programs consequently know what subtle details to look for, and where to look, without being burdened with the combinatorially hopeless task of finding *all* the lines (light intensity boundaries), however faint, and then sorting out the "significant" ones. Similarly, MARGIE's ability to complete syntactically elliptical remarks on the basis of semantic inferences could in principle be provided to a speech interpreter to help it recognize acoustically unclear words—much as one sometimes works out what a word must have been before realizing that one did, in fact, hear it.[7]

Roberts's general paradigm has greatly influenced later efforts in scene analysis. Yet some appear rather different. Indeed, one of the best known, Adolfo Guzman's "SEE," neither resorts to mathematical number-crunching, nor relies on explicit models (theories) in order to pick out the 3D bodies depicted in the 2D image.[8] Accordingly, Guzman's program is both less and more powerful than Roberts's.

As the next section will show, SEE cannot assign precise dimensions to and construct alternative views of the bodies it identifies, nor even classify them as cubes or wedges without the aid of additional programs.[9] But it can recognize most line drawings of *any* convex, and some concave, flat-surfaced object (polyhedron) as depicting a single body, whereas Roberts's program can see only objects constructible out of the specific models previously supplied to it (and viewed without perspective effects). Again, the cues relied on by Guzman are not those used by Roberts, for SEE sees vertices as particularly significant. And SEE's spatial knowledge is "knowledge how" rather than "knowledge that,"[10] being comparatively hidden or implicit in the pragmatically derived rules of thumb (heuristics) built into the program instead of being explicitly represented in a data structure provided for the program's use.

Nevertheless, SEE's knowledge is largely equivalent to knowledge possessed in a different way by the earlier program. And, as consideration of SEE's intellectual descendants will make clear, the theoretical justification of Guzman's choice of cues and interpretative heuristics raises issues that give a rationale also for Roberts's more explicit theory of 2D/3D mapping.

## HOW SEE SEES

To get an idea of what SEE does, you should look at Figures 8.5 and 8.8: how many objects are shown in these pictures? How can you tell?

SEE's purpose is not to describe or classify 3D bodies (as small cubes, for example), but simply to discover *how many* are depicted in line drawings, assigning the various picture regions to individual bodies. For instance, in Figure 8.5, regions 3, 21, 22, 23, 24, 28, and 29 are interpreted as parts of the surface of a single body, while regions 1, 2, and 33 are assigned to another; 34, 36, and 35 are identified as background. The target features represented in SEE's theory of mapping are thus very few—background, body surface, same body, different body; but the (intentionally) relevant picture features are more numerous.

Pictures are seen by SEE as consisting of a set of regions (arbitrarily numbered for ease of reference), each having several vertices associated with it. A vertex is a point where picture lines meet, and SEE classifies vertices into eight types. If you look at Figure 8.6 *without* reading any of the captions you will doubtless agree that these eight vertices "look" different, and you will be able to pick out examples in the two complex scenes mentioned above. In fact (as the captions make clear) this intuitive classification depends upon prior description or articulation of the picture in terms of concepts such as number of lines, meeting of lines, collinear, same side, opposite side, and—for complex patterns like Figure 8.7,i–l—parallel and leg-of-L. These structural concepts, and the vertex concepts derived from them, are richer (can say more things about the items and relations in the picture) than the picture-description concepts needed for addressing theoretical models specified in terms of projective geometry. This is why I earlier described Roberts's program as "comprehensive" *only within the limits of the expressive language available to it.* A fully automated system (such as the M.I.T. robot, of which SEE forms an integral part[11]) of course requires special subroutines for finding lines and regions and classifying vertices in terms of such concepts, much as Roberts's program has to find the approved polygons for itself. I shall ignore these initial cue-finding procedures and assume that SEE starts work already possessing a description of the picture in terms of vertices and regions.

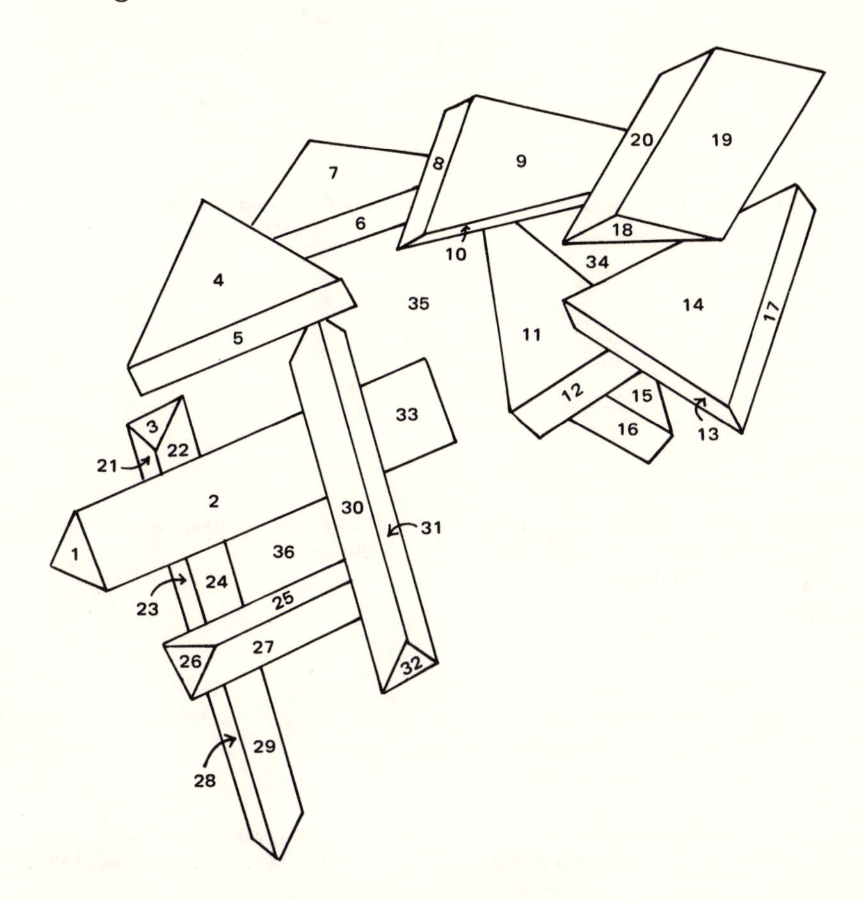


FIGURE 8.5

Guzman's "HARD"—All bodies are correctly identified by SEE, even though one (regions 6:7) has no "useful" visible vertices. The background (34:35:36) also is correctly found.

Source: From Antonio Grasselli, ed., *Automatic Interpretation and Classification of Images* (New York: Academic Press, 1969), p. 273. This and following figures from this source reprinted by permission.

An automated system needs also to be able to locate the background for itself. Sometimes this is easy: light-intensity measures suffice if brightly-lit white objects appear on a uniformly black background. In other cases the figure-ground articulation of a picture requires knowledge of 3D geometry—and more. SEE has a daughter program, "BACKGROUND," that uses geometrical knowledge to pick out the background in line drawings of the scene.[12] BACKGROUND implicitly tends to assume that bodies are convex, that objects do not have bites out of

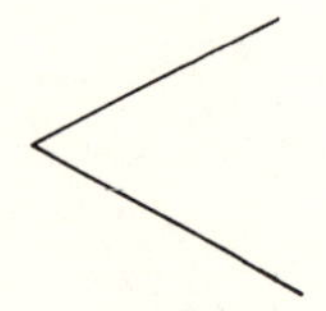

L — Vertex where two lines meet.

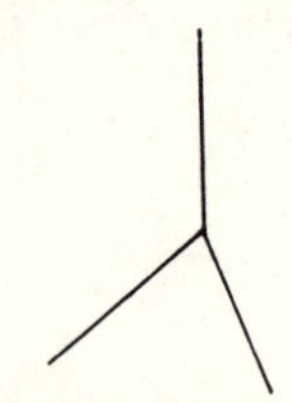

FORK — Three lines forming angles smaller than 180 degrees.

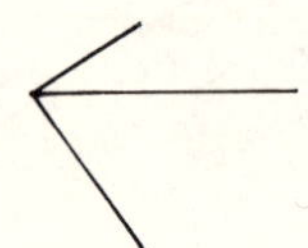

ARROW — Three lines meeting at a point, with one of the angles bigger than 180 degrees.

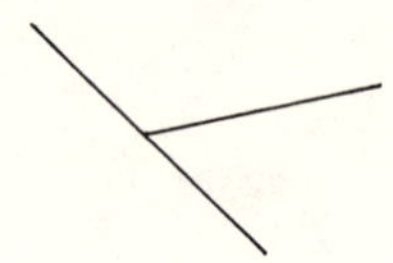

T — Three concurrent lines, two of them collinear.

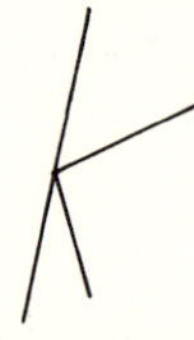

K — Two of the lines are collinear, and the other two fall on the same side of such lines.

X — Two of the lines are collinear, and the other two fall on opposite sides of such lines.

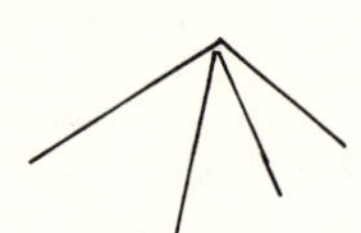

PEAK — Formed by four or more lines, when there is an angle bigger than 180 degrees.

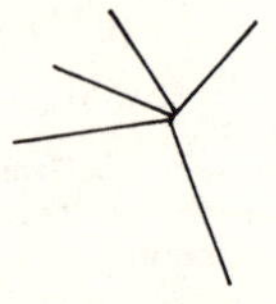

MULTI — Vertices formed by four or more lines, and not falling in any of the preceding types.

FIGURE 8.6

The definitions are Guzman's. FORK, ARROW, and PEAK can all be defined *without* reference to "180°"—e.g. ARROW is a 3-line vertex, two of whose lines have the other two lying on the same side of it.

Source: Grasselli, ed., *Automatic Interpretation and Classification of Images*, p. 251.

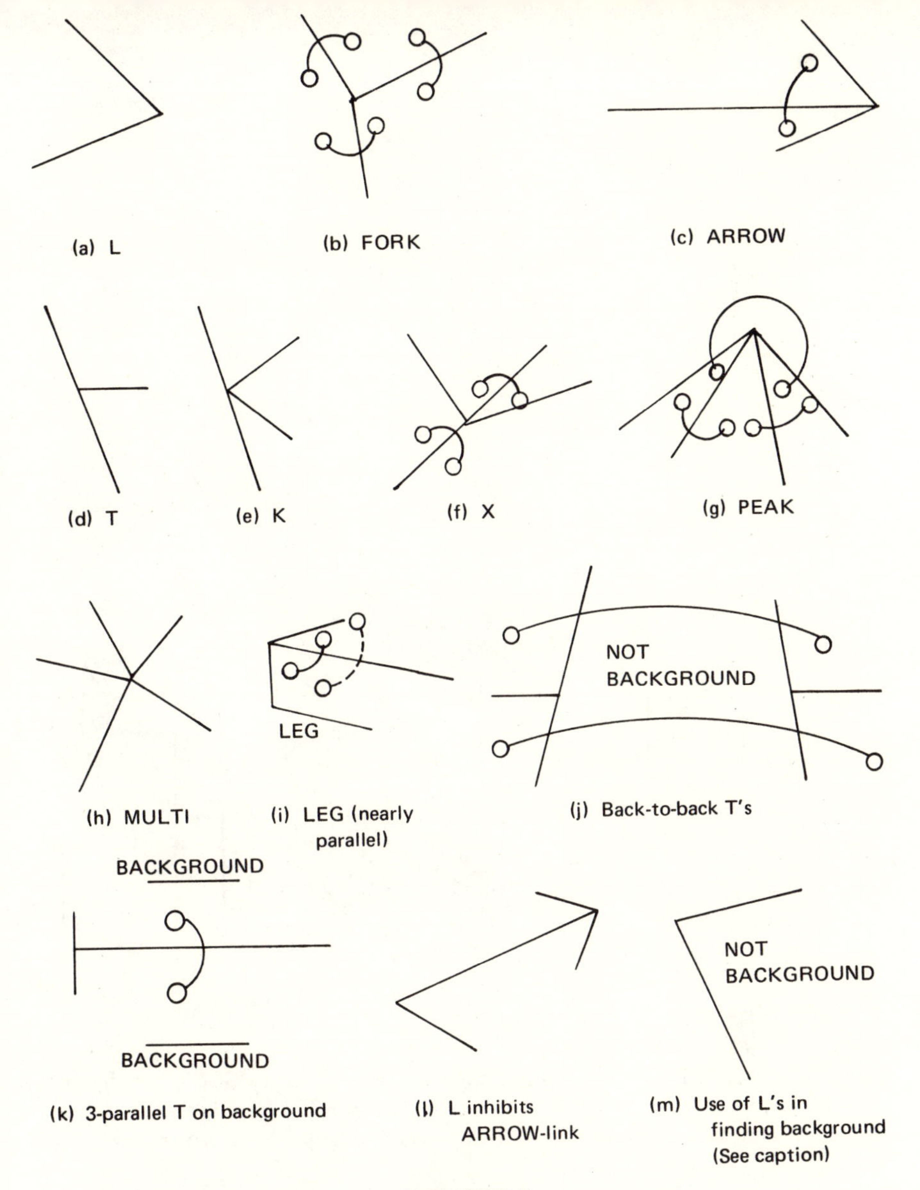


FIGURE 8.7

(a) to (h) show SEE's vertices, with strong region-links placed;
(i) shows weak link placed on ARROW by adjacent LEG.
(j) and (k) show examples of strong links placed on complex patterns;
(l) shows example of link-inhibition by an L-vertex (assumes convexity);
(m) shows embodiment of convexity assumption in use of L's when identifying background. Note that this interpretation of L's is an "ideal" one: it can be overridden in contexts where L's are caused by the overlapping of two bodies—see vertices of region 35 in Figure 8.5. Compare also Figure 8.9a: the BACKGROUND program would misinterpret this picture of a concave object, because it would assume the LEGS of the L to enclose a body region, *not* the background entering into a "bite" in the object.

Source: In addition to my own data, adapted from Grasselli, ed., *Automatic Interpretation and Classification of Images*, p. 258 and Adolpho Guzman, *Computer Recognition of Three Dimensional Objects in a Visual Scene.* (Ph.D. thesis). (Cambridge, Mass.:, 1968).

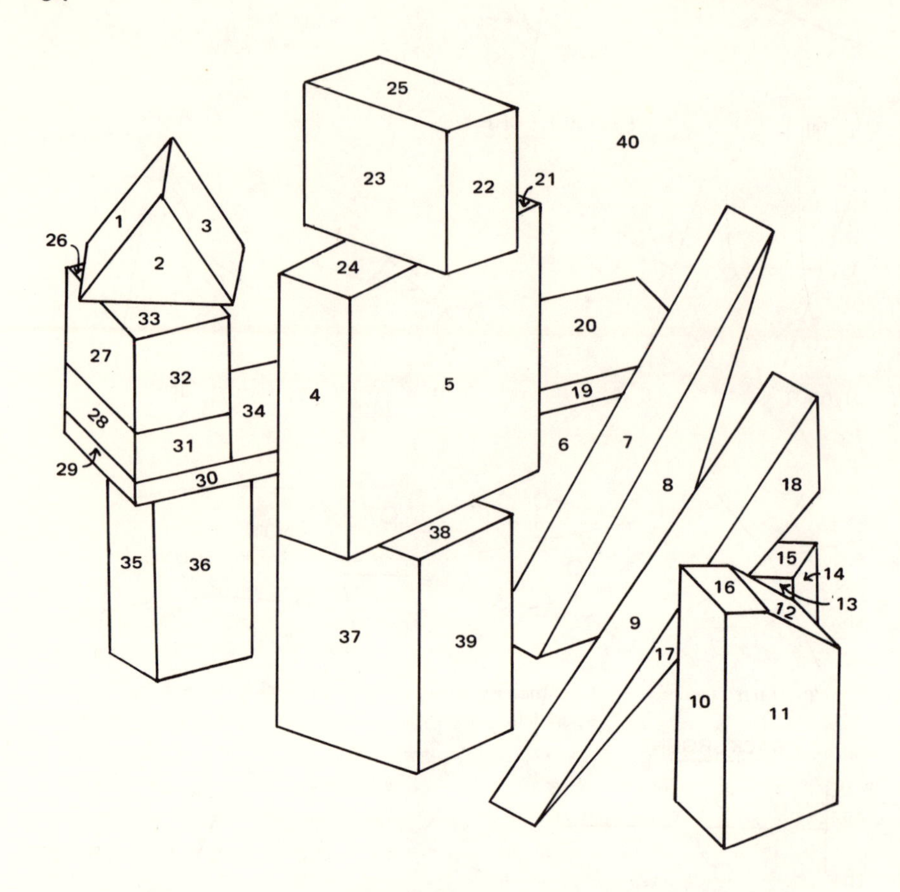


FIGURE 8.8

Guzman's "MOMO"—All bodies are correctly identified by SEE when figure-ground information is provided. Otherwise, BACKGROUND (correctly) labels regions 6 and 40 as background, but also claims region 31 instead of linking it with 28.

Source: Grasselli, ed., *Automatic Interpretation and Classification of Images*, p. 267.

them: this assumption is embodied (for example) in BACKGROUND's use of L-vertices, shown in Figures 8.7,m, and 8.9. However, even if all bodies in the scene are convex, the program often makes mistakes; in Figure 8.8, for instance, BACKGROUND claims not only regions 6 and 40, but also 31. Since Guzman's prime interest is in separating individual bodies, and since this problem remains even when one knows which regions are the background, Guzman normally supplies figure-ground information with the input picture.

SEE starts off, then, knowing which picture regions are background and what types of vertex are associated with each region. Its task is to identify individual bodies, interpreting each (nonbackground) region as belonging to one of the surfaces of a particular body. It does this in two stages: first, it looks at each vertex in turn, asking if it provides evidence for placing "links" (either weak or strong) between regions that probably depict surfaces of the same body. Second, it groups together regions linked by two or more strong links into "nuclei," which then compete with one another for the remaining regions. The resulting groups of regions are then listed, each group being interpreted as representing a separate body.

SEE proceeds bottom-up rather than top-down, in that the interpretations of local picture cues determine the interpretation of the picture as a whole, not vice versa. Although the second stage can decide to ignore linking evidence provided by the first (regions connected by only one link are conservatively assumed to be separate), it cannot influence the first stage in any way, for the first is completed before the second is started. Indeed, there is no way in which the second stage could help the first, since it has no detailed knowledge of the mapping relation between the target and representational domains: it merely knows that the more links there are between the regions in a group, the more likely they are to belong to one and the same body, isolated links being quite possibly spurious. Virtually all of the relevant knowledge is (procedurally) possessed by the first stage, contained implicitly in the various linking rules. No part of the program has an explicit theory of 2D/3D mapping against which to measure local cues, as Roberts's program has.

But bottom-up functioning may be more or less intelligent, because the heuristics or algorithms involved may implicitly contain more or less knowledge. In discussing Colby's neurotic program we saw that his FINDANALOG routine (which searches for a person sharing a certain number of properties with another) to some extent embodies the psychoanalytic insight that like anxieties attach to like objects: if I can't admit to hating my father, I may be almost equally incapable of consciously hating my boss, so I take it out on Ms. Smith instead. We saw also that closer attention to the semantics of neurotic symbolism could have increased the power of the heuristic. If Guzman himself had had a systematic theory of 2D/3D mapping he might have built more procedural knowledge into the heuristics of SEE. In fact, however, he did not, and this accounts not only for most of SEE's mistakes, but also for the bewildering complexity and seeming arbitrariness of the linking procedures.

For instance, consider the basic linking rules and a few of the many additional "rules of thumb" employed by SEE that are shown in Figure 8.7. Of the eight types of vertex, four are interpreted locally, as furnishing links when looked at in isolation (FORK, ARROW, X, and PEAK); two provide contextual evidence for, or *against*, links at closely neighboring

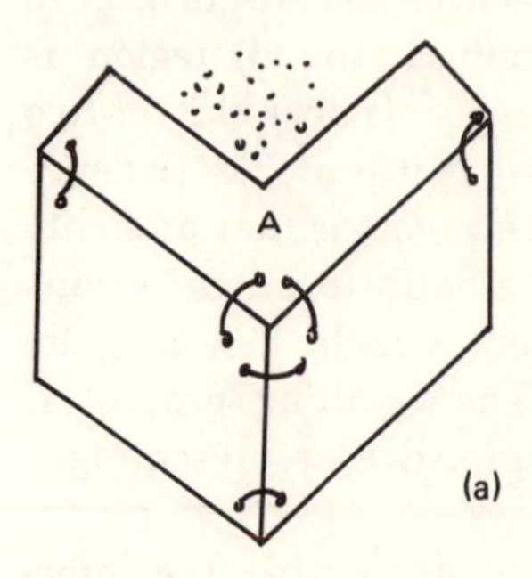


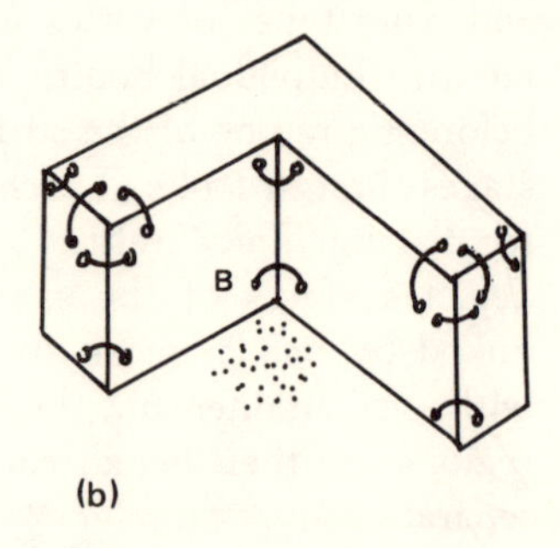


FIGURE 8.9

Concave bodies correctly linked by SEE. BACKGROUND would regard the L-vertex at A and the fork at B as evidence *against* the dotted regions' being background.

points (L and T); and two are not much used (K and MULTI). Sometimes linking depends not only on vertex type but also on other pictorial features, such as parallelism and collinearity (Figure 8.7,i–j). And sometimes it depends also on *target* features—such as background—posited by prior interpretation (Figure 8.7,j–k). In the latter case one might almost say that SEE has a tiny example of top-down functioning built into an apparently bottom-up heuristic: this mixture of picture and target features within a single heuristic occurred because Guzman, as we shall see in the next section, did not distinguish clearly and consistently between the representational and target domains.

Because Guzman had no overall rationale for the various linking procedures, he had no way of understanding why they are so often successful. Nor could he understand why particular mistakes were made, or why whole classes of bodies (those containing holes, for instance, like Figure 8.12) cause trouble for SEE.[13] In short, he had not clearly grasped why the numerous link-addition and link-inhibition rules (which modify the local evidence according to the neighboring context) are helpful yet not foolproof.

For example, Guzman initially suggested placing a link between the two regions on either side of an ARROW shaft (Figure 8.7,c), presumably because he intuitively sensed that an ARROW vertex often represents the outside (convex) corner of an object (see the four linking ARROWs in Figure 8.10,a). When he found that sometimes it does not (as in the *spurious* ARROW vertex of Figure 8.10,a), he added a link-inhibition rule to cope with the difficult case—but this heuristic in turn leads to other difficulties (Figure 8.10,b). Similarly, the 3-linked FORKed is often reliable, but provides only *one* genuine link for inside (concave)

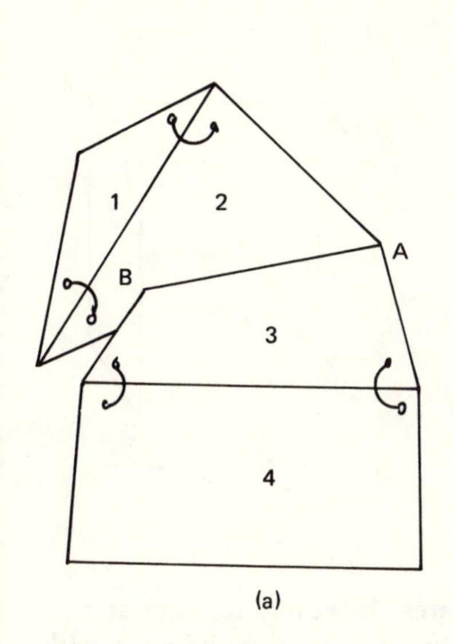


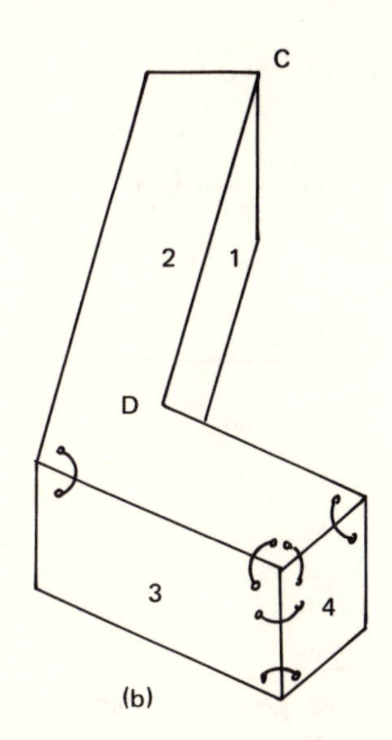


FIGURE 8.10

Both (a) and (b) show inhibition of arrow links (at A and C) by the arrow shaft's forming the leg of an L (at B and D). In (a) this is correct, the most natural interpretation of A being that it is a spurious vertex formed by the accidental alignment of two separate bodies (1:2 and 3:4). But in (b) the inhibition results in region 1 being separated off from 2:3:4 as a second body, whereas this is in fact a picture of one, concave, body. Only in concave bodies could the regions on either side of an L-leg depict surfaces belonging to the same body. Compare Figure 8.7 m and the L's in Figure 8.8.

corners where the base surface is invisible, like B in Figure 8.9. In general, Guzman developed his heuristic rules in an ad hoc and localized manner, based partly on his own unorganized intuitions and partly on continual trial and error with the program: he did not derive them systematically from an explicit 2D/3D representational theory taking account of the overall geometry of polyhedral bodies.

A picture interpreter that itself has recourse to such a theory has already been described, namely, Roberts's program. Indeed, Roberts's most informative cues—for which the program looks before taking account of any others—correspond directly to Guzman's most useful vertices. As Figure 8.1 illustrates, an ARROW vertex often corresponds to a point surrounded by three approved polygons, and a FORK to a line with an approved polygon on either side. This correspondence between Roberts's cue-searching heuristics and Guzman's linking rules, though unplanned, is not fortuitous: as we shall see in the next section, projec-

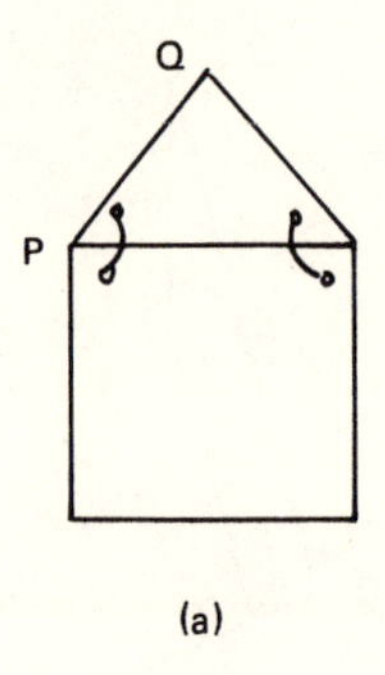


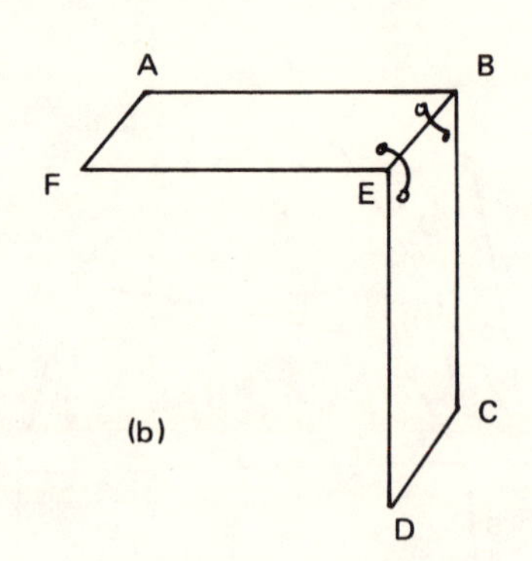


FIGURE 8.11

tive geometry can explain it. One hesitates, however, to suggest that Guzman's program should be improved by somehow providing it with the geometrical knowledge used so successfully by Roberts's program, because that program is in some ways more limited than SEE in its visual abilities. SEE can see polyhedral bodies *in general,* whereas the earlier program can see only bodies conforming to its specific models.

Is there any way in which one might preserve SEE's generality, but increase its understanding of what it is up to, thereby preventing mistakes like those just described or like SEE's "impossible" interpretation of Figure 8.11,b as a single solid body?

## MAKING SENSE OF PICTURES

Increasing SEE's understanding of what it is up to is a matter of giving it more knowledge about polyhedral bodies, and the way they appear in line drawings—but *what* more? There was a clue in the previous section, where reference was made to convex and concave corners, and Guzman's presumed intuitions concerning them. The target features known about by SEE, as was noted earlier, are very few: they do not include items such as convexity, concavity, corners, edges, visibility versus invisibility of surfaces, and "behindness" or occlusion. None of these concepts ever enters SEE's head. But if one asks oneself *why* Figure 8.11,a depicts a body, while Figure 8.11,b (with the same number of regions and links) does not, or *why* FORK and ARROW correspond to

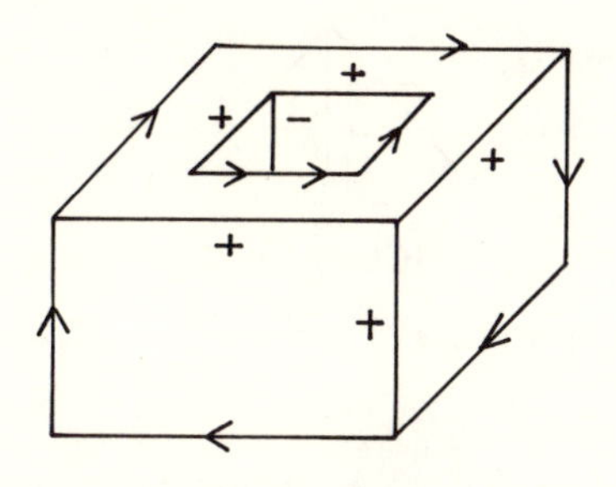


FIGURE 8.12

Source: ***Machine Intelligence 6*** (Edinburgh: Edinburgh University Press, 1971), p. 305. This and following figures from this source reprinted by permission.

Roberts's preferences for picture cues, the answer must be in terms of target features like these.

This was realized independently by two people puzzling over how SEE might be improved—and indeed why it works at all. Both D. A. Huffman and M. B. Clowes stressed two essential points: that a line or vertex considered in isolation may in principle represent one of several states of affairs in the (3D) scene, which can be specified in terms of the features just mentioned; and that neighboring picture fragments must be interpreted coherently, so as to "make sense" in these 3D terms.

Huffman, for instance, pointed out that there are exactly four possible interpretations of a line in pictures like those used by Guzman and Roberts.[14] It can depict either a convex or a concave edge with both associated body surfaces visible, or it can depict an occluding or "hiding" edge that obscures a surface to one or other side of it (*behind* it from the observer's point of view). Figure 8.12 shows examples of these types of edge, labeled respectively by "+," "−," and an arrow (the visible surface is always to the right of an ant crawling in the direction shown by the arrow). Obviously, a line cannot be labeled in different ways at its two ends, because a straight edge cannot be convex at one end and concave at the other, or occluding at one end and nonoccluding at the other.

Guzman's vertices themselves can depict only a few physical possibilities. That is, the lines comprising a vertex of a given type can be sensibly labeled only in a limited number of combinatorially possible ways. These correspond, in the case of FORK, ARROW, and L (and excluding spurious vertices such as the pseudo-ARROW at A in Figure 8.10,a), to corners of solid bodies. Once one realizes this, one can extract more information from a picture articulated in terms of vertices (of which there are now not only classes but subclasses), as well as understanding why they are informative in the first place.

For example, please look at Figure 8.13, which shows the three

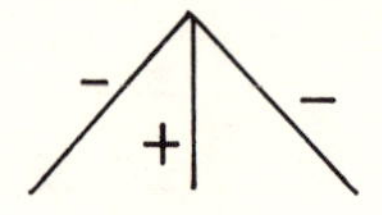

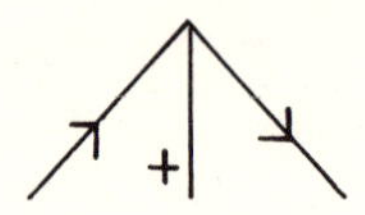


FIGURE 8.13

Source: *Machine Intelligence* 6, p. 304.

"meaningful" possibilities for an ARROW. If you try to see an ARROW in a way corresponding to different labelings (as a corner made up of three convex edges, for instance), you will find that you cannot. The reason for this—as introspection on your own attempt to see such an impossibility may suggest—is that the plane surfaces of a polyhedron can meet at a corner having three convex edges with both sides of each being visible *only* if the corner appears to the observer as a FORK, not an ARROW. Imagine (or try the experiment) how one corner of a cube changes its appearance as you move it up, down, or sideways and as you tilt it toward or away from your eye: at the very moment when you hope to see an ARROW with one "+" label become an ARROW with three "+" labels, the ARROW you have been so breathlessly watching changes into a FORK. (At the same moment, of course, the second of Roberts's preferred cues is replaced by the first: the shaft of the ARROW, which had an approved polygon on either side, suddenly "grows" at its tip a point which, being the central point of the newly arrived FORK, has three approved polygons around it.)

Figure 8.13 can show why (nonspurious) ARROWS are so reliable for planting links: in every possible case, the shaft is labeled as an edge with *both* adjacent surfaces visible, from which it follows that the two picture regions must belong to the same body—whether the shaft edge is convex or concave. Moreover, since Huffman gives *six* possible labelings for L vertices, it is not difficult to see why Guzman found them so unhelpful at the initial stage of local link placing (*cf*. Figure 8.7,a).

Huffman's approach shows why SEE was wrong to interpret Figure 8.11,b as a solid body, for there is no way of labeling the picture consistently. If a picture cannot be consistently labeled, it cannot possibly depict a real 3D object (although Huffman points out that the converse is not true).

There is a foolproof (algorithmic) pencil-and-paper method for arriving at *all* possible labelings of a picture, and therefore for showing what pictures are *im*possible to label. One tentatively assigns to a particular vertex (chosen at random or chosen from a class like ARROWs, which have relatively few possible interpretations) each of its legal labelings in turn, in each case working through successive vertices in a similar manner

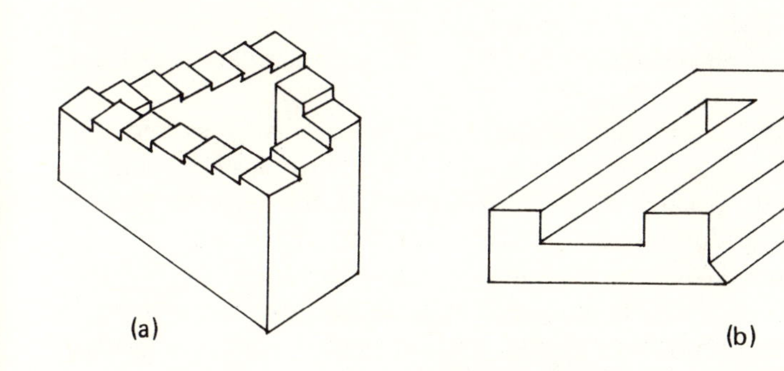


FIGURE 8.14

Source: (a) adapted from *Machine Intelligence* 6, p. 29; (b) from *Artificial Intelligence* 2 (North Holland, 1971), p. 105. Reprinted by permission.

but always bearing in mind the coherence restrictions on the labeling of lines joining neighboring vertices. When one reaches a vertex to which *none* of the legal interpretations can be given, because of coherence problems, one backtracks to the immediately preceding vertex, since the labeling previously assigned to that must be impossible in this context; if that labeling had seemed at the time to be the only possible one, it follows that there is no sensible interpretation of that vertex either, so backtracking must go one stage further. If backtracking reaches as far back as the initial vertex for every legal interpretation of it, then there can be no physical interpretation of the picture as a whole.

This procedure is of a "depth-first" type that is implemented in many game-playing and problem-solving programs, although Huffman himself did not express it as a computer program. It not only rejects Figure 8.11,b (which no one would take to represent a solid object, according to the particular interpretative scheme we are presupposing), but also both items in Figure 8.14, which at first sight appear to represent real things.

Clowes embodied insights essentially similar to Huffman's in a working program called "OBSCENE."[15] His method of mapping picture fragments onto features in the target scene was rather different, but he too subclassified Guzman's FORK, ARROW, and L in terms of the convexity, concavity, and visibility versus invisibility of the edges, corners, and surfaces that could make sense of these vertices. Since in addition he utilized T-vertices as cues, which arise from one body (or part of a body) passing behind another rather than from a genuine corner, OBSCENE was able to analyze pictures of scenes containing several partially occluded objects—a problem Huffman had not tackled. (Notice that the cuboid in Figure 8.3, which you earlier confidently assumed to be sitting *behind* the other two bodies, has no less than six Ts associated with it, the crossbar of the T always depicting an edge of the body in front. Since the crossbar is always an occluding edge, the set of Huffman labelings for Ts always assigns arrows to this line.)

By using vertices to address its knowledge of the types and compatibilities of corners, Clowes's program, like Huffman's labeling algorithm, finds all possible 3D interpretations of the relevant class of pictures (though it does not do so in a depth-first, but in a breadth-first way: this distinction will be explained in Chapter 12). And OBSCENE has a deep enough understanding of solid bodies not to be fooled by some pictures of "impossible objects," and can specify precisely *why* it is not so fooled. For instance, it rejects the "Devil's pitchfork" of Figure 8.14,b, identifying the nature of the pictorial contradiction involved. However, it cannot reject the "never-ending staircase" of Figure 8.14,a, because this would require additional concepts—like height in 3D space.

The details of Clowes's 2D/3D mapping are less important here than the fact that he made an explicit and systematic distinction between the target and representational domains, which he called *scene* and *picture*, respectively. Correlatively, he distinguished clearly between picture features (such as regions, lines, and vertices) and scene features (such as surfaces, edges, and corners).

Guzman did not do this; for instance, he used the terms "surface" and "region" interchangeably, and employed the term "scene" indifferently to refer to both the target scene and the representational picture. This doubtless contributed to the previously noted muddle, in which the background (scene surface or picture region?) was included as a preinterpreted feature within an interpreting heuristic. Of course, there is nothing against letting one's interpretation of a picture fragment (or a word in a sentence) depend on prior interpretation: one of the advantages of top-down processing is that it enables one to do precisely this. But if one does this, one should be clear that one is doing it, if one is not to be confused about the nature of the interpretative process. (OBSCENE does not do this, although it of course employs compatibility restrictions while considering simultaneously all legal interpretations of vertices. Like SEE, OBSCENE functions bottom-up from the picture rather than top-down from interpretative models, and its superiority over SEE lies in the relatively deep, systematic 2D/3D knowledge implicitly contained within its cue-interpreting procedures.)

Another consequence of Guzman's unclarity about the two interpretative domains is his attempt to prevent SEE's being made a fool of by defining "illegal scenes" (*sic*) and "impossible scenes" in terms of purely *pictorial* considerations. Thus Guzman instructed SEE to reject as illegal all pictures in which lines end "nowhere," or do not form part of the boundary of a closed region; accordingly, SEE rejects the pictures in Figure 8.15, as "impossible." However, SEE does not understand, nor did

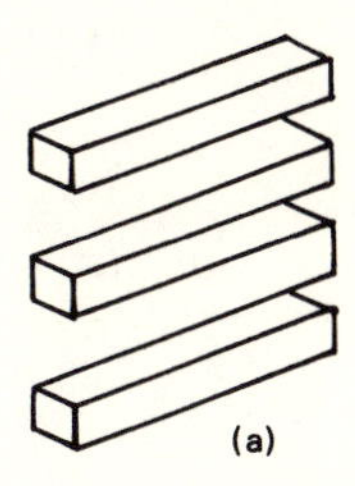
(a)

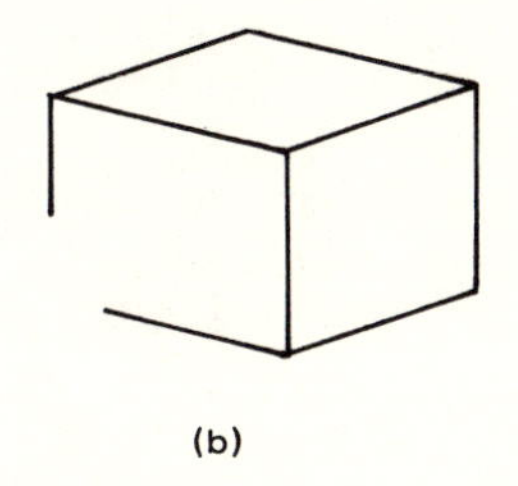
(b)

FIGURE 8.15

Source: Guzman, *Computer Recognition of Three-Dimensional Objects in a Visual Scene*, p. 198.

Guzman clearly state, just *why* they are anomalous (as opposed to what it is in them that is illegal). And SEE is unprotected from the traps offered by Figures 8.14 and 8.11,b, which are perfectly acceptable in pictorial (though not physical) terms.

SEE's acceptance of pictures depicting impossible objects is comparable to the translation of "Our village policeman is a good sport" into a (syntactically perfect) French sentence implying that he is a recreational organization: like the Devil's pitchfork, this just does not make sense. And just as one provides a semantic theory to a machine translator to save it from such absurdities, so one must provide knowledge of the 3D properties of bodies to enable a visual program to make sense of its world. The more powerful the semantics, the better the comprehension of sentences in natural language; similarly, the more extensive the knowledge of 3D space, the more intelligent the vision of things seen within it. OBSCENE sees what it sees better than SEE does (for instance, it can infer the existence of some hidden surfaces and edges), and is not so subject to "stupid" visual illusions, because it knows more. Indeed, what OBSCENE knows explains much of what SEE knows, it explains why the unorganized spatial intuitions that Guzman incorporated in his linking rules are so often reliable.

A recent scene analysis program has a still more extensive knowledge of 2D projections of 3D objects built into it, and consequently sees even better than OBSCENE does. A. K. Mackworth's "POLY"[16] is not limited (as OBSCENE is) to pictures in which no more than three lines meet at a vertex, and so it can deal with PEAK or MULTI configurations. It can reject pictures (like Figure 8.16) that are impossible because of *skewed* surfaces, which neither OBSCENE nor Huffman's basic labeling rules can. And it has the potential to exploit information about the tilt of surfaces (relative to the picture plane) so that the tilt of other surfaces can be immediately inferred. Furthermore, POLY has to consider fewer

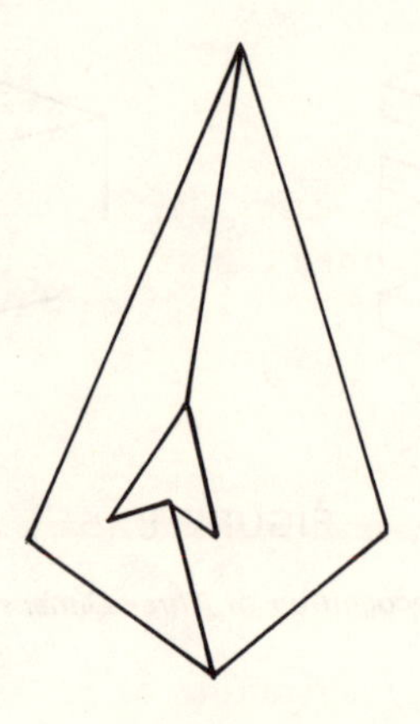

FIGURE 8.16

Source: From *Artificial Intelligence*, 4 (North Holland, 1973), p. 134. Reprinted by permission.

alternatives in deriving all possible interpretations of a picture, basically because it concentrates on lines rather than vertices (a line being less prolifically ambiguous than a vertex).

POLY uses a precise formalism for representing the relative slopes of two surfaces meeting at an edge, and employs this to guide the interpretation of picture lines as depicting edges, or picture vertices as depicting corners where three surfaces meet. By including explicit constraints relating relative slopes to visibility or invisibility of surfaces, POLY articulates intuitively obvious general facts that explain the equally obvious facts about lines and vertices exploited by Huffman and Clowes. In constructing such programs, the achievements of the programmer are: first, finding a precise, explicit way of formulating facts that we normally use unconsciously, and second, devising heuristics and procedures for activating and deploying that knowledge when it is relevant. Some of the facts and heuristics expressed in one program may explain facts and heuristics in another.

POLY's representational theory shows why, in the cube-watching experiment I earlier recommended to you, the ARROW *must* give way to a FORK at the crucial moment. The necessity here is not specifically human, since it derives from the geometrical optics of the method of viewing concerned: even our friend HAL, were he to try it, would find the "all convex" ARROW similarly elusive.

Mention of HAL, strangely enough, brings us back to the real world, for it raises the question of how HAL would fare, were he to be provided with the visual faculties described so far. The answer is that for all practical purposes he would be as blind as a bat.

The programs I have described can see only polyhedra—and only POLY can see all of those. (Even POLY would explode combinatorially if given a complex polyhedral picture, since it makes too much use of undiscriminating depth-first search.) If you show a picture of cubes to SEE, OBSCENE, or POLY, they can see how many there are but cannot tell that they are *cubes.* All four programs can cope only in perfect lighting conditions, where all lines represent edges and all nonoccluded edges are depicted by a line. Indeed, only Roberts's system can start off in a real-world situation at all, instead of being restricted to the contemplation of line drawings, or even of lists of abstract line descriptions. And Roberts's system (like the others) is limited to situations where there are no shadows.

None of the four has any notion of physics. To be sure, Roberts's interpreter can see the cuboid of Figure 8.3 as sitting on (supported by) the ground plane. But this is a geometrical matter of position, rather than a matter of physical support. Roberts's program knows nothing of physical properties like solidity and stability, for example, which are important even in simple practical situations like stubbing one's toe or building with children's bricks. Nor do these programs take account of movement (either in the target domain or on their own account), for they work from a static picture of the world with no sense of how the world might have got that way or what might happen to it next. (This is not quite true of Roberts's program, which can imagine how the image would change according to a change in viewpoint—but it has no appreciation, as MARGIE has, that movement requires a cause.) And how, one may ask, could these programs enable HAL to see that his fellow astronauts intend to betray him—which the fictional HAL is able to do?

The virtual equivalence in the two previous paragraphs of the word "see" and cognitive expressions like "tell," "cope," "know," and "take account of" is significant, for it underlines the fact we have already noted: that knowledge is an integral part of sight. The reference to MARGIE suggests that the understanding of pictures, like the interpretation of language, involves general semantic knowledge as well as expertise in specific domains. We saw in Chapter 3 that a program competent to manipulate symbols like *love* and *hate* in a superficially rational manner may in fact have very little understanding of what love and hate actually are, and how they connect with other concepts. Similarly, a program like OBSCENE, which apparently can tell that one thing is *behind* another, really has rather little appreciation of what "behindness" is: there is more to seeing behindness than just counting T-vertices and inferring that there must be some hidden edges. OBSCENE does not know, for example, that if you start walking toward two objects you will have to get past the one in front before you can reach the one behind. Someone who had only OBSCENE's understanding of behindness would therefore be totally mystified as to *why* one commonly uses the word "behind" to mean "later in time," or even "having learnt less."

It follows that enabling HAL to see, no less than enabling him to speak, would involve providing him with knowledge of many other matters besides the projective geometry we have discussed. Even R. P. Abelson's theory, which might seem to be wholly irrelevant to vision, would be necessary if visual cues were to lead HAL to suspect an alliance between the astronauts aimed at betraying him. The indispensability of rich background knowledge for the apparently effortless achievement of plainly seeing a pikestaff should become increasingly evident as one considers increasingly "realistic" programs. Let us look, then, at some aspects of the fearsome computational *rites de passage* leading from the toy world of cubes and perfect lighting to the real world of shadows, curves, faces, and movement.

# 9

# Glimpses of the Real World

OPERA GLASSES, microscopes, and ordinary spectacles undeniably have the effect of making the invisible clearly visible. But instruments like these would be no help to the purblind HAL, to whom the real world (even the polyhedral parts of it) is almost totally invisible, given only the visual programs discussed so far. If anything, they would make matters worse, by adding new sources of mystification and confusion. HAL's problem is not lack of visual acuity, but incapacity to deal with the visual information he receives in the context of real life.

Real-world vision differs from the "toy" vision so far described in two basic ways: it sees what it sees despite great imperfections in the input; and it sees more things, irrespective of the quality of the input. Each of these differences rests on the wider range of knowledge possessed by the visual system, and its more intelligent deployment in interpreting the visual cues discerned by the eyes and mind.[1]

## HALLUCINATION IN THE SERVICE OF TRUTH

Sunshine and electric lighting can be cruel not only to ageing film stars, but also to "perfectionist" visual programs like those so far described. A photograph (or a retinal image) of a brightly lit white object on a white background, for example, will probably not contain light-intensity gradients corresponding to every edge. Even if it does, some of these will be so faint as to be physically indistinguishable from other intensity gradients in the same picture that do not correspond to genuine edges. This is true even if there are no shadows in the scene. In consequence, a line finder working on a photograph of a cuboid may return a picture with crucial lines missing—like Figures 8.11,b and 8.15,b.

These two pictures were earlier described as having no possible interpretation *with respect to the particular interpretative scheme we were then presupposing*. But you yourself may have had to look twice at Figure 8.15,b, since at first sight it may have appeared to you to be a "straightforward" picture of a cuboid. Clearly, with respect to an interpretative scheme that knows about cuboids and can "hallucinate" missing lines in the picture accordingly, this picture is *not* impossible to interpret. Even Figure 8.11,b might be a cuboid viewed in highly abnormal lighting conditions—and if you think of it in this way you may (whether voluntarily or not) project ghostly images of the two missing edge-lines onto it. The fact that people often see incomplete pictures like these as complete, at least on first viewing, suggests that the distinction between hallucination and veridical perception may not be so clear as is often assumed. Even voluntarily aroused visual images are not always phenomenologically distinguishable from genuine visual perceptions, as was shown long ago by the psychologist C. W. Perky.[2]

Individuals probably differ in the extent to which they can project "convincing" images onto visual arrays, as opposed to being able intellectually to see that a certain pattern *could* represent a particular target object. For instance, please look at Figure 9.1: your knowledge of the world presumably enables you to interpret it as a picture of a soldier walking through a door followed by his dog. But whether you can actually *see* the soldier, even if you try, is less predictable. (This picture, like any other, could be many things besides: how many alternative interpretations can you justify? And how does your interpretative activity here compare with your response to Y. A. Wilks's claim that there is never any "fail-safe" interpretation of a sentence, since with sufficient ingenuity one can always imagine a surrounding context that would force an alternative meaning?) In any event, the important point is that the phenomenological distinction between your seeing *that* a property is depicted by a

FIGURE 9.1

picture, and your seeing *it* as depicted by the picture is not a sharp one. This intuition tallies with theoretical discussions of vision that take a constructivist approach. For instance, experimental psychologists such as R. L. Gregory, and workers in machine vision such as M. B. Clowes, have pointed out the *necessity* of perception's going beyond the evidence presented to the senses.[3]

Given that visual programs functioning in the real world are not going to be able to rely on perfect input like that assumed throughout the previous chapter, they should be able to share your ability to interpret incomplete pictures of cuboids as, nevertheless, *pictures of cuboids.* And much as you automatically complete (at least some) defective pictures, so that you do not even realize that they are defective, so a program should be able intelligently to hallucinate lines (and other pictorial properties) within the representational domain if it is not to be continually plunged into a state of doubt arising from *explicit* recognition that the picture is "imperfect." (Analogously, "Is it or isn't it?" may be an appropriate reaction to the joke picture in Figure 9.1, but would be an absurd waste of mental energy if applied to every slightly imperfect alphabetic character printed on this page.) In this section, then, I shall concentrate not on the "voluntary" hallucination that arises *after* the overall interpretation of the picture has been decided, but on "involuntary" visual hallucination functioning as an integral part of the interpretative process.

L. G. Roberts's program knows enough about cuboids to add the missing edges *after* interpretation in cases of occlusion (see Figure 8.3), and we have seen that he suggested a line proposer and line verifier that could help in finding faint edge-lines without simultaneously overburdening the system by returning *all* the faint lines. As it stands, however, Roberts's program could not see either of the incomplete pictures in Figures 8.11,b and 8.15,b as cuboids. But Roberts's models already predict and advise the interpreter to search for unseen cues that may corroborate the interpretative hypotheses already activated. So there should be no difficulty of principle in enabling a program sensibly to see something in the picture that isn't there at all—even if this runs the risk that it may sometimes mistakenly, and regrettably, see something in the target domain that isn't there either. (Have you never regrettably missed a misprint, because you understood the text too well? Only in proofreading, of course, does one actually want to see the misprints.)

Such a program already exists, and may be thought of as a hybrid of Roberts's and Adolfo Guzman's systems. (It follows that the improvements to SEE elaborated by D. A. Huffman, M. B. Clowes, and A. K. Mackworth could in principle be incorporated into a program of this general type.) Gilbert Falk's "INTERPRET" was developed as part of the Stanford hand-eye project.[4] Its purpose is to construct an exact 3D map of the relevant scenes (polyhedral building blocks placed on a table), making the best of the inconvenient fact that the line finder may have missed out some edge-lines.

INTERPRET has six modules, which function sequentially: first, SEGMENT uses Guzman-type vertices to find the individual bodies in the scene, making allowance for the fact that lines may be missing (and therefore one region may depict two different surfaces). SUPPORT then decides, for each individual body, which (if any) other bodies might be supporting it at its base surface, on the assumption that bodies are *never* suspended in mid-air but are always either on the ground or on some other body.

Next, COMPLETE fills in "obviously" missing lines so that the next module, RECOGNIZE, can do its job more easily. RECOGNIZE (which does not insist on an absolutely perfect line drawing but cannot tolerate gross imperfections) matches up each separate body with one of the nine fixed-size models or "prototypes" it knows about and expects to see, and also (via knowledge of the picture-taking process) locates the bodies in space. PREDICT then "imaginatively" draws a perfect picture of the scene posited by RECOGNIZE, deleting hidden lines appropriately.

Finally, VERIFY takes over and checks whether the predicted picture that "ought" to fit the original input actually does so; if it does not, RECOGNIZE is recalled and asked to try again. (VERIFY is not so pedantic as to insist on an exact match; consequently, sloppy line drawings can be corrected, or cleaned up, by substituting the "imaginary" perfect prediction for the original picture, provided that the discrepancy between the two is not too great.) Since VERIFY cannot ask any module preceding RECOGNIZE to think again, mistakes made at the initial stages cannot be rectified.

The hallucinatory powers of COMPLETE would enable Falk's system to recognize Figure 8.15,b as a cuboid. For COMPLETE has three ways of seeing things in a picture that are not actually in it: JOIN, ADDCORNER, and ADDLINE.

The first of these three subroutines joins up gaps between collinear line fragments that appear (on the basis of the previous segmentation) to depict an edge of a body face. The second—which is the one that could cope with Figure 8.15,b—looks for a face with two lines ending "nowhere" that would join at a corner if extended. And the third looks for evidence that an entire line has been missed, and fills one in accordingly. Since the requisite evidence is parallel-sided L-vertices (between the center points of which a new line is filled in), ADDLINE would not be able to prompt RECOGNIZE to see Figure 8.11,b as a cuboid; nor could ADDCORNER do so, since it needs two dangling lines to work from. This conservatism of COMPLETE, whereby it adds only the most obvious missing lines, is related to the previously noted fact that VERIFY cannot recall modules preceding RECOGNIZE when INTERPRET makes a mistake. Even so, COMPLETE is sometimes overhasty and adds "meaningless" lines to the picture; access to OBSCENE or POLY could, of course, prevent this.

In terms of the previous distinction between "voluntary" or interpretation-*driven* imagination, and "involuntary" or interpretation-*driving* hallucination, one might want to classify PREDICT as being in the former category and COMPLETE as in the latter. However, this would not be quite correct, since the dependence of VERIFY on PREDICT shows that the system's final interpretation of the picture depends in part on the activity of PREDICT. That is, Falk's PREDICT contributes to the interpretative process, unlike Roberts's picture-drawing procedure, which takes over only *after* the interpreter has decided what bodies are present in the scene. But it is clear that *more* interpretation precedes the activation of PREDICT than precedes the functioning of COMPLETE, since PREDICT's imaginative powers are driven by the specific prototype hypotheses produced by RECOGNIZE, whereas COMPLETE relies only on the general polyhedral knowledge implicit in the Guzman-like SEGMENT. One can make these distinctions clearly here (and in the example involving INTERPRET that will be mentioned in Chapter 14) because the control structure of the program is evident—an advantage that is not available in our introspective musings on our own visual abilities.

PREDICT's mistakes are corrigible (by way of VERIFY recalling RECOGNIZE), but COMPLETE's are not. Corrigible anticipation directing attention to the relevant class of cues may be thought of as the key to "top-down" understanding, whether visual, grammatical, semantic, political, or whatever. Hypothetical constructions that go beyond the current evidence are not only useful but harmless, provided that they can be corrected when they are mistaken. But hallucination that (as in the case of COMPLETE) is not open to any possibility of correction is dangerous. Accordingly, INTERPRET would be improved by a more flexible flow of control whereby COMPLETE could be made to think again, on the basis of insights provided by higher levels of the system.

Even so, confirmatory or disconfirmatory evidence would not always be available (the picture might be "a real mess" at the relevant points), and in these situations an intelligent system has to commit itself to a particular perceptual construction and run the risk of being mistaken. We have already seen a similar interpretative process at work in the case of understanding stories, where one sensibly assumes the storyteller to have said that a dime came out of a piggy bank unless specifically informed to the contrary. In short, even a potentially corrigible COMPLETE would be required to hallucinate bravely on occasion.

A visual system's perceptual acuity could in effect be increased with the help of INTERPRET, without thereby running the risk that the system would be unable to see the wood for the trees. For if VERIFY were to fail to find a PREDICT-ed line in the picture initially provided by the line finder, it could call a more sensitive line finder to look again at the scene (or the grey-scale photograph used before), seeking much fainter edge-lines in the appropriate positions only. But of course, the system would be aided only with respect to instances of the prototypes specifically anticipated by RECOGNIZE. What would be more generally useful would be a program functioning as "selective spectacles" for lines having significance with respect to a much wider target domain.

A program that aids line-finding in relation to convex polyhedra *in general* has been implemented by Yoshiaki Shirai, and used by the M.I.T. robot.[5] Shirai's line finder does what seems the intelligent thing to do when in difficulty: it uses a partial interpretation based on the most obvious data to guide it in interpreting the rest. As the interpretation proceeds, it becomes progressively built into the line-finding itself, so that there is no clear distinction between line-finding and higher level recognition except at the very earliest stage of the process. Consequently, the final "line drawing" is not an uninterpreted 2D representation, like the line drawings considered in the previous chapter, but already carries its 3D interpretation with it. The program is hallucinatory in the weak sense that it continually proposes the most plausible lines according to context, and actively searches for them by peculiarly appropriate methods (which may or may not involve using a more sensitive line-finding technique to capture the faint lines). It is also strongly hallucinatory, in that it sometimes extends a line across a gap, with no verification from the input except the fact that the extension "makes sense" by connecting with a previously located line.

The most obvious data, in the particular lighting conditions implicitly assumed by Shirai's program, are usually the "contour" lines, that is, the lines depicting body edges separating the (white) body from the (black) background. The next most obvious intensity gradients are the "boundary" lines, which separate one body from another; and the least clear are the "internal" lines, which separate one face of a body from another face of the same body. The line finder looks first for contour, then for boundary, and last for internal lines, using context-sensitive heuristics which embody general knowledge about bodies but whose specific application may depend on earlier decisions of a more global character. (If the lighting conditions were varied, this strategy would have to be altered: but the lines likely to be the most conspicuous should still be sought first. Deciding between strategies, which requires foreknowledge of the likely effects of different lighting conditions, would have to be done by the program itself were it to be let loose in the real world.)

To find the contour lines, Shirai's program scans the picture for points of high light-intensity contrast, and connects them into lines that are provisionally taken as contour lines. Or, rather, the program in effect scans a small-scale version of the picture for these contour points; instead of examining each one of the 100,000 points in the original picture, it samples one in every 64 ($8 \times 8$ array). Using the approximate contours found in the sampling situation, the program then returns to the original picture and looks in the appropriate places (which it did not know before the sampling procedure) for a more accurate contour map. We shall see in Chapter 12 that this is a simple example of a generally useful strategy in problem solving: the use of an oversimplified plan of the problem to guide one in seeking a solution that, when finally achieved, will specify matters to a greater degree of detail than does the plan itself.

Clearly, this contour-finding procedure depends on there being reliable light-intensity contrasts between the objects and the background. But it is crucial to realize that Shirai's program does not approach the most obvious lines first merely in order to have achieved something—as one might attack a pile of neglected correspondence by getting the easily answered letters out of the way first, so that one's morale may be high enough to wrestle with the others. The program uses its initially acquired knowledge of contour lines to provide cues suggesting where to look for boundary lines, cues that otherwise would not have been available. Similarly, boundary lines aid in seeking internal lines.

For example, in Figure 9.2 the contour is found first, by way of light-intensity measures. Concave points (such as B) on it are then noted. The reason why concave points are useful cues is that, if no concave bodies are allowed, they must depict places where two bodies (convex polyhedra) meet. It follows that the lines contributing to a concave point can be interpreted as boundary lines of two different objects. Since contour lines depict body edges, they are necessarily also boundary lines, or segments thereof. Accordingly, lines AB and PB are tracked from point B to see if either of them continues: AB does, terminating at C. (Notice that this has given us a T-junction, with the usual interpretative significance.) Having found the boundary line ABC, the program then uses a circular search around point C hoping to find an adjacent boundary line

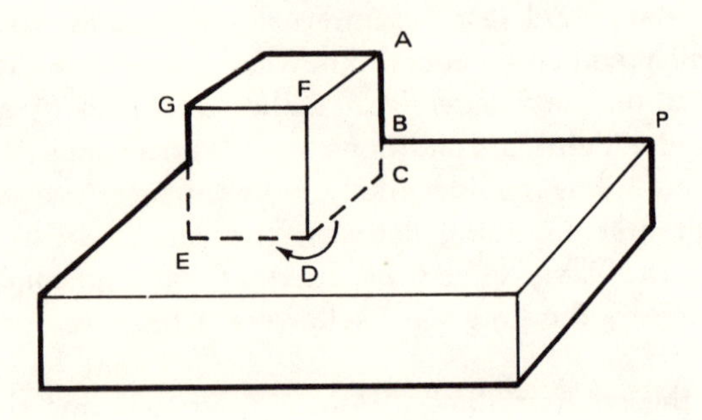


FIGURE 9.2

connecting with it. It does *not* know whether to start searching to right or left of C. Fortunately, this does not matter. It finds line CD, immediately interpreting this as a boundary line.

Now the line finder starts a second circular search, originating this time at point D, and seeking a further boundary line connecting with ABCD. But this time it directs the search in a specific manner, namely, in the direction of the arrow shown. The reason for this is that by now the program has provisionally interpreted ABCD as part of the boundary of a body; since only convex polyhedra are allowed, the continuation at D will therefore be found by searching round the "outside" of the corner at D. If the circular search from D were to proceed in the opposite direction it would find a line, but this would be an internal, not a boundary, line. Having found boundary line DE in this way, the program is now in a position to say to itself, in effect, "I wonder if CDE is part of an ARROW vertex with the shaft missing?"; accordingly, it looks *inside* the corner at D for a line, and if it finds one interprets it as an internal line (you will remember that Huffman's labels show the shaft of all nonspurious ARROWS to be what Shirai calls an "internal" line). DF is found, and joined up later (at a FORK) with internal lines GF and AF, which spring from the discontinuities in the contour at G and A.

This example has mentioned only three of the ten heuristics in Shirai's program. But it should have shown both how they incorporate general 2D/3D knowledge, and how their particular application (like circular search from the unconnected end of a boundary line) may depend on previous interpretation, so that the program becomes increasingly canny as processing continues.

Like Roberts's four cue-searching heuristics, Shirai's ten heuristics are ordered with respect to their likelihood of success in finding useful cue lines (where the ordering depends on Shirai's knowledge of 2D/3D geometry). And just as their specific application varies with the interpretation evolved so far, so their results are continually tested for interpretative consistency with previous results, so that the program is less likely than it otherwise would be to be confused by imperfections in the input.

For instance, whenever a line fragment is found (that is, a line whose termination appears *not* to be a vertex, since no connecting lines can be found by circular search around the terminal point), the program checks to see whether it could be extended to connect with another fragment already found, or whether it could be a continuation of an existing line fragment. (This is comparable in intent to Falk's module COMPLETE, but is more general in application.) In such a case the line finder can if necessary be made more sensitive, the threshold of "significant" intensity gradients being set at a lower level than usual. Similarly, circular search around a point can be made both more sensitive and more wide-ranging in area. Consequently, the program is likely to find the significant lines it hypothesizes, even if they are very faint.

Moreover, Shirai's line finder can ignore irrelevant lines, providing that they do not confuse the contour—as shadows often do but reflections typically do not (see Figures 9.3 and 9.4). It recognizes them as irrelevant because they do not make sense as representations of edges in the 3D scene posited by the interpretation as a whole. INTERPRET, by contrast, can cope with input that has some lines missing, but cannot intelligently ignore any extra lines in the input. This achievement of Shirai's program is analogous to one's ability to ignore the background noises of bottles and breakages when listening to conversation at a party. Information with irrelevancies in it, some of which may obscure evidence that might have been useful, is therefore generally called "noisy" information, whether it is acoustic in nature or not. "Noise," like its complement "cue," is an intentional concept that can be defined only by reference to the particular notion of interpretation, or significance, assumed. To PARRY, for example, every word in the input that does not match one of the expected patterns is merely verbal noise, and as such is ignored by him (although we saw in Chapter 5 that if a high proportion of the input is unrecognizable he is more cautious in his interpretation of the rest, as one usually is in very noisy conditions).

One may be able to compensate for visual noise by referring only to a very low level of interpretation, as Shirai's line finder shows. But just as party conversation is easier to hear if one knows what the topic of conversation is as well as what phonetic cues to expect, so vision in real-world contexts may require access to higher levels of interpretation than are employed by Shirai's program, if noise is to be effectively screened out at the line-finding level. An intelligent line finder written by G. R. Grape, in connection with the Stanford hand-eye project, utilizes abstract models of the expected visual themes (convex polyhedra again) to help it make sense of the chaotic input it receives.[6]

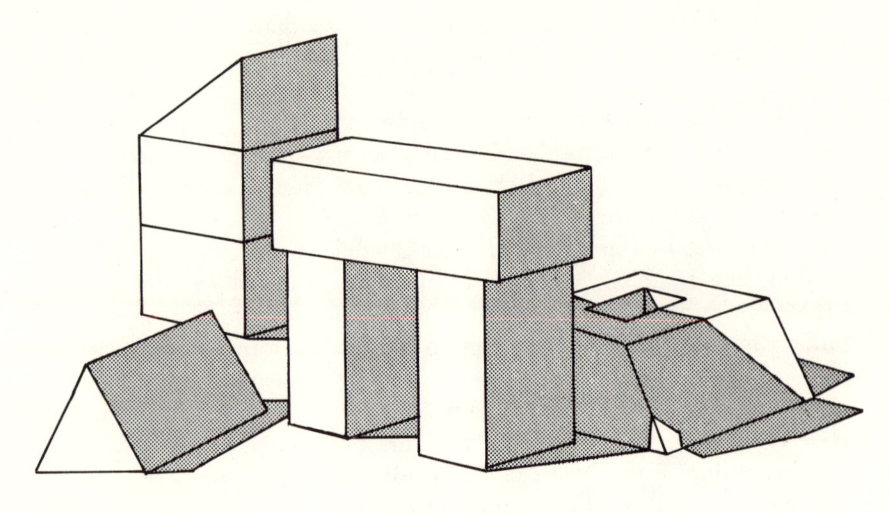

FIGURE 9.3

Source: P. H. Winston, ed., *The Psychology of Computer Vision* (New York: McGraw-Hill, copyright © 1975), Figure 2-1, p. 20. This and following figures from this source reprinted by permission of McGraw-Hill Book Company.

FIGURE 9.4

Source: Adapted from Winston, ed., *The Psychology of Computer Vision*, Figure 2-2, p. 20.

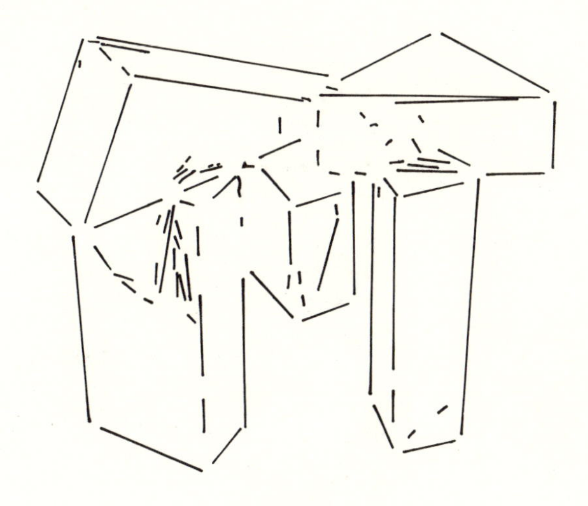

FIGURE 9.5

Grape's system starts off with a digitized TV image, as does Shirai's. The messiness of the data is indicated by the initial line drawing of Figure 9.5, which is produced on the basis of light-intensity gradients in the TV picture. INTERPRET, clearly, would be quite at a loss here. And even Shirai's line finder would get hopelessly confused: what, for instance, would it make of the muddle of little lines in Figure 9.5 at the position corresponding to the top nearside corner of the larger cuboid shown in Figure 9.6?

One might even ask what *you* made of Figure 9.5, before you looked at Figure 9.6. You probably saw some polyhedral bodies in it, but you may have been puzzled by the muddle of lines just mentioned. (It is only fair to say that, judging by the few TV images reproduced in grey-scale by Grape, you would probably have been less puzzled by the TV picture of this scene than by Figure 9.5.) In view of the previous discussion of visual interpretation, the hypothesis is irresistible that you were able to recognize cuboids and wedges in Figure 9.5 at least partly because you knew already what these things look like, and in the context of this text you were specifically expecting to see pictures of polyhedra. Irresistible or not, this hypothesis explains the performance of Grape's system, which converts the relative chaos of Figure 9.5 by stages into the clean line drawing of Figure 9.6. (Grape does not bother to eliminate hidden lines; but, at the cost of added computation, this could readily be done by adapting the line-deleting procedures embodied in Roberts's or Falk's program.) Comparison of the two line drawings suggests that the final one owes a good deal to intelligent hallucination triggered by the fragmentary evidence contained in the first.

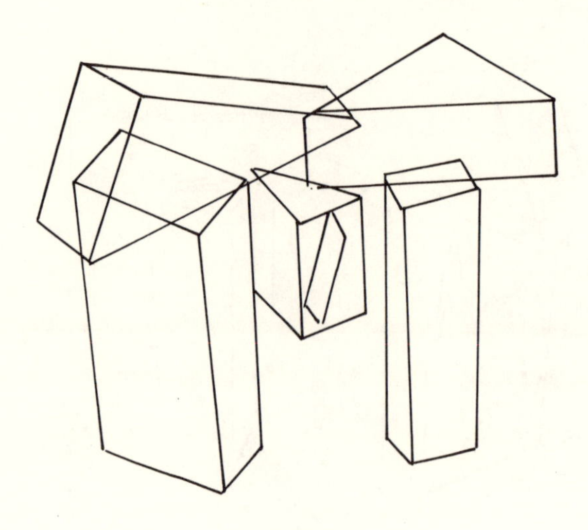

FIGURE 9.6

The reason why Shirai's line finder would not be able to cope with Grape's scenes is that its decisions about which lines connect with which are based on local, rather than global, context. Although these decisions do depend on heuristics embodying general knowledge about polyhedral bodies, no particular bodies are searched for. For instance, Shirai's program looks for a third line at the intersection of two internal lines, to make a FORK (*cf*. vertex F in Figure 9.2), and then follows the line with no particular expectation about what it will find. Suppose it ends "nowhere": the program, in desperation, initiates a circular search around the end of the "dangling" line, and increases the radius of the circle until it finds something. The idea of looking *elsewhere* in the picture for a specific type of help does not occur to it (although some global knowledge is of course already contained in its distinction between contour, boundary, and internal lines, and it could connect the line with a collinear fragment that had *already* been found).

However, as Figure 9.7 shows, the more imperfect the input is, the less appropriate a local strategy for line-finding becomes. The two collections of line fragments inside the circles would most likely be interpreted by Shirai's program as the (imperfect) boundary of a triangular region, and as an (imperfect) L-vertex. In fact, as *only* inspection of the whole picture can show, neither of these interpretations is correct in this case, although they might be in some other picture.

Grape's program actively searches the data for evidence of lines that, however ambiguous it may be in the local context, makes clear sense according to more global considerations. It correctly interprets the two lines in the right-hand circle of Figure 9.7 as contributing to *different*

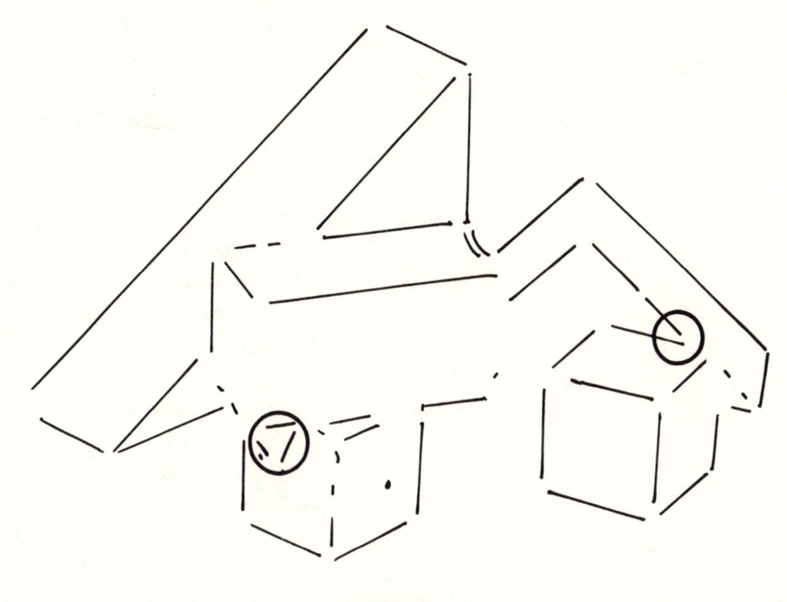

FIGURE 9.7

Source: Figures 9.5, 9.6, 9.7, from G. R. Grape, ***Computer Vision Through Sequential Abstractions***, Stanford AI Memo, Stanford University, 1969. Reprinted by permission.

vertices, being edges of different bodies; and it rightly categorizes as noise all but the top line in the left-hand circle. It does this by matching its abstract models, or prototypes, of cuboids and wedges to the line diagram, searching first for conspicuous features acting (directly or indirectly) as cues to the prototype concerned.

Each of the basic cue features is a line with a particular vertex at each end of it, while compound features are built up out of two simpler ones. The compound features activate prototypes directly, but the basic ones do not: basic features merely suggest the presence of polyhedral bodies, whereas compound cues suggest specific types of polyhedra (wedges, cubes, etc.) In other words, there is a hierarchy of interpretative levels, with the intermediate ones being neutral as regards any particular model; and hallucinatory line-finding can be guided by insights drawn even from the highest level of all. For example, consider the base contour of the right-hand cube in Figure 9.7: tentative joining up of lines whose end points are close to one another gives an L-vertex, connected to an ARROW, connected to another L, where the barbs of the ARROW are legs of the two Ls. The hallucinations involved are guided by the program's need to find a line structure that can be described in terms of the basic features. In this case two basic features are found: each is an L connected to an ARROW. But these two are connected by way of the barbs of the ARROW, giving a compound feature that at first sight strongly suggests a cuboidal body. Following up this hunch, by way of the top-down guidance provided by the cuboidal prototype, all of the remaining visible cube vertices are found—and the circled line set is disambiguated by the way.

First impressions are sometimes misleading, of course. But this need not matter if the system has a way of recognizing and recovering from its mistakes. The advantage of having several interpretative levels is that information gained at an intermediate level can be retained even if the "high level" hunch turns out to be wrong. For instance, the compound feature described above (L-ARROW-L) might *not* turn out to be part of a cube, but alternatively could be part of a wedge (see the base of the central wedge in Figure 9.6). If the prototype-driven cuboidal hypotheses are not confirmed, the information that there is an L-ARROW-L feature is retained, and some other polyhedron is searched for that is consistent with this evidence, namely, a wedge. Sometimes, very *low* level hunches may have to be rejected; for instance, an overhasty identification of the circled lines in Figure 9.7 as constituting an L-vertex. In a case like this, the radical misunderstanding can be corrected only by way of knowledge functioning at higher levels, and there is no useful intermediate level interpretation to be saved.

Continual reference to the prototypes in deciding whether or not to hallucinate two line fragments as being connected (in a genuine vertex or a continuous edge-line) means that high levels of interpretation are repeatedly interacting with lower levels. There is thus a clear contrast with FALK's INTERPRET, whose modules SEGMENT and COMPLETE commit themselves to a list of vertices and line connections with no help whatever from the prototypes known about by RECOGNIZE, and with no hope of correcting any mistaken first impressions. As we have seen, the rigidly sequential flow of control from lower to higher levels of interpretation in Falk's system forces him to make COMPLETE more conservative in its hallucinatory activities than Grape's program is. Conservatism in hallucination is not necessarily a virtue: seeing pink elephants is universally frowned upon only because there are never any pink elephants actually there. Not until workers in artificial intelligence tried to derive line drawings from real-world data did anyone realize quite how messy "raw" visual input is: for one's intelligent assignment of clear outlines to things in the field of view is normally immediate, and not open to introspection.

When one *is* introspectively aware of actively trying to sort out genuine outlines (as in looking at Figures 9.5 and 9.7), one often "imagines" the more obvious bodies to be out of the way while one puzzles over the rest. Similarly, Grape's program successively eliminates from its internal representation of the original picture, all the evidence that makes global sense in terms of its cue-prototype matching criteria. So, by the time that the more obscure parts of the picture are looked at carefully, much of the confusing detail has already been removed.

The final output of the program is not just a clean drawing, like Figure 9.6, but also a separate picture of the scene showing only those lines that the program cannot relate to any model. For example, most of the lines in the previously noted "muddle" on the left of Figure 9.5 are reproduced at the final stage as being uninterpretable by the system, and presumably of no significance in terms of the relevant target domain. It is natural (to us) to categorize these as visual noise, and to assume them to have been caused by shadows, reflections, flies trapped in the camera, and so on. But Grape's program knows nothing of noise and its causes: it just leaves *some* of the data unaccounted for.

Just as humans may sometimes mistake a shadow, or spots before the eyes, for real things in the world, so Grape's program sometimes wrongly finds a body outline in data that in fact arise from some other cause. For instance, the trapezoidal body depicted in the center of Figure 9.6 is a misinterpretation of a shadow at the corresponding position in Figure 9.5. However, the fact that hallucination may sometimes lead one away from the truth is not a good reason for outlawing it. Only in a world that consistently offered perfect visual data would hallucination of missing lines be totally out of place in the everyday functioning of visual systems. And predictive hallucination, or what I have called hallucination in the weak sense, is of course necessary for intelligent interpretation even of perfect data. Knowledgeable hallucination, in short, is an essential part of sight.

## MUFFLING THE COMBINATORIAL EXPLOSION

I would be prepared to wager a tidy sum that you would finish a jigsaw puzzle showing a horse and cart more quickly than one showing a proton accelerator. But if you happen to be an experimental nuclear physicist, I might lose my money. Your extra knowledge of what to expect, and what not to expect, would help you in solving the puzzle. So, provided that one can use it in a disciplined fashion, extra knowledge may speed thought up rather than slowing it down.

Critics of artificial intelligence commonly cite the "combinatorial explosion" as potentially destructive of any really complex and interesting programming project.[7] As Chapter 12 will show in more detail, exhaustive search through all logical possibilities becomes too time-consuming even for the fastest computer when only a few extra dimensions of variability are added to a manageable problem. The number of possible combinations of differing values of the various dimensions "explodes" to an astronomical quantity, so that examination of all of them is quite out of the question. If workers in artificial intelligence cannot defuse this time bomb satisfactorily, there is no chance of producing a HAL— whether in the year 2001 or any other. Insofar as R. P. Abelson's "themes" and "scripts" enable one to ignore inessential details, they are relevant to this problem. And some work in vision is relevant also, in that it shows how additional knowledge can be used to cut down, rather than to increase, the amount of computation involved in visual interpretation. Having more cues to consider, in short, does not necessarily lead one helplessly into a combinatorial minefield.

None of the visual programs yet mentioned treat shadows as cues. Indeed, with the exception of Grape's system, which to some extent can avoid being misled by them, none can tolerate shadows at all. If the lighting is not carefully "fixed," they become unable to see things that otherwise would be clearly visible to them. Apparently feeling shadows to be a regrettable nuisance, Richard Orban wrote a Guzman-like "trial-and-error" program to identify shadows in pictures, so that they could then be blithely ignored by the programs concerned with the main business of seeing.[8] But had he never been to a horror movie, where the menacing approach of some grisly creature is betrayed by its shadow falling on the wall?

Melodrama aside, shadows can often *help* one to see, as well as being useful cues for telling the time. For instance, they can often tell whether or not two objects (one of which is behind the other) touch, information that is not available in shadowless pictures like Figure 8.3. They can show whether an object actually is supported by the ground plane or some other object (as invariably assumed by INTERPRET), or whether on the contrary it is suspended in mid-air. Assuming the illumination, whether sunlight or artificial, to be above the scene rather than below it, they can serve to orientate the whole picture in terms of *up* and *down*, thus enabling one to tell whether a gymnast is hanging from a bar or doing a handstand on top of it. And they are partly responsible for people's finding Grape's TV images less puzzling than Figure 9.5. Consequently, visual programs ought to be able to utilize shadows as cues, instead of ignoring them as visual noise or fondly trusting that they will not occur.

This principle was followed by D. L. Waltz, whose program can not only give a 3D description of the objects in shadowy scenes like that (perfectly) depicted in Figure 9.3, but can also recognize the very different picture in Figure 9.4 as representing the same scene.[9] This ability to perceive essential similarity despite superficial diversity is comparable to the recognition of *West Side Story* as "really" a twentieth-century version of *Romeo and Juliet.* In discussing the possible implementation of Abelson's theory of themes and scripts, we noted that one of his greatest difficulties would be to state explicitly what details at lower structural levels are relevant to particular instances of questions expressed at the higher levels. That is to say, lower level matters must be identified that can act as cues to higher level schemata, whether only in certain circumstances or in all. This is the role played by Roberts's references to approved polygons and by Clowes's references to vertices of various types. Waltz based his own work on the Huffman-Clowes paradigm, and his program offers an example of an even richer 2D/3D interpretative system than theirs. Since it is guided by geometrical insights of the type formalized in POLY, the cue-interpretation mapping relation is much more reliable and systematic than would be any "*Romeo and Juliet*–recognizer" based on Abelson's semantic theory.

Lines in pictures of shadowy scenes may depict not only edges but also shadow boundaries (with the shadowed region being to one side or the other). Two "shadow labels" can thus be added to the Huffman list of four edge-types. Waltz actually used eleven distinct labels for lines. The extra five depict two sorts of "cracks," or *flat* separable edges at which two bodies could be pulled apart (like the radiating lines on top of a presliced cake), and three sorts of *concave* separable edges, where one body could be lifted off the one supporting it (as in Figure 9.2). Separable edges had already been worked on by Martin Rattner: his program "SEEMORE," as the name suggests, was SEE with knobs on—and so was open to the objections already noted.[10] Waltz's program, by contrast, has power comparable to that of OBSCENE within a more richly specified target domain. It can use its label cues (even if a few lines are missing) to infer a description of the scene in terms of concepts like behind, in front of, rests against, shadows, is shadowed by, is turned away from the light, is supported by, is capable of supporting, leans on, and others. (As by now you will expect, its understanding of these concepts is much less rich than ours.)

The purely combinatorial possibilities for vertex labelings, given eleven labels to play with, are enormous: PEAK, for instance, allows of *six million* possibilities. PEAK was not much used by Guzman, but Figures 9.3 and 9.4 show that it is a very common vertex in Waltz's scenes, because a base corner that would otherwise appear as an ARROW may sprout a shadow-line turning it into a PEAK. However, just as knowledge of 3D geometry eliminates the "three-convex" ARROW, so knowledge of the principles of shadowing and illumination allows pruning of the Waltz permutations (PEAK, believe it or not, reduces to *ten*). Waltz describes at some length both how he did this and how he used the relevant knowledge to help him choose particular labels as fruitful cues in the first place. His detailed discussion of these matters illustrates how cues must be identified and assessed by reference to a particular interpretative system.

Nevertheless, despite the impressive reduction of the purely combinatorial possibilities that was achieved by Waltz's understanding of the physical realities involved, the sets of *actual* possibilities for vertices are still large. PEAK is unusual in having only 10 sensible interpretations: FORK and T each have 500; MULTI, X, and K each have 100; and ARROW and L have about 75 apiece.

Consequently, Waltz was unable to search through *all* the possibilities at once, whether in a depth-first way (like Huffman) or breadth-first (like Clowes). To be sure, some of the meaningful possibilities can be immediately excluded for a given picture on the basis of evidence present in the initial input to the program: for instance, if the region on one side of a line is clearly brighter than the other, then the line can be labeled as a shadow-line in only one way (although of course it may not be a shadow-line at all). Waltz made the most of information like this in his initial, tentative label assignments. But even so, the remaining possibility sets of the vertices in any particular picture are so large as to make systematic consideration of all physical possibilities impracticable *even* for a fast-thinking and capacious computer of the type he had available. One of the most significant features of Waltz's program is the way in which he contained this combinatorial explosion within manageable limits.

Waltz uses the analogy of a jigsaw puzzle to explain his strategy. No one with any sense would attempt to do a jigsaw puzzle by systematically trying to fit every piece with every other, watching out for failures-of-fit on the way. Nor would one try to reach an immediate decision as to what the overall design will turn out to be. Rather, one looks for local constraints first (blue pieces with red knobs on fit into blue pieces with red "holes" in them) and builds up from there.

Similarly, Waltz searches for a 3D interpretation of the picture in two stages. The first is a "filtering" stage whereby a pair of neighboring vertices is inspected for mutual consistency of line labels: just as an edge-line cannot be convex at one end and concave at the other, so a shadow-line cannot have the shadow on different sides at either end. (The pair of vertices may be selected at random; but we shall see shortly that certain subclasses of vertices are more useful because they have fewer interpretative possibilities and so are more choosy about their neighbors.) Given that the second vertex, in this instance, is present in the context of the first, some of its initial (independent) physical possibilities are therefore eliminated. The same procedure is now repeated between the already pruned vertex 2 and vertex 3—and so on, once round the whole picture, the adjacent-neighbor constraints being continuously propagated until no more local inconsistencies remain. (Guzman's link-inhibition rules also took account of local context; but quite apart from being less reliable in any case, they were not used to propagate inhibitions through successive vertices in this way.)

Waltz reports his amazement on finding that this simple iterative (repetitive) procedure, when applied to the first simple scenes he experimented with, found a unique label for every line. Even in complicated scenes, many lines receive a unique label at this stage. And those that do not, receive such a small assignment of possible labels that the second, global, stage of interpretation can proceed in a familiar systematic manner without being impossibly bogged down. (Also, of course, the points of ambiguity—and the precise nature of the ambiguity—have been quickly identified, so that the system can now concentrate entirely on them.)

We have already seen, with reference to the programs of Roberts, Shirai, and Grape, that it may be helpful deliberately to seek the most useful cues first. This is what one does with a jigsaw puzzle, when one starts by sorting out all the edges. This strategy not only gets some of the pieces out of the way, but provides a general framework that tells one something about the overall structure of the picture (the red patch is in the lower left-hand corner, and only the middle of the top border has open sky). Waltz employs a similar strategy to cut down the computation that has to be done by his filtering program. Although this program *can* use any two adjacent vertices, those that represent the boundary of the scene with the background are the most useful. The reason for this is that only about one-tenth of the physically possible vertices can occur on this boundary, whereas all of them may be found inside the scene. For example, only two of the ten PEAK interpretations make sense on the scene-background boundary.

Consequently, Waltz's program locates this boundary *before* attempting any labeling. This can often be achieved very simply, either by finding all the regions that touch the edge of the field of view and running them together, or by finding the contour that has the property that every junction lies on or inside it. (Because of the presence of insubstantial shadows in the scene, many parts of this contour would not be "contour lines" in Shirai's sense. Nor could his light-contrast method be used for finding the scene-background contour, since its assumption that all bodies are white and the background is black applies neither to shadow-background contours nor to lines separating white body surfaces from a white ground.)

All the junctions lying on the scene-background boundary are labeled first, using the filtering procedure between adjacent junctions described above. Next, the program labels the junctions that bound regions that share an edge or junction with the background, since these will be more constrained by their "background" neighbors than by their internal neighbors. Finally, the more central junctions are labeled. This is analogous to finding jigsaw edges (which normally *must* have at least one straight side), and building the puzzle onto them until one reaches the middle. (It is also analogous to the fact remarked by the Gestalt psychologists, that people need longer to see "inner" details in a picture than they do to see the figure-ground boundary.)

As a result of his filtering procedure, preferably supplemented by the "boundary first" strategy just described, Waltz is able enormously to reduce the computation involved in recognizing the scene shown in the picture. Indeed, instead of increasing geometrically with the number of lines (as it would for a traditional "depth-first" or "breadth-first" search), the time needed to interpret the picture is roughly proportional to the number of lines in it.

Waltz's achievement here has relevance for artificial intelligence in general. We have seen in earlier chapters that as soon as a program promises to get really interesting it raises the spectre of combinatorial problems. For the more a program knows, the greater is the potential search space involved when it asks, or is asked, a question. (The concept of "search space" will be clarified in Chapter 12.) But think of a jigsaw again: the really difficult jigsaws are those that have *few* shapes for the pieces, not many; or that have only a *few* colors, concentrated in different areas of the picture. A puzzle of this type is irritating rather than entertaining, because it reduces one to the stupid procedure of systematic search; normally, one has to fall back on this only for the sky. In short, intelligent use of the varied constraints between the many different features of the puzzle helps one to finish it quickly. At first, the relevant features are simple colors and shapes; but soon one is in a position to hypothesize "That must be a horse—so where are the ears?" and to direct one's search top-down accordingly.

The deeper and richer one's understanding of the way in which various target features can co-occur to make sense in the target domain, the more potential there is for utilizing this knowledge in identifying and interpreting cues in the representational domain. A corollary is that rich domain-specific knowledge is essential to intelligence, artificial or not, although this does not preclude the necessity for general principles of thinking (like depth-first search, or the classification of problem bugs to be described in Chapter 10) that are useful in many domains. In Waltz's case, for example, his knowledge of geometry and illumination determined his choice of cues, his distinction between sensible and meaningless vertex interpretations, his simple filtering procedure, and his strategy of labeling the scene-background boundary first. All of these contributed to the interpretative constraints without which the program would have been hopelessly bogged down in endless computation.

FIGURE 9.8

Source: ***Proc. AISB Summer Conference, July 1974.*** University of Sussex. (Society for the Study of Artificial Intelligence and Simulation of Behavior), p. 246.

The necessary computation varies with different domains, since they require different cues and/or interpretative "jigsaw" constraints. For instance, if one allows *curved* objects into the target domain, one may find new cues—such as radius of curvature—to be useful, as well as having to relax the constraints that a line cannot be interpreted as convex at one end and concave at the other, or as being illuminated in different ways at opposite ends. Huffman discussed briefly how his labeling scheme might be adapted to pictures of smooth curved surfaces like saddles (which are curved from back to front and from side to side), and to folds of cloth like those in a full skirt or an Elizabethan ruff. And K. J. Turner has recently programmed an extension of Waltz's system, which (starting either from a noisy TV image or from a perfect hand-drawn line diagram) can interpret pictures of shadowy scenes containing both curved and plane objects, as in Figure 9.8.[11]

Turner's basic approach was to use polyhedral knowledge to generate knowledge about the intersections of curves and planes, by letting lines tend to curves. For example, you will see from Figure 9.9 that a 3-positive FORK (depicting the corner of a cuboid with all the surround-

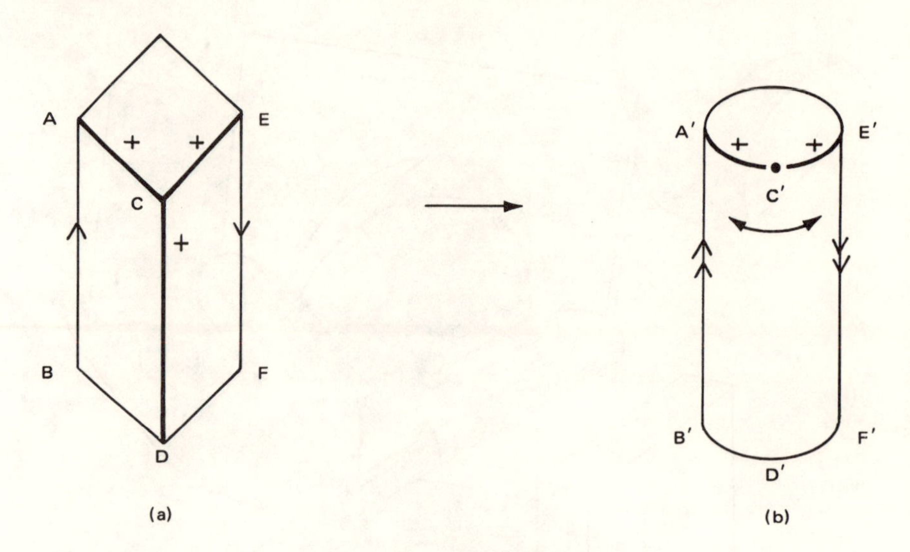


FIGURE 9.9

As the cuboid is transformed into a cylinder, the actual junction at C becomes the notional junction at C′. Line CD disappears. Lines AB and EF, which are both genuine edge-lines depicting physical discontinuities in the cuboid, are transformed into lines A′B′ and E′F′, which (as the double arrow labels mark) do *not* correspond to "real" edges in the cylinder, but merely to the visible outline of it.

ing faces visible) can be transformed into a 2-positive curve with a *notional* junction in the middle, the stem of the fork having got lost as the cuboid is transformed into a cylinder with no vertical edges. In its analysis of the picture into parts, therefore, Turner's program has not only to identify and label vertices composed of lines that are explicitly (objectively) present in the picture, but has also to "see" junctions that are not there as such.

Since the physical nature and illumination of a curved edge can each vary at different points along the edge (for example, on either side of a notional junction), Turner had to replace the straightforward Huffman-Waltz "same-at-each-end" constraints by transition rules specifying how particular illumination and edge labels can sensibly transform into others along a given line. Largely because of the complexity of these rules, compared with Waltz's more straightforward filtering rules, Turner's program takes about four times as long as Waltz's to interpret pictures of comparable complexity. A second reason for this increase in computation is the relatively large set of possible labelings for each vertex: since the component lines may be straight *or* curved, and if curved may be convex *or* concave, each type of vertex can have more variants than in purely plane-surfaced domains.

Sometimes computations one assumed to be necessary, or at least helpful, turn out not to contribute valuably to interpretation after all. For instance, Turner found that illumination cues were not very useful in dealing with the curved objects he considered, since ignoring them entirely gave rise to only a few ambiguities. Superfluous computation could in these cases be avoided, therefore, by not bothering with illumination labels. But quite apart from the fact that this neglect of illumination would not be so intelligent if the program were to be presented with one of Waltz's wholly *planar* scenes, so that it should be made to depend on initial judgments (by the line finder?) about the general nature of the lines in the picture, it might be inappropriate for some *curved* surfaces also. Turner considered only smoothly curved surfaces, without bumps, wrinkles, and indentations. Irregularly curved surfaces would be more difficult to cope with, and illumination might come into its own again as a highly significant cue. Other cues also would be relevant; for instance, Huffman remarked that the line patterns around picture points that are to be interpreted in terms of a "crumpled paper" target domain are constrained by the fact that the angles in the scene that are represented at the picture point must sum to 360°.

A program that could see not only polyhedra but also paper (and pikestaffs, and piggy banks, and people . . .) would have to possess the requisite knowledge (including jigsaw heuristics) for each class of objects in the target domain *and* be able to activate it selectively in appropriate circumstances if it was not to be overcome by the combinatorial explosion. Writing such a program would involve gaining general insights into the representation and use of knowledge that would be useful to Abelson, in his attempts to prevent the IDEOLOGY MACHINE from foolishly approaching "Berlin Wall" questions by inquiring into the chemical composition of the bricks. For just as illumination cues are sometimes useful and sometimes not, or as contour lines are sometimes conspicuous and sometimes not, so knowledge of chemistry is sometimes helpful and sometimes not. Indeed, visual programs of the perceptual power fictionally attributed to HAL would embody general principles of intelligence that are relevant to complex knowledge of all sorts. However, the next section will show how little of the path from polyhedra to people has been laid down: the problem still is how to formulate the relevant knowledge, never mind how to find it on realizing that it is needed.

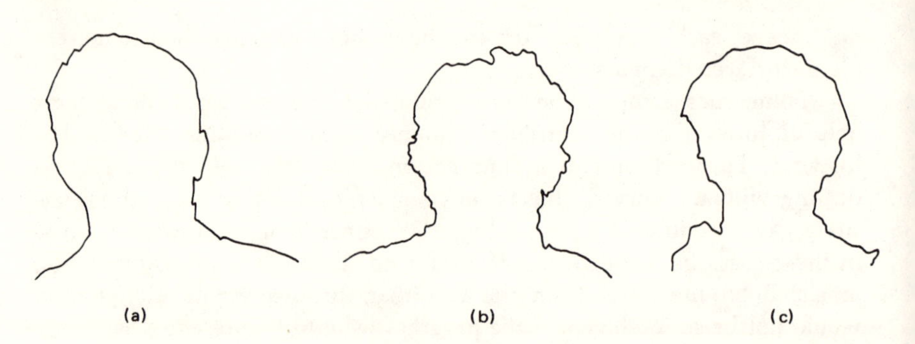


FIGURE 9.10

Source: Adapted from *Machine Intelligence 6*, eds. Bernard Meltzer and Donald Michie. (Edinburgh: Edinburgh University Press, 1971), pp. 397-410. Reprinted by permission.

## OF MEN AND MOVEMENT

Humpty Dumpty complained to Alice that he wouldn't know her again if they were to meet:

> "You're so exactly like other people. . . . Your face is the same as everybody has—the two eyes, so—" (marking their places in the air with his thumb) "nose in the middle, mouth under. It's always the same. Now if you had the two eyes on the same side of the nose, for instance—or the mouth at the top—that would be *some* help."

He obviously assumed that one could recognize eyes, noses, and mouths without any implicit reference to their usual relations to one another. Grape's work on locally ambiguous evidence, such as the circled patterns in Figure 9.7, implies that this is not so. And a person-recognizing program written by M. D. Kelly[12] suggests that HAL, no less than Humpty Dumpty, would have to rely on his global knowledge of what people usually look like, if he were to be able to tell Alice and astronauts apart.

Kelly's program is a line finder specially sensitive to the outlines of people's heads. Its task is to extract an accurate outline of a person's head from an ordinary grey-scale photograph of the person standing in front of various backgrounds. The output of the program is a line drawing of the edge of the head alone—like what you would produce if you were asked to trace the outline of a head in a snapshot, ignoring both the edges of background objects against which the head is viewed and the details of the face (see Figure 9.10).

Like Shirai, Kelly uses a simple form of "planning" (a concept to be further discussed in Chapter 12). First, a smaller, less detailed, photograph is produced by physical reduction of the original. Next, the desired edges are located as lines in the small picture. And finally, these are used as a plan for finding more nicely specified lines in the original picture. If there is a straight line between points A and B in the plan, the program concludes that there must be an approximately straight line between points A′ and B′ in the original photograph. The plan-following line finder therefore searches only a relatively narrow band between A′ and B′, so that false trails are quickly recognized as such. For instance, if the person is standing in front of a door, the line depicting the edge of the door—which may in objective terms be continuous with a head-line—will not be followed, or not for long, by the program intending to trace only the head. The program thus backtracks appropriately if it is temporarily led onto a false trail, but is able to preserve smaller irregularities, like the contour of the hair.

You may already have guessed, as Humpty Dumpty presumably would not, that significant line-finding in this task of tracing head outlines depends upon prior assumptions about the usual shape of human heads. Thus the outline of a lion's or a unicorn's head would not be found by Kelly's program, which (unlike you) does not have any alternative head models that it can match up with the input. Nor would the program find the devil's slender horns—and you, too, might fail to notice these if the current embodiment of the devil (or as much of it as was shown in the photograph) lacked the traditional "key" features of goatee beard, cloven hoofs, and forked tail. In short, this is a case of hallucination again, where what is seen is what was expected: Kelly's program does not function in an epistemological vacuum, but knows something about what it is looking for, namely, human heads.

Like the polyhedral knowledge in Shirai's line finder, this knowledge is procedurally embedded in the program, which includes various heuristics implicitly defining an acceptable head shape. For instance, if at the stage of plan production the program finds a vertical line in the small-scale photograph among a row of horizontal lines that have already been interpreted as the top of the head, this line is rejected as noise. And vertical lines in the equivalent position in the original photograph are only accepted as genuine head (hair) edges if they turn back toward the top of the head *before* passing out of the narrow band currently being searched under the guidance of the plan. It is the combination of these two heuristics that would prevent recognition of the horned Mephistopheles, should the program be so unfortunate as to encounter him.

The first cues that Kelly's program looks for (in producing the plan from the small-scale photograph) are three short line segments that might be part of the top and sides of the head. Just as the spatial relationships between the cue features in Grape's pictures have to be "reasonable" in terms of the polyhedral prototypes, so for Kelly's line finder the top of the head must be *above* the sides. (Consequently it could not

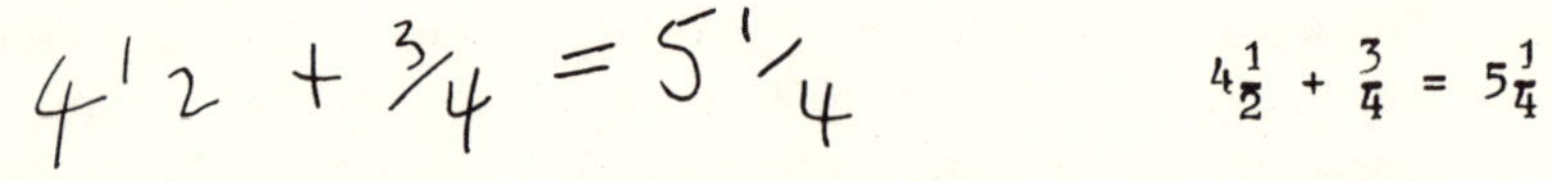

$4^{1}2 + {}^{3}/_{4} = 5^{1}/_{4}$

(a)

$4\frac{1}{2} + \frac{3}{4} = 5\frac{1}{4}$

(b)

FIGURE 9.11

recognize circus artistes hanging upside down from the trapeze.) The program tries to connect these line fragments together to make a "head" shape, by searching between them for a roughly semicircular set of lines depicting the crown of the head. Once the top half of the head has been found, the program searches below it for inward curves (depicting the neck) followed by outward curves (toward the shoulders). This often achieves good results like those shown in Figure 9.10—but Alice's flowing locks would confuse the program, unless they were swept back so as not to obscure the outline of her neck and shoulders.

Clearly, Humpty Dumpty's thumb-in-air sketch of the relative positions of eyes, nose, and mouth could fairly easily be communicated (as additional heuristics) to the machine, to enable it to search for lines depicting these target features also. But one could not improve Kelly's program by telling it, "You'll easily recognize Alice: she has almond-shaped eyes, a snub nose, and a pretty rosebud mouth." Nor could one generalize its visual abilities by describing the appearance of the houses, doors, and trees shown in the backgrounds of the pictures it looks at. The reason for this is that there is as yet no general language for describing picture structure that is powerful enough to communicate the relevant pictorial concepts to the machine.

The general problem of describing picture structure was discussed by M. L. Minsky in the early 1960s, and since then has formed the focus of much of the work on machine vision.[13] We saw in Chapter 8 that the symbolic description of new picture cues (such as Guzman's various vertices) may require new concepts, as well as new combinations of old concepts. But providing even such apparently straightforward concepts as *above*, or *to the left of*, to the machine is not an easy matter. The normally unsuspected complexity of concepts like *above* becomes apparent as soon as one tries to provide a computational definition of them, whether descriptive or procedural in form.

For example, in Figure 9.10 it is of no significance whether or not any part of a line that is interpreted as the crown of the head (and which must therefore be *above* the sides) is *above* the side-of-head lines in the sense that some part of these would intersect if it extended upwards; correlatively, a picture is not a "better" picture of a head because some parts of the top *are* "above" the sides in this sense. By contrast, the fact that in Figure 9.11a, the "1" is not *directly above* the "2" constitutes prima facie evidence against interpreting these two figures as meaning "the fraction, one half" rather than "the integer, twelve." Only a program (or a person) with mathematical knowledge could correctly interpret *both* this sloppily handwritten fraction *and* its superior typed equivalent, by recognizing that the handwritten "1" should in this context be regarded as *above* the "2" rather than to the left of it. If you place your hand over all but the first three figures in the handwritten formula, you will be more likely to interpret these as meaning "four hundred and twelve" than "four and one half." This is closely comparable to the interpretation of the circled lines in Figure 9.7 as a triangle and an L-vertex, when they are seen *without* reference to the global context.

One might argue that highly abstract descriptions of pictures, such as those cited so far, are very unlike our usual ways of describing visual features. For we commonly describe appearances in a more concrete way, by reference to familiar examples: "almond-shaped" eyes and "rosebud" mouth are descriptions of this type. Whether one could convey the appearance of a cat to anyone who had never actually seen a mammal of any type is doubtful. Indeed, we shall find in the next chapter that some workers in scene analysis are trying to teach visual programs perceptual concepts by way of showing them concrete examples, rather than relying on abstract descriptions referring to shape alone. For instance, the concept "door" can be introduced to the program either in terms of a definition like "rectangular object with long axis vertical," or by showing it a (photograph of a) real door. The program represents the concept of door by means of two interrelated data structures, "semantic" and "iconic"; the iconic structure stores all the visual information picked up by the program from the sample photograph (including color, brightness, and so on), and can help the program to search for door*like* things if necessary. (Although the semantic data structure may be thought of as relatively "abstract" and the iconic as relatively "concrete," *both* involve symbolic descriptions of the door itself. All learning by example requires the storage of symbolic representations and the matching of new instances against these representations, as we shall see in Chapter 10, so in this sense the abstract-concrete distinction breaks down.) However, this approach also cannot escape problems of local ambiguity, and so needs to know how to apply knowledge of the global context.

It follows that if one were to program a general purpose line finder, which could find the significant lines in pictures of polyhedra *and* people, on the basis of descriptions of their visual appearance, one would have to provide two things. First, one would have to supply (or help the program to learn) computational definitions, whether "abstract" or "concrete," of many different picture-description concepts (ARROW, semicircle, above, snub, rosebud-shaped) in terms of which to tell the program what to look for. Second, one would have to back up these descriptions of pictorial structure by domain-specific knowledge restricting the allowable range and application of the visual concepts in particular cases. For instance, in handwritten maths (*directly*) *above* can look very like *to the left of*; and since two real rosebuds on the same tree can be seen to be above each other but two mouths (in the same face) cannot, a pictorial "rosebud" may or may not sensibly have another one above it in a picture.

Some recent work of Guzman's illustrates the necessity of developing new picture-description concepts and formulating new domain-specific knowledge, if HAL is to be able to spy on the astronauts as they go about their tasks.[14] Guzman gives a theoretical discussion of the problems involved in interpreting line drawings of everyday scenes like those found in children's painting books. Like Grape, he points out that pictorial evidence may be locally ambiguous (as indeed his own ARROW and other vertices are), and that hierarchical sets of features must be specified that suggest interpretation in terms of one conceptual model rather than another. Thus a line configuration that might depict a hand, *or* a horse's tail, *or* a girl's hair, could be disambiguated by seeing whether or not it connects with something that could be an arm. (One's knowledge of anatomy would probably be *consciously* set to work in understanding a photograph of two circus contortionists performing their act.) Guzman takes the abstract rather than concrete approach to picture description; an abstract description of "something that could be a hand, or a tail, or hair" would have to represent the relevant kind of wiggly line in a way that distinguishes it from "something that could be a flower or a puddle." Description of the picture in terms of straight lines and polyhedral vertices is obviously inadequate, and ways of describing curved line configurations (more varied than those considered by Turner) would have to be found before pictures like these could be analyzed into the relevant cues.

However, whether one relies purely on abstract definitions or resorts also to concrete pictorial examples, there are insuperable difficulties involved in specifying *spatially* characterized models of classes of everyday objects that can enable a program to recognize instances of those objects. This is partly because shape can be extremely variable, but also because other (nonvisual) factors enter into our concepts of many things. Consider, for instance, what we understand by "hat." Guzman tried to define *hat* for the purposes of interpreting children's pictures: although his abstract spatial definition of *hat* would indeed enable a visual program to recognize indefinitely many hats, he found it necessary to include a specific provision for a dent in order that the model should mediate recognition of a Homburg as well as of a bowler. Not surprisingly, then, there are many "normal" hats that would not be accepted as instances of the model, because their shape is so different. And if one passes on to fancy-dress hats, or the millinery confections of Royal Ascot or Broadway's

Easter Parade, one loses any possibility of classing these as hats in terms of shape alone.

Someone might wish to object here that many of these are not really hats at all, or that they are hats *only* on the relevant ceremonial occasions. These reservations depend not only on knowledge about the usual shape of hats, but also on knowledge of their typical and occasional functions. Hats shaped like full-rigged galleons, which cost the earth and which have to be constantly held to stop them falling off, may attract the attention of a crowd but admittedly are not useful as protective head-coverings. However, since the *normal* functions of hats include decorative and symbolic aspects, it seems somewhat arbitrary to insist that these extravagant objects are not "really" hats.

Even so mundane an object as a hat, therefore, cannot be visually recognized *as a hat* on all occasions without the implicit use of pictorial cues to access a great deal of "nonvisual" knowledge, including sociological knowledge about the institutions and annual festivals of a particular country. Correlatively, closer attention to the visual image may show that the woman is not wearing a hat at all: rather, she is a shop-assistant selling model galleons, one of which has just fallen onto her head, and she is raising her hand to remove it. (Why does seeing a cat walking along the shelf above her encourage this interpretation?) It follows that the notion that hats—and countless other things of human interest—can be seen by visual systems knowing only about 2D/3D shapes, however complicated and subtle their spatial descriptions may be, is absurd. Visual interpretation is no less dependent on semantics and common sense than is the understanding of language.

The problems involved in enabling computers to see people could temporarily be made to appear more tractable by eschewing relatively rich contexts like ordinary photographs, or even children's painting books, and concentrating instead on highly stylized pictures drawn by the same hand and having a fairly constant content—like the "Peanuts" cartoons.[15] But even seeing "Peanuts" cartoons, if it is to progress beyond the mere identification of the characters and their environment (Snoopy, baseball bat, tree, kennel), requires additional computation capturing further significance in the pictures. In particular, it needs an appreciation of the meaningful relations between successive frames of the strip-cartoon: that is, it demands an understanding of both spatial movement and psychological change.

One can always make a conceptual distinction between seeing movement and seeing the cause of the movement, much as R. C. Schank always distinguishes between verbs of movement and their causal inferences in his Conceptual Dependency analysis. And sometimes this is an appropriate distinction at the phenomenological level also: it is easy to recognize that seeing that the model ship has fallen is not the same as seeing that the cat dislodged it, and that seeing that the woman's hand has been raised is not the same as seeing her intent in raising it.

But even these examples suggest that the conceptual distinctions may not always be phenomenologically evident, since a cat's presence in the picture may make it "obvious" that the galleon has fallen without one's being aware of noticing the cat, still less of making any inferences concerning it. And the Gestalt psychologist A. E. Michotte showed long ago that whether or not one sees movement, as well as what type of movement one sees, may depend upon visual cues implicitly activating causal theories about the sorts of changes that occur in the world.[16] Michotte concentrated on the intuitive attribution of *physical* causes to movement, but the psychologists Fritz Heider and M. L. Simmel have since generalized his experimental method in investigating how movement is interpreted in *psychological* terms, as being caused by particular motives or intentions.[17] "Seeing movement," then, normally involves not only seeing that something has changed its spatial location, but also seeing this change meaningfully, as being attributable to a cause of a certain kind—whether a clumsy jolt from a cat or a woman's deliberate intent to remove an unwelcome object from her head.

Michotte's experiments required people to watch two "meaningless" spots changing their relative positions at varying speeds. However, the spots typically were not seen as meaningless 2D stimuli, but as representations of 3D objects (such as billiard balls). This "effort after meaning" on the part of the perceiver involved the interpretation of their movements by implicit reference to the relevant domain-specific knowledge, namely, the laws of physics.

Similarly, Heider and Simmel's subjects saw "mere" triangles, circles, and squares (shown to them in cartoon films) as animate objects chasing, fleeing, threatening, fighting, and embracing one another. This progression of psychological concepts calls to mind Abelson's *rescue* script, and one of the experimental cartoon films was indeed experienced by most subjects in terms of a rescue scenario. The larger and more pointed 2D figures were more likely to be seen as the aggressors, particularly if the point pointed in the direction of movement; clearly, the point's direction of movement functioned as a cue activating the "head-first" feature of the general "animate being" model, while its shape and relative size helped selectively to activate interpretative models like Abelson's themes, which specify good or ill intent regarding the other person's plans and which assume particular power relations between the two actors. (A dwarf father *could* abandon his immature giant daughter, but in watching a film with this theme one would have constantly to check a sense of absurdity by reminding oneself that the usual correlation between physical size and overall competence did not hold.) It is significant that when the "thematic" interpretations assigned at different points in time by Heider and Simmel's subjects did not cohere into familiar "scripts," they were both less intuitively compelling and less uniform across subjects.

A programmed "Peanuts"-reader would therefore have to be able to see that Snoopy has retired to his kennel in a sulk because his human friends have abandoned him to go off and play baseball—and this requires *more* than seeing that Snoopy was pawing at their car door in frame *n* of the strip-cartoon, but is sitting in his kennel in frame $n+1$. (The progression of tenses here marks the interpretative convention that successive frames denote successive times.) HAL's seeing the astronauts plotting to betray him would naturally require even more. Before such programs could be written, one would have to identify the relevant visual cues calling the interpretative schemata involved. One could tell Kelly's program, for instance, that tears (small pear-shaped contours directly below the eyes, preferably in two parallel columns) suggest unhappiness. But since Snoopy does not always cry when he is sulking (do you?) other cues, like "droopy ears" or "downcast eyes," would be involved.

The computational description of different facial expressions would be considerably more taxing than the description of tears, and would in addition have to be related to global context specified in terms of psychological rather than purely visual features: a shout of laughter and a grimace of pain may look the same out of context, so only psychological context could disambiguate them. The relevant jigsaw heuristics would have to rely on assumptions about what successions of psychological states *make sense.* Correlatively, the strip-cartoonist could not express very unconventional mental reactions without using speech-bubbles—and even these could not be confidently interpreted if, in context, they seemed highly bizarre (like someone's laughing or crying at the most *in*appropriate moments in a Greek tragedy).

The problems involved here have hardly been touched upon, at least as far as "motivational" visual interpretation is concerned. But Sylvia Weir has written a program that to some extent captures the psychological processes with which Michotte was primarily concerned, namely, the attribution of physical causes to perceived movement and the effect that causal theories have on the perception itself.[18] And she has done this with the domain of strip-cartoons in mind, asking how her methods might be generalized to deal with the perception of human actions and the attribution of intentions.

Weir modeled some of Michotte's experimental situations in which one thing is seen as hitting another and somehow setting it in motion. Her program in effect watches successive pictures ("frames") representing consecutive instants in time, like the "stills" making up a cartoon film ("in effect," because—like Guzman's SEE—Weir's program accepts *de-*

*scriptions* of pictures as input rather than pictures themselves). It interprets the visual changes in terms of physical movements and causes of various kinds.

Michotte found that the visual stimulus itself was not the only significant factor: his experimental instructions about where to look within the picture somehow played a part in encouraging one perception of it rather than another. Analogously, Weir's program can have its attention drawn to particular features of the picture, with the result that specific expectations are aroused that differentially determine the interpretation of a given visual cue. For instance, the same sequence of pictures is seen by it either as one object hitting another and making it move, or as one object passing over another (stationary) object, depending on what type of change was expected.

The perception of change in general requires that changes (of whatever character) in the representational domain be noticed before they can be interpreted in terms of altered target realities. This is as true of perceiving that someone has changed her mind or her mood as it is of perceiving that she has changed her *bodily* attitude or position. Seeing that movement of some kind has occurred is, then, an interpretative gloss on noticing that something in the picture has altered, so that the first requirement of Weir's program is that it be able to detect change *in the picture*.

This is achieved by a "differencing" method whereby the structural descriptions of successive pictures are compared, resulting in a list of differences for each pair of adjacent frames. For present purposes, the way in which the program does this may be ignored: it is essentially similar to the differencing methods used by P. H. Winston and T. G. Evans, whose work on learning and analogy, respectively, will be discussed in Chapters 10 and 11. For example, Frame I of Figure 9.12 is input to the program as an assertion that there is a white square on the far left and a black square at the center; Frame II is input as the statement that there is a white square rather to the left of the picture and a black square at the center. (Actually, positions are specified by numbers like those on the horizontal axis of a graph.) The difference list drawn up by the program marks the fact that in the earlier frame there is a white square at position 3, whereas in the later frame there is a white square at position 4.

If you ask yourself why all the descriptions (including the difference list) in the previous paragraph refer to "a" white square rather than "the" white square, you will see that speaking of "the" white square already involves an interpretative process, since two pictorially distinct regions are thereby assimilated in terms of the general category of "identity." And of course, as soon as the (*sic*) white square is referred to (or seen) as "the white object," or "the white cube," it is being further interpreted

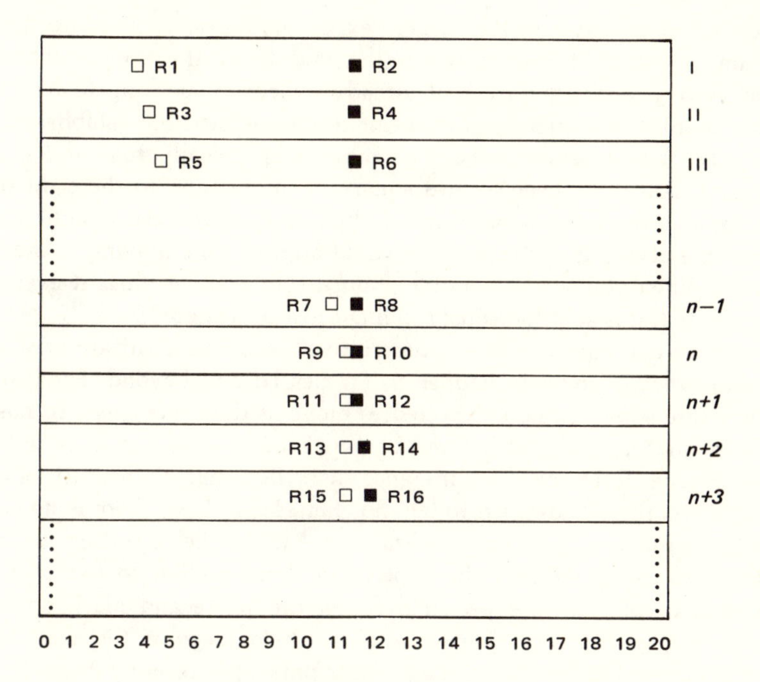


FIGURE 9.12

Source: Adapted from *AISB Conference Proc.*, 1974, Figure 2, p. 250. Reprinted by permission.

in terms of *bodily* identity. Weir points out that there is in general more than one way of pairing picture regions in successive frames, and that a set of interpretative rules is thus needed for deciding which of the possible pairings corresponds to an enduring object in motion. She also points out (what is implied by Michotte's experimental results) that the application of these rules depends on prior expectations set up in various ways by the particular context of perception.

For example, let us see what the program makes of the change between Frames I and II of Figure 9.12. Comparison shows that regions R2 and R4 have identical descriptions, which constitutes good evidence for pairing them (that is, for regarding them as "the same" pictorial and target item). R1 and R3 have different descriptions on account of their positions, so they cannot be so straightforwardly paired; but their color is identical and differs from the color of R2 and R4. Accordingly, the most sensible pairing is deemed to be R1 with R3 and R2 with R4. The latter pair is then interpreted as a stationary object at midscreen, while the former is seen as a moving object traveling from left to right at fast, slow, or medium speed (calculated from the size of the position difference).

Further, if in this case the earlier "experimental instruction" to the program was that it should watch the middle of the screen, then (like Michotte's human subjects) it already expects something to happen at that point. Consequently, a frame of reference has been established such that the white square is seen as moving specifically *toward the black square*; had there been a third square in the picture, to the right of the black one, the white square would objectively have been moving toward that one also, but subjectively it would appear to be moving to the *black* one. (When you see one marble about to hit another, does it ever occur to you that it is simultaneously rolling toward the door?)

Part and parcel of these "descriptive" interpretations are *predictions* about what is going to happen in Frames III and beyond. For instance, the white square is now expected to move at the same speed to position R5, so that its doing so will *not* be perceived as a change to be noted on the Frame II–Frame III difference list. Between Frames II and III, therefore, there is deemed to be "no change"; indeed, there is no change until Frame $n-1$ gives way to Frame $n$. Meanwhile a further prediction is made, encouraged by the experimental instruction to fixate at midscreen: a collision is expected between the white and black objects at midpoint, with the moving white object being seen as the agent of the collision and the stationary black one as passively suffering it.

To say that the program is "expecting" a collision at midscreen is not just to say that it predicts that the two squares will touch there. It is actively on the lookout for this to happen, and when it does it has an interpretation ready. This is achieved by programmed "demons" similar to Eugene Charniak's story-understanding demons, which look out for remarks praising pencils when pencils are first mentioned in a "trading" context. As soon as the expected event occurs, the demon automatically assigns the relevant interpretation. (Charniak's demons are PLANNER antecedent theorems, whereas Weir's demons are CONNIVER if-added methods: CONNIVER and PLANNER will be discussed in Chapter 12.)

When Frame $n$ is finally reached, and the expected adjacency description is noticed by the watching demon, it has more work to do than just to say "Ah! A collision!" It modifies the region-pairing process and it sets up a search for consequences of the impact. Thus instead of continuing with the region-pairing rules that have been assuming that the white square will move steadily across the screen, the program reverts to the position of having an open mind: anything can happen, including movement, explosions, and disappearances. Consequently in Frame $n+1$, R9 and R11 are paired with *no* sense of surprise, as would have arisen if the white square had unaccountably "stopped moving" (*or* "disappeared"?) in Frame IV. Also, the demon predicts that there will be an effect on the *black* square, and watches out for a relevant item in the difference lists comparing successive frames after Frame $n$. When the black square is indeed seen to move, the interpretation of a bodily collision (as opposed to a mere pictorial adjacency) is confirmed and the white square is seen as having caused the black square's movement.

There are a number of subtleties involved that parallel Michotte's results to some degree. For example, you will see from Figure 9.12 that the expected movement of the black square does not happen at once: in Frame $n+1$ it is still adjacent to the white one. But by Frame $n+2$ it has moved. Weir's collision demon is a "suicide" demon, which kills itself if it does not find what it is seeking within a certain time; in other words, if the black square had not moved until Frame $n+7$, the collision demon would already be defunct and so the movement would *not* be interpreted as a consequence of the collision with the white object, but as an autonomous movement attributed to the black object itself. Analogously, Michotte found that the greater the latency between touching and movement, the less likely the white object was to be seen as the *agent* causing the black object's motion.

In this simple context, and with the advantage of Michotte's results, Weir could give "reasonable" lifetimes to her demons, based on the latency times observed to be significant for human beings. But we saw in Chapter 6 that it is less easy to decide how long a "trading" demon should be on the lookout for remarks praising the vendor's possessions: if I offered you some pencils last month and you refused them, should you be wary of my praise of them today? And if it is my house that I am trying to sell you, does this make a difference to the trading demon's longevity?

Similarly, if the speed of the black square is much greater than that of the white, it is seen (by the program as by many of Michotte's human subjects) as having been "triggered" or "launched" by the white object, instead of being merely pushed and so set into motion by it. In short, one's implicit knowledge of physical dynamics affects one's perception of what one sees. No wonder Alice found it difficult to regard the Queen of Hearts' curled-up hedgehogs as croquet balls—for croquet balls do not move on their own account, or run quickly away uphill when gently nudged by a mallet. The circles and squares in the Heider and Simmel experiments were seen as animate largely because their movements were either autonomous or "triggered" in some way, rather than conforming to the physics of billiards and croquet. Weir does not distinguish in her program between different sorts of triggering, but she is currently working on this problem (which requires a complex semantic representation of the various causes concerned).

A further subtlety is that different experimental instructions make a difference to the perception of what is going on in the frame sequence. For example, if instead of fixating the midpoint, the program is told initially to watch a point above the screen, it does *not* expect anything to happen at the midpoint. So it does *not* see the white square as moving toward the black one, but merely sees it as moving from left to right; similarly, no collision is expected between the two bodies. Consequently, at Frame *n* there is *no* demon waiting to change the region-pairing rules. According to the region-pairing rules set up at Frame II, R9 is expected to move rightwards but R10 is expected to stay still. Consequently, at Frame $n+1$ region R12 is paired with R9, and R11 with R10; but in the following frames, odd-numbered regions are paired together as before, and even-numbered regions are paired as before. In other words, the picture sequence is interpreted as one object *passing* over another, stationary, object. The change in color at Frame $n+1$ is noted by the program, but treated as less significant than the fact that movement apparently is as was to be expected.

Analogously, Michotte's human subjects saw the passing effect rather than a collision when they fixated a point above the screen. Sometimes they noticed the "change in color" of the squares at the moment of "passing," but sometimes they did not; in the latter case, their normal assumptions of color consistency overruled the evidence of their senses, so that they hallucinated constant coloration in the picture regions that they took to depict single enduring objects. Michotte remarks in a footnote that some subjects who "apparently observe in a particularly analytical way" see a small *retreat* of the stationary object as the moving object passes over it. As Weir points out, this phenomenon is neatly explained by her approach, since it corresponds to the pairing of R10 with R11 instead of with R12 (see Frames *n* and $n+1$).

In the experimental situation just described, two squares are always visible. This accounts for the fact that one object is seen as passing *over* another, rather than behind it. But just as *invisible* edges and corners of blocks can be "seen" by sufficiently knowledgeable programs, so invisible movements could be inferred by systems having a more comprehensive representation of pictorial and physical change than Weir's program. Accordingly, a visual stimulus that had already been interpreted as depicting an enduring object could be seen to pass behind another object, its hidden movement being inferred partly from extrapolation of its earlier path and partly from looking to see at what point on the far side of the occluding object it eventually reappeared. One way of providing this facility to Weir's program would be to allow for what Charniak calls "demon-demon interaction," whereby (for instance) one demon passes information about the *consequences* of an event to the demon who had been unsuccessfully watching for that event: the latter demon then infers that the watched-for event must indeed have happened, even though it did not have direct evidence of it.[19] This method was suggested by Charniak as a way of filling gaps in stories, but is equally appropriate to gaps in other conceptualized sequences. In general, changes other than movement often have to be inferred on the basis of their consequences; if Shakespeare's script did not actually show Iago lying to Othello about Desdemona, the audience would nonetheless know that he had carried out his evil plan off stage, when Othello's treatment of his wife was observed to have changed accordingly.

Weir is currently working on the inference of invisible movement, and on the seeing of actions—such as a person's *picking up* a ball and *walking away* with it. Her results will probably illuminate the perceptual interpretation of visual change that goes on when we see movement. But there is a general feature of her approach that limits its plausibility as a model of human perception. At the beginning of the "experiment," she has only a small number of possibly appropriate interpretative schemata waiting in the wings for their cues, and these are selectively activated fairly early. For example, between Frames I and II in the center-fixation condition, a collision demon is immediately called onto the stage. But this does not seem an apt model for all cases of human reaction to visible change.

For instance, Weir herself has shown the Heider and Simmel films to her colleagues, one of whom saw a particular sequence as "like the first act of *Figaro*."[20] The notion that this particular interpretative scheme was already aroused, with its demons watching for their respective cues, is not convincing. To be sure, Grape's work showed that if one is specifically looking for wedges, one may reasonably interpret ambiguous data as evidence of a wedge; and if one were primed with several operatic scenarios to be matched up with the Heider and Simmel films, then a close cousin of Weir's program might be an appropriate model. But the "spontaneous" assimilation of geometrical figures to specific operatic characters such as Figaro or Brunhilde is a more mysterious matter than this. Is it mediated first by general psychological schemata (like Abelson's themes and scripts) and only then by specific dramas conforming to these abstract scenarios (such as *The Marriage of Figaro*, or *Romeo and Juliet*)? And why is one specific drama taken as the exemplar rather than another? The way in which particular experiences are made sense of by being assimilated to general cognitive frameworks will be further discussed in Chapter 11. But one should certainly expect the epistemological processes in the *Figaro* interpretation to be more complicated than those modeled by Weir.

More could be said in comparison of Weir's program with human perception. Indeed, more could be said about the extent to which any of the visual programs so far described actually simulates vision in live creatures. How limited are they, for instance, by the fact that even the "movement-oriented" programs do not utilize movement *of the visual system itself* as a way of generating useful cues?[21] And has research on the "toy worlds" of simple polyhedra *inhibited* progress in machine vision, where concentration on more realistic domains (as in the program of J. M. Tenenbaum to be described in the next chapter) would have led to more efficient artificial seeing, in addition to approximating human sight more fully? I shall postpone closer comparison of machine vision and human sight to Chapter 13, when I shall discuss the general psychological relevance of machine models of the mind.

Meanwhile, the question inevitably arises whether visual programs must always be presciently provided with the representational theories they will use to interpret their world. A person can learn to tell a hawk from a handsaw, or become expert at using a microscope, largely through her own efforts: could a program do the same? These are special cases of the general problem of learning, where one asks whether (and how) a program can acquire new knowledge of cues and models, new knowledge of facts, and new skills, without having this knowledge specifically spelled out for its benefit by the programmer. Let us turn, then, to ask whether any current work in artificial intelligence casts light on this problem.

# *Part V*

# NEW THOUGHTS FROM OLD

# 10

# *Learning*

AUNTS AND UNCLES of Victorian damsels sometimes tried to improve their niece's mind by writing suitably moral verses in her autograph book. No doubt they occasionally succeeded, for there are many different ways of achieving valued and lasting psychological changes, of generating new thoughts on the basis of old ones.

In both common-sense and theoretical psychology, these are usually thought of as falling into three distinct classes: learning, creativity, and problem solving. Broadly, learning is then seen as the improvement of general cognitive capacities under outside influence; creativity, as the spontaneous generation of new representations of the world; and problem solving, as the achieving of new knowledge by way of thinking things out in a specific situation.

But it should be evident that these phenomena are very closely connected. General learning may derive from specific experiences, and may involve creative thinking; "spontaneous" construction of new representations may be elicited by a particular need or problem, and may be aided by environmental cues; and solving a specific problem may require creativity and lead to general learning. So the distinctions should be regarded as a matter of emphasis rather than an expression of rigid partitions within mental reality. Indeed, we have already seen that everyday psychological categories can mislead through obscuring unsuspected similarities underlying superficially disparate phenomena: "vision," for instance, is not so different from "memory" and "problem solving" as is normally assumed.

Accordingly, the topic common to the whole of Part V is the generation of new thoughts from old, by the intelligent development of representations already in the mind. In the next three chapters we shall see how this topic underlies and connects psychological changes variously ascribed to learning, creativity, and problem solving.

Learning itself is often subdivided into various types, including learning by example, learning by being told, and learning by doing. Roughly, these provide, respectively, new knowledge of cues and models, new knowledge of facts, and new skills. We shall see that these are computationally similar, in that each involves the development of inner representations of whatever it is that is learnt. So, as John McCarthy forecast in the early days of artificial intelligence,[1] the problem of how to enable a system to learn is inseparable from that of how to represent the knowledge concerned, and how to transform the representation in case of error so that it becomes increasingly adequate to the task in hand.

Sometimes what is learnt is stored explicitly, in the form of "facts" or data that can be used and examined by many different procedures; but sometimes, the new knowledge is stored implicitly in procedures, that may be more or less general in application. Sometimes, learning is primarily a process of accepting new information, and integrating it into a preexisting structure; but sometimes it is more a matter of reorganizing the information that is already in the mind. These varying aspects of learning are exemplified in the programs described in this chapter. To the extent that programs can be enabled to learn for themselves, new information will not have to be provided to them by laborious step-by-step programming: one will be able, for instance, to *tell* them something in English, and rely on them to include it in their inner models of the world in an intelligent fashion.

## LEARNING BY EXAMPLE

Humpty Dumpty and Bertrand Russell have more in common than meets the eye, their striking differences in physique being offset by shared (and widespread) assumptions about the nature of learning by example. A common-sense cousin of Russell's technical distinction between "knowledge by acquaintance" and "knowledge by description" is implicit in Humpty Dumpty's answer to Alice:

> "And what does '*outgrabe*' mean?"
>
> "Well, '*outgribing*' is something between bellowing and whistling, with a kind of sneeze in the middle: however, you'll hear it done, maybe —down in the wood yonder—and, when you've once heard it, you'll be *quite* content."

A similar distinction was made in Chapter 9, in contrasting "concrete" with "abstract" visual descriptions. In each of these cases, the implication seems to be that learning a concept by meeting an example of it is not only significantly different from learning it by hearsay, but somehow better. Indeed, Russell regarded knowledge by acquaintance as independent of and prior to knowledge by description, because of its directness: according to him, genuine learning by example is free of all taint of abstraction or description, involving the direct apprehension of reality without any intervening interpretative activity.[2]

But is learning by example really superior to mere hearsay? And if so, why? Is it because it is wholly direct, thus avoiding the fallibility inherent in interpretation? Or is there some other reason for its so commonly being regarded as a Good Thing?

Alice's contentment on hearing outgribing would indeed be justified —but not because the actual example somehow obviates all need for description. Rather, it allows for a check, and perhaps an improvement, on the conceptualizations previously provided to her. After having heard it she would know *what* kind of a sneeze is in the middle, and *which* bellowing-cues are combined with *which* indicators of whistling. She might even learn that outgribing is really more like hissing than whistling, and revise Humpty Dumpty's description accordingly. In short, acquaintance with the target domain increases knowledge of it because it suggests and refines specific interpretative (cue-schema) connections between it and the representational domain, not because it makes the representational domain in general superfluous.

Artificial analogues of Alice, too, can learn from example (and counterexample) by generating and improving representations of the target realities concerned. Consideration of programs that can learn by example helps clarify some of the difficulties and complexities behind the question how such learning is possible: that is, what kind of computational mechanism is required?

Several common instances of learning by example were mentioned in the previous chapter, namely, learning what cats (and catlike things), tables, and telephones look like by seeing some. J. M. Tenenbaum's program has been taught to recognize the door, chairs, table, pictures, floor, wall, and telephone in grey-scale photographs of his office at the Stanford Research Institute.[3]

In view of the discussion in Chapter 9, you may be wondering how the complicated shape of a telephone is communicated to the program in abstract terms. The answer is that it is not. Rather, the program is *shown* a telephone and learns to pick out telephones in the relevant class of pictures. Similarly, it learns to recognize table-tops by way of ostensive definitions. But, as Wittgenstein pointed out in criticizing epistemological theories essentially similar to Russell's, ostensive definition itself involves abstract descriptions, or concepts.[4] And Tenenbaum's program, accordingly, relies on abstract interpretative schemata some of whose cues are provided in a "concrete" manner by way of an actual visual image.

Tenenbaum's program uses two data structures to represent the concepts learnt: semantic and iconic (image-storing). The semantic representation of a telephone, for instance, carries the information that a black rectangular prism of a certain size (supported by the table) itself supports a black rectangular wedge with a grey area on its sloping surface, smaller circles being arrayed round the edge of the grey (dial) area. This abstract description is not merely "spoon-fed" to the machine, but is developed in an increasingly discriminatory way in conjunction with an iconic representation of the object concerned. Since the semantic network includes pointers to the iconic storage, concepts are largely defined in "concrete" pictorial terms even though they include abstract descriptions like the one just given. Because of the links between the two data structures, the program can learn what something looks like either by being shown an example, or by being told that it resembles something seen earlier—much as Alice can learn the meaning of "outgrabe" either by hearing outgribing, or by being told that it is rather like bellowing.

Given that it starts off knowing nothing about table-tops, for instance, how does the program learn to recognize them? In particular, how does the iconic data structure interact with its semantic counterpart so as to generate a useful description of a table-top?

The program has primitive visual operators that can detect intensity gradients, colors (both hue and saturation), height, and surface orientation. First, the intensity-gradient operators produce a representation of the photograph in which different *regions* are provisionally distinguished (much as line finders find lines). Next, Tenenbaum encircles the table-top region in the picture with a light pen. He does not have to trace the outline of the region, merely to draw a rough circle around it; the program finds the "outline" region within the circle. This region is thereupon stored in the iconic data structure with the label "TABLETOP." At this stage, the program has an "inner picture" of a table-top, but is unable to use it to find or describe table-tops since it is not yet connected with any particular visual operators (nor with any semantic information).

Tenenbaum now guesses which operators may turn out to be useful in discriminating table-tops, and asks the program to ascertain their values for the particular example (image) stored. For instance, he inquires as to the height (average, maximum, and minimum) of the region in question, and its hue. The relevant values are computed and stored in the semantic net as cues for locating "TABLETOP," together with a pointer to the image stored in the iconic data base. Next, Tenenbaum experiments to see if the current description is sufficient to find the table-tops, exclusive of other things in the picture. This is done by instructing the program to illuminate all the regions of the whole picture that fit the description evolved so far. The result is that the table-top is found, but so too are parts of the buff floor and brown door. Perhaps color saturation or surface orientation might help? Tenenbaum chooses the latter, asks the program to ascertain the orientation of the TABLETOP region, and the program adds this (horizontal) to the description. On retesting, this three-cue description (height, hue, orientation) is found to be sufficient to find the table-top in this scene and to exclude the counterexamples that were mistakenly identified earlier.

The semantic net, and its linkages with the iconic data, are thus grown by the program in interaction with its teacher. Colors can be defined pictorially (BUFF is the color of the chair seat) or in terms of the abstract "color triangle" that codes their physical properties. Extra information can be added to a concept symbolically, using already defined concepts; for instance, the program can be *told* that table-tops are buff, if it already knows what buff is (the color of the chair seat, namely, the region arbitrarily tagged "3" in the region map of the scene). Compound concepts can be defined symbolically too, or they can be built up by judicious questioning about pictorial examples. For instance, the compound concept TABLE could be learnt by first learning TABLETOP (as described above) and then being told that it is supported by TABLE-LEGS, which are vertical and cuboidal, and which in turn are supported by the FLOOR. Alternatively, the program could find this out by inspection of the (compound) region made up of the table-top and the four legs at its corners. Analogously, Alice may be *told* that outgribing is something like bellowing, or she may notice this for herself. And, like Tenenbaum's program, she may find that a description covers counterexamples as well as examples, in which case she (like it) will refine her initial concept accordingly.

In addition to providing a check on previous verbal descriptions, the experience of an actual example can enrich knowledge in another way. For instance, when Tenenbaum's program is given a symbolic definition of a concept, it stores only those features that are explicitly mentioned. Thus if it is told that tables are buff, it adds this information to its semantic storage, but cannot answer the question "How bright are tables?" By contrast, once it has seen a table it can access its stored image to answer questions like this, even though it may not use brightness as a cue for identification or location. That is, the particular characteristics of the iconic example are assumed (at least provisionally) to be representative of the class. It may be that similar processes occur in more complex cases of learning by example—for instance, in K. M. Colby's neurotic patient's taking her father's irresponsible behavior to be typical (though not definitive) of all male authority figures, and perhaps of social relations in general. (The use of particular examples as paradigmatic frameworks for relatively wide-ranging knowledge will be further discussed in Chapter 11.)

It is important to notice that Tenenbaum's program needs to start off with a great deal of knowledge, in order that it should learn about table-tops and the like. For instance, it needs to know something about what a region is, what colors are, what surface orientations are, and so on. It needs to know how to store and quickly find such information, and how to transform very specific descriptions or representations into more general ones. The implication is that learning can be achieved only by a cognitive system that has the ability to construct, analyze, and manipulate complex symbols. This was pointed out in different terms by Kant, in his account of the a priori prerequisites of empirical knowledge.[5] The psychological purport is that human babies (and the young of many animals) are born with powerful computational abilities—although precisely which these are is not known. (Kant's views about the specific nature of the a priori "forms" of empirical knowledge were incorrect, since he regarded the categories of Newtonian physics as inescapable by human thought.)

As we have seen, it is Tenenbaum who directs the program's refinement of its initial concept. The program has to wait for him to nudge it into building appropriate descriptions of table-tops and the like, since it is he who inquires about the specific properties (such as height and hue) of the example shown. Similarly, on meeting a counterexample (part-of-door, not the desired table-top), it has to wait for him to suggest orientation as a discriminatory cue. Moreover, Tenenbaum stresses that his approach is empirical rather than theoretical, in that he has no explicit theoretical preconceptions about which primitive visual operators will turn out to be useful to the program; he sees himself as experimenting with the program, much as one might do to discover what visual discriminations are easy for a newly discovered animal.

But if programs are to learn from examples by building the initial descriptions themselves, and locating the relevant (discriminatory) refinements themselves, some theoretical decisions about which attributes are likely to be relevant must be implemented in the system. In the case of a *general* concept learner—as opposed to a system capable of learning only a few restricted types of concept—this would involve considerations of general semantics and of previously accumulated knowledge of the world, including sociocultural roles. And, in any event, the particular sequence of examples encountered in the learning process is likely radically to affect the success of the whole procedure.

These points can be seen by reference to P. H. Winston's program, which learns to recognize structures such as tables, arches, pedestals, and arcades by being shown examples and counterexamples of them.[6] The program's teacher has to tell it which is which, but does not have to say why: the program itself spontaneously searches for the difference that makes the difference.

For instance, instead of being told what an arch is (what is it?), Winston's program learns what arches are by being shown the sequence of pictures in Figures 10.1 to 10.5. Similarly, it can then learn about arcades from Figures 10.6 to 10.9. By "learning what arches are" I mean coming to be able to interpret correctly an infinite number of pictures of individual arches, not just those exactly like the specific examples encountered in the teaching process. How is this possible?

The generality of the program's concept of ARCH, like the generality of the visual concepts described in Part IV, rests on its use of an abstract model or interpretative schema that is activated by observable cues. On first being shown a picture of an arch, Winston's program builds a structural description of it, in terms of objects (such as BRICKS and WEDGES) and their relations (such as SUPPORTS and MARRIES), which is provisionally taken as a representation of the concept concerned and is gradually refined in light of later experience. This obviously requires prior articulation of the picture into objects and relations, which is done by a series of programs (including Adolfo Guzman's SEE) that interpret the line drawing in 3D terms. Less obviously, it requires decisions—or provision for subsequent decisions—about the importance, or conceptual salience, of the various features observed in the examples being compared. It is these decisions and comparisons that form the heart of Winston's program, and that give it whatever interest it has as a representation of intelligence in general, as opposed to a mere "arch recognizer."

This question of salience can be illustrated by Figures 10.1 to 10.5. In Figure 10.1, one of the observable relations is the alignment ("marrying") of the side faces of the upper brick with the outer faces of the two supporting bricks. Being observed, it is recorded in the initial description built by the program—but is this feature merely accidental or is it essential to the concept ARCH? The program (or a person in like case) has no way of knowing, until an example of an arch is encountered that does not have this feature. Thus Figure 10.2 enables the program to amend its previous description so that this alignment is allowed for but is not insisted upon.

Correlatively, suppose that all the examples encountered were to share a certain feature: should an intelligent system assume it to be necessary to the concept, or could it be regarded as merely accidental (though universally present)? The observational evidence provided by the actual examples cannot decide here, since whether or not the feature is essential it is *ex hypothesi* found in all the known examples. However, observation of a *counter*example could settle the matter: for instance, Figure 10.3 provides the information that the SUPPORT relation be-

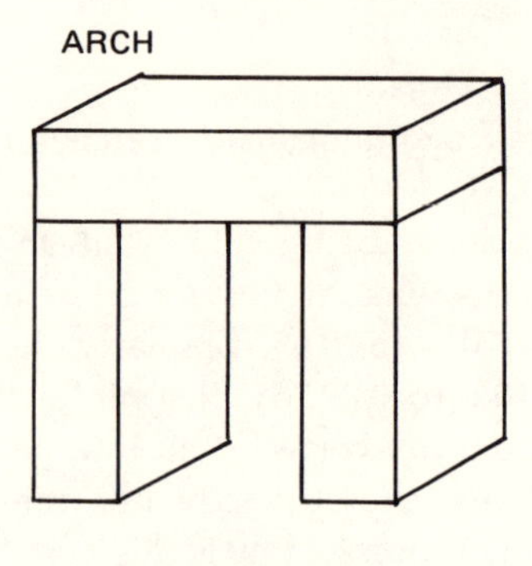


FIGURE 10.1

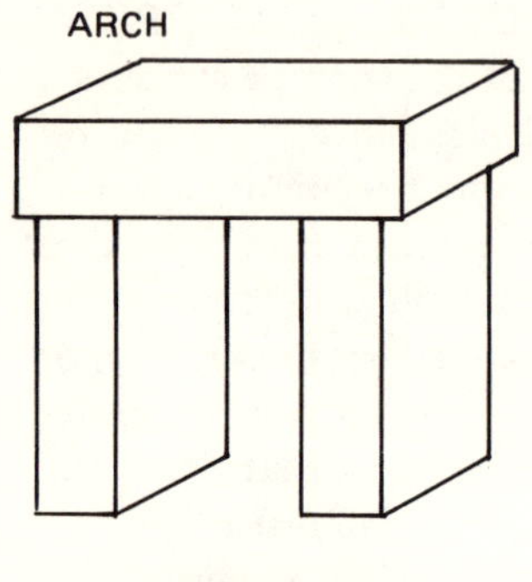


FIGURE 10.2

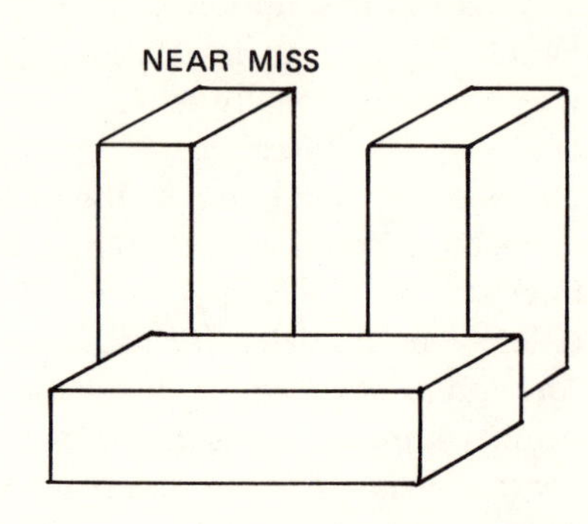


FIGURE 10.3

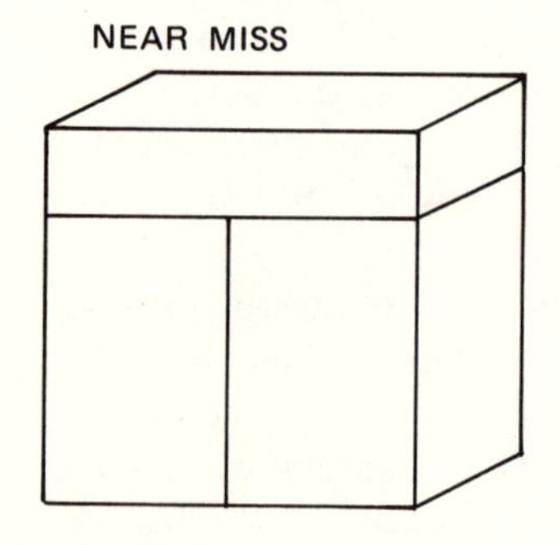


FIGURE 10.4

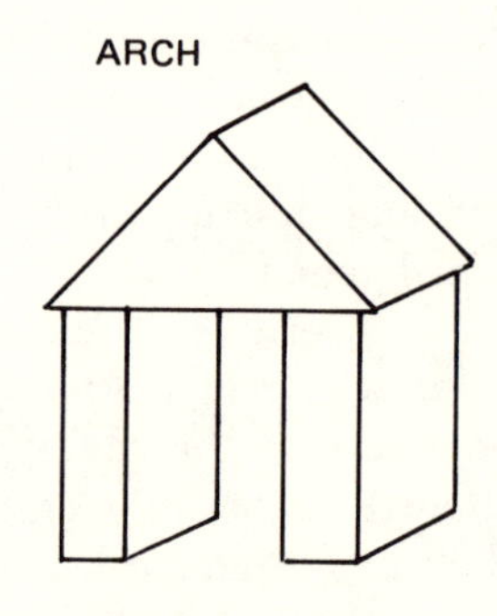


FIGURE 10.5

Source: Figures 10.1-10.5 adapted from P. H. Winston, ed. *Psychology of Computer Vision.* (New York: McGraw Hill, Copyright © 1975), Figure 2.1, p. 20. This and the following figures from this source are reprinted by permission of the McGraw-Hill Book Company.

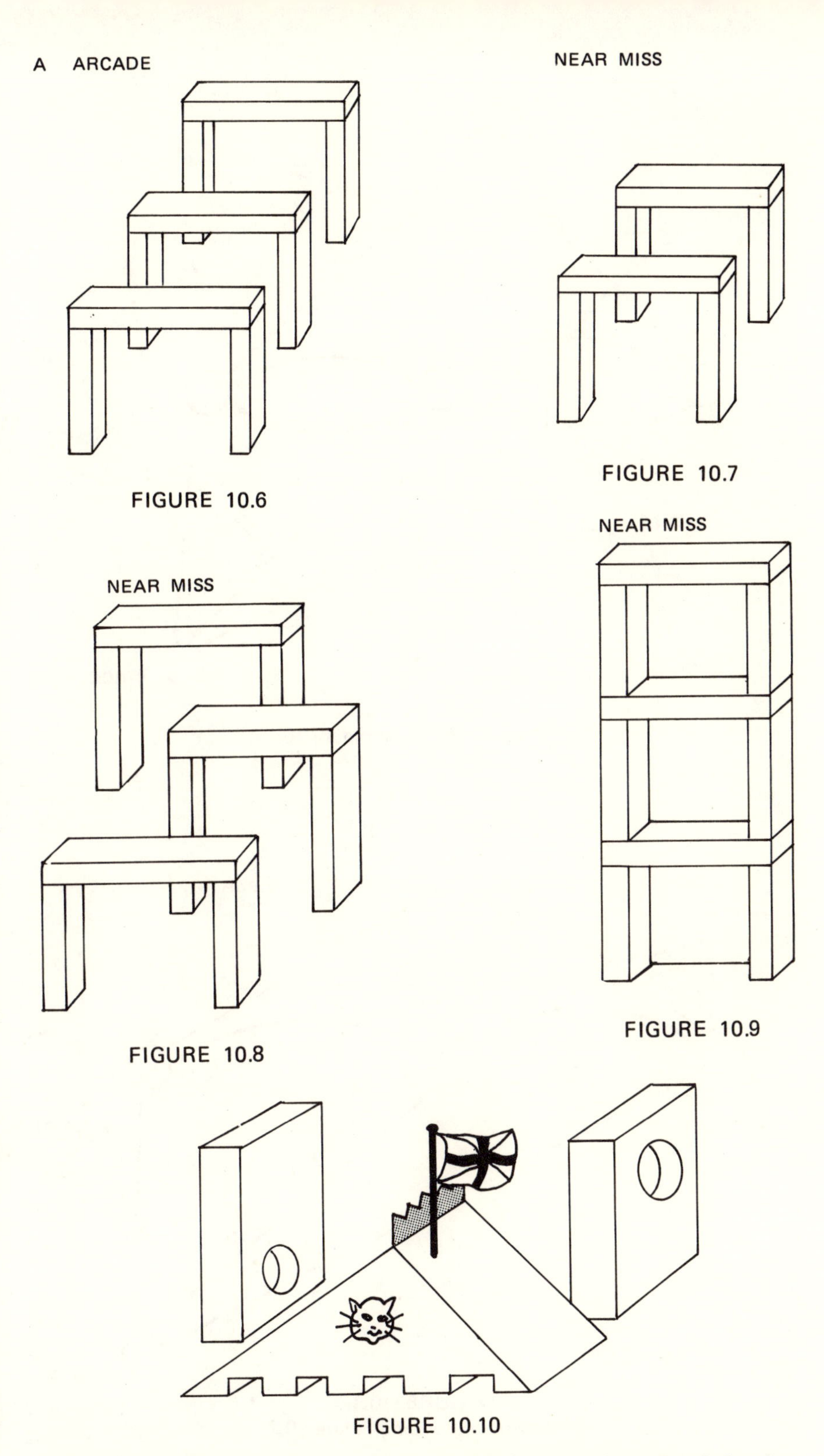


FIGURE 10.6

FIGURE 10.7

FIGURE 10.8

FIGURE 10.9

FIGURE 10.10

Source: Figures 10.6, 10.7, 10.9 from P. H. Winston, *Learning Structural Descriptions from Examples* (Ph.D. thesis). (Cambridge, Mass.: AI-TR-231, MIT AI Lab. 1970). Reprinted by permission.

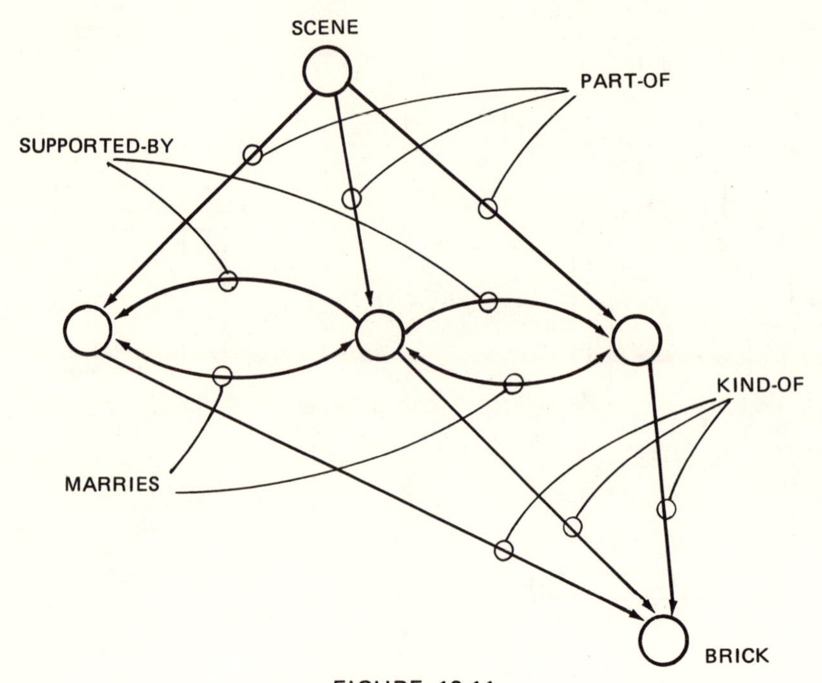


FIGURE 10.11
Network describing Figure 10.1.

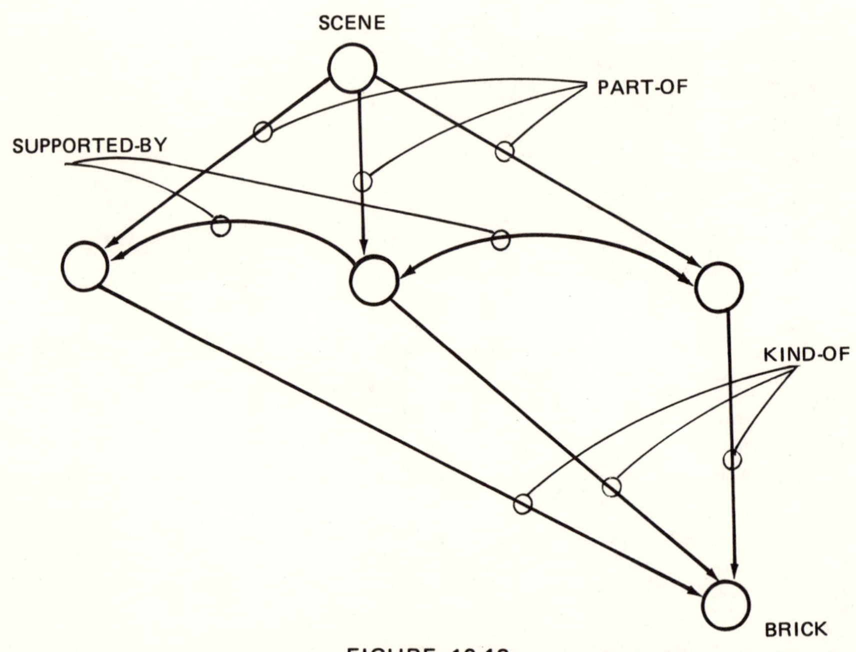


FIGURE 10.12
Network describing Figure 10.2

Source: M. L. Minsky & S. Papert, *Artificial Intelligence*. (Eugene, Oregon: Condon Lecture Publications, 1973), Figure 65, p. 47. This and other figures from this source reprinted by permission.

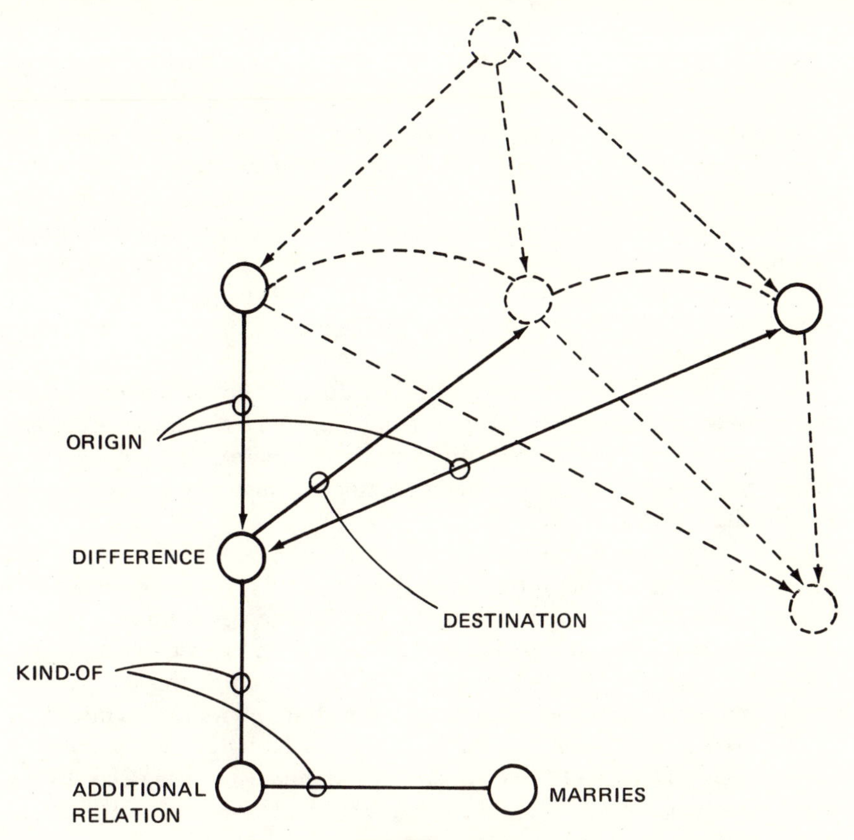


FIGURE 10.13
Network describing difference between Figures 10.11 and 10.12.

Source: Adapted from Minsky & Papert, *Artificial Intelligence*, Figure 69, p. 47.

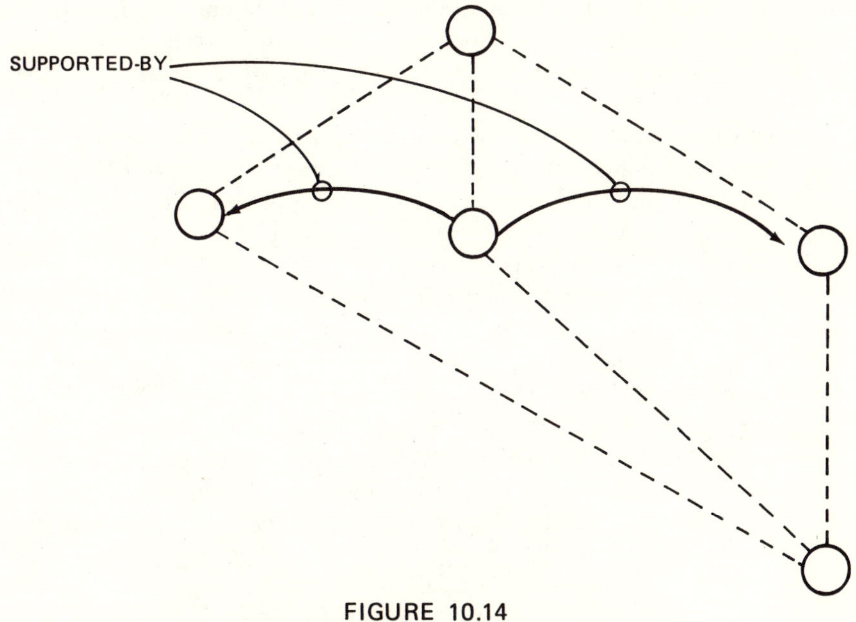


FIGURE 10.14
Amendment to model after seeing Figure 10.2.
(The MARRIES pointers have been dropped.)

tween the upper and the two lower bricks is essential. (Why would not Figure 10.10, in *this* position in *this* sequence of examples, provide this information so reliably?) In another sort of situation, too, one counterexample may be worth a thousand additional examples: if *no* example has been observed to possess a certain feature, should one conclude that the feature is forbidden or merely that it is not, in fact, found in any examples likely to be met with? A judiciously chosen counterexample could give the answer.

Precisely how counterexamples can do this, and what one might mean by a "judicious" choice of example or counterexample, are questions that can be clarified only by considering more closely the way in which the concept learner generates, compares, and progressively modifies the descriptions involved.

On being shown an example, Winston's program builds a structural description of the scene in the form of a semantic network, in which the nodes represent component objects and the links represent their relations. For instance, the network of Figure 10.11 is built on encountering Figure 10.1. On being told that this is an ARCH (the first time this concept has been mentioned), the program stores the description as its model of arches. The way in which this initial model gradually evolves is outlined in Figures 10.11 to 10.17, which show the refinements triggered by the series of examples culminating in Figure 10.5. The procedural heart of this evolutionary refinement is the comparison of each successive example with the current version of the model, as follows.

If you look at Figures 10.11 and 10.12, which respectively show the networks for the first two examples in the series, you will notice that the only difference between them is that the first contains two MARRIES relations, whereas the second does not. Winston's program has a special network-matching module that is able to compare networks and that expresses its discoveries in the form of further networks. Thus the difference between the networks of Figures 10.11 and 10.12 is represented as the network of Figure 10.13. (We shall see later that this fact is significant, since it allows the same network-matching procedures to be used to compare "second order" networks so that the program is sensitive to differences between differences. . . .)

Figure 10:13 can be thought of as identifying a mistake in the initial model, or at least as locating a point at which the model might be misleading. For although the supporting bricks *may* be aligned with the upper one, they *need* not. The program amends its model accordingly, by dropping the MARRIES pointers to give the structure shown in Figure 10:14. If Figure 10.2 had been the initial example with Figure 10.1 as the second, the program would simply have ignored the difference, so that the model prior to meeting Figure 10.3 would be the same as in the alternative sequence. In short, *contingent* features—once they are recognized as such—are excluded from the model.

In similar fashion, the refined model of Figure 10.14 is compared with the network describing the counterexample of Figure 10.3, and is modified as in Figure 10.15 so as to include a MUST-BE pointer as a satellite on the SUPPORTS relation. The next counterexample leads to the addition of the MUST-NOT pointer of Figure 10.16. Finally, the last example of the series (namely, Figure 10.5) prompts generalization of the class of the upper object from BRICK to PRISM, this being the smallest category that the program knows about which includes both bricks and wedges. (What other amendments would have been consistent with the last example?)

There are a number of general points to be made about this learning process. First, you may have wondered why Figure 10.2 caused the program to *drop* the MARRIES pointer, thus forgetting about it entirely, instead of, say, putting a MAY-BE satellite onto it as in the hypothetical model of Figure 10.18. The basic reason is that if one were explicitly to include a MAY-BE pointer for every contingent feature of a concept, the concept would become unmanageably complicated. Imagine all the different sorts of marble, or stone, or wood, or metal . . . from which arches *may* be constructed. Only if there are special reasons why the system might expect a feature to be salient that in fact is not, would it be worth including a MAY-BE pointer in the model: for instance, it might be worth including "may be a woman" within the representation of "chairman" if one is to eschew the abomination of "chairperson." (However, the program does sometimes make use of MAY-BE pointers in its description *of the scene*, as when it is specifically looking for an ARCH and sees something that could be a half-hidden ARCH.)

Winston's program cannot be described as having any "expectations" about what particular features may be salient, except in the sense that it knows about very few features and doggedly records all of them as observed in the samples. I oversimplified in describing the ARCH model, for the final model of ARCH actually looks like Figure 10.19, in which are included additional observable features like LYING, STANDING, LEFT-OF, and RIGHT-OF. Quite apart from the fact that many familiar "perceptual" concepts (like *chair*, for instance, or even *arch* itself) are not purely structural as is Winston's ARCH but primarily functional in meaning, we would not normally consider including each and every observable feature within a potential perceptual schema. A *general* concept learner, as opposed to one primed with a tiny subset of features to play with, would need to be sensitive not only to abstract semantic relations like negation, or contrast, but also to a provisional epistemological structuring of the world in which particular features were expected to be

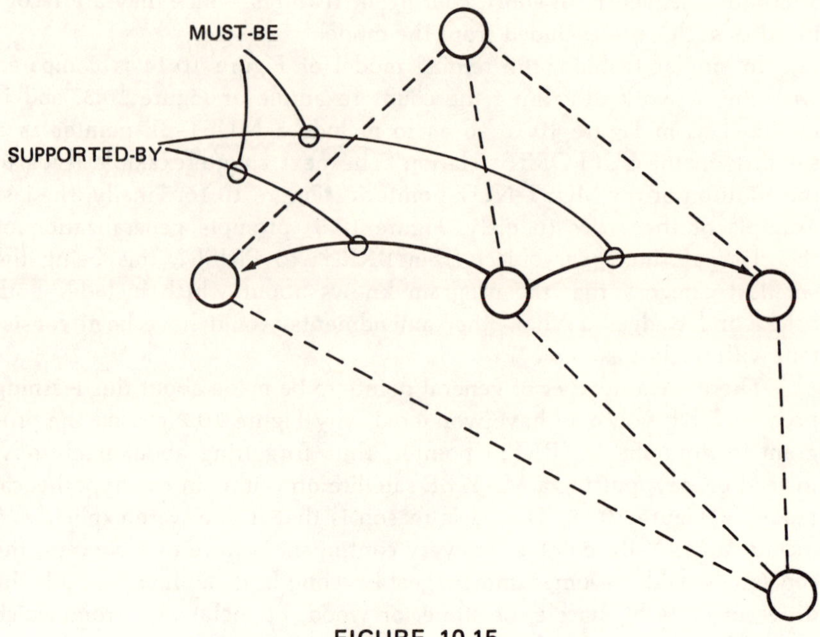


FIGURE 10.15
Amended model after seeing Figure 10.3.

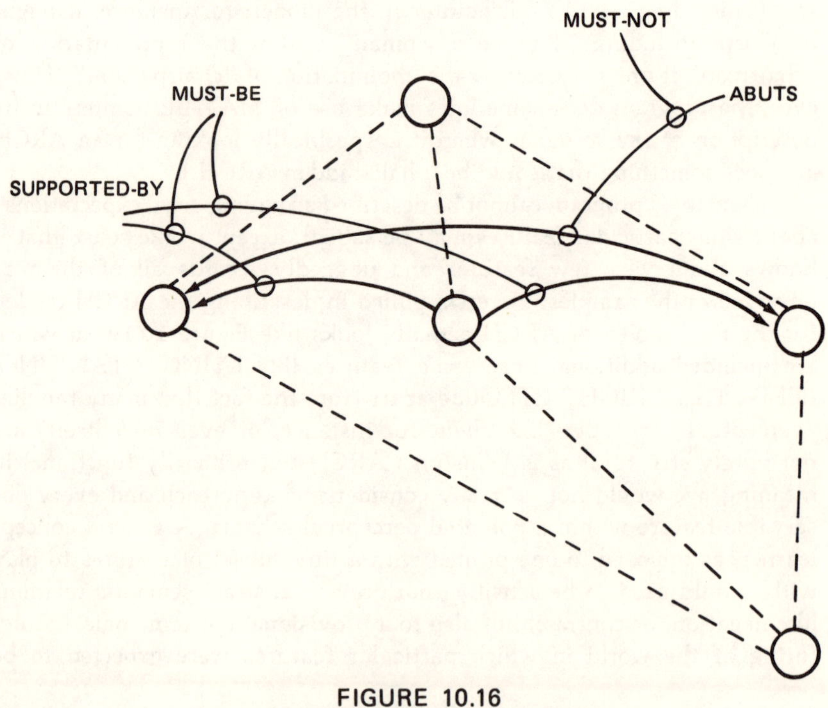


FIGURE 10.16
Amended model after seeing Figure 10.4.

Source: Adapted from Minsky & Papert, *Artificial Intelligence*, Figure 70, p. 47.

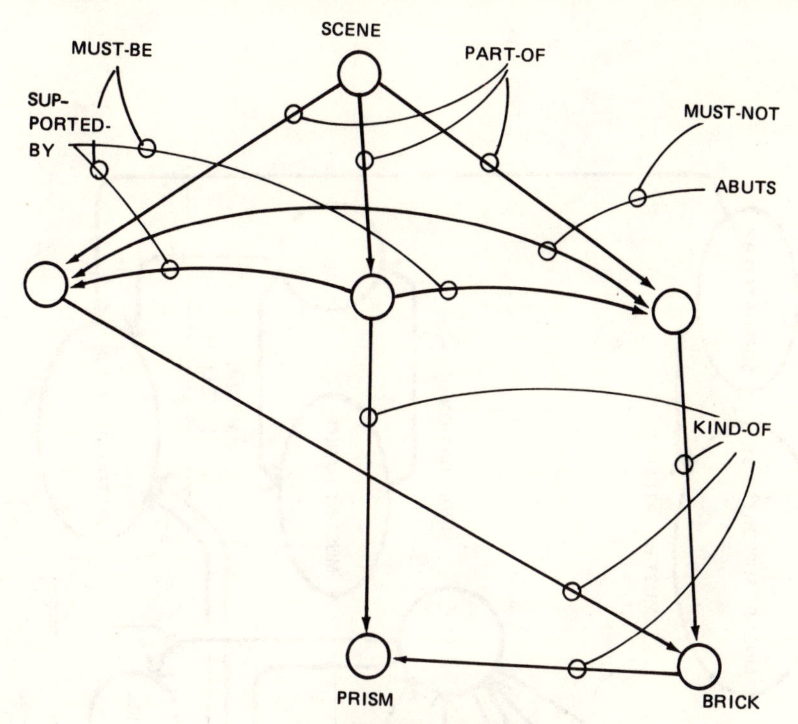


FIGURE 10.17
Amended model after seeing Figure 10.5.

Source: Adapted from Minsky & Papert, *Artificial Intelligence*, Figure 74, p. 49.

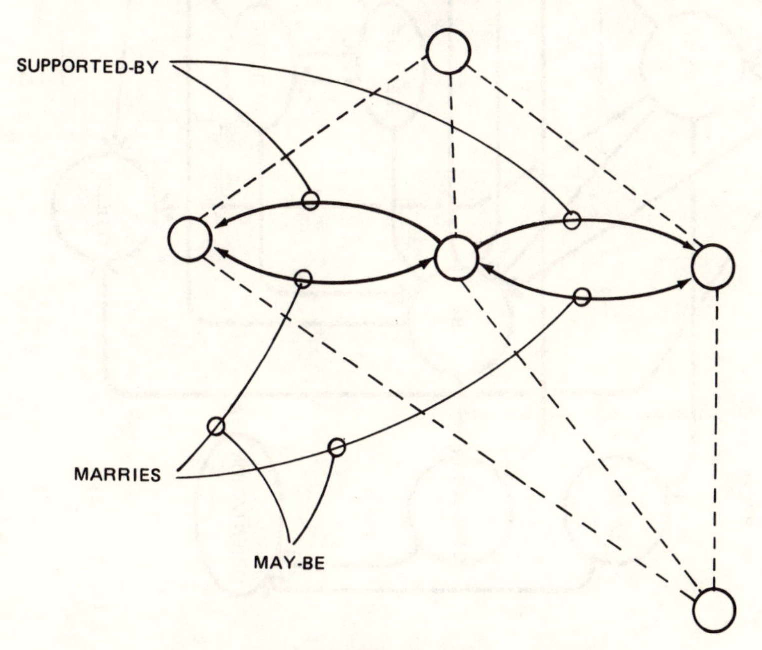


FIGURE 10.18
Hypothetical amendment to model after seeing Figure 10.2.
(The MARRIES pointers have been retained,
and a MAY-BE satellite pointer added.)

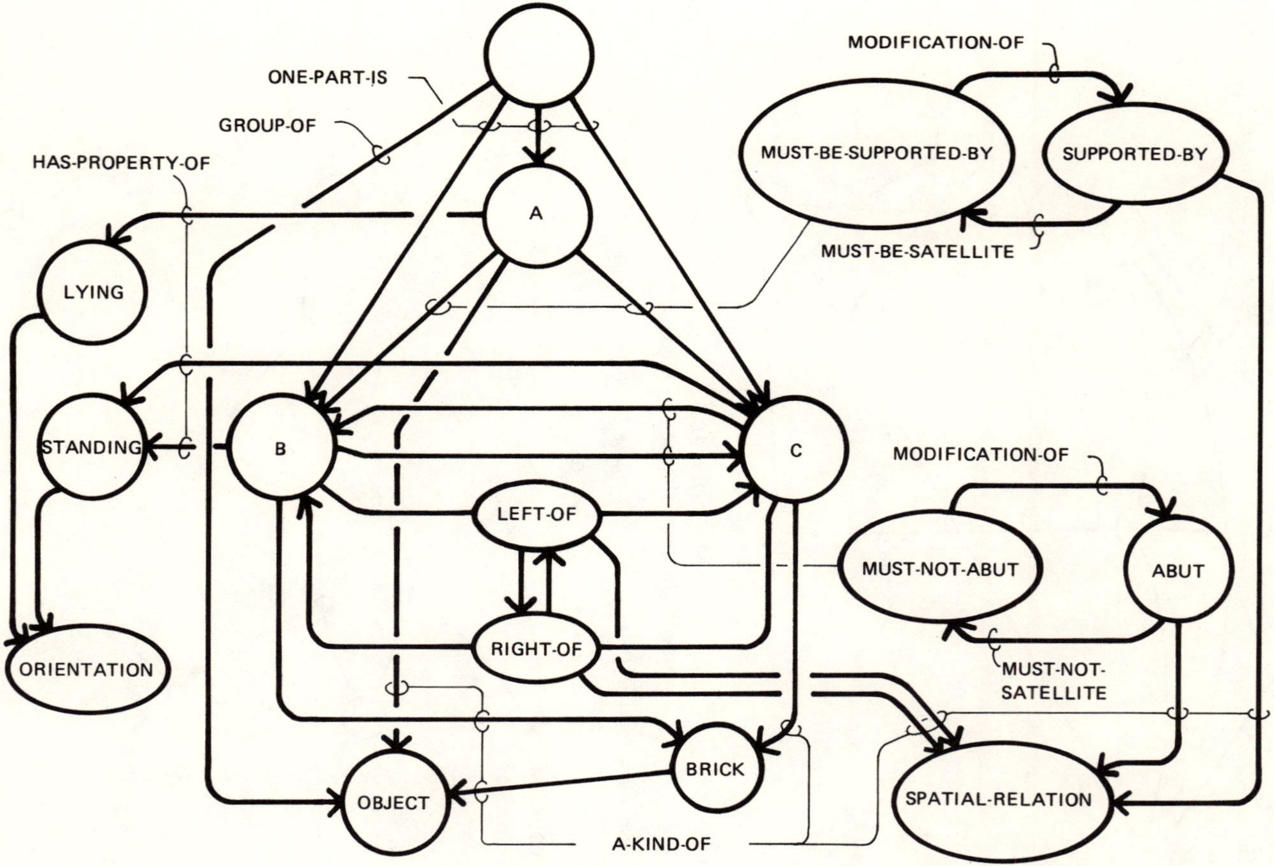

FIGURE 10.19

Source: From Winston, ed., *The Psychology of Computer Vision*, Figure 5.62, p. 198.

salient with respect to given interests and to be related in certain broadly specifiable ways.

Such expectations would play a role analogous to D. L. Waltz's jigsaw heuristics, in limiting the combinatorics involved. As the example of "chairman" suggests, they would themselves be heavily dependent on past experience within a particular physical environment and social milieu. Learning within a strange environment (such as a very different human culture, or the planet Mars) would very likely be hampered by the inappropriateness of the background expectations, since there is no guarantee that the sorts of categorizations previously found useful would still be salient. Insofar as such expectations have been "built in" during evolution (for example in the form of physiologically-based feature detectors like those to be described in Chapter 13), discounting their influence is probably more difficult. For Winston's program it is quite impossible, since the program has no way of learning that any of the features it knows about is *in general* unlikely to be salient. ("Salience" for Winston's program relates not at all to its interests, for it has none: rather, it relates to categorizations made by its human teacher for human purposes.)

Even Winston's program, however, can economically capitalize on the structures exemplified in past experience. If shown an arcade as in Figure 10.6, it can immediately recognize that the scene consists of three arches, since its network-matching module identifies substructures within the new description as corresponding to the schema of ARCH it already has. In consequence, it can learn from the counterexamples in Figures 10.7 to 10.9 without being overwhelmed by detail.

The role of counterexamples is crucial, since MUST-BE and MUST-NOT pointers are inserted at particular points in the model only at the instigation of certain types of NEAR MISS. (The preliminary discussion of counterexamples showed that this is necessarily so.) In general, the *sort* of difference between sample (whether EXAMPLE or NEAR MISS) and model directs the *sort* of modification employed.

Winston lists 18 distinct difference-modification pairs, which are applicable to a wide range of concepts since they are expressed in the abstract structural terms used by the network matcher.[7] Put informally, the matcher comparing Figures 10.11 and 10.12 notices that there is *a pair of relations* present in the current model though not in the example, but it is not interested in whether the relation is MARRIES or SUPPORTS . . . or MISTRUSTS. The formally defined "comparison notes" recorded by the network matcher include such members as: a-kind-of-chain; a-kind-of-merge; negative-satellite pair; must-be-satellite pair; and supplementary-pointer. Most of these can occur either in a comparison of the model with an example, or in a comparison of the model with a near miss; and the supplementary pointer can involve four specifically different structural contexts in either case, such as a negative-satellite in the model or a fundamental pointer in the near miss.

The crucial point that justifies this careful classification of types of conceptual difference is that each comparison note, according to the specified structural context of comparison, has its own particular significance in aiding development of the model. Whether or not the program inserts a MUST-BE satellite, and if so on which part of the network, is determined by the specific nature of mismatch between model and sample. In the final section of this chapter, we shall see that learning by doing also involves the intelligent classification or diagnosis of one's mistakes so that procedures may be modified accordingly.

The importance of the near miss (as opposed to the counterexample *simpliciter*) is that in such a case the comparison procedure discovers only a few differences between model and sample (with luck or good management, perhaps only one), rather than many. If there are several comparison notes recording differences between the two networks, then any one, or more, of these could be salient. For instance, if the network of Figure 10.14 (which corresponds to the system's concept prior to meeting Figure 10.3) were to be compared with the network describing Figure 10.10, there would be a plethora of comparison notes, and the problem would arise of sorting out which was relevant. Is it the presence of the Union Jack that prevents this from being a genuine arch? Or the holes? Or what? Clearly, nothing definite can be concluded from this sample (in *this* position in the training sequence) about how the concept should be modified. An attempt on the program's part to bear in mind all those differences for future reference, so that they might be gradually confirmed as salient cues or eliminated as irrelevant, would involve horrific combinatorial problems.

Psychological research on concept-learning confirms that people likewise become hopelessly confused if too many differences are presented at once.[8] The experimental method characteristic of natural scientific inquiry is itself designed to get round these combinatorial problems. Learning by example, then, depends upon a gradual improvement of representations of the world that is instigated by training sequences best structured in particular ways, and that utilizes appropriate symbol-manipulating mechanisms within the mind. If Humpty Dumpty's explanation of "outgribing" was deficient only in its inaccurate reference to whistling, then Alice would later be able to amend it by substituting "hissing" instead: but if his description hardly matched up at all to the sound heard in the wood, she would be at a loss what to conclude.

Even in carefully planned training sequences, however, it is often impossible to avoid introducing more than one difference at a time. Thus the progression from Figure 10.1 to Figure 10.2 in fact involves two differences, since *two* MARRIES pointers disappear. Winston's program therefore has to be able to cope with multiple differences to some extent, bearing in mind alternative hypotheses where necessary. In outline, it produces a set of alternative refined models, each of which embodies the hypothesis that a particular interpretation of a particular comparison note is salient. Then it ranks these hypotheses in order of priority, so that those most likely to be useful are considered first. As it encounters further samples in the training series, it gradually weeds out the models found to be incompatible with the incoming evidence (backtracking to follow up the remaining alternatives), so that the "tree" of potential models is successively pruned.

Multiple comparison notes generated by an example cause relatively little difficulty, since there is no problem of deciding which is most important. Because none of the observed differences debar the sample from being a member of the category, they are assumed to be equally relevant. Consequently, each is independently allowed to prompt amendment of the model in its appropriate fashion. This is why both the MARRIES pointers are dropped together in generating the second-stage model shown in Figure 10.14. Only if one of the comparison notes allows of multiple interpretations (as does a-kind-of-merge) are several alternative models generated.

The case of a near miss is more tricky, as we have seen, since the program has to figure out which (one or more) of the several comparison notes, in which interpretation, is salient. The program puts each comparison note onto one of two lists, according to the nature of the model/near-miss difference concerned. The primary list includes those differences most likely to be salient. For logical reasons, some differences can never appear on this list so need never be considered. Others are more variable: for instance, if a brick in one scene is replaced by a wedge in another, this difference may or may not be a likely candidate. Imagine a NEAR MISS to an ARCADE in which three ARCHES are piled one on top of the other as in Figure 10.9, but the top object of all is a WEDGE: even without knowing that it is arcades that are in question, one would probably be more impressed by the difference of orientation that is expressed at the level of *groups* of objects, than by the difference of shape affecting only a *single* object. In general, Winston's program opts for the least detailed differences in assembling the primary list; the relatively niggling differences are not totally ignored, but are placed on the secondary list from which they can later be promoted if necessary.

In forming and testing hypotheses on the basis of the primary list, Winston's program makes a very important, though not infallible, assumption, namely, that each difference functions *independently*. This does not mean that it assumes only one of the several comparison notes to be salient, but rather that it takes each difference that is salient to be salient irrespective of the presence of any other differences. In thus assuming a *linear* conceptual model, Winston avoids the combinatorial problem of simultaneously exploring all the possible permutations of differences. He points out that this assumption of linearity is a very common, and highly fruitful, strategy in scientific inquiry (as well as other types of learning from experience of examples). And it is useful also in learning by doing, as we shall see. No concept learner that is forced always to assume linearity could achieve the conceptual schemata possessed by people, for we are able sometimes to relinquish the assumption; but this does not prevent its being a reasonable first step, or heuristic strategy. The one exception to linearity allowed by Winston's program is when several differences gleaned from the near miss have virtually the same description, in which case they are all allowed to prompt amendment of the model. For instance suppose that a NEAR MISS to an ARCADE were shown as three ARCHES in a LYING position: it would be counterintuitive to ignore the symmetry or treat it as a mere coincidence, and more reasonable to conclude (like the program) that all three ARCHES must be STANDING in concert.

You may feel that it is not merely "reasonable," but *right* to interpret the hypothetical near misses to ARCADE in the ways described in the two preceding paragraphs. If so, you are probably intuitively relying on the fact that arcades are usable as passageways through which one may move from one space to another. With respect to this human interest, the shape of a top object is indeed irrelevant, and three fallen arches are clearly more obstructive than one. Winston's program has no sense of function, merely of structure. But a recent program of M. J. Freiling can recognize some of the functional possibilities inherent in structures like those considered by Winston.[9] Specifically, Freiling's system can recognize and express the way in which a wide variety of three-dimensional structures limits or enables the movement of a walking animal, a flying creature, or a rolling inanimate object. It would therefore have even more reason than Winston's program to interpret the hypothetical comparison notes regarding arcades in a sensible fashion.

But the important point is that one cannot hope to pass from hazy notions like "sensible," "reasonable," "counterintuitive" to straightforward *right* and *wrong*. As the long history of debate between essentialism and nominalism in philosophy shows, our concepts do not necessarily (if at all) follow "natural kinds" of things, and are more or less obviously constructed with human interests of differing types in mind.[10] Similarly, philosophical discussions of induction (such as Nelson Goodman's papers on the disjunctive predicate *grue*—"green or blue"[11]) show the difficulties facing anyone who tries to say that the previously cited response of Winston's program to Figure 10.5 is *right*, as opposed to intuitively reasonable. There is nothing in logic to prevent the program's amending its previous model to include BRICK-OR-WEDGE rather than PRISM (at least while there is no independent evidence that "prism-ness" is often important); the psychological fact[12] that people find disjunctive concepts very difficult to learn merely underlines the "acceptability" of the program's generalizing strategy, without showing it to be correct in absolute terms.

My reference to essentialism may have revived some readers' doubts about the wisdom of attempting to develop *any* computational representation of concepts, doubts that may have arisen when I first described the MUST-BE and MUST-NOT pointers inserted in the program's models. For Wittgenstein showed that one typically cannot assign rigid essences to concepts. He likened conceptual classifications rather to a system of "family resemblances," in which the similarities between individual examples of the concept overlap in an untidy fashion instead of a clearcut, all-or-none manner.[13]

However, this criticism cannot be leveled at Winston's program. For if it is asked what a particular structure is, or whether it is an X, it is able to recognize borderline cases as such—and, more importantly, it is able to describe the relevant similarities and differences. Wittgenstein's famous dictum, "Say what you choose!" assumes that one's choice is made *after* carefully considering the particular case concerned.[14] In short, the program has network-matching capabilities that enable it to specify the nature and closeness of analogies between different things. This ability will be further discussed in Chapter 11; for the moment, the point is that computational models of concepts do not necessarily assume an essentialist epistemology.

Learning by example, then, is not the "direct" apprehension of reality, unsullied by any intermediary interpretative activity. It involves the discerning development of descriptions, or interpretative schemata, representing the target domain, which are constantly checked by reference to example and counterexample so that salient cues are identified (what Piagetian psychologists term "assimilation and accommodation"). The ability to compare concepts with one another in a discriminating fashion is the basis of the recognition of analogy, in terms of which concepts can be intelligently used to describe things that were not included within the target domain as initially conceived.

In the next section we shall see that learning by being told also involves the active manipulation of inner representations of the world, and the comparison or knitting together of descriptions arising from different sources. Equally, a program that could learn a wide range of things "simply" by being told them would need considerable knowledge in the form of jigsaw (compatibility) heuristics, just as would one that could learn about things "simply" by being shown examples. For one has to know a fair amount about something already if one is to profit by being told new facts about it.

## LEARNING BY BEING TOLD

It is tempting to think of learning by being told as a mere passive reception of new information, requiring no thought on the part of the learner. And some programs do indeed "learn" in this way: thus I pointed out in Chapter 3 that Colby could have "taught" his artificial neurotic to believe *My friend will betray me* simply by adding this item to the Belief Matrix. But the scare-quotes here suggest that this passive acceptance of information hardly qualifies as learning at all. Careful consideration of cases where a person learns something by being told it will show that a good deal of thought actually goes on.

For instance, if Colby's human patient had been told "Your friend will betray you," she would (whether consciously or not) have assessed the credibility of the remark, by reference to such matters as her opinion of the speaker, her friend's past behavior, and her views on the trustworthiness of social contracts in general. Simple examples of these types of credibility tests were included in R. P. Abelson's earliest computer simulations of belief, described in Chapter 4. And many examples of spontaneous thought on being told something were mentioned in Chapter 7, in connection with R. C. Schank's Conceptual Dependency analysis of natural language. If one is told that "Ms. Brown pushed the table" one normally infers that the table moved, and one may be quite unaware that this was not explicitly stated.

A more complex example is contained in the following story. A mayor was delayed on his way to open the town fete, and the local priest was asked to speak to the crowd meanwhile. He described some aspects of his pastoral experience, including his distress and embarrassment when, many years before, his first penitent had confessed to a particularly nasty murder. Eventually the mayor arrived, and in thanking the priest for entertaining the audience in his absence, he said: "I am very glad to see Father Brown here today. He and I are old friends—indeed, I was his very first penitent." Naturally, many of the listeners immediately realized that the mayor was the murderer.

This realization counts as a case of "learning by being told" because it was seemingly immediate, involving no introspective difficulty and probably no consciousness of inference at all. Yet inferences obviously had to be made here, since neither speaker explicitly stated that the mayor was the murderer. This would have been so even if the priest had indiscreetly said: "My first penitent was a murderer; actually, the mayor was my first penitent." Moreover, we shall see presently that the inferences involved are more complex than the automatic assumption that the table moved, given that "Ms. Brown pushed the table," involving the creation of symbolic structures that are specifically suited to the assimilation of information presented after the structure is first set up.

In other words, learning by being told often requires inferential thought processes classifiable as "problem solving," so that there is no clearcut distinction between the two psychological phenomena.[15] Indeed, some members of the audience may not have drawn the obvious conclusion on the spot—only to be "struck" by it later in their bath, or to have it explicitly spelt out to them by a friend. Since what one is told is more usually a sequence of statements, as in a story or conversation, rather than one isolated sentence, learning by being told typically requires the comparison and knitting together of inferences initially triggered by many different sentences.

The "demon-demon interaction" sketched by Eugene Charniak for the purpose of understanding children's stories represents this implicit inferential thinking, and his most recent work on making sense of what one is told will be mentioned later in this section.

D. V. McDermott's program "TOPLE" employs inferential strategies rather similar to demons on being told such things as: "The banana is under the table by the ball. The monkey goes over to the table and picks up the banana. She eats it."[16] Before actually being told that the monkey picked up the banana, TOPLE had predicted that she would pick up either that or the ball; similarly, it had assumed that the monkey was hungry before receiving the confirmatory evidence of the final sentence. TOPLE's understanding differs from SHRDLU's (which also makes inferences on being told things) in that it can hypothesize alternative possible worlds and bear them in mind simultaneously, gradually weeding out those that conflict with the information as it comes in; SHRDLU, by contrast, has to plump for *one* prediction, backtracking if it turns out to be false. (TOPLE's ability to weigh alternatives depends on McDermott's use of CONNIVER instead of PLANNER, as we shall see in Chapter 12.) McDermott's program inhabits a strictly limited and well-defined microworld (comparable to SHRDLU's environment), within which it is able to make clearly reasoned choices between specific potential inferences.

Perhaps the most extensive theoretical model to date of learning by being told is the MEMORY module of MARGIE, the system of Schankian programs described in Chapter 7. As its name suggests, this module is intended as a simulation of human verbal memory, which is studied experimentally by cognitive psychologists.[17] C. J. Rieger's MEMORY program represents the world in a way that enables it to make many simple conceptual inferences justified by Schankian semantics, such as that a pushed table has probably moved. More important, it is able also to interrelate these inferences so as to interpret sets of sentences (sets of Schankian "conceptualizations") in terms of scenarios expressing cause and motivation, and so as to establish identities of reference that are not explicitly stated. For example, MEMORY could learn that the mayor is a murderer by realizing that "Father Brown's first penitent" refers to the very person who is also referred to by "the mayor" and "the murderer." As we shall see, this realization is a complex matter, despite its introspective immediacy. Mapping linguistic cues (such as referring phrases) onto target realities (such as individual people) is an interpretative activity requiring a great deal of knowledge, both general and specific.

On being told something, MEMORY stores it and then asks if there are any significant inferences (or interactions between inferences) relevant to it. The "inference evaluator" searches for mutual confirmations and contradictions between the spreading inferences, and for interesting new information of various types (including the identity of items referenced by referring phrases).

One of the criteria of "interest" or "relevance" is the degree to which new information is surprising. But what is perceived as surprising depends upon background knowledge and expectations: a confirmed cynic, or someone who had already decided that the mayor was a nasty piece of work, would be less astonished than her neighbor on hearing his unwitting confession. Accordingly, MEMORY's ideas about what is or is not normal play a large part in the comparison and evaluation of inferences, as well as affecting the content of inferences drawn in the first place.

Since MEMORY's inferences are applied to information represented in a Schankian fashion, they naturally recall the conceptual relations marked by Schank's semantics and Abelson's theory of belief. For example, types 7, 8, and 9 (below) are concerned with the enabling and interfering influences on purposive action that are represented by Abelson at the levels of *plans, themes,* and *scripts*; correlatively, MEMORY embodies highly general knowledge about the abstract psychological structure of human action and intention, as well as culture-specific knowledge about motives and "normality," which were less systematically represented by the "implicational molecules" of Abelson's earlier work. The sixteen classes of MEMORY inferences (of which you may like to think up some examples) are as follows:[18]

1. *specification inferences*: what are the missing conceptual components in an incomplete conceptual graph likely to be?
2. *causative inferences*: what were the likely causes of an action or state?
3. *resultative inferences*: what are the likely results (effects on the world) of an action or state?
4. *motivational inferences*: why did (or would) an actor want to perform an action? What were her intentions?
5. *enablement inferences*: what states of the world must be (must have been) true in order for some action to occur?

6. *function inferences*: why do people desire to possess objects?
7. *enablement-prediction inferences*: if a person wants a particular state of the world to exist, is it because of some predictable action that state would enable?
8. *missing enablement inferences*: if a person cannot perform some action she desires, can it be explained by some missing prerequisite state of the world?
9. *intervention inferences*: if an action in the world is causing (or will cause) undesired results, what might an actor do to prevent or curtail the action?
10. *action-prediction inferences*: knowing a person's needs and desires, what actions is she likely to perform to attain those desires?
11. *knowledge-propagation inferences*: knowing that a person knows certain things, what other things can she also be predicted to know?
12. *normative inferences*: relative to a knowledge of what is normal in the world, determine how strongly a piece of information should be believed in the absence of specific knowledge.
13. *state-duration inferences*: approximately how long can some state or protracted action be predicted to last?
14. *feature inferences*: knowing some features of an entity, and the situations in which that entity occurs, what additional things can be predicted about that entity?
15. *situation inferences*: what other information surrounding some familiar situation can be imagined (inferred)?
16. *utterance-intent inferences*: what can be inferred from the *way* in which something was said? Why did the speaker say it?

What conclusions would MEMORY draw if, like the audience at the town fete, it were to be told by the priest that his first penitent was a murderer, and by the mayor that he was the priest's first penitent?

The initial input to MEMORY is a semantic representation of the speaker's words in the form of a conceptual dependency diagram (such as could be generated by C. J. Riesbeck's analysis program). The first thing MEMORY does is to store this information—but it does so in a rather different form, called a "conceptual graph." The reason for this is that MEMORY, like people, is interested not so much in abstract meanings as in their application to actual individuals. In other words, it wants to know *who in particular* had a murderer as his first penitent, and *who in particular* was that penitent. So MEMORY always tries to link the general meaning (or "connotation") of a sentence to its particular point of reference (or "denotation").

In our imaginary example, the concept "my" would therefore be connected by MEMORY to a special label (say, C2493) denoting the speaker, namely, Father Brown. But the concept "my first penitent" has to be linked to a dummy label (say, C2517) because at this stage MEMORY has no knowledge of who that person is. The concept "murderer" is also attached to this dummy label, so that the conceptual graph stored by MEMORY represents the information that someone or other (tagged

C2517) is *both* a murderer *and* Father Brown's first penitent. The graph also stores the information that the second predicate applied to the unknown individual at some time in the past.

For the moment, let us ignore the diverse inferences that MEMORY immediately draws from this conceptual graph, and the ways in which the program then links it to neighboring conceptualizations (such as the priest's admission that he was both frightened and embarrassed at the time). Let us "skip" to the point at which the mayor declares "I was Father Brown's first penitent." On being told this, MEMORY generates a conceptual graph showing that someone—C2534, namely, the mayor—was Father Brown's first penitent. The question now is, how does the program come to realize that C2534 and C2517 are labels denoting one and the same person? (How do you think you achieve this insight?)

The basic requirement is that the descriptions associated with the two numerical labels be compared: do they match sufficiently closely for the system to conclude that they refer to identical individuals? The criteria of closeness of match are crucial. The more restrictive they are, the less prone the system will be to jump to unwarranted conclusions; but the required match must not be so close as entirely to rule out intelligent intuitions about matters that have not been explicitly stated. What is needed is reasonable insights as opposed to wild imaginings.

In effecting references, the first step of Rieger's MEMORY is to reorder the descriptions within the "floating" descriptive set, so that the ones listed first are those that are more likely to be useful in this enterprise. For example, proper names (like BROWN), sex (like MALE), and class memberships (like ISA-PERSON or ISA-MAYOR) are rather more likely to be useful than predicates denoting temporary states and acts (like HASA-REDSETTER or IS-SITTINGDOWN). MEMORY uses a heuristic measure of likely significance that places "useful" predicates at the top of the list. If you try to work out what these heuristics might be you will find that they do *not* necessarily order from most specific to least specific, that they depend on common-sense knowledge of the world as well as "general" semantics, and that they are not foolproof. (Consider how HASA-REDSETTER might have contributed to the discovery of the mayor's guilty secret, had Father Brown spoken of "*one of* my first penitents" and remarked on the magnificent beast left waiting at the confessional door.)

MEMORY's reference mechanism searches for a referent in a way that exploits the fact that significant predicates are listed first. It starts with the first feature on the list and locates all the entities in the memory that share this feature. The process is then repeated on this subset until *either* no more features are left on the list, *or* only one candidate remains that shares the features examined so far, *or* the next feature listed would eliminate all the surviving candidates at a stroke. If you try simulating these rules with pencil and paper, you will see that there will rarely be a situation in which no possible candidate is found (although there may be no way of picking out *one* of the possible candidates, in which case the question of referential identity must be left open, at least temporarily). You will see also that the function of the third "stoprule" is to prevent idiosyncratic, and so probably previously unknown, information on the new list from blocking identification with other descriptive sets on which this item does not occur.

For example, in the case we have been imagining, the floating descriptive set for which a referent is required is the one tagged C2517: ISA-PERSON, SEX-MALE, ISA-MURDERER, IS-FATHER BROWN'S FIRST PENITENT (TIME PAST). This description is originally introduced into the audience's minds by Father Brown himself, in his speech pending the mayor's arrival. All persons known to MEMORY are immediately located, and as a second step the males are separated out as the class of potential felons. But the process has to stop here, since the next feature would eliminate all candidates at a stroke, there being *no* entity described as a murderer (or, for that matter, as the priest's first penitent).

I remarked in Chapter 7 that among the "interesting information" dumped by Rieger's memory module onto N. M. Goldman's response generator is the existence of "dummy" markers; accordingly, were MARGIE to be present at the scene, she could ask Father Brown at this point "Who was your first penitent?" Unfortunately, for MARGIE if not for the mayor, this admirably straight question is not one likely to elicit a straight answer.

However, MEMORY (and therefore MARGIE) has the inner resources to keep in mind the descriptive sets whose referents have not been successfully identified, so that information received later in the discourse can be knitted into these latent threads to produce the answer required. Thus when the mayor casually remarks that he was the priest's first penitent, MEMORY recognizes this as a concept that has already been "activated" by previous thinking on the program's part. A quick check shows that the context was one of attempted reference, and the newly introduced descriptive set (attached to the mayor) is compared with the one that proved so puzzling before. The match of the two sets is not perfect, since ISA-MURDERER occurs only on the first. But since one and only one of the (male) persons previously located as possible candidates shares the feature of being Father Brown's first penitent, that one—*viz.* the mayor—is deemed to be the villain.

It should be clear that the procedure I have described is very crude, compared with what probably goes on (*what* probably goes on?) in the human mind. Even allowing for the brevity of my remarks, and the attendant failure to do justice to Rieger's actual program, there are obvious points at which further subtleties could be introduced.

For example, suppose that TIME PAST were accompanied by a numerical index—"twenty years ago," or "nearly thirty years ago": how close should the two indices be for the match to be regarded as satisfactory? Any answer would not only depend on one's views about what types of time perception are normal, but would doubtless vary with the context: a discourse on dinosaurs should provide little room for quibbling about a mere ten years either way, whereas an apologist for the mayor might conceivably seize upon this discrepancy in arguing a case of mistaken identity. Moreover, the "first penitent" story implicitly draws on our understanding of ordinal numbers: only one individual can possibly be truly described as Father Brown's *first* penitent. But MEMORY does not know this and so cannot exploit it in evaluating a suggested reference. Nor would its limited understanding of superlatives lead it to baulk at the cosmetic advertisement I once saw assuring me that "You, too, can be the most beautiful woman in the world."

In these ways, and many others, MEMORY is not so clever as it seems. We have seen already that a system's "reasoning power" depends crucially on the richness of its semantics and the extent of its knowledge of the world. Making MEMORY less stupid would be a nontrivial task.

Even apparently simple inferences (such as the mayor-murderer case) require computational resources that far exceed the complexity of what psychologists usually put in their theories. In writing an artificial intelligence program to perform the equivalent tasks, one must specify precisely the symbol-manipulating mechanisms and knowledge concerned. SHRDLU, for instance, could not recognize the fact that the mayor was the murderer. The reason is that SHRDLU has a relatively *closed* data base, in the sense that it needs to *find* a referent already in its world for each definite noun phrase in the sentences it interprets. In discussing MEMORY, we saw that Rieger's program can create a dummy label for the concept "my first penitent" if it does not yet know who the referent of that concept is. It later attaches the concept "murderer" to the same dummy label—and only later, if at all, does it establish the actual referent (named as the mayor). Any system capable of realizing that the mayor was the murderer would have to possess computational mechanisms defining a data base that is "open" in this sense, although they need not be identical to those embodied in MEMORY.[19]

It would similarly be a nontrivial task to improve MEMORY's "inference evaluator." This is a computational mechanism that detects points of contact (confirmations, contradictions, and augmentations) between the conceptual inferences of various types that spread out like ripples from the original conceptual graph representing "what MEMORY is told." Again, assumptions about what is or is not normal are crucial in assessing the plausibility of a single inference and the compatibility of two inferences (is it normal to be embarrassed by a confession of murder, or should we regard the priest as either oversensitive or a liar?). Sherlock Holmes solved one of his cases only because he noticed how strange it was that the dog did *not* bark in the night. MEMORY, too, notices prima facie incompatibilities and actively looks out for further information relevant to their resolution, much as it watches for future facts (inputs or inferences) that will establish the identity of an unknown referent.

Rieger stresses that in making sense of what one is told one typically has to deal with "fuzzy" matches between descriptions—fuzzier than the matches considered by Winston's program, and fuzzier even than that central to the "first penitent" story. In brief, the question asked by MEMORY is more often "Is it likely?" or "How likely is it?" than "Is it definitely so?" what is deemed most likely being taken as true in default of specific information to the contrary. We have already noted that Charniak's demons fulfill a similar function (the dime "must" have come out of the piggy bank), and in Chapter 11 we shall see that intelligence in general requires the use of representations (variously termed "models," "schemata," "scripts," and "frames") that can provisionally supply default values of specific variables as well as defining the broad outlines of the phenomenon being thought about.

Rieger regards the 16 types of inference as conceptual "reflexes"—that is, as thoughts that occur spontaneously in the absence of special inhibiting or directive conditions and that form the essential basis of any higher level interpretative process. Like Schank, he sees his work as an exercise in computer simulation, claiming that the processes of human memory (and of any intelligent machine interpreter of natural language) must be broadly similar to those implemented in his program. Though human understanding undoubtedly employs more inferential types than MEMORY does, he believes that these must include those he has specified, and in any case probably amount to no more than 30 classes in all. He sees these reflex inferences as exploratory in nature, aiming to expand the semantic content of the input "bottom-up" as fully as possible to see if anything of interest should turn up (which of course does not prevent the superimposition of high level "top-down" thought processes searching for specific implications).

For instance, consider Rieger's account of what goes on in someone's mind (and what should go on in MEMORY) on being told that John McCarthy went to Boston:[20]

> "He went to Boston, eh? That means he was in Boston, and he probably wanted to be there. Why would he want to be there? Probably to do something which requires his personal presence, like talking to some other high-up. Oh yes, he probably went to talk to someone like Minsky at MIT . . . about grant money or new research proposals, or something

> like that. That's understandable. Of course, it could be just a vacation. How did he go? Probably by flying. That's OK, he has the means, and there's no air strike on. Wait a minute. I thought he was giving a talk here tomorrow. Either he'll be back then, or it's been called off, or something. Better find out. . . ."

The interplay of general and specific knowledge here is evident. Someone who did not know that McCarthy and Minsky are both "high-ups" in artificial intelligence laboratories on opposite coasts of the USA would learn much less than this on being told "John McCarthy went to Boston." Even so, she would be in a position to ask the same sorts of question as are posed in the above passage. Indeed, there are many other questions (based on the 16 inference types) she could ask, and additional knowledge of specifics allows—potentially—for even more.

Clearly, then, Rieger is not to be interpreted literally when he speaks of MEMORY (or memory) drawing "all possible" inferences. He assumes[21] that with greatly increased inferential power, *fewer* inferences will be recognized as applicable to any given meaning structure. But the heuristics for stopping and organizing inferences have yet to be worked out. The concept of *frame systems* to be discussed in the next chapter is centrally relevant, though as yet relatively ill-formulated.

Rieger has not yet had to face this problem in practice, primarily because current computers are too small to allow MEMORY to run in more than one inference mode at a time (although sets of diverse conceptualizations can of course be fed in by hand as material for the inference evaluator, whose function in principle is to compare inferences arising spontaneously from many different sources), the "depth" of inferences being arbitrarily limited to ten. A further reason is that he often concentrates on one sentence only, such as "John McCarthy went to Boston," so that the problem of knitting together many sentences in an intelligently selective fashion does not arise. But in more extended discourse, like Father Brown's speech or a paragraph of text, this question becomes crucial if the combinatorial explosion is to be avoided.

To some extent, one is helped by the fact that compatibility heuristics have usually already influenced what one is told in the first place, due to the (conversational and literary) convention that one only tells someone things one regards as relevant. This is why it is so difficult to follow someone whose speech is rambling and parenthetical.

This fact has been noted by Schank, in a discussion of how one might go about paraphrasing (expressing the gist of) entire paragraphs or stories.[22] He assumes that one may take what one is explicitly told as clues to important themes that are likely to be developed later. His basic strategy is to identify the underlying causal chains expressing the semantic relation of ACT-results in-STATE-enables-ACT. These can be chained and nested, and obviously bear a close relation both to Rieger's inferences and to Abelson's plans, themes, and scripts. Schank shows how "dead-end" chains are to be identified and then abandoned, how an appreciation of necessary conditions for certain actions guides thinking, and how "peculiarity markers" are used to sensitize the search for unusual inferences that are likely to represent the *point* of the story. Schank's account is very sketchy, but it is an initial attempt to identify some of the heuristics that one might use to understand Father Brown, to fathom what *Romeo and Juliet* is all about, or to draw a moral from verses written in one's autograph book.

However, even if adequate compatibility heuristics were available, a great deal of inferential "waste" would be inevitable in systems that could learn by being told. Rieger's basic tenet is that one typically does not know just what will turn out to be interesting, so that many inferences must be made initially in case a few of them should turn out to be useful later. He thus contrasts his model of memory with more "problem-directed" approaches like SHRDLU, which draw an inference only in response to a specific inquiry, and likens it rather to "demon-based" approaches like Charniak's.

Rieger's reason for preferring spontaneous inferences to demons is that the interpretations effected by demons are normally represented as *potential* rather than *actual*, so that while they are lying in wait for their conceptual pattern, the information that the alerted demons contain is not available to other parts of the system that might need it for some purpose of their own. For instance, Sylvia Weir's "collision-expecting" demon (described in the previous chapter) lies in wait for something to happen at a particular point in the visual field that it will then interpret as a collision; but until something actually does happen there, no use can be made by the system as a whole of the insight that a collision is likely. But, depending on what was expected to collide with what, it might be useful for other procedures, such as hand-raising or child-comforting modules, to know of the expectation. This difference between Rieger's and Charniak's approach is not radical, however, and demons could readily be made to express their information more openly (though the problem of *finding it* intelligently would remain).

The broad similarity between Rieger's and Charniak's accounts of what one thinks when one is told something is shown by some of Charniak's more recent work.[23] He has identified a general psychological assumption that is closely comparable with MEMORY's *motivational* and *enablement* inferences and that he regards as a universal (culture independent) feature of human understanding.

Informally, his "R + SSA" rule can be stated thus: if you are told something that makes it plausible to infer that a person has a reason for wanting action A to be performed, and that person does S—where S is a means to (a significant subaction of) A—then you may infer that the person is doing A. For example, if you are told that Guy Fawkes has criticized Parliament and has been seen purchasing gunpowder, you may draw the obvious conclusion about what he is up to.

Charniak shows how this rule, with the requisite knowledge of reasons and significant subactions, can be employed in one of the two possible paths to understanding the final sentence in this story:

> Today was Jack's birthday. Janet and Penny went to the store. They were going to get presents. "I will get a top," said Janet. "Don't do that," said Penny. "Jack has a top. He will make you take it back."

Difficult as it may be to believe, Janet has to deploy a great deal of knowledge, some of which is highly culture specific, in learning with apparent ease from Penny's words that Jack will make her take the new top back to the toystore to exchange it for something else.

Charniak's lengthy discussion of this deceptively simple passage raises even more questions than it answers, many of which are relevant to a wide range of cases of learning by being told. These include questions about what is to count as "evidence" for an assumption, how one is to decide that someone has a reason for doing something, and how one is to represent "biplausible" conceptualizations that could (at least provisionally) equally well be regarded as true or false. As one might expect, preferential notions of probability and normality, with all their attendant risk of error, are crucial to Charniak's account: insistence on *rigorous* thinking (in the sense defined in Chapter 1) would buy infallibility only at the price of gross stupidity and an inability to learn from communications with one's fellows.

In short, any system (whether HAL or human) that can learn "simply" by being told must be able to tolerate these fuzzy epistemological criteria, as well as having a richly equipped memory and a powerful inferential (problem-solving) competence in terms of which to read between the lines. And learning by being told (like other forms of learning) in a sense involves *doing* something, namely, thinking. But "learning by doing" is more commonly understood as learning how to do something by trying to do it, or learning how to do it better by repeated practice. These types of learning are discussed in the next section.

## LEARNING BY DOING

Practice, they say, makes perfect—but how? Learning how to do something better by doing it is sometimes merely a matter of learning to do "more of the same." For instance, Colby's neurotic program develops an idiosyncratic manner of coping with anxiety-ridden beliefs over a given series of runs, by noting at the early stages which defense mechanisms are most successful in reducing emotional charges and making them more readily accessible thereafter. The program has no notion of *why* a particular mechanism is or is not helpful, but relies apparently on a blind allegiance to the motto that "nothing succeeds like success."

A. L. Samuel's checkers (draughts) player—the most impressive of the early learning programs—also relies relatively blindly on its memory of past successes to improve its current performance.[24] It does this in two ways, which Samuels called "rote learning" and "learning by generalization."

In the first of these, the program stores all past positions together with the evaluations it made of them initially. Since the stored record of a given position can be found quickly, this saves time that can be used for further computation. (Samuel stressed the importance of indexing, and fast sorting and searching procedures: without such procedures, searching a very large store of records for the one required might take longer than working out the evaluation again.) Moreover, because the evaluation concerned is a decision based on a "look-ahead" of a fixed depth—for example, three moves ahead—the saving of evaluations can increase the effective depth of the look-ahead. Thus if the program looks ahead three moves, and finds that the third move is one for which it already has an evaluation, this is equivalent to its looking *six* moves ahead. And since the process can be repeated indefinitely, the program can "pull itself up by its own bootstraps" so that the more recent evaluations rest on an effective search that is considerably deeper than the official limit of three.

Samuel's "generalization" procedure is a way of improving the evaluation decision itself, by continuously adjusting the weighting of the test parameters involved according to their success in actual performance. Thus Samuel provided the program with a list of 38 parameters marking strategic features of the game (such as "threat of fork" and "center control"); starting with a subset of 16 of these, the program experiments to see which are most useful, the least helpful ones being replaced by others drawn from the "reserve" pool. The actual evaluation procedure is fairly complex, and allows for different mathematical weightings of individual parameters. Since different opponents have different weak spots, a parameter that is normally not much help can be very effective against a particular player. (Since the program treats each board position as a new problem, it cannot sustain an overall strategy that is developed throughout the game.)

These types of learning, then, enable a game-playing program to take advantage of its experience. And Samuel's program (which beat him regularly even in its early stages) improved with practice to such a degree that it once beat a checkers master, the champion saying later: "In the matter of the end game, I have not had such competition from any human being since 1954, when I lost my last game."[25] He added that, from a position roughly halfway through the game, "all the play is original with us, so far as I have been able to find." Evidently, *more of the same* can produce some surprises.

But you may feel, even so, that increasing one's skill by practice can be a decidedly more intelligent matter than this: that Samuel's program, for instance, should understand why center control is useful, and should be able to think up new strategic parameters for itself, instead of being confined to permutations within the set provided by the programmer. You may feel also that one ought to be able to learn something positive from one's mistakes, as well as from one's successes, instead of merely refraining from repeating the unsuccessful activity or decreasing its numerical weighting.

These intuitions about learning inform the work on two recent programs, "STRIPS" and "HACKER," which increase their skill by generalizing solutions in a more intelligent way than Samuel's checkers player. Each of these programs relies crucially on knowledge of the purposive structure of the task. As we shall see, this knowledge is represented rather differently in either case, but STRIPS and HACKER use it in broadly similar ways so as to learn from their experience.

STRIPS (so called because it is the STanford Research Institute Problem Solver) can solve a wide range of different problems, and is used, for instance, to produce plans of action for execution by the SRI mobile robot, "SHAKEY."[26] SHAKEY spends much of his time in a world consisting of seven rooms, variously interconnected by eight doors, and containing several large boxes which he can push from one place to another. If he is asked to move Box 3 from Room 5 to Room 2, SHAKEY first calls on STRIPS to work out a pathway, and a sequence of pushes, that will achieve this goal. For present purposes, we need not ask *how* STRIPS solves such problems (a question taken up in Chapter 12), but rather how it can learn to do so better.

The crucial point is that STRIPS produces a means-end analysis of the task, and expresses the plan as a series of actions that cumulatively establish the preconditions necessary for the final, consummatory action. For instance, if the robot is required to open a window, he must first get himself into the relevant room and ensure that there is a box under the window to climb on, bringing one into the room if necessary; then he must climb on it; and only *then* is he in a position to open the window. (The current version of SHAKEY does not have the motor abilities of climbing or of opening windows; but this does not affect the present point, which is that STRIPS can tell the robot what would need to be done to achieve the end.)

STRIPS's representation of the plan not only lists the relevant actions in order of their execution, but also clearly shows *which* actions (or subsets of actions) satisfy *which* preconditions for *which* other actions (or subsets of actions). It also distinguishes the important effects of actions from their side effects. And it shows how the state of the world changes progressively at different steps of the plan. In short, the inner purposive structure of the plan is made apparent by the representation, which can consequently direct flexible use of it in varying circumstances *without* the need for a new plan to be worked out. The representation that fulfills these functions for STRIPS in an economical and perspicuous way is called a "triangle table," and its manner of construction will be explained presently.

Suppose that SHAKEY (sitting in Room 7) has been asked to open Window 2 in Room 3, and that STRIPS has produced a plan accordingly. This plan is expressed as a triangle table that tells SHAKEY how to get from Room 7 to Room 3, where to pick up a particular box (Box 1 in Room 4, say), how to push it into Room 3 and under the window in question, and how to get into position for opening the window. In order to produce this plan, STRIPS relied on specific information such as where suitable boxes were located at the time. Like Samuel's program, STRIPS *learns* from this particular problem-solving experience in two ways, by storing its decision (plan) and by generalizing it.

The benefits accruing to STRIPS by storing a decision for future reference are the same as for the checker-playing program: it saves computation time and it increases the system's effective computational power. Thus if STRIPS has already thought out how to open a window, this plan can be inserted *as a whole* (as a complete triangle table) into future plans that would otherwise be too complicated for STRIPS to work out. Plans expressing complex actions are called "MACROPS" (macro-operations); since STRIPS can combine MACROPS into strings that are themselves (higher level) MACROPS, its experience cumulatively increases its ability to cope with new, complex, problems.

The generalization available to STRIPS, by contrast, is considerably more powerful than that blindly carried out by Samuel's system. STRIPS can be said to generalize intelligently in two senses. First, it can convert a plan initially produced with a specific problem in mind into a plan schema covering an indefinite number of plans. For instance, the triangle table expressing the plan "From Room 7 to Room 3 via Room 6, to open Window 2, using Box 1 fetched from Room 4" is converted into a generalized table expressing the plan schema "From any room to any other, via some other if necessary, to open any window, using any box found in any place." For this schematization, the descriptions in the original triangle table have to be made more general in logical type. In outline, what STRIPS does is to replace terms that refer to specific individuals ("logical constants") by terms for which one can substitute any constant of the relevant type ("variable parameters"). During subsequent execution, then, STRIPS merely substitutes the appropriate room, box, and window constants for the variables in the plan schema, and goes ahead without having to work out afresh how to open a different window in different circumstances.

The second sense of generalization is even more interesting, in that it provides for a flexibility of behavior that allows SHAKEY to adapt himself to changed situations on the spur of the moment—and which is not available to the dungbeetle. The dungbeetle is a creature that, at a certain stage of its lifecycle, executes a dungball-rolling plan. It grips a ball of dung between its feet, rolls it along the ground to the top of a burrow at the bottom of which it has already deposited its eggs, and taps it into place so as to seal the entrance to the burrow. "How clever!" you may say. But if the dungball is very gently removed from the beetle's grasp at an early stage of the plan, the remainder of the plan is executed *precisely* as before—except, of course, that there is actually nothing to roll, nothing to tap, and nothing waiting for the hungry larvae when they hatch.

Clearly, the beetle is not so clever as at first it seemed. It can execute its plan, certainly, but it cannot monitor this execution and vary it accordingly. The dungbeetle cannot ask questions like these at every step of its plan:[27]

(a) Has the portion of the plan executed so far produced the expected results?
(b) What portion of the plan needs to be executed next so that after its execution the task will be accomplished?
(c) Can this portion be executed in the current state of the world?

SHAKEY, however, has a monitor program (called "PLANEX") that can do this, by reference to the information about the purposive structure of its action that is stored in the triangle table provided by STRIPS.

To take a concrete example involving these three questions, suppose that SHAKEY approaches Window 2 only to find to his surprise that there is already another box there, so he cannot push Box 1 under it: what should he do?

He could move the intruding box out of the way, so that the world conforms to his previous expectations, and then push Box 1 up as originally planned—but this would be stupid. The point of having Box 1 there at all is merely to establish a precondition of his reaching the window, and *any* box would do. For similar reasons, it would be absurd to push Box 1 right up to the unexpected box so as to clamber over both of them before reaching the window. If SHAKEY has any sense, then, he will abandon Box 1, climb onto the serendipitous box, and open the window—thereby saving himself some trouble. And this is indeed what SHAKEY's monitor program advises, since the schematized triangle table provided by STRIPS represents the fact that Box 1 was fetched purely to fulfill a particular purpose, and is no longer needed if that purpose is prematurely achieved in some other way. It clearly shows, also, that the tail portion of the plan (climb onto box, then open window) can be executed in the current state of the world. So, on this occasion, the plan requires no "patch": in other words, STRIPS need not be recalled by the monitor to work out how to adjust the unexpected situation so as to make the tailplan feasible.

Bearing this example in mind, please look now at the schematic triangle table of Figure 10.20. The numbers 1–4 represent the names of individual actions in the order in which they are to be performed. So "1, 2, 3, 4" is itself a superficial representation of the plan, which states what has to be done when, but affords no inkling as to why. The "why's," the "how's," and the "what happens when's" are all contained within the body of the triangle.

STRIPS's understanding of the "why's," the "how's," and the "what happens when's" is rooted in its background knowledge of the preconditions and results of actions. The knowledge of results is stored in the data base in the form of a set of newly true (and another of newly false) facts for each primitive action, to be added to (or deleted from) STRIPS's overall world model whenever the action is performed. For instance, the execution of "Action: climb onto box" adds the fact "SHAKEY on box" and deletes "SHAKEY on floor." If a fact added by one action forms part of the precondition set for another, then the first action can in principle be used to help achieve the second. In short, STRIPS shares the sort of common-sense knowledge about actions that is embodied in Rieger's MEMORY program by way of the resultative, action-prediction, and three varieties of enablement inferences. And STRIPS uses this knowledge to fill each cell of the triangle table with a list of descriptions, as follows.

Let us consider the cells in Column 1 first. (Please note that this is *not* the leftmost column; each column *except* the leftmost is labeled with a specific action name.) In the top cell STRIPS lists all the facts that are normally newly true after doing Action 1, the first action of the plan. In the next cell down, STRIPS enters all of the facts listed in the previous cell that will *remain* true after (i.e., are not deleted by) the second action, and so on for the next two cells. Columns 2, 3, and 4 are constructed in the same way. Each numbered column, then, maps the survival through time of the positive effects of the action concerned.

The cells of the leftmost column cannot be filled in this way, since it bears no action name. This column represents the state of the world prior to the first action. Or, rather, it represents those facts about this world state that are relevant to the plan, namely, the preconditions for the plan. Going down the leftmost column, the top cell is filled by the list of preconditions required for Action 1; the next cell lists those preconditions,

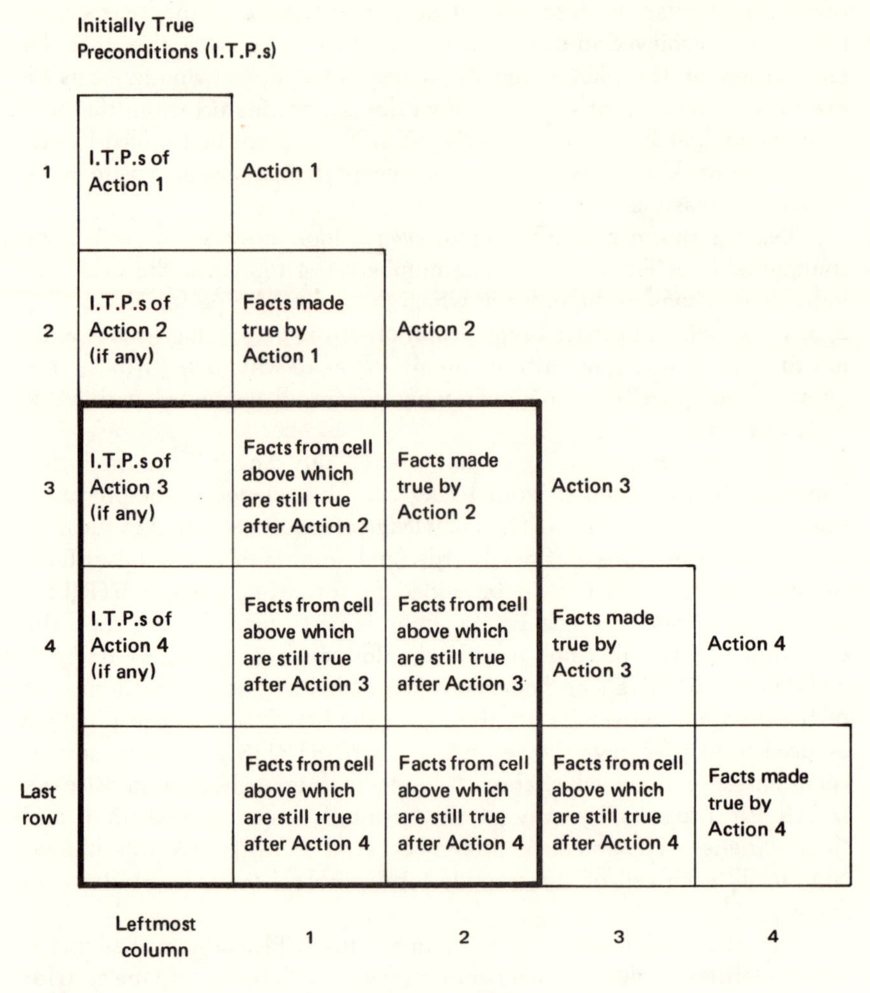


FIGURE 10.20

Source: Adapted from *Artificial Intelligence, 3* (North Holland, 1972), Figure 1, p. 259. Reprinted by permission.

if any, of Action 2 *that are not brought about by Action 1,* so which have to be already satisfied before the plan gets started; similarly, the third cell lists those preconditions, if any, of Action 3 that are not made true by either of the first two actions. The bottom cell must be left empty, since there is no Action 5; but if this triangle table were later to be inserted as a MACROP unit in the beginning of a larger scale plan (such as "open the window and turn on the light"), any relevant preconditions would have to be entered to achieve purposive continuity.

It follows that each entire numbered *row* lists all the preconditions for the action that labels that row (whether these are established by previous actions or are satisfied before the plan gets going), and also any side effects of previous actions that have survived to the point at which the row's action is taken. The last (unnumbered) row must contain a full description of the overall goal state—"the" purpose of the plan as a whole—together with any surviving side effects.

STRIPS marks the preconditions with an asterisk, thereby distinguishing them from the side effects. Obviously, all facts listed in the first cell of each row are so marked, because the leftmost column lists *only* preconditions. If a particular asterisked fact occurs for the first time in the third (*n*th) cell of a row, it must result from the second (*n*—*1*th) action; so, unless it appears independently as a newly true fact added by some *subsequent* action, one may assume that the point of the second action was to establish this fact (or facts).

This last observation is reminiscent of the previous example of SHAKEY's abandoning Box 1 because he finds another box already in position. Indeed, questions (a), (b), and (c) may each have been brought to mind during the description of STRIPS's construction of the triangle table.

For instance, consider question (c). Each action column is preceded by a "kernel" rectangle whose right-hand side is immediately to the left of the column in question (such as the heavily outlined rectangle to the left of Column 3). The construction of the table implies that all the preconditions of the action are contained within this rectangle. Indeed, the kernel rectangle contains all the preconditions for the entire tail portion of the plan, except for those which are to be established during execution of the tail portion itself. So, to answer question (c) with respect to the last two actions of the plan, all the monitor program need do is check to see whether each of the asterisked descriptions within this kernel is currently true. It does *not* have to ask whether the first two actions have been accomplished. Nor need it assume, if they have been, that the relevant descriptions *must* therefore now be true. (If one is to attribute assumptions to the dungbeetle at all, it seems that this last is the assumption it makes.) In other words, any tail portion of a plan can be treated as a whole unit with regard to which question (c) can be answered. The execution of MACROPS, therefore, can be intelligently monitored by SHAKEY: given that he has begun to perform some complex procedure, he is not committed to performing *all* of it, come what may. (How do you think the triangle table might be used to answer questions (a) and (b)?)

SHAKEY's monitoring ability contributes to his learning power in several ways. If he is executing a stored MACROP and something untoward happens, he does not have to start planning again from scratch.

He may not even have to call STRIPS to work out a patch for the plan—but if he does, the patch is stored as a new MACROP and is available for future use. The sort of purposive analysis used by SHAKEY to answer questions (a), (b), and (c) during execution can also be used by STRIPS when puzzling out new plans, so that MACROPS can be broken up and the relevant portions selectively incorporated into new plans. Finally, this purposive analysis is also used in the first type of generalization I mentioned. I oversimplified when I said that STRIPS replaces constants by variables, for if this were done unthinkingly it would sometimes produce a less general schema than one would like and it would sometimes produce an incorrect schema. If James Bond's views on the making of martinis are to be taken seriously, then an example of the latter case would be STRIPS's substituting "shake *n* times" for "shake 3 times." As an example of the former case, suppose that SHAKEY starts off in Room 3, to which he returns after fetching a box, to open Window 2: if STRIPS were to assume that the plan schema for opening windows must start and finish in the same room (even though this could be any room), SHAKEY would often have to make an unnecessary journey to get to that room so that his window-opening procedure could get started. Intelligent use of the triangle table prevents STRIPS from making these mistakes, so that its learning of plan schemata is rather more subtle than I earlier suggested.

So far, I have implied that SHAKEY's plan is ideally correct, but that it may go astray in execution due to unforeseen accidents. Accordingly, it may have seemed to you that plan-patching is a mere afterthought to problem solving, required only to cope with irritations such as an unauthorized person's putting boxes in unexpected places, or unkindly stealing dungballs from beetles. In SHAKEY's case (as currently implemented) this is almost true, since STRIPS foresees all known consequences of a plan when formulating it, and the "add," "delete," and "precondition" lists are well-nigh exhaustive with respect to SHAKEY's grossly restricted world of action. But plans are not always perfect, so that failure sometimes arises because of a faulty plan; and the more unfamiliar and ill-specified the task, the more likely this is to be so. Consequently, in a rich and changing world, plan-patching is essential, and the more intelligently it can be carried out, the better. If one can learn something useful about the task in hand *from one's mistakes*, then —with practice—clumsiness and error should progressively give way to skilled insight into what one is doing. Correlatively, the translation from strategy into tactics should be increasingly smooth and trouble-free.

These points have been stressed by G. J. Sussman, whose computational model of skill-learning is called "HACKER," a name often applied to people who spend a lot of time programming.[28] The human hacker constantly has to locate and eliminate bugs in her program. As the contemptuous term suggests, "bugs" are annoying, and often essentially trivial, mistakes made by the programmer in writing the program. By a trivial mistake, I mean something like forgetting to close a parenthesis, or negligently omitting a line of code that one knows perfectly well to be essential. Such mistakes are mere annoyances, and finding them teaches one nothing of any theoretical interest. But other bugs are considerably more significant, which is why Sussman insists on the *virtuous* nature of many bugs, meaning that these bugs are "manifestations of powerful strategies of creative thinking," such as the assumption of linearity in problem solving. Accordingly, in creating and removing bugs one can learn from one's mistakes, so as to refine one's understanding of the purposive structure of the task that the buggy program (faulty plan) was intended to achieve. As the discussion of triangle tables should already have made clear, this understanding is crucial to intelligent, flexible action.

Like STRIPS's, HACKER's performance improves with practice because the program learns general lessons from specific experiences. But HACKER does not try to get everything right first time, being content to "have a go" using a rough and ready method that may be only broadly relevant. When he[29] makes mistakes, he analyzes them to see what went wrong and why. This may suggest a way in which the previous method (problem-solving plan, program) can be adjusted. If so, HACKER tries to classify his mistake in general terms, to be put on a list of traps to be deliberately avoided in future. Correlatively, he generalizes the adjustment as a subroutine that can be called on in all cases where this type of trap may arise (that is, in all problems of the same pattern). In order to do this, he criticizes faulty plans in terms of very general purposive concepts defining the teleological structure of action, such as "unsatisfied prerequisite," or "prerequisite clobbers brother goal."

HACKER inhabits the simulated BLOCKS world that is the home of Winograd's SHRDLU. In other words, insofar as HACKER has knowledge of a specific problem domain (as opposed to general knowledge about problem solving) it is the BLOCKS world that he knows about. But, at the level of effective achievement, HACKER starts off knowing rather less than SHRDLU.

For example, whereas SHRDLU can immediately *pick up a big red block* on being asked to do so, even though this is the very first request made of it, HACKER cannot. The reason is that (as you may remember from Chapter 6) the top of the red block has to be cleared off first, since the green one is sitting on it. SHRDLU is provided with a program for PICKUP that includes CLEARTOP in such a way that SHRDLU automatically asks whether an object's top is already clear before trying to pick it up, and automatically clears the top if necessary. HACKER has to learn this procedural strategy for himself. To begin with, all HACKER knows about picking up is that the relevant primitive action allows only one object to be moved at a time. In other words, although he knows that it is a theoretical prerequisite of this action that the top be clear, HACKER does not know that *in practice* a preparatory action of clearing must therefore often precede the action of moving. How does he learn this, so as to acquire the practical knack that is innate to SHRDLU?

HACKER learns by experience, and the *order* of experience is important, as it is (for basically similar reasons) for Winston's concept-learning program, whose constructive use of the NEAR MISS corresponds to HACKER's building on an almost-right plan. Let us see, for instance, how a training sequence that carefully puts first things first can bring HACKER to realize the general importance of a CLEARTOP subroutine.

Please look at the scene in Figure 10.21. If HACKER is asked to get A onto B, he can put A on B without hesitation, using the primitive action for moving blocks. (At this initial stage, the only performance programs he has are the BLOCKS world action primitives.) But if he is now given Figure 10.22, and asked to get B onto C, he cannot. The existing performance program that is most nearly relevant fails, because it is just a call to a primitive and cannot move B without moving A. The failure of the primitive sends an error message to HACKER, who investigates and classifies the error, and accordingly produces a patch for the performance program. The patched program is now rerun, and it works: A is put on the table, then B is placed on C, so the scene looks like Figure 10.23.

That HACKER has learnt a skill of some generality here can be seen by looking at Figures 10.23 and 10.24. If he is given the first of these scenes, and asked to get C onto A, he can *immediately* do so, without the false start that was made in the earlier case. Moreover, given Figure 10.24 and requested to get A onto B, HACKER again unhesitatingly does the right thing. That is, the patch produced earlier was sufficiently general to direct these steps:

```
Wants to put A on B
   Notices C and D on A
      Puts C on TABLE
      Wants to put D on TABLE
         Notices E on D
            Puts E on TABLE
         Puts D on TABLE
Puts A on B.
```

This results in Figure 10.25.

The patch was generally useful because the error was assessed in general terms. That is, right from the start HACKER thought about what he did, and what he did wrong, in terms of what he was trying to do and

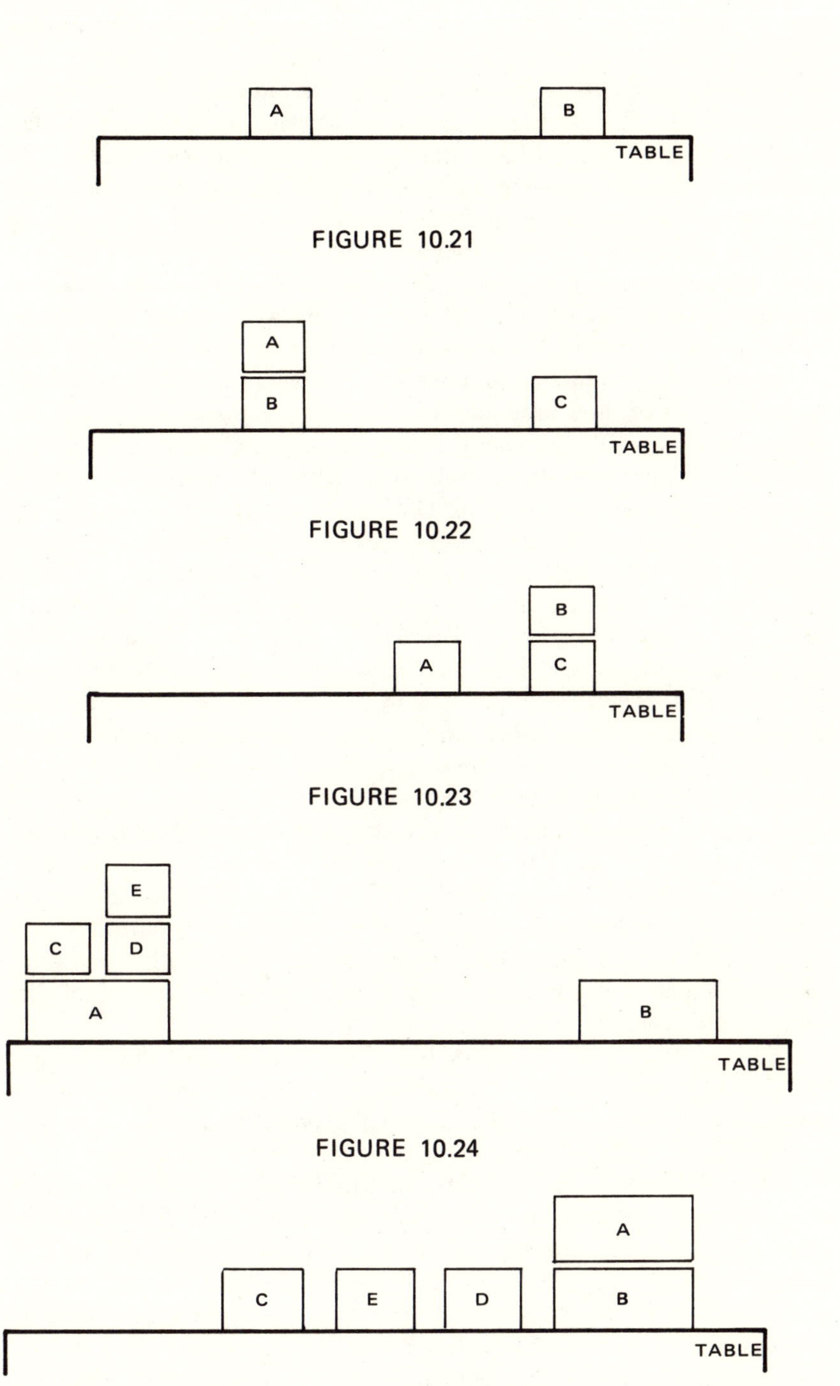


FIGURE 10.21

FIGURE 10.22

FIGURE 10.23

FIGURE 10.24

FIGURE 10.25

Source: Figures 10.21–10.25 from G. J. Sussman, *A Computer Model of Skill Acquisition* (New York: American Elsevier, 1975), pp. 9-11. Reprinted by permission.

how he was trying to do it. For example, instead of blindly using the primitive action PUTON (as a "Samuelish" program would have done), HACKER entered this initial performance program in a notebook with comments explicitly showing that he was calling the primitive action PUTON in order to get A onto B. Similarly, comments on the patch showed that putting A on the table was done purely in order to clear B in preparation for moving it onto C. In other words, the sort of structural information that is implicit in triangle tables is made explicit by HACKER at all stages of evolution of the performance program.

Human programmers, too, have been accustomed to "commenting" particular parts of their programs so that debugging is easier to do later. But before Sussman's automatic HACKER, these comments had been extrinsic to the program itself, being mere nonfunctioning labels that the *human* programmer finds helpful. HACKER's comments, of course, are also extrinsic to the execution of the performance program: in the scene of Figure 10.21, it is sufficient to *put* A onto B in order to *get* A onto B, and HACKER (like the dungbeetle) needs no explicit knowledge of means-end relations to succeed. This knowledge is needed, however, for the program-writing part of the HACKER system to patch this primitive program so as to be able to put B onto C, given Figure 10.22. In short, HACKER produces a detailed purposive commentary on each stage of the evolving performance program that can be consulted to clarify the relation between *intention* and *execution*.

Different types of mistake can be classified in terms of this relation, as the philosopher J. L. Austin suggested in his discussion of excuses.[30] Sussman classifies bugs (of which he has so far identified five main types) in these terms, and outlines *general* patching subroutines appropriate to each type of bug. (Compare the general rules used by Winston's program to amend its conceptual models, according to the type of discrepancy discovered.)

For example, in the case we have been considering, the primitive PUTON tries to put B onto C, given Figure 10.22, and finds that it cannot. That is, HACKER tries to run the primitive performance program PUTON, and finds that it doesn't work. An error message is recorded by PUTON, saying that it cannot do its job *because* A is onto B, in other words because a prerequisite of action (that B be clear) is unsatisfied. HACKER thereupon calls one of his store of debugging programs, namely, the one for UNSATISFIED-PREREQUISITE bug manifestations, of which there are several possible underlying causes. By investigating the purposive structure of the buggy program (that is, by a general analytic method independent of any specific problem domain), HACKER decides that this bug manifestation at runtime is caused by an underlying PREREQUISITE-MISSING bug in the program's structure, and that the type of patch appropriate is therefore one that supplies the missing prerequisite. Accordingly, a patch to achieve CLEARTOP-B is written for insertion into the performance program. This patch is converted into a general subroutine that does not confine itself to B and A, but uses variables rather than constants so that it can clear *anything* off *anything*. Moreover, it is a recursive program that can be used on several levels if necessary—which is why the complex structure of Figure 10.24 can be immediately unpacked by HACKER as required.

It is important to be clear about the distinction between a manifested UNSATISFIED-PREREQUISITE bug and a causal PREREQUISITE-MISSING bug. These two correspond respectively to mistakes discovered on trying to run the program (at "runtime"), and mistakes discovered by a critical analysis of the program's structure before any attempt is made to translate the program into action (at "compile time"). That is, the first is a hitch that becomes evident during execution, whereas the second is one possible underlying cause in the structure of the program being executed.

In our example, the prerequisite (that B be clear) is manifestly unsatisfied at runtime because it never was satisfied, so some entirely new action (the CLEARTOP subroutine) must be inserted into the plan at compile time to provide for it. But sometimes a prerequisite is manifestly unsatisfied at runtime because it has previously been clobbered: that is, an early action satisfied this prerequisite, but a later one (whose point was to establish some other state of affairs) unfortunately reversed the earlier condition as a side effect. In such a case, the program patch required is one that avoids the clobbering. Often, this can be achieved merely by reordering the two actions concerned, so that the one that originally came first now comes second. This is the sort of general problem-solving (programming) knowledge that is embodied in HACKER's bug-classification and debugging routines. This knowledge itself evolves into more powerful forms as HACKER gains experience in patching buggy programs; and it is also used by HACKER *while* writing programs (as opposed to *after* writing them) so that bugs that have been met once are appropriately anticipated and avoided in future.

If you think back to triangle tables, it is clear that these two causes of an UNSATISFIED-PREREQUISITE bug manifestation would correspond to different structures in the corresponding table. Indeed, SHAKEY's monitor program in effect watches for these hitches during execution and can either adjust matters itself or call on STRIPS to do so. For example, the monitor continually checks to see whether any kernel can be found such that all the prerequisites within it are true. If so, then the relevant tail portion of the plan can be executed. If not, then STRIPS is recalled to produce a new MACROP that establishes the preconditions required, given the current situation (much as HACKER writes the CLEARTOP routine). Since the checking of kernels is done by working backwards from the goal, SHAKEY can take advantage of serendipitous situations where he finds himself unexpectedly close to the goal. For the same reason, superfluous subsequences within the plan—i.e., "self-defeating" sets of actions that leave the world exactly as they found it—are never actually executed by SHAKEY (although the monitor cannot recognize them as such, and so does not tell STRIPS to remove them from the plan). For this reason also, if SHAKEY were to find a prerequisite suddenly clobbered by outside influence (as when one removes the dung from the dungbeetle) he could repeat his previous actions accordingly.

If SHAKEY were to be provided with a faulty plan, containing a clobbering bug, he could call on STRIPS to produce a new MACROP to restore the deleted state of affairs. While this has the same immediate practical effect as HACKER's reordering the actions in a buggy program, it is important to realize that SHAKEY's monitor does not actually recognize the causal bug, but merely notices the bug manifestation. In consequence, the original plan would remain in store, with the bug still in it. This is somewhat less disastrous than it sounds, since STRIPS's method of problem solving implies that a plan with this type of bug inherent in it should never be produced. However, it remains true that SHAKEY (by way of STRIPS) learns patches for relatively small classes of bug manifestation whereas HACKER learns to forestall a much wider class of problem. The reason is that types of (manifest and causal) bug are explicitly represented in HACKER, whereas they are not so represented in STRIPS. Correlatively, SHAKEY is content to work with a buggy program if the bug manifestation can be avoided by appropriate monitoring at execution time (like the self-defeating actions previously mentioned), whereas HACKER tries to eliminate causal bugs as well as bug manifestations.

The five main types of bug so far identified by Sussman as underlying causes of manifest failure are abbreviated as PCB, PM, PCBG, SCB, and DCB. These are, respectively, PREREQUISITE-CONFLICT-BROTHERS; PREREQUISITE-MISSING; PREREQUISITE-CLOBBERS-BROTHER-GOAL; STRATEGY-CLOBBERS-BROTHER; and DIRECT-CONFLICT-BROTHERS. As these names suggest, the bugs concern the interactions between the various actions within a plan. These underlying bugs manifest themselves during execution in a number of ways, in the form of unsatisfied prerequisites, double moves, or failure to protect a condition that must continue to exist until a specific point in the plan.

The patches required to remedy a buggy program may require not only "brand new" main steps, like the CLEARTOP subroutine in the previous example, but also actions required merely to interface one main step with another, like cleaning a saucepan one has used to cook peas in order to use it again for custard. Sussman points out that most interesting and difficult bugs result from unanticipated interference between steps in a proposed solution of a problem. When this interference is particularly complex, so that a bug manifestation has more than one underlying cause, or the manifestation of one bug hides or cancels the manifestation of another, HACKER—like many people—cannot cope.

Nor can HACKER handle things when the plan required is *non*-linear. HACKER assumes when writing programs that all plans are linear, in the sense that each subgoal is independent, so that the overall goal is achievable by successively achieving its separate components. (This is strictly analogous to the assumption of cue linearity made by Winston's program.) This is a sensible, and very common, assumption; but sometimes it is wrong (although even then it may help in suggesting a first approximation to a suitable plan). For instance, if one wishes actually to build an arch (like those defined by Winston's program) one can*not* do so in a linear fashion whereby the cross-piece is first placed on the left-hand upright and then on the right-hand one—or vice versa. One has to use a nonlinear strategy such as placing it first on a supportive *central* upright that is removed after the two side uprights have been added. However, we shall see in Chapter 12 that problem-solving programs already exist that can handle this case, and whose representation of the problem does not assume linearity. Consequently, it is not impossible that some future version of HACKER might be able to cope with it also; indeed, Sussman points out that the bug involved in a linear arch-building plan like the one I just outlined is of his fifth (DCB) type, and that this type of bug is characteristic of problems requiring nonlinear plans.[31]

HACKER constructively criticizes its own efforts at doing something, and learns accordingly. One of the functions of the human teacher is to help her pupil to do the same, and for this she needs to be aware of what it is that the pupil is trying to achieve, and how current performance matches up with the goal. A preliminary example suggestive of how the teacher's role might be simulated in artificial systems is Ira Goldstein's "MYCROFT" program.[32] As devotees of Sherlock Holmes will know, Mycroft was the elder brother who advised the detective on particularly difficult cases.

Goldstein's MYCROFT may be thought of as complementary to HACKER, in that it applies general knowledge of debugging like HACKER's to a specific domain, namely, children's drawings. Or, more accurately, it assesses and corrects *drawings done by computers programmed to draw by children.*

The children used the LOGO programming language, which is a very simple one: its instructions tell a pen-carrying mechanical "turtle" to go FORWARD for a specified number of unit steps, to turn RIGHT or LEFT through a specified angle, to lift the pen UP (so that it no longer marks the paper), and to drop the pen DOWN (so that as the turtle walks, the pen draws).[33] By way of this language, young children wrote programs for drawing schematic people, faces, trees, houses, and so on. Usually, they did not succeed at the first attempt: like HACKER's efforts, their initial procedures contained bugs of various kinds. Figure 10.26 shows some examples of drawings done by their corrected and their buggy programs.

A child interacting with MYCROFT tells it what she was intending to draw, and a little about how she was intending to draw it (for example, head first or feet first); she also shows it her program and the drawing done by that program. If these three representations (intended plan, turtle program, and actual drawing) do not match, MYCROFT's job is to diagnose the bugs in the program and help the child to put it right. That is, MYCROFT does not merely (*merely*?) correct the program and spit it out at the child. The child probably would not learn much if it did. Rather, MYCROFT analyzes the faults in terms of specific bugs, which the child learns how to correct and how to avoid in future. The result is a child with a greater understanding of what she is up to, as well as a more satisfactory drawing.

For example, MYCROFT points out to the child that if she starts to draw a man (as in Figure 10.26,b) by drawing the two legs splayed out from her starting point, she has to put in an *extra* orienting step before starting on the body. Otherwise, as in this buggy picture, the body will be aligned with the second leg instead of being at an angle to it. Reorientation bugs in general involve the failure deliberately to *restore* some previous state of affairs so as to continue with the next *main* step of the procedure. They occur in all sorts of contexts, not just in drawing. For instance, my seven-year-old son was trying to blow up a paper bag: after half-filling the bag with his first puff, he inhaled deeply *with the bag still at his mouth*, so that in filling his lungs he emptied the bag. Intelligent teaching that combines correction with explanation not only develops the learner's understanding of the specific domain in question, but improves her thinking *in general*, as she learns about various types of interaction and equivalence between different information-processing procedures that are used in all manner of situations. MYCROFT, accordingly, combines general (HACKER-like) knowledge about bugs with specific expertise in the domain of LOGO drawing programs.

The type of learning-by-doing modeled by HACKER, and assisted by MYCROFT, is often thought of as basic to human intelligence, and incapable of formal expression. The philosopher Michael Polanyi refers to the "tacit" knowledge embodied in functional skills and creative thought, which only experience can teach and which no textbook can satisfactorily supply.[34] Polanyi and his disciples (including H. L. Dreyfus) regard this nonfactual knowledge as a fortiori incapable of simulation in digital computers. Their general philosophical position will be discussed in

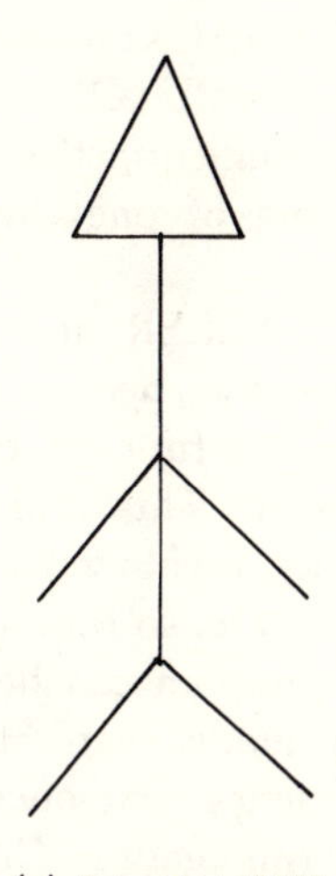

(a) Intended MAN

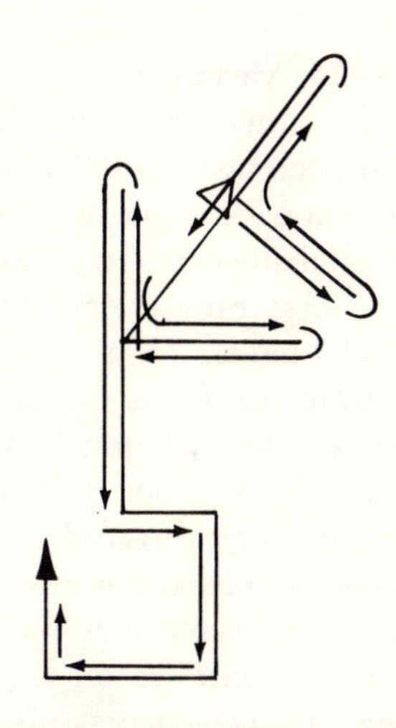

(b) Picture drawn by buggy MAN-program. (The small triangle shows starting position of the turtle.)

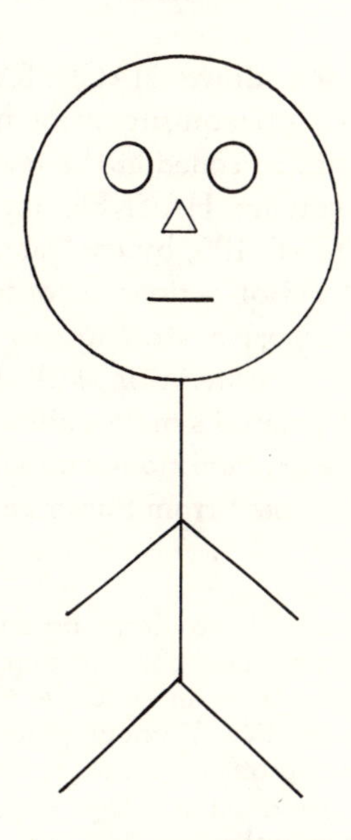

(c) Intended FACEMAN

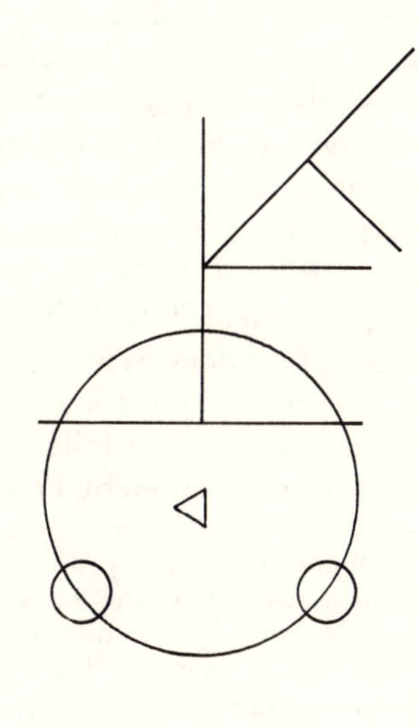

(d) Picture drawn by buggy FACEMAN-program.

FIGURE 10.26
Drawings done by children's LOGO programs.

ource: *Artificial Intelligence 6* (North-Holland, 1975), p. 250. Reprinted by permission.

Chapter 14. Meanwhile, it is worthy of note that HACKER, too, starts with merely theoretical knowledge (that the primitive PUTON requires a clear top, for instance) that is complemented through practice by procedural knowledge of how to achieve the putting of one thing onto another in many different situations.

The first time a new program is run by HACKER, it is run in CAREFUL mode. That is, each step is individually checked as it is executed, to see if it is fulfilling the overall purposive function coded by the relevant line comment in HACKER's notebook. This typically involves a large number of checks and cross-checks, together with a detailed chronological record of the changing world state, so that any bugs that have escaped the previous criticism of the program can be located and identified as early as possible. In CAREFUL mode, then, HACKER haltingly concentrates on the level of detailed tactics. But once a new program has been successfully executed, it is not run again in CAREFUL mode unless it actually gets into trouble, in which case CAREFUL running is reinstituted. A smoothly functioning program thus implicitly embodies a good deal of knowledge that is made explicit only when necessary. This characteristic of HACKER's skilled behavior is reminiscent of Polanyi's "tacit" knowledge.

Moreover, it is only practical experience that draws HACKER's attention to the "clear top" prerequisite: if this prerequisite were never unsatisfied, then although it would of course still be coded at the level of the primitive action PUTON, the problem-solving HACKER himself would seem to have no special cognizance of it. STRIPS, by contrast, has to list explicitly all the preconditions and effects of actions in order to construct a triangle table representing the purposive structure of any plan. To this extent, HACKER satisfies Polanyi's criteria of skillfulness more closely than does STRIPS, for it is one of Polanyi's main points that one does not have explicit knowledge of all the preconditions and effects of what one is doing. The following passage is quoted from Sussman, but it could have been written by Polanyi:[35]

> An important property of skill is *effectiveness.* It wouldn't be enough to memorize all of the facts in the plumber's handbook, even if that could be done. The knowledge would not then be in an effective, usable form. One becomes skilled at plumbing by practice. Without practice it doesn't all hang together. When faced with a problem, a novice attacks it slowly and awkwardly, painfully having to reason out each step, and often making expensive and time-consuming mistakes.
>
> Thus the skill, plumbing, is more than just the information in the plumber's handbook; it is the unwritten knowledge from practice which ties the written knowledge together, making it usable.

Whether or not one regards Sussman's model as an analogue of Polanyi's tacit knowledge, it is clear that HACKER's learning is a creative matter. For HACKER is a system for automatic programming, a problem-solving program that itself writes and improves programs and that learns to do so better with practice. In other words, just as learning can only arbitrarily be distinguished from problem solving, so it is often intimately connected with creativity. Even Winston's program automatically evolves representations that can be used to assess degrees of analogy—a typically "creative" activity. It must be admitted, however, that neither Winston's nor Sussman's program can step right back from its initial hypothesis (whether "concept" or "plan") and switch to a radically new way of looking at its problem—which is why its first thoughts have to be *almost* right for it to succeed. In the next chapter I shall ask to what extent, and in what way, creativity is rooted in novelty.

# 11

# *Creativity*

CYNICS may deny that one ever gets anything for nothing, but most people apparently assume that one can get something *from* nothing. Indeed, the first definition of "create" listed by my dictionary is "to bring into being or form out of nothing." The medieval theologians of Islam, Jewry, and Christendom showed that this assumption can be questioned at the metaphysical level. What is more to the point, it can be queried at the psychological level also.

Perhaps the new thoughts originated in creative thinking are not wholly novel, in that they have their seeds in representations already present in the mind? And perhaps they are not wholly inexplicable, in that something can be said about ways of manipulating familiar representations so as to generate others that are somehow fresh, or original? In this chapter I shall ask whether creativity, like learning and problem solving, involves the production of new thoughts from old.

In the first section, I describe various failings of a story-writing program, and argue that an improved program would need a semantic base incorporating greater knowledge of shared assumptions about psychological matters than the program has at present. The second section continues this discussion, with reference to the notion of "frames" in intelligent programs. In the third section I discuss computational models of analogical thinking of various types, including scientific creativity. And in the last, I ask what artificial intelligence has to say about the generation and use of radically different representations for problems.

## CREATING WITHIN A CONVENTION

Rumor has it that a popular novelist, asked by a beginner how to write a detective story, advised combining a glimmering of religion, a touch of class, a soupcon of sex, and a strong aura of mystery. The resulting story ran as follows: "My God!" said the Duchess, "I'm pregnant. Who done it?"

This literary effort undeniably shows a certain stylistic economy, but is unlikely to be a bestseller. Yet it followed the advice: does this show that in creative writing there are no rules, that the novelist was foolish to be seduced into giving advice in the first place? Surely not, for her counsel was reasonable enough. Indeed, literary critics concerned with rather more elevated genres than the detective story have claimed that there are no new themes in literature, that human relations allow of only a limited number of plots, which can be given new settings and superficial variations but whose basic motivational structure is invariable. Even Goethe thought it possible that there are less than forty distinct tragic situations. What is required, then, is not the rejection of all rules but the intelligent interpretation of them.

Good literature, to be sure, is distinguishable from hack writing largely by its ability to surprise, to extend the range of the reader's experience by giving a series of shocks to her preconceptions that alert her to human possibilities of which she was formerly unaware. But this achievement would be impossible had she no preconceptions to begin with. Moreover (as we shall see more clearly in the third section) the new insights have to be intelligible in terms of the basic cognitive structures that underlie the preconceptions concerned: this is implied by the reference to "human *possibilities.*" Granted that fine literature avoids being stereotyped and cliché-ridden, its power depends on a shared understanding of normative assumptions that it is able somehow to transcend. More popular forms of writing, like the mystery one buys to while away the train journey, exploit these assumptions less critically.

A psychology of creativity, then, needs to be able to handle these common assumptions and conventional stylistic forms before it can fruitfully approach questions about their transcendence. Correlatively, critics of artificial intelligence should avoid what has been called the "Superhuman-Human fallacy," which is to demand of a machine model that it parallel the highest flights of human thought—flights which most people cannot achieve, still less explain. Initially, at least, it is exacting enough to ask that a computational theory represent the mediocre thinking of the average person on an off-day.

But even this apparently modest requirement has not yet been met (nor is it satisfied in "ordinary," verbal, psychological theory). For example, the 2,100-word murder mysteries generated by Sheldon Klein's "automatic novel writer" are only marginally more satisfactory than the snippet about the Duchess.[1] The unlikelihood of their being nominated for any literary prize will be evident from this excerpt:

> The day was Monday. The pleasant weather was sunny. Lady Buxley was in a park. James ran into Lady Buxley. James talked with Lady Buxley. Lady Buxley flirted with James. James invited Lady Buxley. James liked Lady Buxley. Lady Buxley liked James. Lady Buxley was with James in a hotel. Lady Buxley was near James. James caressed Lady Buxley with passion. James was Lady Buxley's lover. Marion following them saw the affair. Marion was jealous.

Despite the touch of class and the soupcon of sex, you may be wondering how the program's stories are any better than the ten-word saga about the Duchess. The answer is that they have an embryonic plot, represented in terms of the motives of conflicting personalities; and the Duchess's final query is answered, in that the murderer is named.

The most basic shortcomings of Klein's program are not apparent in a short extract like the one just quoted. That this passage is unlikely to set anyone's blood afire is due to its stylistic crudity, rather than its subject matter; a novelist could even use this paragraph as a semantic skeleton for one of her more breathtaking chapters. But no purely expressive skill could render the story as a whole acceptable, even by the jaded standards of a hack writer. Aside from the lack of detail in the narrative, the program's mysteries have three radical weaknesses: the stories are shapeless and rambling; the specific motivational patterns are relatively crude and unstructured; and the identification of the murderer comes as a statement rather than a discovery, there being no step-by-step detection —still less any deliberately planted false clues.

Curing these faults would require that the program's semantic data base be enriched by representations of the relevant structures, and also that the system be given ways of manipulating these structures so as to generate sensible stories. The epistemological issues involved are currently ill-understood, so that a successful cure is not in sight—nor will it be a reasonable practical research project for many years (Klein's efforts to generate automatic folktales are somewhat less ambitious, but hardly more satisfactory[2]). Nevertheless, some work relevant to the first of these requirements has already been done by workers in artificial intelligence— "relevant" in the sense that it raises some of the essential questions even though it provides no more than the sketchiest outlines of the requisite answers.

For example, D. E. Rumelhart has written a "story grammar" that might be used in expressing one's intuitive sense of the *shape* of stories in computational terms.[3] Currently, the grammar consists of only 11 rules, which cover a wide range of simple stories (such as Aesop's fables, many folktales, and the sort of children's stories discussed by Eugene Charniak) independently of their specific content, or plot. By this I mean that Rumelhart, and his human readers, can often successfully "parse" stories by reference to the grammar, not that any programmed implementation exists that can do the same. For just as "ordinary" syntax can be intelligently used to parse sentences only with the help of background semantic knowledge (as the heterarchical organization of SHRDLU makes clear), so story syntax can be applied to particular cases only by one who understands what the story is about. However, a story writer certainly needs an appreciation of the shape of stories, such as Rumelhart attempts to make explicit in his story grammar. Its nature can be briefly illustrated by considering the first four rules.

Rule 1 says that every *story* consists of a setting and an episode. Since this gets us nowhere without knowing what a setting or an episode is, we are given Rule 2: a *setting* is a series of statements indicating the time and place of the story and introducing the characters, and Rule 3: an *episode* is an event followed by, and understood as causally initiating, a reaction of one of the characters. Finally, Rule 4 tells us that an *event* may be one of four things: an episode, a change-of-state, an action, or a sequence of events; in the latter case, the sequence is to be understood as a set of enabling or resultative events relative to some action within the story. You can see from Rules 3 and 4 that stories may involve hierarchical nesting to an indefinite degree of complexity, since every episode involves an event, while some events are themselves episodes; also, each event within a high level "sequence of events" may itself be a complex structure, so that the narrative dealing with the enablement or result of the corresponding action may be not only lengthy, but complicated too. Rules 5–11 introduce further concepts such as reason, emotion, desire, plan, and consequence, all of which have a clearly defined place within the overall narrative structure.

The shapelessness of the literary efforts of Klein's program can be thought of as a failure to conform to Rumelhart's narrative grammar (though this is not to say that the grammar captures all the architectonics of the typical detective novel). To be sure, a well-formed story can be distilled out of the ramblings of the novel writer: early sentences mention the setting (an English country-house party) and characters (Lady Jane, Lord Edward, and Clive, the lusty butler); and the main part of the composition is a sequence of actions, one of which contributes to the crucial episode in that it is followed by a murder performed in reaction to it. But many of the events recounted in the story are actions to which there is *no* reaction, and which therefore do not contribute to *episodes* in

Rumelhart's terms; even when they do, the episodes are undeveloped in the sense that there are no sustained hierarchical structures of episodes understood in light of a succession of higher level episodes. Nor are most action sequences interpretable as coherent sets of enablements or results of other actions.

Thus the little drama concerning Lady Buxley and her lover James is not used to further (or intelligently to conceal) the main plot. Marion's jealousy has no outcome; even when similar incidents have an outcome (such as "Lady Jane yelled at Lord Edward"), this has nothing whatever to do with the performance, concealment, or detection of the murder. In the particular story I have been quoting, the amorous James ends up being poisoned by his poor relation, the butler, who hopes to inherit his money. The notion that Marion might cooperate with the butler for reasons of her own is not one that the program's semantics can handle.

As regards the second shortcoming of the automatically generated stories, the crudity of the genuine episodes involved, R. P. Abelson's theory of *themes* and *scripts* could provide for more richly structured plots than those currently dreamt up by Klein's program. The social encounters that occur prior to the murder are composed with only minimal concessions to psychological coherence. For example, the witness of an illicit love affair can react by blackmail, fighting, or forgiveness, but within these limits responds virtually at random. It is assumed that deceived wives never blackmail their adulterous husbands, but apart from dubious constraints such as this, almost anything goes.

But consider, for instance, the various possibilities inherent in the family of *betrayal* themes, including the relevant restrictions on social power, that were sketched in Chapter 4. It is probably Klein's intuitive knowledge of these psychological structures that explains his program's reluctance to portray a betrayed wife threatening her unfaithful spouse: mere infidelity could so easily be transformed into abandonment. Abelson's theory could represent the fact that a wife of independent means is less open to abandonment than her unfortunate sisters, and correspondingly more able to resort to blackmail in response to marital treachery. Given such a representation, together with the knowledge it already has that Lady Jane is independently wealthy, Klein's program could sensibly allow Lady Jane the option of reacting to Lord Edward in a more effective way than yelling at him. And assuming that several distinct scripts had been identified as having betrayal as a thematic unit, the thematic structure of the story prior to the disloyal act could selectively affect the likelihood of different reactions.

The importance of motivational context in determining plausibility is relevant also to the third weakness of Klein's program, its failure to allow for detection within the story. In the construction of the narrative, the slaying is not foreseen and prepared for: indeed, we have seen that action prior to the homicide is well-nigh random. Rather, when the story has reached a length that guarantees that some of the characters will have reason to be at loggerheads, the program looks for such a pair, kills off one of them, and proclaims the other to be the murderer. To be sure, the police and houseguests are described as "looking for clues," and one character is announced to have found one; but there is no genuine or developed detection involved. Nor are any false clues planted, nor real clues slipped in unobtrusively or rendered ambiguous or downright misleading by the local context.

If you remember the account of G. R. Grape's vision program in Chapter 9, you will recall the way in which cues can attach to different schemata or, in the local context, can appear not to be significant cues at all. If a novel-writing program had a computational theory connecting specific psychological cues with appropriate motivational schemata, it might plant both true and false clues throughout the story such that people's actions would be open to ambiguous interpretations, which would allow for an element of "real" detection on the part both of the fictional detective and of the reader.

As the difficulty of recognizing even cubes and pyramids suggests, this is a big if: and it gets bigger the more one demands subtlety and surprise, as opposed to a mere wooden plausibility. The interpretative key to Henry James's brilliant story *The Beast in the Jungle* is a single motivational schema that many readers realize only on the penultimate page, although on rereading one sees that cues to the schema concerned had been presented throughout the narrative. Without emulating such a superhuman human as James, however, one may reasonably hope to provide an automatic novel writer with a sensitivity to psychological cues and schemata that would outlaw the grosser absurdities and allow for some degree of motivational development. It should not, for instance, ape the author of the Duchess story, in stupidly assuming that the words "My God!" always merit a religious interpretation instead of sometimes displaying emotion.

Abelson has recently said more about the "intermediate" levels of belief previously termed *molecules* and *plans*.[4] He gives detailed analyses of mundane social actions that exhibit their purposive structure. As well as distinguishing the conditions required to start and to achieve various primitive actions (knowledge of which underlies the detective novelist's fondness for locked doors and barred windows), he notes the points at which an adversary might intervene to frustrate the person concerned. This sort of intervention is not something that HACKER or STRIPS has yet had to worry about, but it is common both in everyday life and in fictional contexts. (Frederick Forsyth's *The Day of the Jackal* is a prime example of a story constructed with such interventions in mind, the pressing question being not "Who done it?" but "Will he succeed in doing it?") Alternatively, someone might intervene at these points to help rather than frustrate: Marion may leave the poison cupboard unlocked to enable the butler to open it.

Abelson mentions various types of coordinated sequences of behavior, such as PLAN, SERVICE, DISPLAY, and STYLE schemata, constituents of which can function as cues to the overall schema. (He calls these schemata "scripts," but they are at a lower level than the "scripts" defined in Chapter 4.) PLANS and SERVICES are each directed to attaining a premeditated state of affairs, but someone performing a SERVICE is acting as an agent for someone else, having no interest in the goal for its own sake. DISPLAY and STYLE schemata are expressive rather than goal-oriented, relating respectively to temporary emotional states and enduring attitudes.

In attributing motivations to people one often wonders whether they are carrying out a PLAN or a SERVICE: was Marion cooperating with the murderous butler under duress, or not? And the cue to the answer often lies in a DISPLAY of emotion: did she smile gleefully as James swallowed the poison? My remarks in Chapters 3 and 4 about self-esteem, shame, and vanity suggested that many distinct reactions and feelings may be involved in such emotions, and that specific semantic relations must be satisfied to merit the interpretation of any of these as a cue expressing one emotion rather than another. Expressive and goal-oriented schemata are thus closely interlinked within many psychological concepts, and a program competent to use these concepts either in understanding or in composing stories (as also in plausibly modelling neurosis) would need to know about the linkages concerned.

In general, concepts at the lower levels of a Schankian representation can function as cues to concepts at higher levels. Misleading cues would lead the reader to expect specific consequences that are never realized, much as dead-end chains complicate the central development of the Schankian "story skeletons" mentioned in the previous chapter. In principle, then, plot, subplots, and red herrings could all be generated by reference to a psychological theory structured in this way. And the interpretative work of detection—which aims to discover which of these is which—would be intelligently guided by knowledge of the various cue-schema relations concerned.

## FICTION AND FRAMES

As you may have muttered on reading the previous paragraph, it is a long road from principle to practice. Even if the relevant conceptual relations had been clearly spelled out, the problem of expressing them in computational form so that a program could manipulate them creatively would remain. One aspect of this problem is the familiar hurdle facing any attempt to make intelligent use of a large data base representing complex world knowledge: how to organize it in an economical fashion so that the combinatorial explosion is avoided. The stylistic crudity of Klein's automatically generated stories referred to earlier is, in part, a poverty of specification at the lower Schankian levels. To overcome this one would need not only to tell the program more things about the world, but enable it to find its way quickly through the epistemological maze without losing the thread of the story.

Abelson's plans, themes, and scripts are themselves suggestions as to how this could be done. And so also are his recent remarks on "situational" schemata, which represent stereotyped action scenarios relevant to a familiar situation, or social context.[5] When discussing *Romeo and Juliet* and *West Side Story* in Chapter 4, we saw that circumstantial details are only broadly determined by the exigencies of the plot: fencing foils or flick knives would do, but feather dusters would not. Given that only the first two of these three instruments could fill the instrumental case slot in conceptual dependency representations of *fight* and *kill*, the choice between them depends on the setting: medieval Venice or present-day New York. The important point is that it is not necessary to be a learned medieval scholar to opt for fencing foils, any more than it is necessary to be an expert on the domestic arrangements of the aristocracy to know that Lady Jane is likely to have a butler, or that he spends some of his time serving drinks on a silver salver. In short, we possess a number of paradigmatic schemata representing specific settings, situations, and roles, which incline us toward certain details in default of more accurate knowledge, and on which we draw in an economical fashion when writing and understanding stories.

M. L. Minsky has offered an influential theoretical discussion of such conceptual schemata, which he calls *frames*.[6] He refers to the use of frames (generally conceived as inner models of the world, or systematic epistemological representations) in widely varying domains, such as visual, problem-solving, and semantic contexts. Within each of these there are an indefinite number of levels of systematically interrelated subframes. Thus even a comparatively simple visual schema like P. H. Winston's *arcade* described in Chapter 10 may be seen from many viewpoints, and involves arches, blocks, vertices, and lines—while Schankian representations in general, like the family of betrayal themes, may be of great complexity. Minsky's discussion is an attempt to give some theoretical unity to a wide variety of particular hypotheses about knowledge-using systems in artificial and natural intelligence, and as such it inevitably blurs some of the distinctions already made within the relevant literature. But, while there may be many differences of detail between different types of epistemological schemata, or frames, a number of broadly similar questions arise with respect to their manner of functioning in the information processing of the system concerned.

The issues raised by Minsky include "conceptual" questions about what knowledge an intelligent system needs and how it should be interrelated, and "methodological" questions about how a computational system could be enabled to use this knowledge efficiently. The equivalent psychological questions concern the organization and activation of human memory, including how memory contributes constructively to perception, how we know what information we need when we need it, and how we find it smoothly accordingly. (For instance, how does a physician quickly achieve a reasonable diagnosis of a disease?[7])

Such questions were asked in the 1930s by the psychologist Bartlett, whose theory of "memory schemata" postulated stereotyped representations creatively interacting with specific details in intelligent thinking.[8] But Bartlett was self-confessedly unclear about precisely how an epistemological particular (such as a visual image or a word) can access the relevant general schema, and how other details can be immediately recalled by way of the schema. He was also puzzled about how one's memory of an event or complex episode may be aided (and often located at a particular point in time past) by sensory images, such as the taste of a madeleine that proved so evocative for Proust.

These sorts of question about the structure and function of memory are highlighted by considering how frames might be computationally implemented. One needs also to inquire how frames are interrelated and compared, how schemata are initially learnt and afterwards refined, how the necessity for such refinements—or for a new schema altogether—is recognized,[9] and how "default" assignments can be replaced when appropriate by details drawn from examination of the particular situation.

Learning by example, for instance, can be thought of as the acquisition of frames that thereafter function as differentially modifiable paradigms guiding one's thinking. Thus I suggested in the previous chapter that K. M. Colby's neurotic patient somehow used the example of her father's callous behavior as a relatively inflexible schema representing social relations in general, one which radically affected her perceptions of and reactions to the world. And one of the aims of Colby's research is to clarify how particular interventions in psychotherapy can fruitfully modify the neurotic thinking concerned, so that individual people can thereafter be judged for reliability according to their own actions, instead of being automatically represented "by default" as fickle, irrespective of particular circumstances.

There can be no *general* answer to the question how typical the first example should be assumed to be, or how easily "detachable" the default assignments should be; these matters depend on the purposes for which the exemplary frame is being used. If the purpose involves life and death, it is better to be safe than sorry: even the subtle distinctions between very closely related frames that an expert could draw may excusably be regarded with reserve when one's life hangs in the balance. So Colby's patient might defend her undiscriminating suspicion of her "untrustworthy" fellows in the same way that one defends one's refusal to walk on the "unsafe" ice *irrespective* of what anyone says about its safety. If she persists in this attitude, her analyst might hope at least to limit the range of applicability of her mistrust, by somehow getting her to regard a smaller class of personal characteristics as automatically included within the neurotic frame (all men rather than all people, or older men rather than every man . . .), much as Winston's program comes to see some observable properties of arches as salient while others are not.

As in this case of the neurotically skeptical woman, the nature of the lower conceptual levels of a frame (which Minsky calls "terminals") crucially affect its application to the world and its accessibility to other frames. In his discussion of thinking by means of frames, Minsky asks what sort of factual and procedural information is stored at the terminals, and how the restrictions and advice within such terminal knowledge guide thought so that control is passed from one frame to another (whether within the same conceptual system or a different one). In outline, then, Minsky describes a frame as a collection of questions to be asked about a hypothetical situation, together with recommended answers in default of contrary information; the frame also prescribes certain methods for dealing with unexpected information of particular types, including switching to a different frame system altogether.

Before considering any of the work in artificial intelligence that has followed on Minsky's paper, you can get a feel for the complexity of real-life frame systems, and for the way in which they are selectively cued and switched by specific details of information, by trying out a version of a well-known parlor game. Find a friend or group of friends, and tell them you are going to describe a situation; they must then ask you questions to which you answer either "Yes" or "No," in order to discover the one plausible explanation, or background, of the situation. This is the description you should give them:

> There is a room. In the room is a bed. On the bed is a man. Under the bed is a small pile of sawdust. On the floor by the side of the bed is a piece of wood, 2′11½″ long [or 99 cms, if your friends are accustomed to the metric system]. The door opens, and another man comes in. He sees what is in the room, looks pleased, and walks out.

Notice what questions you are asked, and in what order; notice what (often misleading) assumptions are made at particular points in the unravelling of the puzzle, and what are the characteristic points at which questioning switches from one frame into another; notice what clues you need to give if your questioners get stuck (but try to give as few as possible, or the point of the exercise will be lost); and notice when the "Yes / No" format is unhelpful, in that it forbids your mentioning something that would prevent your friends from making an assumption that leads them away from the solution in a predictable manner.

The solution is this: The man on the bed is dead. Before his death he was the star of the circus, billed as the shortest man in the world. He was exactly one yard tall, and had a yard-long stick (*not* a graduated yardstick) in his room against which he measured himself on ceremonial occasions. The second man, also a dwarf, persuaded him that he had grown half an inch, so that he himself (at one yard and one quarter of an inch) was now the shortest man. While the first man was absent from his room, the second man sawed half an inch off the stick and hid the sawdust. When the first man checked his height and found that he had, apparently, grown half an inch, he understandably committed suicide since his supreme place in the world's circus community had seemingly been irredeemably lost to his hated rival.

(You could, of course, check the *Guinness Book of Records* to find the actual height of the smallest person, which is less than a yard or a meter; but if you adjust the length you give in the initial description to accommodate this item of real-world knowledge, you will find that the specific cue to "graduated yardstick" is eliminated and the pattern of solution is altered. Similarly, if you refer to women instead of men, or to a woman and a man, your non-Soviet questioners will be markedly less likely to suggest that the relation between the two people is one of professional rivalry, a useful fact that is commonly elicited long before the nature of the profession is guessed.)

As you would expect, the psychological complexity that is intuitively tapped by this parlor game is not approached by current formalizations of frame systems, where every epistemological relation must be explicitly marked if it is to be given any function at all. For example, Charniak has pointed out that one could express much of the knowledge underlying his interpretation of "He will make you take it back" (cited in Chapter 10) in terms of frame statements rather than demons. Thus one could construct a *birthday party* frame that records the fact that one is expected to take a present to the person whose birthday it is; and in most cases (what are the exceptions?) the appropriate present will turn out to be something the recipient does not already possess. If the present is a top and it turns out that Jack already has one, the frame provides the information that the usual way of righting matters is to take it back to the toyshop and exchange it for something else. Charniak has not spelled out this example in any detail, but he has recently done some preliminary work on formalizing a *supermarket* frame, one which would express the mundane knowledge of how one goes about shopping in this type of emporium.[10]

Charniak shows how even such an apparently simple matter as mastery of the use of a shopping-cart, or trolley, requires one to know about the purposes and preconditions involved, what to do if something goes wrong with the normal procedure, and how the use of the cart relates to the overall context of purchase. Correlatively, of course, understanding stories about supermarkets involves this knowledge also. In outline, his supermarket frame (like the cart frame that shares "terminal" information with it) consists of a set of statements summarizing the scenario concerned.

Charniak compares frame statements with demons, regarding the former as in several respects the more economical way of organizing common-sense knowledge. For instance, many inferences connected with a given frame statement are stored implicitly in the structure of the overall frame (set of frame statements), whereas since demons are conceived of as independent facts any relevant inferences have to be stated explicitly within each demon. Thus if the sentence "Jack got a cart" occurs in a story and is recognized as an instantiation of the general frame statement "SHOPPER obtain CART," then the fact that this frame statement is followed within the frame by "SHOPPER obtain PURCHASE-ITEMS" suffices to establish a context whereby the sentence "Jack picked up a carton of milk" will be appropriately interpreted in turn as the story unfolds. (The step-by-step method by which a SHOPPER goes about obtaining ITEMS is itself stored as a subframe in Charniak's representation, and can be accessed if necessary to interpret sentences like "Jack dropped the milk.") By contrast, a CART-demon invoked by "Jack got a cart" would have to specify explicitly every aspect of the context that it could watch out for, even if no relevant instances were actually to turn up; and so also would the independent SHOPPER-GET-ITEM-demon. Although Charniak concentrates on frames made up of statements rather than questions, he stresses the overall similarity of his supermarket schema to Minsky's conception of frames.

Abelson's account of the schema (or "script") representing the *restaurant* situation similarly prescribes expected roles and scenario, together with *what-if* methods for getting back on track if things go wrong.

If you can't read the menu because the print is too small or it is in a foreign language, you don't have to rack your brains furiously to decide what to do next: there are familiar algorithms which immediately spring to mind suggesting ways to overcome the obstacle. Sometimes, of course, you do have to rack your brains; Abelson characterizes the sort of problem solving involved in such cases as guided by *planning* scripts, each of which involves specific knowledge about the purposive structure of a given domain, much as HACKER knows about how to generate plans to move blocks. So planning frames can be embedded in situation frames when something genuinely unexpected happens; if it becomes routinized, the new instantiation of the planning frame may become an additional *what-if* within the situation frame. (Let us hope, for the sake of the other guests, that the butler's carefully elaborated plan to slip the poison into James' nightly milk-drink remains unroutinized.)

Although somewhat different matters arise in connection with the writing and the understanding of stories, each involves the intelligent use of schemata expressing world-knowledge (and interpretation in general, as we have seen, is largely a creative activity of filling in the gaps). Accordingly, the notion of frames is relevant to the problem of how Klein's program might be improved. Charniak's and Abelson's accounts, like Minsky's more wide-ranging discussion, are theoretical explorations of the nature of frame schemata as opposed to descriptions of functioning programs. But the ideas of Schank and Abelson are being applied in programs designed to understand as well as to generate stories. The Yale artificial intelligence group is developing "SAM" (Script Applier Mechanism), which they regard as a natural successor to MARGIE in the sense that it deals with texts rather than isolated sentences, and addresses the problem of how one should organize and select some from amongst the many *possible* inferences.[11]

Wendy Lehnert has contributed a question-answering module to SAM, which shows its understanding of stories involving the restaurant script by answering and sensibly establishing the focus of questions that were not explicitly answered in the text. SAM has five basic question types, each of which has a distinctive Conceptual Dependency Diagram. These concern *why* someone does something (Why did Rosie leave the restaurant without eating anything?), *how* she does it (How did she get the menu?), *whether or not* she does it (Did the waitress give Rosie the menu? If not, who did?), *what occurred* (What happened next when Rosie sat down at the table?), and *complementations* of explicit statements (Who gave Rosie the menu?). SAM assumes that the restaurant script guided the storyteller in her implicit structuring of the story, determining what she decided to mention and what to leave unsaid. The program maps the story accordingly onto a conceptual representation of it, a representation that may involve "ghost paths," which are generated by SAM to represent *things that might have happened, but didn't.* By noting the points at which these ghost paths (which are contradicted by the actual story) branch off from the central story line, SAM can locate the "interesting," or "significant" events among those mentioned by the storyteller.

Another program inspired by Schank and Abelson's recent work is Jim Meehan's "TALE-SPIN." As its name suggests, TALE-SPIN actually generates stories. It does so by way of planning structures (like those described by Abelson) which represent goals and methods of achieving those goals. Decisions have to be made at various points as to whether the ongoing method will succeed or whether it will encounter some obstacle. For instance, TALE-SPIN starts telling a story about a famished Joe Bear, who asks Irving Bird where he can find some honey—but will Irving Bird cooperate? If not, Joe Bear may use bribery (in the form of a juicy worm) to persuade Irving Bird to agree to tell him the location of the honey. Supposing that Irving Bird promises to give the information, TALE-SPIN then has to decide whether the promise will in fact be kept, and what Joe Bear will do if it isn't. Even if it is, Henry Bee may complicate the storyline by objecting to Joe Bear's pillaging of his hive. Clearly, knowledge about the interpersonal relations of the various characters (as well as their dietary preferences) will affect TALE-SPIN's selection and instantiation of planning scripts. These matters are currently represented by having each character described (relative to some other character) by a point on each of three scales: COMPETITION, DOMINANCE, and FAMILIARITY. This crude model of interpersonal relations might be improved (for instance) by drawing on the theory of reciprocal interests and roles represented in Abelson's earlier account of themes and scripts. (Why might Irving Bird be *expected* to betray the starving Joe Bear by breaking his promise to lead him to the honey?) Although the antics of James and Lady Buxtley are rather more enthralling than the interchanges between Joe Bear and Irving Bird, the conceptual structure of TALE-SPIN's stories is richer and more plausible than that of the yarns spun by Klein's program.

Finally, C. J. Rieger has recently written a program illustrating the use of a simple system of frames (which he terms "conceptual overlays") to interpret stories. Rieger's "EX-SPECTRE" (so called because it expects things—please don't blame me!) employs an interpretative theory, underpinned by his notions on inferential memory that were described in Chapter 10, to make sense of sentences in varying contexts.[12] EX-SPECTRE uses a number of stereotyped conceptual overlays, together with "common-sense algorithms" that are computational versions of the purposive structures represented by Abelson's primitive actions and PLANS, to respond to text in the sort of way required of the reader of a detective story.

For example, mention of certain actions in the text triggers overlays anticipating other broadly defined action possibilities; if one of these is confirmed by later input, the appropriate detailed common-sense algorithms are assumed to be relevant. Various alternative action overlays can be simultaneously kept in mind by EX-SPECTRE, which can switch from one to another as the evidence comes in (compare: did she fall, or was she pushed?). Rieger compares the "cloud" of activated overlays, certain aspects of which will have been weighted for saliency by specific information gleaned from the text, with the subliminal expectancies that incline a human reader to one interpretation of a story-line rather than another.

The *theft* context, for instance, is triggered by certain cue words (such as "steal"), and activates a range of broadly defined expectancies as indicated in Figure 11.1. If one reads that "Mary stole Lady Jane's jewels," one naturally then interprets "Lady Jane snatched up her car keys" as a detailed step in a familiar GOTO algorithm whose overall purpose in this case is probably to do something about the felony—though whether Lady Jane intends to drive to the police station, to attack Mary, or to get her valuables back by gentle persuasion is not yet clear. The more one knows about the personalities of and relations between the two protagonists (not to mention the gems themselves), the more one is inclined to expect certain reactions: the "universe of expectancies" and "expectancy selectors" of Figure 11.1 serve to establish these anticipatory interpretations.

Much as a vertex in a picture can suggest a wedge but turn out to belong to a pyramid instead, so an activated action overlay can be abandoned in favor of another, given appropriate inputs. Thus on being told that Mary apologized, or that Lady Jane smirked, the reader may switch from the *theft* context to the *atonement* or *practical joke* overlays, with correspondingly altered interpretations of the subsequent sentences (perhaps the "diamonds" are really paste, and the joke is on Lady Jane's insurance company: if so, Lady Jane is more likely to be driving to a celebration than a confrontation). Sometimes the termination of an overlay itself activates the beginning of others, as the return from the funeral commonly makes way for the reading of the will. In sum, EX-SPECTRE cannot write stories as Klein's program can; but EX-SPECTRE knows more about the likely scenarios relating what a person might do were she to discover that someone had stolen her jewels.

Klein's program draws quite unwittingly on the knowledge of psychology and the conventions of detective fiction that are implicitly embodied in it. Its mysteries are set in a house-party, the characters include a baroness and a butler, and the small set of possible actions include playing tennis, cheating at bridge, and flirting. Klein chose this stereotyped setting because it provides a felicitous context for a detective story:

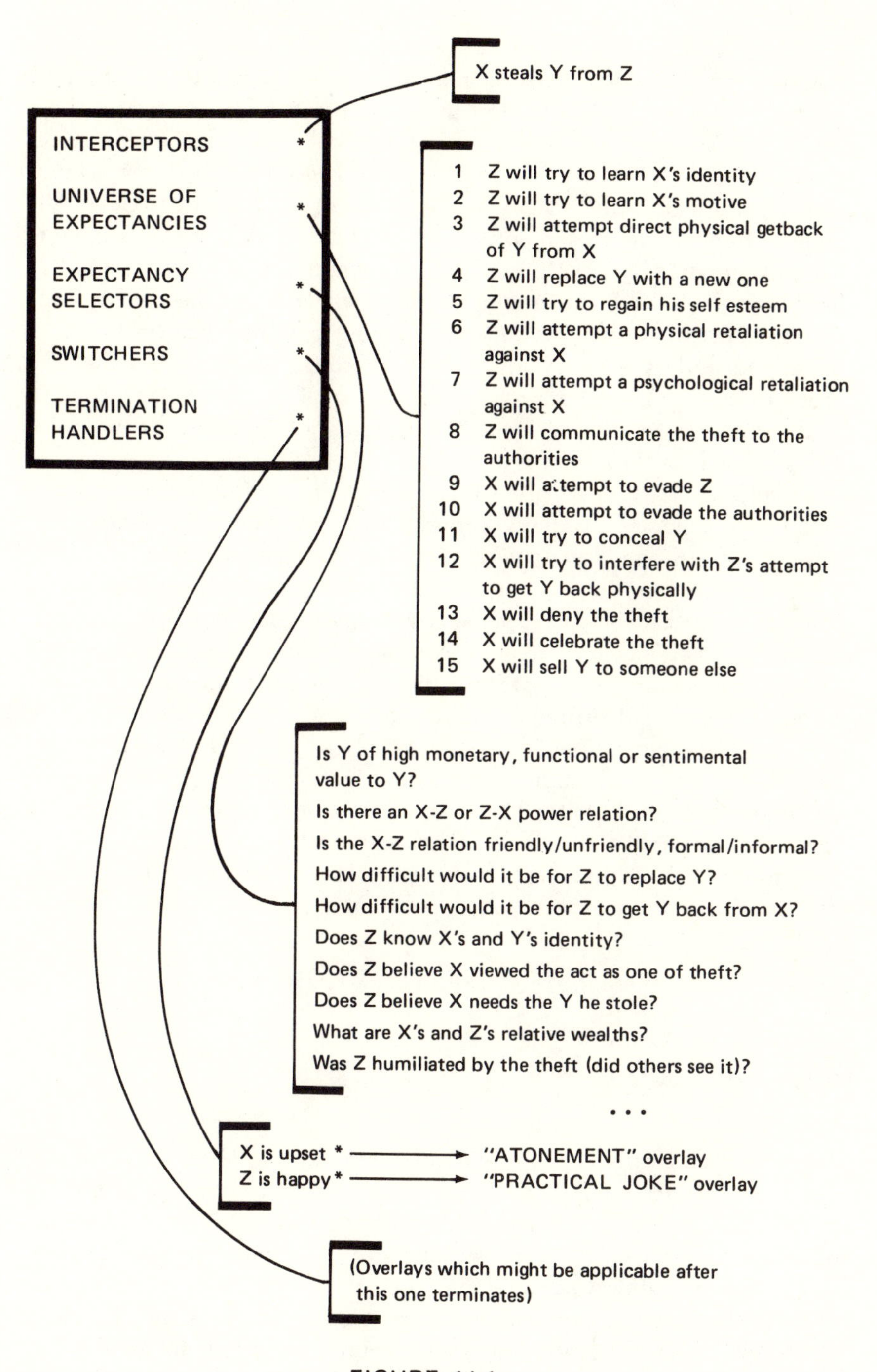


FIGURE 11.1

Source: From *Proc. Fourth International Joint Conference on Artificial Intelligence*, Tbilisi, USSR, 1975. Figure 1, p. 145. Reprinted by permission.

it allows for ill-assorted personalities plausibly to encounter one another (and overhear crucial encounters) in a concentrated fashion over a short period of time, and dictates also that the killer be one of this select company; it even allows for an observant guest to outwit predictably stupid policemen in the strictly localized search for clues (it is clever Dr. Hume who finds the poison bottle). Agatha Christie was adept at exploiting this situational frame, but Klein's program exploits its possibilities hardly at all. Indeed, the automatic novel writer would be equally content to spin yarns about astronauts bringing poisoned goblets to duchesses, and butlers fencing with New York cops. The reason is that the program itself has no structured representation of the house-party context to guide its construction of the narrative, relying blindly on the programmer's dictat as to the concepts available for its use.

Quite apart from its gross ignorance of the world, and consequent inability to generate ingenious stories, Klein's program lacks creativity in other ways. For instance, its vocabulary is not only limited but inflexible: it cannot use a fresh analogy or metaphor to express what it wants to say (though Klein can of course include analogical uses in the dictionary). Similarly, it is doomed to accept the authorial conventions provided to it by Klein: it cannot even appreciate the peculiar suitability of the house-party setting for an orthodox murder mystery, still less dream up new compositional conventions for itself. These two types of literary creativity have their parallels in mathematical and scientific thinking, where someone may intelligently extend the use of a familiar proof or theory to a somewhat unfamiliar problem, or may conceive a fitting method of approach that is novel in a significant sense. The next two sections will be concerned with these aspects of original thinking.

## ANALOGICAL THINKING

To think analogically is to see one thing *as* another, not in the sense that someone mistakes the one for the other, but that she conceives of the one in terms of the other. To perceive an analogy, it is necessary that one recognize an agreement or correspondence in certain respects between things that are otherwise different. The "things" may be anything you like: words, pictures, physical objects, problems of many kinds. Indeed, I challenge you to name two things that are not analogous in some sense: when the Mad Hatter asked Alice "Why is a raven like a writing-desk?" he had no solution to offer her, but many ingenious answers to the riddle have been suggested since. Accordingly, some conception of the degree, or closeness, of analogy is in order. (Even Colby's crude FINDANALOG routine embodies a simple criterion of the closeness of analogy it seeks.) And the "certain respects" should preferably be specifiable, if the analogy is to be intelligently exploited rather than vaguely remarked. For most analogies that people take the trouble to draw are not mere riddles or idle fancies: on the contrary, we continually make use of analogy in approaching the serious matters of life.

The question therefore arises whether analogical thinking is a necessary aspect of intelligence, whether it enables an intelligent system to do something that otherwise it could not have done so economically, or perhaps could not have done at all. A psychological theory of analogy should address this issue, as well as asking how it is that analogical thinking is possible at all.

It is often assumed by people having no familiarity with artificial intelligence that a basic defect in comprehension limits the power of all possible language-understanding programs, namely, the inability to interpret *new* word senses. More comprehensive dictionary facilities could not eliminate this defect, since words are continually being employed in new senses, both in casual conversation and formal writing. What is required, then, is not a post hoc dictionary definition, but a way of correctly assigning meaning to the word on its first occurrence in its new sense. Correlatively, one requires the ability to use words in extended senses in expressing the meanings one intends. No language user unable to cope sensibly with such cases could be held to possess a really satisfactory linguistic understanding: a glance at the daily newspaper should suffice to show the *normality* of analogical usage.

Moreover, the skeptic might object that to say that any word can always be used in a new sense is to say that no set of rules, however complex, can separate the meaningful from the meaningless. What is apparently nonsense today may assuredly be good sense tomorrow. Consequently, no talking machine could ever achieve the power of comprehension enjoyed by human speakers. And a fortiori, it seems, no set of computational rules that—like Y. A. Wilks's system, for instance—functions by way of programmed assumptions about what is or is not a meaningful message, or "gist," could possibly overcome this basic limitation on linguistic creativity. It would follow that HAL must forever remain a charming figment of the fictional imagination; for, in the event that the human astronauts of 2001 coin a new metaphor or subtly extend the usage of a word, their mechanical traveling companion would surely not be able creatively to perceive their locution as a meaningful English sentence.

However justified this skepticism may be when directed to programs such as Terry Winograd's SHRDLU, in which the relations between underlying semantics and surface vocabulary are rigidly fixed, it cannot properly be applied to more flexible systems. Even when a program relies on rule-bound assumptions about what is or is not meaningful, it may nevertheless be able to cope intelligently with language used in newly extended senses or unusual contexts.

For instance, in an early discussion of the limits of meaningfulness, Wilks programmed a system that would locate the problematic word in a chunk of text (itself not a trivial task), and then use semantic insights to choose some other term or phrase within the passage with which to identify its meaning.[13] The semantic structure of the text as a whole, not merely of the sentence containing the offending word, is taken into account by the program in divining the appropriate extension of sense. Crude as this procedure is (what if there is no other suitable word?), it achieves acceptable readings of some notoriously tricky texts, including passages drawn from Wittgenstein's *Tractatus Logico-Philosophicus* and Spinoza's *Ethics*.

Indeed, Wilks's basic set of rules (excluding the meta-rule required to enable the program to cope with new senses) suffices to give a coherent semantic representation of test paragraphs from the philosophical writings of Descartes, Leibniz, and Hume. This is worth remarking in light of the positivist attack on metaphysics launched, for instance, by Rudolf Carnap, an attack that uses a set of empiricist rules to draw strictly defined limits between what is meaningful and what is not. Wilks agrees that a Spinozist axiom may be meaningless *with respect to* a set of rules adequate for more matter-of-fact contexts, but uses his EXPAND algorithm to show that these rules may be developed in a rational fashion so as to make sense of the axiom in its proper textual setting.

As its name suggests, the EXPAND meta-rule can cope with a semantic oddity only if the anomaly involves a genuine extension of sense, relative to some interpretation already stored in the dictionary. If the programmer has omitted any major sense of a word, then EXPAND will not enable the system to understand the word when it is used in that sense. But what is a "major" sense of a word? Is it precisely one that cannot, in fact, be derived from other senses by way of EXPAND or something similar?

For example, consider this exchange between Humpty Dumpty and Alice:

> "In winter, when the fields are white,
> I sing this song for your delight—
> Only I don't sing it."
> "I see you don't" said Alice.
> "If you can *see* whether I'm singing or not, you've sharper eyes than most!"

One might say that Humpty Dumpty apparently lacks a major sense of "see" in his dictionary. On the other hand, someone of his intelligence ought to be able to work out the expansion of sense whereby "see" as "gain information via eyes" is also used for "gain information" *tout court*. These senses of "see" are much closer than the senses of "bank" that connote the riverside and the financial institution; even though these two meanings of "bank" are etymologically connected, the analogy is so tenuous that a person (or a program, or Humpty Dumpty) may be forgiven for regarding them as different senses that should be separately indexed in a semantic dictionary.

Wilks's EXPAND procedure could not make sense of Alice's claim in the quoted conversation, assuming that the meaning of "see" had been represented as specifically visual. The reason is that there is no other expression in this paragraph with which her use of the word can sensibly be identified. But his more recent system of "semantic preferences" (which was mentioned in Chapter 7 with reference to "The car drank the petrol") is not similarly limited, and can assign new senses to words in a more subtle way than the EXPAND procedure. Also, it can access the previously described probabilistic common-sense inferences to aid this assignment, the shortest possible chain of inferences being preferred in interpreting analogy just as in puzzling out the likely referent of "it."[14] It is characteristic of Wilks's approach to natural language, as against Rieger's or Charniak's, that the program draws such inferences (and deepens the representation) *only* when they are necessary to resolve the sense of a word.

Much as Rieger's MEMORY tries to locate fuzzy matches between descriptive sets when dealing with problems of reference, so Wilks attempts fuzzy matches of semantic templates at various conceptual levels in order to interpret not only referential phrases but also cases of anomalous use of words. He opts for the least possible change of meaning (the least fuzzy match) that makes sense of the whole, changes at more superficial semantic levels being more acceptable than changes at deeper levels.

Intuitively, you may feel that to speak of a car as "drinking" is somehow less creative than to say one can "see" that Humpty Dumpty isn't singing. In other words, the analogy seems closer in the first instance than in the second, though both are commonly accepted figures of speech. Analysis in terms of Wilks's theory of semantic preferences would show that more assumptions can be correctly carried over from case to case for "drink" than for "see," where only the deep concept of *acquiring information* can be reliably matched. Thus the car has an aperture through which is passed fluid that is essential to its proper functioning, and which has to be continually replenished. Accordingly, one might even describe the sputtering of the engine as it runs out of petrol in terms of "gasping." You may want to object that Alice's auditory "seeing," like vision itself, involves also the use of sense organs, which would not function if covered or if placed a long way away from Humpty Dumpty; but even this semantic match is forfeited if one describes her as "seeing" Humpty Dumpty's point of view (*sic*) or argument.

Analogical uses of language that (like "see" or "point of view") are very widespread might profitably be separately coded within the dictionary of a program or a person, especially if they involve comparisons at relatively deep levels of meaning: it would be absurd to have to work out the semantic match every time one encountered the word "see." Indeed, the fact that one continually employs the word "see" when talking to someone congenitally blind, usually without realizing that one is doing so, suggests that "visual" and "intellectual" senses of the word are separately accessible in our memory. Similarly, a dead metaphor is one that has lost its capacity to surprise us and that is used as an idiom rather than a metaphor.

But the point of importance is that a semantic system like Wilks's exhibits the conceptual rationale underlying these common everyday expressions, and can be used for interpreting less familiar uses also. And in the more surprising cases, the textual setting has to force the novel meaning more strongly than in a relatively unadventurous extension of the usual sense. The distinction between more and less close analogies is made in terms of a measure of "semantic density," which the program computes with regard to semantic preferences on various levels and common-sense inference chains of varying lengths.

In its current implementation, Wilks's system can interpret many of the analogies it finds in the input text, but cannot itself project semantic preferences so as to create new senses of words. However, this would in principle be possible: for instance, Wilks suggests that "Britain escapes from Common Market" could be formed by projecting the normal preferences of "escape" for a human agent and a prison source onto "Britain" and "Common Market" respectively, thus creating a MAN-headed formula for the nation and a "prisonlike" formula for the political institution.[15]

Supposing it had found this sentence in the text being translated, it could thereby learn for itself (without having to be specifically told by Wilks) that Britain—and perhaps all nations or human groups?—can sometimes be analogically assigned to the category of MAN without violating the rules defining meaningful messages, or gists. The problem would remain of determining when the extended sense can sensibly be assumed and when it cannot: to revert to the sentence first discussed in Chapter 7, one would not want an undiscriminating use of analogy to garble "Our village policeman is a good *sport*, he captains the cricket team every Saturday" by deciding that in this case FOLK can be assimilated to MAN, so that the policeman is seen as a recreational organization. On the other hand, "Sport escapes tax" would merit such a semantic

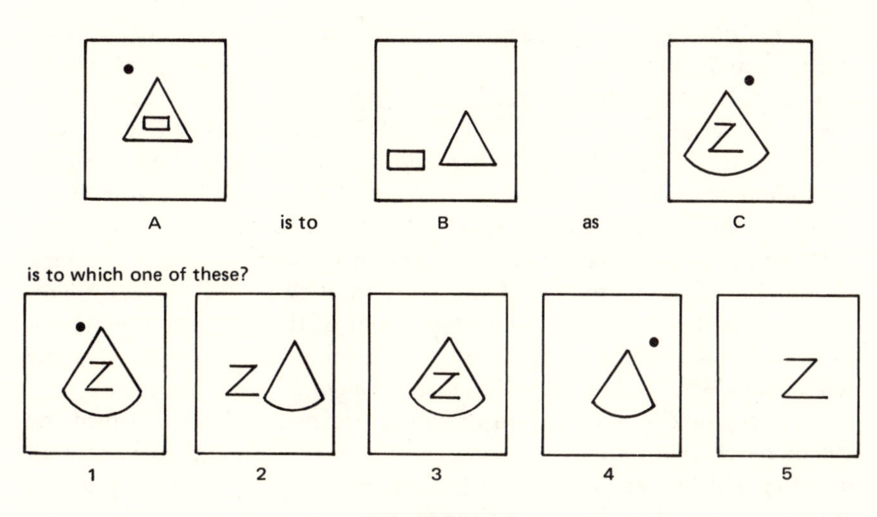


FIGURE 11.2

Source: Adapted from M. L. Minsky, ed. *Semantic Information Processing*, (Cambridge, Mass.: MIT Press, 1968), Case 1, p. 328. This and following figures from this source are reprinted by permission.

assimilation. The more subtle the analogy, the less simple it would be to enable a computational system to create or interpret it appropriately. But this is because the rules involved are complex, not because there are no rules involved at all: analogies are not drawn by magic.

Just as linguistic analogy happens not by magic but by subtle comparison of the semantic representations of words, so other varieties of analogical thinking depend on the intelligent comparison of descriptions of the things between which an analogy is perceived. This point was first clearly made within the artificial intelligence literature by T. G. Evans, whose ANALOGY program could appreciate structural likenesses between geometrical patterns of the type commonly included in IQ tests.[16]

Presented with a set of items such as those shown in Figure 11.2, Evans's program uses descriptions, and descriptions *of* descriptions, to search for deeper similarities underlying surface disparities. For example, ANALOGY concludes in answer to this particular test problem that "A is to B as C is to 2." Its rationale is that the transformation between A and B involves both the removal of the dot and the shifting of the figure inside the larger figure to the left of it, and that the same two rules if applied to C would give pattern 2 alone. In order to arrive at this rationale, the program had first to describe both A and B; next to describe the difference between those descriptions; and finally to apply the difference as a transformation rule to C, to find out whether it would produce a pattern having the same description as any of the five numbered patterns.

This procedure clearly depends for its success on the way in which patterns and differences are described. In the initial description phase, ANALOGY has first to articulate each pattern into "separate" figures, and then has to apply concepts like INSIDE, ABOVE, LEFT OF, BIG, SMALL, ROTATED, REFLECTED, and so on. The articulation is relatively simple in the example just given, but less so in other cases where two individual figures overlap as in Figure 11.3. And notice that the matching procedure depends on the appropriate level of generality being chosen for the descriptions: if the path from A to B had been represented in terms of shifting the *rectangle* out of the *triangle*, it could not have been applied successfully to C, which contains neither a rectangle nor a triangle but rather a Z and a *circle-segment*.

Psychological investigation of concept formation has confirmed the common-sense insight that human beings do not attempt to handle all possible matches at once—indeed, they can keep in mind only a few features at a time if the comparison is arbitrary with respect to conceptual schemata they already possess.[17] When dealing with a problem like that in Figure 11.2, people are not always so lucky as to light immediately on a useful description of the difference between A and B; if they do not, then they change the description of this difference and try again. For instance, in the case of a failure caused by overspecific description, merely generalizing the description may suffice to find the analogy: thinking in terms of shifting "the figure inside the larger figure" allows one to transform C into 2 despite the fact that these patterns contain a zigzag and a circle segment while A and B do not. If generalization does not help, then a different difference entirely has to be discovered and tested. Clearly, this unsystematic procedure cannot guarantee that one will find a match allowing C to be transformed into one of the five numbered patterns, still less that one will find the *best* match, which picks out one of the five as being the most suitable of several possible candidates.

ANALOGY does not suffer from these drawbacks, since it computes *all* matches defined by the class of generalized Euclidean similarity transformations that underlie this type of analogy problem. More complex problems, of course, could not be handled in this "parallel" fashion because they would run into the combinatorial explosion; but (as we shall see in Chapter 14) the more complex the analogy, the less likely that human intuitions will agree about its precise nature and strength. In general, an analogy is closer the more numerous are the descriptions concerned in the transformation rules (which is why pattern 2 is preferred over pattern 3 in Figure 11.2), and the more specific these descriptions are. So if the top line of Figure 11.2 were to contain C′ as well as C, and the bottom line were to have pattern 6 added (see Figure 11.4), the analogy between the pair (A,B) and the pair (C′,6) should be recognized as stronger than that between (A,B) and (C,2). Correlatively, the first of

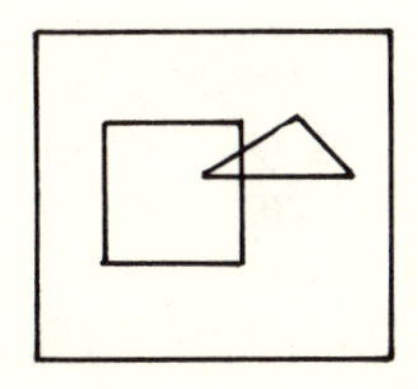

FIGURE 11.3

Source: Minsky, ed. *Semantic Information Processing*, p. 295.

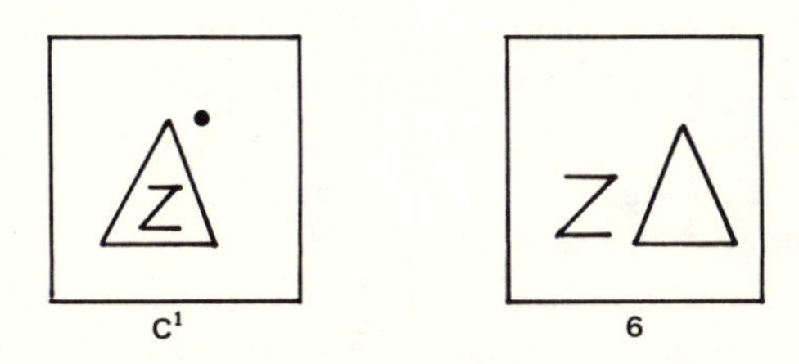


FIGURE 11.4

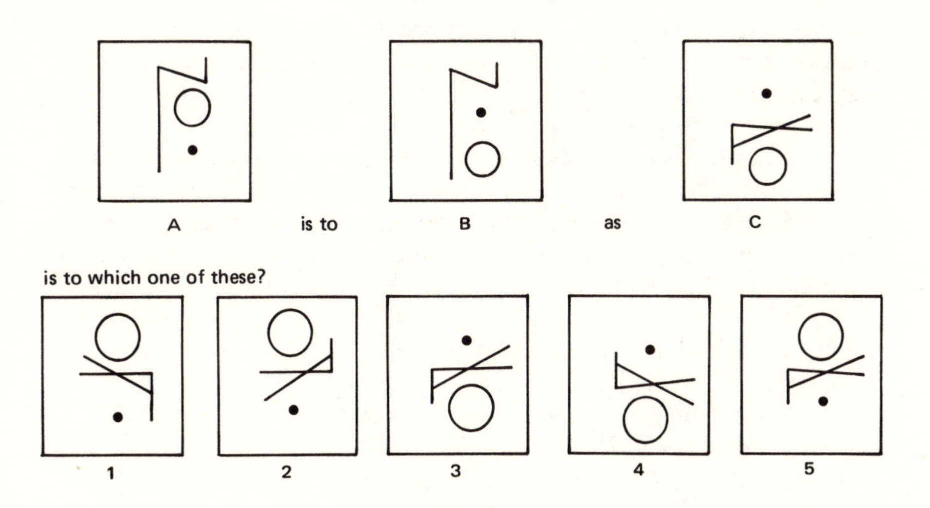


FIGURE 11.5

Source: Minsky, ed. *Semantic Information Processing*, Case 10, p. 332.

these analogies should be stated "strongly" in terms of something being inside a *triangle*, whereas the second would merely justify a vague reference to something being inside a *figure*.

ANALOGY is able to show such preferences for specificity over generality, and for a high number of covarying differences, in choosing the "best" match between C and one of the numbered patterns, while recognizing that more than one of these patterns (usually) can be "correctly" paired with C in that many such pairs are in some sense analogous to the original (A,B) pair. ANALOGY's overall performance on these geometrical problems is comparable to that of a fifteen-year-old child. On only one of the twenty problems discussed by Evans did ANALOGY's answer differ from human intuitions: in Figure 11.5 the program selected pattern 3 whereas you would almost certainly pick 5.

It has probably occurred to you that ANALOGY is distinctly similar to Winston's concept-learning program, which also uses descriptions and descriptions *of* descriptions . . . in comparing specific examples with a preexisting conceptual schema to find the best match. This structural analogy has a historical basis, for Winston built on and adapted the computational methods earlier introduced by Evans. Because the network-matching procedures can be applied to networks irrespective of whether they record properties or differences between properties, Winston's program can recognize analogies on various levels; so it cannot only solve analogy problems, but can answer higher level questions about analogies between analogy problems.

The program utilizes heuristic rules that weight the various differences so that a "best match" is found. These are not confined to geometrical matters, as Evans's rules are, but are defined in terms of the very general semantic notions that make up the various comparison notes. Consequently, the *content* of the concepts is immaterial; provided that the program were given generalized semantic information about a range of predicates (for instance, that abandonment and letting-down are species of betrayal, and that betrayal is opposed to loyalty), it could discover some analogies between conceptual structures defined in terms of those predicates. Similarly, its network-matching procedures could in principle recognize the systematic semantic reversal that generates the diabolical Black Mass, provided it knew such facts as that God and Satan are theological opposites much as LEFT and RIGHT are physical opposites. To appreciate the detailed significance of the Black Mass would of course require the matching of numerous specific items within the Christian and daemonic rituals, which many *people* have insufficient knowledge to achieve: if you don't know that a Black Mass is conducted by an unfrocked priest, you are necessarily unaware of part of the analogy concerned.

World knowledge such as this is essential to the intelligent assessment of analogy. Thus the program's ignorance of the world prevents it from deciding—as people commonly do—that a particular aspect is especially crucial to a proposed analogy, by virtue of the specific implications involved. For instance, in Chapter 14 we shall see that some people object to an analogy being drawn between natural intelligence and artificial "intelligence," on the grounds that a particular feature of the former is lacking from the latter, one so important that any supposed "analogy" must be superficial at best and perniciously misleading at worst. This feature is the possession of *intrinsic* interests, purposes whose appearance cannot be explained by reference to the interests of some other agent. Without anticipating that discussion here, one should note that the *use* of analogy to posit further similarities between the analogous items implies that domain-specific knowledge must be brought to bear in assessing the strength of analogy in specific cases. Since Winston's program is grossly deficient in world knowledge, it can assess analogies only in terms of general conceptual form as opposed to richly specific semantic content.

Winston's program spontaneously utilizes its capacity for analogical thinking to help it identify scenes as examples of one concept or another. Thus it spends its "idle" time in comparing its various stored models with one another. Accordingly, it builds up a similarity network interconnecting its models, which records cases of two models being only slightly different: for instance, it knows that a TABLE differs from a PEDESTAL only in having four supporting objects rather than one. This knowledge is used by the program to suggest what closely analogous model should be tried next, in cases where it has attempted to identify something without success.

It could also be used by a program such as M. J. Freiling's, which could recognize that this particular difference is irrelevant to the usual function of a table, so that the analogous pedestal could be used *as* a table if necessary. For real-world action, considerations of height would need to be included also, in which case some pedestals could not function as tables for normal human beings despite their close structural analogy to tables. Freiling notes that recognizing the functional significance of a staircase for a walking animal, for example, depends upon having a concept of vertical heights that are "suitably small" for the animal concerned. This concept is vague because Freiling can suggest no general, systematic method for handling such a concept (can you?). Heuristics relating the height of the steps to the animal's "footstep size" would doubtless be useful, but many other factors are also relevant, such as whether the legs are rigid or jointed and whether the creature is a large mammal or a tiny spider. If a robot were to be able to recognize the functional significance of a terrain, it would need knowledge (implicit or explicit) of its own motor capabilities in order to see structures *as* pathways, obstacles, or tables.

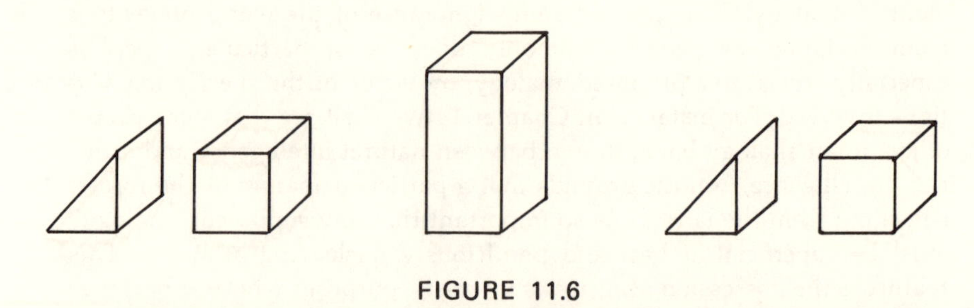

FIGURE 11.6

Source: P. H. Winston, *Learning Structural Descriptions from Examples* (Ph.D. thesis). (Cambridge, Mass.: AI-TR-231, MIT AI Lab., 1970), Figure 7.7, p. 208. This and Figure 11.7 reprinted by permission.

Symmetry is a special case of analogy, in which there is a structural likeness between different parts of a superstructure. The superstructure may be a representation of a physical scene, a logical problem, a story plot, and so on. By taking note of similarities and simple oppositions between comparison notes at various levels of detail, Winston's program can assess degrees of symmetry within scenes. It can notice that there is a symmetry if the objects within a given scene are grouped in one way rather than another; and it can also recognize symmetry within the groupings, so that two symmetrical scenes can be compared with respect to the *depth* of symmetry involved. For instance, Figure 11.7 is more deeply symmetrical than Figure 11.6, since there is not only symmetry with respect to the location of groups of objects, but also in the positioning of objects within the groups. Similarly, a scene that is symmetrical on the three dimensions of LEFT-RIGHT *and* ABOVE-BELOW *and* LARGE-SMALL is "more" symmetrical than one that exhibits symmetry on one dimension only, and Winston's program can recognize this fact. As we shall see in the next section, this is a potentially useful ability, since creative problem-solving strategies often exploit symmetries of various kinds.

Winston's discussion of the varying ways in which the components even of simple scenes may be grouped, according to analogies between the descriptions available, casts light on Gestalt properties of perception. For instance, the organization of the visual field into groups of parts sharing a "common fate" largely predetermines the course of later perceptions, as in the attributions of causality modeled in Sylvia Weir's program: the recognition of analogy, then, is integral to perception no less than to speculation.

That analogical thinking can serve many different purposes is evident from the disparity of aims implicit in asking riddles, taking IQ tests, and participating in a Black Mass. However, one can readily imagine a

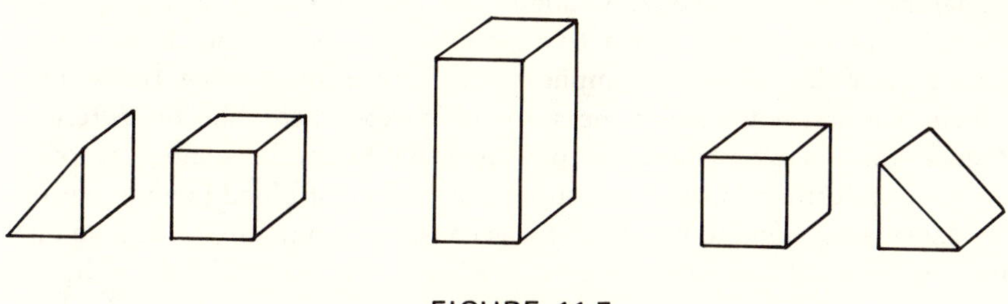

FIGURE 11.7

Source: Winston, *Learning Structural Descriptions from Examples*, Figure 7.8, p. 208.

society of highly intelligent beings in which these forms of frivolous amusement and deliberate blasphemy never took place, and whose armoury of tortures lacked the IQ test. Would such a society be able to dispense entirely with analogy? It is less easy to imagine such a thing, for this type of information processing serves a general purpose with which intelligence is essentially concerned, namely, the creative use of preexisting schemata to guide the exploration of novel phenomena.

In Minsky's terminology of *frames*, one might say that useful analogical thinking involves the comparison of frames, wherein admittedly disparate frames are initially assimilated by virtue of certain points of likeness, and are thereafter scanned for further similarities in terms of which the relatively novel schema may be developed and understood with the help of the more familiar one. The comparisons may be made at the superficial level of the frame terminals, and/or at deeper levels, and the analogy may be more or less detailed and may fail in different ways accordingly.

For instance, Harvey approached the task of understanding the blood system with a well-developed conception of the water-pump already in mind. His "hydraulics" frame apparently had terminals some of which matched specific observations of contemporary anatomists, such as that the blood flowed through tubes, spurted out if the tubes were broken, passed through a specially shaped structure connected to the tubes, and so on. The hydraulics frame dictates that the tubes be a closed continuous system, through which a fixed amount of fluid circulates in a given direction because of the action of a pump.

Observational comparison with the blood system confirmed the analogy at various points. For example, venous valves were found, consistent with only one direction of blood flow; observation of cold-blooded animals with slowly beating hearts showed (what was not apparent in warm-blooded mammals) that the blood did indeed travel in one direction only, and that the *systole* of the heart (its muscular contraction) provided the motive power responsible. (Descartes used a different explanatory frame in theorizing about the circulation, in which cardiac *diastole* was supposedly the active moment: but he did not think sufficiently carefully about the implications for the observable frame terminals, and did not attempt the systematic experimentation on different animals that enabled Harvey to pin down his hydraulic analogy so convincingly.) Further, Harvey measured the volume of blood flowing out of the damaged system, to find that it was constant in amount . . . and so on, and on. Not all the observations prompted by the hydraulic schema found a parallel: Harvey was unable to see the capillaries, since he had no microscope. Undeterred, he posited the existence of capillaries, which were observed some years later by Malpighi.

This example illustrates two important creative uses of analogy, which are particularly clear in instances of scientific theorizing like this one (or like the kinetic theory of gases, in which gas molecules are thought of as tiny billiard balls), but which are relevant also to less systematically disciplined thinking in everyday life.

The first is the use of the familiar frame to prompt inquiry aimed at developing the novel (and initially often more sketchy) frame in an economical fashion. For example, the distinction between the pulmonary and aortic circulations, arising from either side of the heart, was newly made by Harvey because it was suggested by the constraints implicit in his hydraulic schema. Similarly, the kinetic theory of gases was used to predict that the gas laws would break down under conditions of very high pressure: for then the "billiard balls" can no longer move freely, but are pushed up against one another so that one may expect interactions between them to be important that under "Ideal Gas" conditions may be ignored. In this case, the potential breakdown of the analogy was conceptually evident long before the analogy could be tested in practice. The actual (as opposed to the potential) breakdown of the analogy must depend on further empirical investigation. Thus one might use the billiard ball theory to suggest certain sorts of molecular interactive effects, but whether these effects or others are actually present cannot be settled merely by "thought experiments": the terminal slots in the high-pressure gas frame must be filled on the basis of empirical observation, after which they may or may not be found to match closely with the corresponding slots in the billiard ball frame.

The second creative use of analogy enables one not merely to gather new factual knowledge about the novel phenomenon, but correlatively to *understand* or *explain* it, by relating it to the concepts already accessible in the familiar frame. The importance of making something thus intelligible underlies Harvey's confident postulation of the still invisible capillaries—for the general principles of hydraulics can be applied to a closed, continuous, system but not to one in which the tubes peter out allowing the "enclosed" blood to escape. (At the submicroscopic level this is in fact what happens, since the blood escapes from arterial capillaries into the body tissues and has to be reabsorbed into venous capillaries; this reabsorption clearly must be explained in *non*hydraulic terms.) Harvey therefore felt that he could understand the blood system only on the assumption of there being capillaries. By contrast, those observable "terminal" details of the water-pump frame that (like the color of the fluid) had no conceptual connection with the frame expressing the general principles of hydraulics would *not* need to be paralleled in the circulatory system, in order for Harvey to be satisfied with his explanatory analogy.

Few programs exhibit even the beginnings of scientific creativity, since the human thought processes concerned are so ill-understood. The most impressive automatic scientist at present is the DENDRAL system of Joshua Lederberg, B. G. Buchanan, and E. A. Feigenbaum.[18] This set of programs has been in continuous development since the mid-1960s and is still evolving. It combines general principles of problem solving like heuristic search (to be described in Chapter 12) with considerable expertise in organic chemistry.

The "performance program" embodies the process by which an analytical chemist identifies an unknown compound by way of mass spectroscopy, a technique in which the unknown molecule is broken into fragments by an electron beam inside the spectroscope. Like the human chemist, DENDRAL's performance program formulates probable hypotheses about the compound's molecular structure on the basis of its spectrograph, and then tests these hypotheses by way of further predictions. Its performance as a practicing analyst compares favorably with that of expert chemists, for certain classes of compound. And DENDRAL has been of use to chemists also in giving for the first time a complete list of the set of possible isomers of a given empirical formula within several families, including amines and thioethers, because its STRUCTURE GENERATOR, which is called on by the performance program, can in addition be used to generate exhaustively all the molecular structures in (or outside) any given class.

This latter achievement is one of predetermined rule application rather than creative insight, for it depends on the fact that the program (unlike the human chemist) can systematically avoid missing out any of the logical possibilities—though it can also be programmed to draw attention to novel structures that might be expected, on general grounds, to be interesting. (DENDRAL's area of expertise includes the steroids used in contraceptive pills, so the structure generator can be primed in terms of what is pharmacologically useful as well as what is chemically interesting.) A closer approximation to genuine creativity is shown by a recently developed module called "meta-DENDRAL," which is described by its authors as a preliminary exercise in automatic theory construction.

Unlike the performance program, meta-DENDRAL does not merely apply previously articulated rules in a routine fashion to experimental data, but it formulates *new* rules on the basis of observation of the data. Thus it has already discovered mass spectrum fragmentation rules for several classes of compounds (such as aromatic acids) which had not previously been recognized by expert chemists, and which can be provided to the performance program. In outline, what it does first is to inspect the experimental data to see if it can find any regular patterns in it; second, it thinks about these in light of some of the concepts it already knows about, to see whether it can find any explanation of the observed patterns. The latest version of the program can accept "mixed" data drawn from a number of significantly different molecular structures, and can separate out the different subgroups and find a characteristic explanation for each.

The experimental data consist of a series of molecular structures (represented for the purposes of the program in LISP), each with its correlated spectrograph. The sort of rule at which the program can inductively arrive includes "IF the graph of the molecule contains the estrogen skeleton, THEN break the intramolecular bonds between nodes 13–17 and 14–15," which expresses a hypothesis about the chemical process underlying a particular class of spectrographs. The most likely explanatory processes are selected by DENDRAL from the class of a priori plausible ones partly by reference to strictly chemical considerations (such as the number of bonds assumed broken) and partly by reference to logical considerations of simplicity, uniqueness, and evidential strength that are relevant to scientific explanations in general. Various chemical concepts can be applied by meta-DENDRAL to ask what is so special about (say) nodes 13–17 and 14–15. That is, the program attempts to isolate a smaller, more general (*sub*molecular) structure in the immediate environment of the bonds broken in order to say *why* the bonds break. Since it sometimes succeeds, the program is useful in extending the theory of mass spectrometry.

DENDRAL's ability to separate out significant subclasses of molecules from a mixed bag of experimental data may be compared with the ability of Winston's program to group objects together in various ways according to "significant similarities." Insofar as grouping depends on the recognition of similarity, DENDRAL's identification of molecular subclasses may be thought of as a form of analogical thinking. But DENDRAL also embodies knowledge of chemical analogies more naturally so called, such as that amines are something like ethers, which are helpful in reasoning about the behavior of an unidentified compound. As yet, DENDRAL is not able spontaneously to recognize the analogy between amines and ethers, but it has much of the knowledge required to make sense of this comparison. Indeed, the nature and extent of the analogy were made more precise during the writing of the program, since the intuitively recognized similarities normally left largely implicit by expert chemists talking among themselves had to be made explicit so that the programmer could communicate them to the program.

It should be clear from the discussion of "learning by being told" in Chapter 10, that DENDRAL would need to know a good deal about what (chemical) questions to ask and what inferences to draw, if it were to be capable of being *told* that "Amines are like ethers." Buchanan and Feigenbaum aim to enable the program to learn by being told such things, thus bypassing the need for a human programmer to be constantly interposed between program and chemist. The large amount of expert knowledge already possessed by DENDRAL makes this aim more realistic than that of enabling Rieger's MEMORY to make sense of Father Brown's speech.

In everyday thought, as well as scientific theorizing, one constantly seeks to make relatively unfamilar things intelligible in terms of more familiar concepts. Explanation in general involves the assimilation of familiar to unfamiliar, and the discriminating assessment of the inevitable *dis*similarities as crucial or not.[19] Even Winston's program, as we have seen, uses analogies to guide its observations in identifying objects in its visual world. Similarly, HACKER exploits the analogy (common pattern) between the current problem statement and some past problem, in order to pick the most appropriate performance program out of its library, which it then refines as necessary. When a refinement is generalized and stored as a new subroutine, it is indexed by the problem pattern that gave rise to it, so that analogous problems may be recognized in future. And Abelson's IDEOLOGY MACHINE has the basic knowledge required to recognize the structural analogy between Barry Goldwater's conservatism and *Pravda*'s Communism, much as a filmgoer appreciates the similarity of *West Side Story* to its Shakespearian precursor, or Alice notices the likeness of seeing to hearing.

All these achievements require the comparison of descriptions of the analogous phenomena, and the more richly detailed and tightly structured the representations concerned, the more fruitfully the analogy can be extended and explored. The matching and comparison of frames (and even their individual formulation) is a topic on which much work remains to be done, for the computational issues involved are currently ill-understood. However, the difficulty in getting machines to appreciate subtle analogies lies in the complexity of the problem rather than any essential mystery or impossibility in principle.

You may feel that the *really* creative step in analogical thinking is lighting on the fruitful analogy in the first place: anybody—well, almost anybody!—could have deployed the vascular/hydraulic analogy once it had been suggested, but it took a Harvey to suggest it. That this underestimates Harvey's intellectual achievement should be clear from the contrasting example of Descartes, who did not deploy his theoretical analogy of the circulation in a comparably detailed experimental fashion. Still, it is true that greater problems are involved in matching a frame with the most relevant of an indefinite variety of frames, than in matching two *given* frames: how does someone even know which frame terminals will turn out to be the interesting ones, before having any notion of the particular analogy required? For example, how does one know that the color of blood is irrelevant?

Before one can cope with this elliptical question, it must be completed: irrelevant to what? To its function? To whether or not it circulates? One must have at least a vague idea of the sort of question one wants to ask about the blood before one can sense the relevance of its color. Correlatively, one needs at least a vague notion of the sort of answer that might be acceptable, if one is not to be totally at a loss where to look for an appropriate analogy. This latter condition does not preclude finding an analogy "by accident"; but apparently accidental or inexplicable insights into analogy usually turn out to have some degree of intellectual directedness.

We have seen already that someone speculating about the circulation of the blood with a broadly hydraulic model in mind has good reasons for disregarding its color. What if one is interested in the function of the blood? Not two hundred years ago, the Abbé de St. Pierre believed that every natural phenomenon could be best explained in (theologico-teleological) terms of its immediate convenience to animals or people: he actually claimed that God designed melons in segments so that they would be easier to eat *en famille*, and someone capable of saying that is surely capable of believing the blood's red color to be a direct consequence of its function of making the lips and cheeks beautiful.

You are doubtless incapable of such a belief, and of the fancy that there are little bespectacled imps in the body tissues, who collect the life-giving spirits from the blood and who recognize them solely by their red color. Rather, you assume that the *sort* of explanation appropriate describes the "life-giving spirits" in chemical terms in which colors do not figure. Even if you knew nothing of hemoglobin or of oxygen, you might know that chemists and physicists do not explain bodily functions in terms of colors. On the contrary, colors are explained in terms of properties like length, size, or shape.

This *very* general knowledge about the sorts of concepts we expect to be useful in explaining matters like the function of the blood crucially influences our choice of which frame terminals to concentrate on and which to ignore, but it is so "obvious" that it is unlikely to be explicitly mentioned. Only if we come across a cultural anomaly such as the Abbé de St. Pierre, or a person from a wholly nonscientific culture who naturally does not share these expectations, are we shocked into realizing that they underlie our questions about the blood. (Even theological explanations, in our tradition, pay no attention to the colors of things in the natural world.)

By the same token, these implicit expectations may help to explain the famous "creative leap" of the nineteenth-century chemist Kekulé, who reputedly dreamed of a snake biting its tail after he had been puzzling over the behavior of benzene, and accordingly postulated the benzene-ring as the underlying molecular structure. Since chemists already expected a molecule's behavior to depend on its shape, or spatial structure, and since the structures previously envisaged failed to exhaust even a very limited number of topologically distinct types, it is perhaps not so surprising as is sometimes suggested that Kekulé should have had this dream and should have recognized its significance.

Even if you accept Jung's theory that the tail-biting snake is one of the ancient archetypes of the mind, and a common occurrence in dreams (even of dreamers with no interest in chemistry), you may still admit that Kekulé was able to see its chemical significance only in the specific context of his current concerns. For that matter, there may be no *psychological* explanation for the image of the snake in Kekulé's dream—whether in terms of problem-solving or archetypes or any other psychological category: possibly, the cheese he ate for supper upset his cerebral physiology in the sort of way that hallucinogenic drugs do, so that the cause of his dreaming of a snake was physiological rather than psychological. Even were that the case, computational questions remain as to how he was able to recognize and exploit the relevant analogy. In general, during creative thinking a frame may be "accidentally" activated by a (psychological or physiological) cause that is unrelated to the problem in question. But once so activated, the problem context can facilitate its analogical matching with a previously activated frame by way of implicit assumptions about which features are most likely to be relevant, and which are probably of no account.[20]

Since computational knowledge of how to compare two given frames is currently in its infancy, it is not to be expected that artificial intelligence has much to say about precisely how creative achievements like those of Harvey or Kekulé may be mediated. True, some programs can reason "inductively" or "analogically" to draw sensible (though not foolproof) inferences from incomplete data. For instance, the SCHOLAR program answers geographical questions by extrapolating from facts it already knows: in default of evidence to the contrary, it will assume that the climates of Sydney and Los Angeles are similar, since they are both at sea level and share the same latitude.[21] MERLIN is enough of a wizard to use an operation typically employed on one concept on a different, though analogous, concept, by means of "forced matches" between disparate features.[22] M. L. Minsky showed in the 1950s that a set of geometrical heuristics, soon to be embodied in a geometry theorem-proving program, could produce a superbly elegant proof of the *pons asinorum* theorem, a proof superior to Euclid's own in that it does not require any construction.[23] (The mathematician Pappus is credited as the first discoverer of this proof: *six centuries* after Euclid.) And DENDRAL, as we have seen, can use inductive inference to arrive at genuinely new knowledge: "new" not merely in the biographical sense, but in the stronger sense that *no one* was previously aware of it.

But even DENDRAL can function only within a tiny specialized corner of a well-established scientific paradigm, and its "new" hypotheses are generated within strictly defined theoretical constraints. There is no question of its producing a novel convention, a significantly different representational frame in terms of which to think about chemical phenomena. The creative shift to a new scientific paradigm—which T. S. Kuhn has likened to a Gestalt-switch, after which different cues are emphasized or "the same" terminal information is interpreted in different ways—is beyond the scope of DENDRAL's reasoning.[24] To be sure, a learning program such as HACKER can come to interpret an instruction (to put block A on block B, for instance) very differently from the way in which it first interpreted it; even so, we saw in the previous chapter that HACKER's first attempt at solution has to be *almost* right in order that HACKER can improve on it, since the program is unable to "stand back" from the problem and view it in a radically new way.

Before one can say whether (and how) a program might do this, the concept of a "radically new" or "significantly different" representation needs to be clarified. Some relevant questions have been considered with computational models in mind, as we shall now see.

## CHANGING REPRESENTATIONS

I mentioned Spinoza's metaphysics in the previous section as an example of a novel linguistic representation of the world to which a person (or program) can attach some meaning by extending the existing set of interpretative rules, but I did not ask why anyone should be interested in doing so. Doubtless some sense can be made also of "Colorless green ideas sleep furiously," but why bother?

The power and fascination of Spinoza's philosophy is due to the economical way in which his abstract axioms and definitions can be used to generate mutually consistent answers to an indefinite number of particular theological questions that had been hotly debated throughout the Middle Ages (and are still so debated by many people).

In order to appreciate this power, one must be able to map the abstract Spinozist dicta onto the specific questions concerned. One needs to know, for instance, that when he says "Nothing exists from whose nature an effect does not follow," tersely referring to three previous theorems in proof of this one, this should be interpreted as a suggested solution to the perennial problem of whether it makes sense to assume a divine plan behind the world, a solution related in specific ways to alternative answers previously offered by Aristotle, Maimonides, and Descartes. A good Spinozist commentary provides (what Spinoza himself left implicit) a detailed mapping of the novel representation onto a wide range of discussions within medieval Islamic, Judaic, and Christian thought.[25] In so doing, it shows how a Spinozist formulation allows of answers not only to "obvious" questions but to less immediately obvious problems also. To understand Spinoza's thought is to understand how it relates to and integrates these philosophical matters that are commonly expressed in other ways.

The example of Spinoza's philosophy, and also Kekulé's model of the benzene molecule, are special cases of the fact that one representation of a problem may be better than another. Better, both in the sense that it provides a solution relatively economically as regards the amount of processing needed, and in the sense that it can be generalized to other, closely related, problems—and perhaps even to problems that had not previously been recognized as related at all.

Since a representation has to be used by a problem-solving system, a representation that is useful for one problem solver may be less helpful to another. Thus we shall see presently that diagrammatic formulations of a problem may be extremely fruitful for human thinkers—but it does not follow that they would be equally valuable to all problem-solving systems. A reasoner (whether animal or artefact) that was incapable of handling the epistemological structures and mapping rules involved in manipulating diagrammatic representations would not be able to solve problems in this fashion.

Problem solving in artificial intelligence requires that a representation of the problem be available that the program concerned can usefully exploit, whether it is thought of at the level of the machine code, the high level programming language, the details of the specific problem statement concerned, or the overall plan of the problem as a whole that is manipulated by the program in the early stages of the solution. Some of these issues will be further discussed in the next chapter, but for the moment we need to consider what sort of creativity is involved in producing a novel representation for a problem and whether it could be achieved by programs as well as persons.

The work of Saul Amarel focuses on these issues. One of his early papers illustrates the comparative power of different representations by reference to the missionary-and-cannibal problem.[26] This can be stated as follows: Three missionaries and three cannibals seek to cross a river, say from the left bank to the right bank. A boat is available that will hold two people and that can be navigated by any combination of missionaries and cannibals involving one or two people. If the missionaries on either bank of the river, or en route across the river, are outnumbered at any time by cannibals, the cannibals will indulge their anthropophagic tendencies and do away with the missionaries. Find the simplest schedule of crossings that will permit all the missionaries and cannibals to cross the river safely.

If you have not already encountered this puzzle, you might like to try it out yourself before reading further. How do you proceed? Do you use matchsticks? Or pencil and paper? What details of the situation do you explicitly represent—and how? Can you identify significant steps or crucial insights in your approach to the puzzle? What is significant about them? Above all, do you use any *general* procedures in formulating a restatement of this particular problem in a more helpful way?

By a general procedure, I mean one that may be applied to other problems superficially different from this one. The missionaries' quandary is one of a class of problems of reasoning about actions, where the formulation of the poser specifies an initial situation, a terminal situation, a set of feasible actions, and a set of constraints that restrict the applicability of actions. One must find a sequence (preferably the shortest sequence) of permissible actions that can transform the initial into the terminal situation. This obviously requires an appreciation of the sorts of transformation effected by each action—but, as you may already have realized in tackling the puzzle, the level of abstraction at which actions and transformations are represented can vary, with corresponding variation in problem-solving power.

Amarel's discussion shows how the specific details of the missionaries' dilemma can be mapped onto six different abstract representations of increasing power. The sixth can furnish solutions to an indefinitely large class of new problems of the same general type: for instance, where there is a very large number of cannibals (try doing that with matchsticks!), where they are arbitrarily distributed on either bank of the river at the initial and terminal stages of the problem, where the boat's capacity may vary at different stages in the development of the solution, and where a certain number of missionary "casualties" is allowed. His interest lies in making explicit the features of these alternative conceptualizations that render them more or less powerful, and in identifying general principles of representation shift that could function as creative procedures enabling a thinking system to reformulate many different problems in a useful manner. As yet, he has had more success in the first aim than in the second.

The first of his sixfold series of representations of the puzzle is the verbal formulation previously quoted. This does not specify just what are the feasible actions in terms of which one should think about the difficulty. For instance a "crossing" could be represented as the traversal of the river from left to right, by Jane and Lumompo, at 4:25 P.M., by means of a balsa craft rowed with two oars. But does one need to distinguish each person by name? Need one specify the direction of the crossing? Does the time of the crossing matter, or the material from which the craft is made? (In a problem concerning Cinderella, the time of actions involving her pumpkin-coach would be crucial.) From the fact that such information might be specified in a detailed plan representing the *actual execution* of a solution, it does not follow that it has any importance in *finding* the solution.

Assuming that one finds a solution by thinking in terms of rather more abstract frames, decisions have to be made about which details are irrelevant, and at what level of abstraction the actions should be schematized. For instance, is the problem symmetric with respect to time reversal? Can one safely "work backward" to the solution? Can several crossings be combined into one conceptual unit, and if so, how? The previous discussion of STRIPS showed that the combination of elementary actions into macro-actions can facilitate problem solving, and such a combination is a general strategy of creative thinking wherein underlying regularities are more economically expressed in the new representation.

An intelligent system must be able somehow to keep sight of the way in which the more abstract formulation maps onto the real-world details of the case. Many trick puzzles rely for their effect on the two representational levels' getting out of step with each other. For example, the teaser about the frog in a 30-foot well who every morning climbs rapidly up three feet and then exhaustedly slips back two, depends on the fact that one mathematically describes the frog's several actions as simply "one foot per day"—usually forgetting the physical reality that on the *twenty-eighth* day the frog reaches a position from which it will not slip back. In this case, creating a representation of actions in terms of macro-actions can prevent one from seeing the solution. But in practice this may not matter: no sensible frog will climb back into the well, however surprised it may be to reach safety already on the twenty-eighth day; similarly, the robot SHAKEY's execution program can monitor its progress in the real world to eliminate superfluous steps, as we have seen.

In the sixfold series of representations suggested by Amarel, elementary actions are clearly defined (in a way that shows the identification of Jane and Lumompo to be unnecessary), and are variously chunked into macro-actions. In addition, redundant conditions are eliminated, sym-

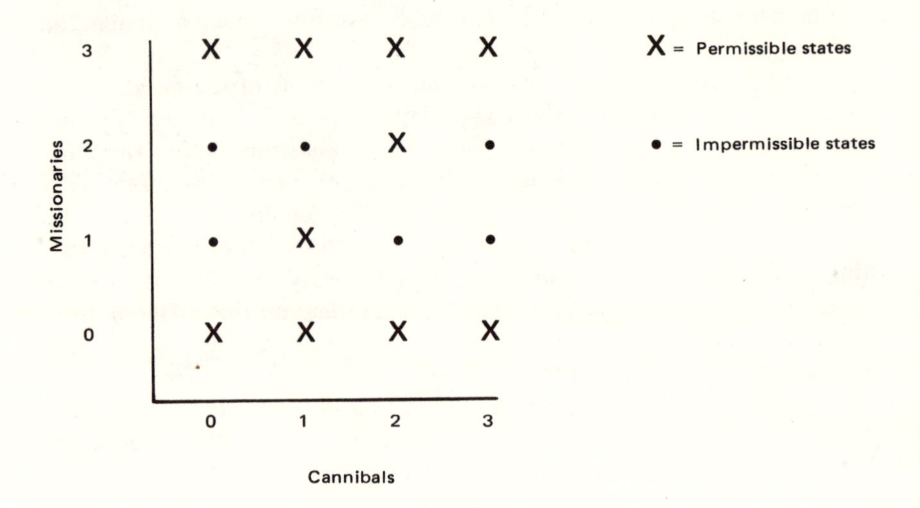


FIGURE 11.8

metries in the problem space are exploited, and critical points are identified forming a high level subspace within which the problem may be schematically solved, the detailed solution being generated later. This latter principle is the approach of *planning,* a general method of creative thinking that will be further discussed in Chapter 12. The combination of elementary actions into macro-actions is itself a contribution to planning, but Amarel is particularly interested in how the structure *of the plan representation* can aid thinking by suggesting methods of inference that lead economically to a solution.

For example, *given* the insight that a particular set of *n* actions can be represented as one macro-action, would it help to express the problem as a visual diagram rather than as a string of symbols? How would the *performance* of the macro-action be represented in either case? One of Amarel's most powerful representations formulates the problem geometrically or spatially as a zigzag pattern, and conceives of macro-actions (combinations of "forward" and "backward" crossings) in terms of sliding along the cross-bars and jumping on and off the diagonal at particular points.

Amarel's fifth representation is shown in Figure 11.8, and if you have not yet succeeded in solving the puzzle you might try doing so by way of this diagram. Each point on the grid represents a certain grouping of people on the left bank (from which the party sets out). The composition of the group for any given point can be read off the axes of the grid: the vertical axis gives the number of missionaries, while the horizontal axis gives the number of cannibals. The complete set of points covers all the logical possibilities for grouping. But if the missionaries are not to meet a ghastly end, some of these possibilities are impermissible. The Z represents all and only the *permissible* groups. Obviously, then, a path (*sic*) has to be found from starting to final point that passes through Z-points only. The starting point is the top right-hand corner (3M,3C), the goal is the bottom left-hand corner (0M,0C).

One's immediate intuition is to go *straight* along the diagonal of the zigzag. But this is impossible, since though you can pass direct from (3M,3C) to (2M,2C) in one move, you cannot get directly from there to (1M,1C)—nor from *there* directly to the goal. Nor can you pass straight along the upper cross-piece from one point directly to its immediate neighbor, for the same reason. (Or, rather, you can—but the next move *must* bring you back to the same position: try it with matchsticks.) In short, as you probably discovered very quickly yourself, this puzzle requires a strategy of *reculer pour mieux sauter*: this is an *abstract* strategical concept that (as the French idiom suggests) can be represented spatially. Figure 11.9 shows the path you have to take to get from start to finish. It does indeed involve using the diagonal as a link between initial and terminal points, but you have to maneuver yourself into a position on the top bar that allows you to jump onto the diagonal, and then slide along the diagonal in the "wrong" direction until you can jump off it onto the lower horizontal.

The pathway shown in Figure 11.9 involves eleven moves. If you solved the problem with matchsticks, you probably found that a minimum of eleven moves is needed. The 3rd to 5th and 7th to 9th are fixed, and are mirror-image sets. The sixth is the crucial move, often not tried because it seems to be going *away* from the goal. And the first two and last two moves each allow of two alternatives, so that there are four possible ways of solving the problem. However, the state resulting from the first two moves is always identical (namely, net transfer of one cannibal, with boat returned to left bank). The strategy of "moving along the upper horizontal" leads one to transfer two cannibals on the first move, and to ignore the other possibility of transferring one missionary and one cannibal as the initial step.

The symmetry of the problem is accounted for by the fact that permissible points that do not involve *equal* numbers of missionaries and cannibals must involve either three missionaries or no missionaries at all. That is, the point (2M,1C), for instance, is *not* shown as permissible because its "opposite," namely, (1M,2C) is not allowed. This is why Amarel can consider only the situation on the left bank, apparently ignoring the goings-on on the right bank.

This form of representation is generalizable to an indefinite class of missionary-and-cannibal puzzles. In most cases, not all of the permissible Z-points need be visited: it is sufficient that one can jump from *some-*

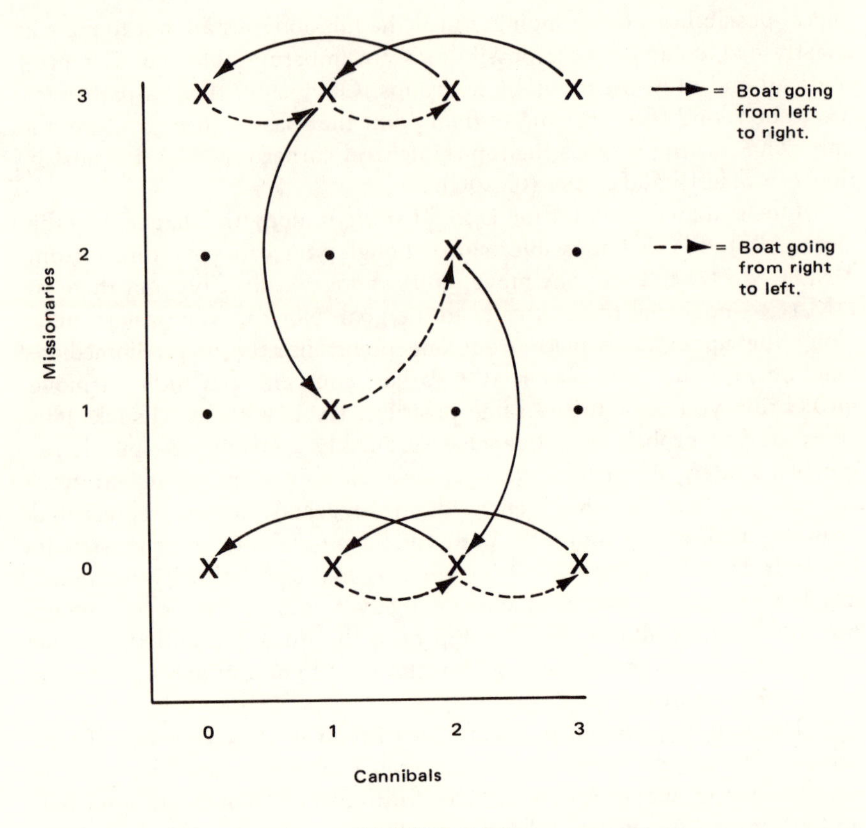


FIGURE 11.9

Source: Adapted from Donald Michie, ed. *Machine Intelligence 3*, (Edinburgh: Edinburgh University Press, 1968), Figure 9.3, p. 154. Reprinted by permission.

*where* on the top bar onto the diagonal, and then move along the diagonal so as to jump onto the lower cross-bar *somewhere.* Moreover, since the solid lines (left to right crossings) can go only in a roughly southwesterly direction, whereas the dotted lines point only in a roughly northeasterly direction, one can see almost at a glance that there can be *no* solution to the case where there are *four* missionaries and *four* cannibals (assuming the boat to hold two, as before). Differently "shaped" paths representing macro-actions in the form of double moves are generated by different boat capacities. And since one can specify that after *n* moves (say) the size of the boat changes, one can readily transfer this constraint onto the manipulation of the relevant zigzag diagram.

The usefulness of this representation clearly lies in our facility to conceive of the zigzag as a pathway along which one might move in various ways. We saw in Part IV that recognition "at a glance" of spatial properties of visual arrays like zigzags is not a simple matter to simulate by current techniques. And Freiling's work shows that appreciation of the functional significance of a spatial structure, in terms of the potential pathways involved, depends implicitly on knowledge of motor capabilities that "comes naturally" to a person, but would have to be provided for (or, ideally, learnt by) a machine. The biological basis of visual-motor coordination may be responsible for the fact that Amarel's diagrammatic representations are more powerful (for us) than are his formulations in terms of strings of symbols. In general, abstract creative thinking that uses spatial analogies (whether zigzags, planetary systems, or tail-biting snakes) may rest on information-processing mechanisms that have been evolved for very practical purposes.[27] The recognition of spatial relations of various types can then be put to use in exploring (*sic*) and exploiting diagrammatic representations of problems originating in widely diverse conceptual domains.

Amarel points out that it is easier to formalize the recognition of features like symmetries and redundancies within a given representation (whether spatial or not) than it is to formalize the choice of "relevant" basic elements and operations, and "sensible" chunks thereof, in terms of which to represent the problem in the first place. That is, the critical structuring of the search space is likely to be more difficult to automate than the recognition of useful properties of the search space, *given* an appropriate representation. Similarly, it is easier to compare two given frames for their degree of analogy than it is to pick a suitably analogous frame "out of thin air."

This is especially true when the representation shift is one in which the novel formulation is superficially very different. Such cases are common in science (you may remember Eddington's famous image of the physicist as someone who thinks a table is like a swarm of bees buzzing in a cathedral), including computer science. Even Colby's numerical matrices fit the bill, as not superficially resembling the psychological phenomena they represent: and programming languages in general also qualify as significantly novel representations of familiar matters, each with its characteristic types of inference procedure. In a later paper on automatic programming, Amarel gave another example of a relatively dramatic representation shift, and (as before) had more success in showing why the novel formulation was helpful than in explaining just how he came to think of it.[28]

In this paper, Amarel assumes that a problem is to be given to the machine in a language *different* from that in which the desired solution will be stated. (He calls such cases "formation problems.") The crucial difficulty, then, is creatively to find a way of moving between these two languages (a mechanism for mapping the relations between them) that can help to locate the solution. In the context of Spinozist philosophy, for instance, the task is to show how an answer to the traditional theological problem of creation *ex nihilo* (could the material world have been made by God out of nothing?) can be found that is expressible as a demonstrable consequence of Spinoza's metaphysical axioms. In the context of machine intelligence, the goal is described by Amarel as a matter of automatic programming: how can one construct by computer a program, in a certain programming language, that satisfies problem conditions expressed in *another* language?

Amarel's main point is that this task may[29] be achieved indirectly, in the sense that candidate solutions are generated within a representation that is other than (and somehow simpler than) the final language itself. In his theoretical case study, Amarel uses a mathematical (algebraic) model for the intermediate representation. He shows clearly how the three languages contributing to his example map onto one another, and what are the specific features that make the intermediate (planning) model helpful. The mathematical model is used in formulating an overall strategy and in building up a repertoire of problem-solving moves. In this way, a radical change in representation of the overall problem is made, with a consequent change in strategy of solution so as to exploit the implicational structure of the model.

In the Spinozist case, the axioms and definitions fulfil much the same function in generating candidate solutions as does the "middle" language in a formation problem. The axioms containing technical terms that are semantically assimilable (analogous) to the everyday concepts involved in the familiar "creation *ex nihilo*" problem are manipulated deductively (or, as Spinoza put it, "geometrically"), to conclude that creation of a material world out of nothing is metaphysically impossible. This obviously requires that the problem-solving system, namely, Spinoza, already have a mastery of the type of inference (formal reasoning) concerned. In Amarel's case, the intermediate model is a modified form of algebra, with a lattice structure of relations that can be manipulated in mathematically familiar ways so as to provide a framework for suggesting sensible solutions and selecting the best one. And in his zigzag example described earlier, the inferences made within the intermediate model or plan are represented as physical movements on the lines.

However, just as it took a Spinoza to find the Spinozist axioms, so it takes an Amarel to suggest a zigzag, or to find (and appropriately to modify) a suitable algebraic inference frame. Not even Amarel understands quite how he did this. It would be easier to enable an automatic programming system to *use* Amarel's algebraic model than to *find* it, because the latter achievement would presuppose a generalized theoretical understanding of the comparative power of different forms of representation.

Aaron Sloman has raised some relevant questions in comparing what he calls "analogical" and "Fregean" representations.[30] An analogical representation of something is one in which there is some significant correspondence between the structure of the representation and the structure of the thing represented. To understand an analogical representation is to know how to interpret it by matching these two structures (and their associated inference procedures) in a systematic way. But in a Fregean representation there need be no such correspondence, since the structure of the representation reflects not the structure of the thing itself, but the structure of the procedure (thought process) by which that thing is identified. To understand a Fregean representation is to know how to interpret it so as to establish what it is referring to, basically by the method described by the logician Frege as applying *functions* to *arguments.* (Sloman points out that his distinction is neither exclusive nor exhaustive, even if it is applied to a relatively restricted range of well-understood representations such as programming languages; but for present purposes this may be ignored.)

For example, a drawing of a cat sitting on a mat is an analogical representation of the relevant aspects of reality since there are corresponding spatial relations of above/below in either case. Similarly a family tree is an analogical representation of hereditary relations since these relations can be read off the corresponding spatial relations in the genealogical diagram. In the latter example, the "reading off" is fairly direct, because the "corresponding" relations, though different in type, are isomorphic. But this need not be so: in Part IV we saw that the interpretation of 2D pictures as denoting 3D scenes may require complex computations for determining what is the target reality equivalent to a particular representational feature. But the point is that even though there need be no identifiable part of the analogical representation that corresponds to a certain part of reality (to a hidden corner, for instance), there must *in general* be parts of the representation that correspond to parts of reality (as vertices correspond to visible corners). Moreover, it must be possible to specify the mapping rules by way of which representational features (such as picture properties) can be interpreted in terms of target realities. Indirect and context-dependent as it may be, there must be some significant (that is, interpretatively relevant) correspondence between the structure of an analogical representation and that of what it represents.

By contrast, sentences like "The cat sat on the mat," or "Jill is Jane's grand-daughter" express the relations concerned in a nonanalogical manner. Understanding these linear strings of words involves thinking like that described in Part III, thinking that does *not* proceed by matching the sentence's structure with feline or genealogical reality. Similarly, predicate calculus formulae and LISP representations of (for instance) chemical structures are interpreted in a Fregean manner.

Sloman's distinction is of interest here because it suggests the possibility of there being very *different* ways of arriving at valid and rigorous proofs in problem solving. He points out that mathematicians, philosophers, and logicians have often assumed Fregean methods of proof (using representations expressed as axioms and formal inference rules) to be superior to "informal" proofs—if these last are to be honored with the name *proof* at all.

For instance, Spinoza felt the axiomatic method to be so intellectually superior to discursive philosophizing that he reformulated Descartes's philosophy in "geometrical" terms, even though he disagreed with its basic assumptions. Similarly, the traditional proofs used in Euclidean geometry, which relied on diagrams, are often regarded by mathematicians as mere heuristic hunches, *real* proofs having been achieved only with the later axiomatization of geometry. The geometry of one's schooldays, on this view, was a mathematical confidence trick in which unsuspecting children were fobbed off with inferior goods: only recently, with the introduction of formalized mathematics (based on set theory) into the school curriculum, have youngsters been offered the genuine article.[31]

Contrary to this attitude, Sloman shows how notions of validity and rigor can apply to informal proofs too, such as those based on diagrammatic representations of the problem. This raises interesting questions about which of the "intuitively obvious" steps in proofs, whether formal or not, need further explanation or justification. When does heuristic adequacy need to be underpinned by explicit logical vindication? No one acquainted with artificial intelligence will be surprised to find that one *can* specify underlying computations in explication of intuitive steps within a proof. The psychologist L. N. Landa has formulated criteria for determining what operations of thought are "elementary" for one person or another, in connection with his project of teaching children of various ages and experience by way of "algorithms" very like computer programs.[32] Clearly, the usefulness of a representation to a particular system (program or person) depends on what may be taken as the elementary mental operations for that system, and what has to be spelled out.

Sloman shows also that analogical representations can sometimes be more powerful than Fregean ones, much as Amarel's zigzags are more heuristically useful (and no less rigorous) than his first four, nonanalogical, representations of the missionary-and-cannibal problem. Sloman cites the example represented diagrammatically in Figure 11.10: what would happen to point B if point A were raised, given that the two horizontal centrally pivoted levers are rigid and the pulley-string is unstretchable? (If you are a masochist, you should try to represent this problem using *only* words.) The diagram can be used to show rigorously—given the constraints of the problem—that point B would fall, whereas a formal proof of this fact would be more convoluted and (for many people) less

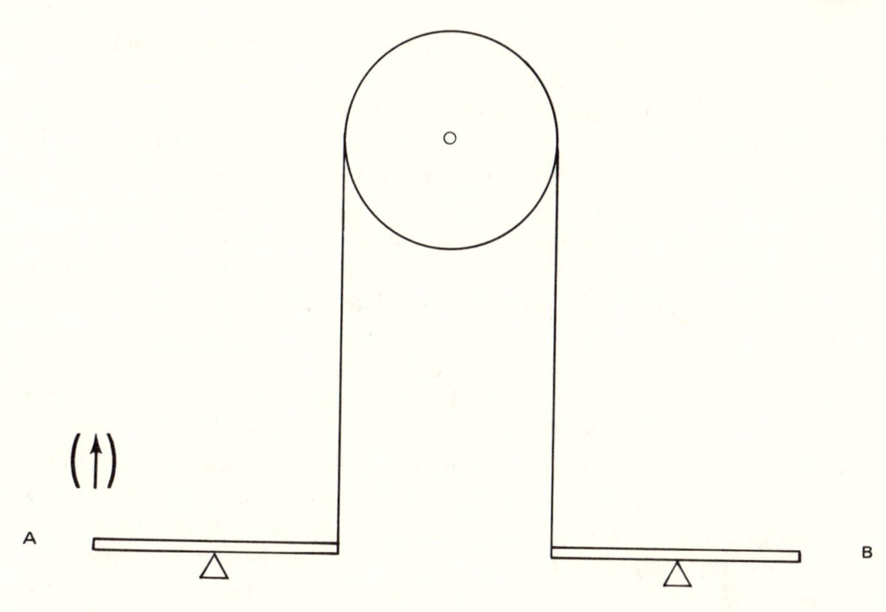


FIGURE 11.10

Source: *Artificial Intelligence 2* (North-Holland, 1971), Figure 21, p. 213. Reprinted by permission.

intuitively compelling. Sloman suggests that problem solving in general may often be aided by analogical representation, because the possible manipulations of (inferences within) such a representation are more tightly restricted by its structure than are the possible manipulations of Fregean representations.

However, the questions involved are still very obscure, and afford no more than hints about how to identify the relevant aspects of the problem and how to pick the most helpful representation of them. There is at present no systematic theory of representation in terms of which to discuss these issues of creative thinking. It is not even clear just what questions such a theory would be required to answer, though D. G. Bobrow has usefully distinguished a number of dimensions along which representations may vary, and P. J. Hayes has raised some relevant theoretical issues also.[33] Amarel suggests that there may be certain classes of forms of rules of action to which there correspond problem spaces with certain special properties, characteristic patterns, and so on. A theory of representation should identify these classes, as well as showing what it is for several representations to be "equivalent" to one another, how novel representations can be appropriately generated with a particular task in mind, and what are the general constraints that decide to what problems a given form of representation is most appropriate. (Is there any *general* characterization of those problems that are suited to zigzag diagrams?) The "rules of heuristic" discussed by the mathematician George Polya and Gestalt psychologists such as Max Wertheimer and Karl Duncker are relevant to these questions, but are suggestive rather than definitive.[34] It is an open question, also, to what extent creative problem solving rests on *general* representational methods, and to what extent it demands domain-specific *expertise* that would not be included within a strictly general theory. Until issues like these have been clarified, there is little hope of enabling a program fruitfully to reformulate its problems in a radically new way.

A general theory of representation, in short, must await fuller understanding of the strengths and weaknesses of specific cases, such as the intuitive hunches used by Polya, the various "languages" of art,[35] and the computational representations currently employed within artificial intelligence. Some relevant examples will be described in the next chapter, wherein I discuss the helpfulness of "branching tree" representations of problems; the increasing heuristic power of four distinct planning programs; and the advantages and disadvantages of three different programming languages.

# 12

# *Problem Solving*

THE previous chapters have shown that all thinking involves problem solving, no matter how simple, immediate, and effortless it may appear. Current artificial systems parallel only a tiny part of the computational complexity implicit in everyday thought. Most thinking relies on specific expertise, whether this is embodied as specialized knowledge in the data base or as domain-specific heuristics and procedures of various kinds.

In this chapter I shall concentrate less on those aspects of problem solving that exploit specific expertise, turning rather to more general methods. These may suffice to solve relatively "contentless" problems, such as logical puzzles and some simple games. But most programs of any interest combine these general methods with domain-specific knowledge, as DENDRAL applies heuristic search to chemical data in a chemically sensible way.

The first section describes some ways in which programs can improve on a "merely mechanical" approach based only on the most general aspects of a problem (for instance, treating it as a combinatorial problem). The second section discusses the use of a *plan* of the problem as a whole: a representation that reflects important aspects of the problem's structure, and within which a solution can be sketched, leaving the details to be worried about later. And the last section indicates some ways in which different programming languages affect the problem-solving power of the system thinking in terms of them.

## THE WEAKNESS OF BRUTE FORCE

You may know the story of the lion saved by the mouse, who delicately gnawed a lion-sized hole in the net in which the raging beast was trapped. It is not only in the pages of Aesop that one finds knowledgeable subtlety and selective planning proving superior to mere arbitrary strength. In artificial intelligence also, brute force approaches have been supplanted by more powerful, more intelligent, procedures.

Brute force thinking is minimally adequate to the problem rather than nicely appropriate to it. It relies on speed of computation, slavish adherence to some exhaustive procedure, and faultless memory, as opposed to intelligent ways of minimizing thinking. A brute force method is not the same thing as an algorithm, although some brute force procedures (such as exhaustive search) are algorithmic. An algorithm is a procedure, brutish or not, that *guarantees* solution.

The prime example of brute force thinking is exhaustive coverage of a set of possibilities (a search space) without any attempt to survey the likely possibilities first. A survey or overview of the possible things to do may enable one to eliminate large sets of alternatives without exploring them individually, or to select the most promising alternatives to explore first. (Sometimes an overview of the possibilities you already know about may even provide clues to new possibilities; for example, Einstein conceived of non-Euclidean space-time by dropping one of the constraints on possibilities within the Euclidean viewpoint.) A police department tracking an elusive murderer, or a frustrated crossword fan tussling with an anagram, sometimes resorts to a similarly plodding approach. But in these cases, usually, searching is not by pure brute force since certain possibilities are outlawed (so you fail to complete the crossword containing a Polish name, "--CZY--"), and others are tried first (males between 18 and 60 before female infants and crones). Similarly, some programs are less brutish than others in that they first reduce the search space in an intelligently selective manner, and only then use exhaustive methods to find the solution. For instance, D. L. Waltz's scene analyzer combines an initial heuristic filtering with exhaustive consideration of the few remaining possibilities (though for many scenes the filtering procedure suffices to give a unique solution, so no further search is needed).

Since computers, unlike people, do have the speed and bookkeeping ability required for successful brute force thinking, it is often assumed that they rely on it whenever they can. We have already seen that they often cannot, that special knowledge must be used by a program as by a person in many situations. But the point of importance here is that brute force computational methods may be insufficient in practice, even though in principle they could solve a particular problem. An example sometimes cited in this regard is the "British Museum Algorithm," a procedure guaranteed to generate Shakespeare's sonnets. In case you have not guessed, it consists in putting a barrow load of monkeys into the basement of the British Museum, together with typewriters and dictionaries, and letting them type indefinitely. With luck, they might produce "Shall I compare thee to a summer's day?" before their first summer is out. Then again, they might not.

Less fanciful examples include problems of a type that many people assume to be readily amenable to brute force computation on the part of current computers. For instance, choosing a move in checkers or chess, or proving a theorem in predicate logic, are activities to which definite rules apply—and in each case one can even state a relevant algorithm of a "British Museum" character. But no computer could have world enough and time to achieve them by means of exhaustive search: the limits on its memory store and on the patience (or even the lifespan) of its interlocutor would each demand a more subtle approach. Even assuming that a game-playing program limits itself to looking ahead for only a few moves, rather than trying to foresee the end of the game, it could not consider individually each of the astronomically large number of possibilities open to it in chess, or *Go*. Similarly, the "keys and boxes" puzzle (which can be solved in 21 steps by a program to be described in the next section) would take on average 8000 years for a program using exhaustive search, even at the rate of one million decisions per second.[1]

Given that a finite intelligence usually can consider only some possible answers to a question, on which ones should it concentrate? Random methods are sensible only when the problem domain (or one's knowledge of it) is virtually unstructured. Thus a hatpin may be useful for placing bets on horses, but for most problems one needs not hatpins but heuristics. A heuristic is a method that directs thinking along the paths most likely to lead to the goal, less promising avenues being left unexplored.

Heuristics are sometimes contrasted with algorithms, it being said that the former are inherently fallible rules of thumb whereas the latter are methods guaranteed to succeed. But if heuristic thinking is understood as *directed* thinking, this contrast is not strictly accurate.[2] Some procedures are algorithmic but not heuristic: for instance, the brute force method par excellence, random exhaustive search. Many are heuristic but not algorithmic: such as the fallible linking-rules in Adolfo Guzman's SEE. A few poor specimens are neither: for example, consideration of a random sample of the search space. And some are both: like certain ordered search procedures described in heuristic search theory.

These examples show that the algorithmic-heuristic distinction breaks down if it is conceived of as *guaranteed versus directed*. Indeed, every program can be shown to be algorithmic if its goal is defined appropriately. Another way of expressing the distinction might be in terms of flexibility due to "postponement of decision making," with *heuristic* programs being relatively flexible in their functioning because they can adapt to the particular situation in hand.[3] For instance, the flow of control may be decided completely by the programmer at program-writing time, resulting in an inflexible program. Or some conditionals may be written into the program, saying "If this happens, then do that." In the latter case, the programmer may decide exactly what conditions to test and what alternative corresponds to each result of each test. Or the programmer can let the program itself take these decisions at runtime. The more decisions are postponed to be taken by the program, the less algorithmic the program is. If even the criteria for taking decisions about taking decisions can be postponed, then one is approaching programs like HACKER—which alters its CRITICS's gallery so that, with experience, it modifies both what it looks for and what it does when it finds it. Another example of postponement is seen in SHAKEY's execution monitor PLANEX, which allows reconsideration of a plan (originally produced by STRIPS) during the actual execution of the program. However, the theory of "heuristic search" focuses not on the issue of postponement of decision making, but rather on that of directing search through a search space in an ordered and economical fashion.

Heuristic search theory has developed from early work on problem solving, and concerns *general* principles of heuristic thinking.[4] Most artificial intelligence programs exploit these principles, and many supplement them by domain-specific heuristics like those described in previous chapters. Basic to search theory is a distinction between two types of thinking, *depth*-first and *breadth*-first search. These methods respectively characterize the mind that follows its first ideas to their limits before turning to consider radically different alternatives, and the mind that skims all possibilities before lighting on a chosen few to be examined more fully. The formal definition of these contrasting approaches that is provided by artificial intelligence enables one to compare them clearly with respect to problem-solving power, and to relate them to the concepts of algorithm and heuristic.

Informally, the difference between depth-first and breadth-first search can be illustrated by reference to Figure 12.1. Imagine that there is an ant crawling along these paths in search of a drop of nectar placed inside a circle or square somewhere on the diagram. The ant must start at the top circle (number 1) and begin by walking downwards, though it can retrace its steps thereafter. A depth-first ant always goes *down* the page as far as it can before going up again to try another route. It can try any path first, so long as it remembers which paths are so far unexplored; so, sitting at circle 1 and facing down the page, it may take any one of

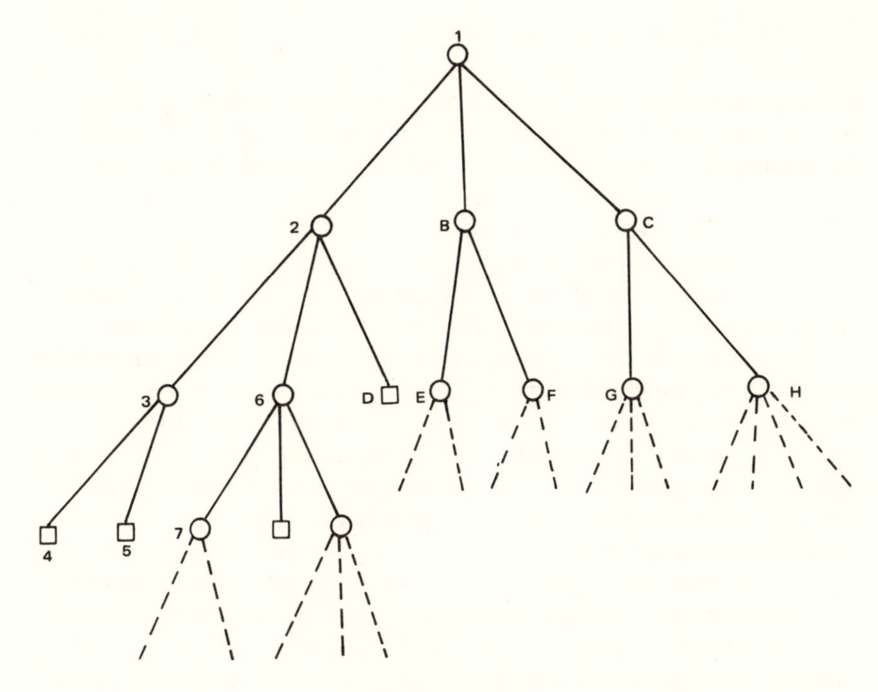


FIGURE 12.1

the three paths available. Let us suppose that it has no particular reason to choose one path rather than another, since it has no idea where the nectar is, and that it wants a relatively easy way of remembering where it has and hasn't been. In this event, it may decide to "follow its right feet," always taking the rightmost path.

Thus its initial itinerary will be: 1, 2, 3, 4. Point 4 is a dead end (represented by a square instead of a circle), so at this point the ant retraces its steps up to the *nearest* circle (number 3) and takes the rightmost path again. When it retreats from the dead end 5 it goes to circle 3 again—but no more unexplored paths remain at that point, so it goes right up to point 2 before starting the whole procedure all over again, via points 6, 7, . . . and so on. Its itinerary then is: 1, 2, 3, 4, 5, 6 7, . . . Clearly, this "right foot first" search strategy is applicable to *any* set of paths of the general form shown in Figure 12.1, no matter how many choices are allowed at the circles and no matter how many circles have to be passed through before reaching the squares.

By contrast, a breadth-first ant goes *across* the page as far as it can, before starting again at the opposite side one level down and working across on that level, after which it passes down to the next. Its itinerary is: 1, 2, B, C, 3, 6, D, E, F, G, H, 4, 5, 7, . . .

Obviously, if the nectar is at point 7, the depth-first ant has more chance than the breadth-first ant of reaching it before collapsing from exhaustion. Contrariwise, if the nectar is at point C, the breadth-first ant will be better off. In principle, breadth-first search will eventually find a finite path to the nectar, if such a path exists at all. But a depth-first search might go down a wrong (and infinite) alley for ever, even though a finite path exists somewhere else.

This diagram represents the *state space* of the ant's problem, namely, the set of states through which the problem solver might conceivably (legally) pass in solving the problem. It is called a "search tree." The *nodes* (circles and squares) correspond to possible states of the problem (starting state, intermediate states, goal state), and the *links* or *arcs* (the lines) represent possible actions or operations that transform one state into another. Any problem that can be represented in this way can be attacked by a general problem-solving program—though whether the program can solve it in a reasonable time is another matter. (Thus even the breadth-first ant will probably reach the nectar at point 7; but if the nectar is at C, the depth-first ant will starve long before it gets there.)

Problems that can be represented in this way include the 3D interpretation of 2D pictures by way of the Clowes-Huffman labels described in Chapter 8 (you may remember that D. A. Huffman suggested a depth-first search, whereas M. B. Clowes's OBSCENE uses a breadth-first procedure). They include also the missionary-and-cannibal problem discussed in Chapter 11, and many other familiar posers like the 15-Puzzle (a 4 by 4 frame within which you must move 15 numbered blocks into the correct sequence, with the empty square ending at the bottom right-hand corner), and the Commercial Traveler problem (how does one plot the shortest route enabling one to visit a dozen scattered towns?). Heuristic search theory concentrates on general principles of searching appropriate to problems that can be represented in this way. And *general* problem-solving programs like Donald Michie and J. M. Doran's GRAPH TRAVERSER, for example, are designed to solve problems so represented, irrespective of specific content. Thus the GRAPH TRAVERSER can solve various different puzzles by growing a search graph from the starting node given by the initial conditions of the problem, and searching this graph structure until the goal node is found.[5]

It is not clear that all problems can be so structured as to conform to this "state space" representation. How about these, for instance—Should I marry him? How can I write a detective story? The trouble with these examples is that it is difficult to know how one can distinguish a solution from a nonsolution (which is not to deny that one can state *some* constraints on what might count as a solution, as we saw with respect to detective stories in the previous chapter). In principle, provided one can tell when one has a solution, exhaustive search through all possible combinations of actions is a way of finding it. But in practice, one may not have a way of ordering one's repertoire of basic actions. What are the basic actions required to win a tennis match? And should devotees of Gamesmanship—who dress formally when their opponent is dressed casually, and vice versa—be thought of as experts in finding surprising new types of basic action, calculated to put their opponent off form?

The affecting vignette of the starving depth-first ant is a reminder that heuristics may be in order in tree-searching, since each of the pure brute force strategies I have described will be inappropriate to (though occasionally successful with) problems having large state spaces. An ant with a sense of smell could use it both in deciding which general section of the tree to explore in the first place, and also in assessing its distance from the nectar at any given node. Similarly, there are two classes of heuristic applicable to tree search: *generator* functions, which influence the order in which the tree is grown (that is, the order in which nodes are visited); and *evaluation* functions (like those used by H. A. Samuel's checker-playing program), which direct the choice at a given node by supplying information about the probable distance from the goal node.

Samuel's evaluation functions are domain-specific, involving concepts like "center control." But heuristic search theory is concerned with abstract properties of heuristics in general. One of the central results of this theory is a mathematical proof that a heuristic using a numerical evaluation function (like Samuel's, for instance) in a specified way will *always* find the shortest path to the goal, given certain relations between the depth of the goal node and of the (intermediate) node at which the evaluation function is applied.[6] A procedure making use of a heuristic in the way described by this mathematical formula is therefore both algorithmic and heuristic, in the sense that it guarantees finding the shortest path while avoiding exhaustive search of the state space. Moreover, it guarantees the *best* solution, where several exist.

Lest you imagine that this abstract formula is a magic wand enabling the programmer instantly to solve any problem, you should notice that the formula tells you how best to use a (numerical) evaluation function once you've found a good one for your purpose, but it tells you nothing about how to find a "good" evaluation function in any particular case. For example, it neither suggests the intuition that center control might be a relevant factor in evaluating a move in checkers, nor gives any hint of how center control might be measured in computational terms. Analogously, the "means-end analysis" strategy to be described in the next section tells you how to use your knowledge of the preconditions and effects of actions, but does not help you to discover what are the preconditions of any particular action that you might decide to consider.

Another general finding of heuristic search theory concerns the efficiency of bi-directional search. This is a way of thinking about a problem in which one simultaneously works forwards from the starting node and backwards from the description of the goal. Ira Pohl has shown that bi-directional search is most useful if the forward and backward search trees meet in the *middle* of the problem space.[7]

A third general result is that depth-first and breadth-first search can be "combined," by specifying a lower-depth boundary within the state-space (such as the third level in Figure 12.1, containing points 3: 6: D: E: F: G: H). The search proceeds depth-first until the limiting level is reached, and then backtracks to the next downward path. When the entire space above the limiting boundary has been searched depth-first, and no solution has yet been found, the boundary can be (successively) lowered by one or more levels and the process is repeated. An ant using this "combined" search-strategy on Figure 12.1 would follow this path: 1: 2: 3: 6: D: B: E: F: C: G: H: (decides to lower the boundary by one level): 4: 5: 7. . . . As you can see, the ant goes *right across* the diagram above the limiting level, even though it is searching in a depth-first way. If a problem-solving program also has an evaluation function available, it may be able to locate the area of greatest promise in an initial breadth-first scan, before attempting to search at deeper levels. This is like the ant (or the person) who ranges over all the broadly defined possibilities first in order to "sniff out" the most likely region. The lower-bound generator heuristic, together with an evaluation heuristic, thus gives the system some of the advantages of each of the brute force methods.

A disadvantage of breadth-first search in general, as compared with depth-first search, is that breadth-first search makes more demands on the searcher's memory. In describing the itineraries of our two ants, I have assumed that each ant always knows where it has already been and where it should accordingly go to next: if it did not, it would lose its place in the search space, perhaps revisiting a former position or missing out the "proper" next node. If you imagine searching through Figure 12.1 bearing this fact in mind, you will find that you need to remember *every node visited on the previous level* when you are searching breadth-first, but merely *every node on the direct path to the top node* when you are proceeding depth-first. For a regularly "double-branching" search tree, which is $n$ levels deep, a breadth-first ant has to remember $2^{(n-1)}$ previous positions, whereas its depth-first cousin needs to remember only $n$ previous contexts. As $n$ increases, the memory-load for breadth-first search soon explodes to unmanageable proportions.

General problem-solving methods such as "mini-maxing" and "alpha-beta pruning" are used by game-playing programs, which have to face difficulties *both* from the game itself (How can I legally move my pawn to the last rank on the board?) *and* from the opponent (If I make that legal move, will I lose my piece?). These assume that the program has a domain-specific evaluation function in terms of which to assess possible moves. Each is a precisely defined computational version of common-sense ways of thinking about chess, for instance.

Thus in mini-maxing, one chooses the move that is *most* likely to lead to good positions for oneself and *least* likely to do so for one's opponent. This involves looking ahead in the search tree, and could in principle involve looking at *all* possible future moves (assuming a lower-bounded search). But the alpha-beta technique cuts down the number of moves looked at, yet is logically equivalent to a systematic mini-max procedure. Alpha-beta pruning was first described by Samuel with reference to his checker player, but is a general method applicable to many problems. It uses the domain-specific evaluation function to assess the value of various nodes in such a way that it identifies—and ignores—subtrees within the state space that *cannot possibly* affect the value of the choice being considered.

Chess-playing programs thus do not use mere brute force methods. They set bounds to the lookahead (which may be relaxed in certain circumstances). They use evaluation functions embodying knowledge about what moves are likely to contribute to king safety, material balance, center control, and so on. They use mini-maxing or alpha-beta pruning to take account of their opponent's likely actions instead of merely the legality of moves open to them. They can sometimes exploit their opponent's weaknesses in much the same way as does the checker player, learning to attack individuals in an aptly idiosyncratic manner. Some of them can benefit from expert advice. And some of them play very respectable chess: the Greenblatt Chess Program, for instance, wins over 80 percent of its games against nontournament players as well as a fair proportion of tournament matches.[8] No wonder, then, that it is an honorary member of the United States Chess Federation!

However, there is no prospect of a chess master being beaten by a program in the near future, as a checkers champion was routed by Samuel's program years ago. Even with the heuristics I have described, there are far too many alternatives to be considered for the program confidently to pick the right one; and as it cuts down the range of alternatives, it is as likely as not to lose the best move, so throwing out the baby with the bathwater. A grandmaster was asked how many moves he considered before choosing, and his answer was: "One—the right one!" H. J. Berliner has pointed out that master chess players develop global perceptual schemata in terms of which they can *see* threats and opportunities on the board much as a lesser mortal can *see* a complex emotional response in a cartoon face.[9] You will remember from Part IV how difficult it is to make our understanding even of *cubes* explicit, so you will hardly be surprised to hear that chess masters are unable to communicate their "intuitions" in a clear fashion to others. Berliner, himself an American chess master, is now tackling this problem in cooperation with other expert players; until chess masters have made more progress in articulating the knowledge they use, there is no question of its being represented computationally.

Like problem solving in other epistemological domains, then, chess playing needs more than quick thinking and a retentive memory: it requires an appreciation of the overall structure of the problem, so that intelligent action can be economically planned. The next section asks how such planning can be achieved in computational terms.

## PLANS FOR PROBLEMS

One of the most effective problem-solving strategies (briefly mentioned in Chapter 9)[10] is planning, in which a simplified version of the problem as a whole is used as a model for solving the problem, details being filled in later as necessary. We saw in discussing Saul Amarel's work that a person, or program, using this strategy successfully must be able to formulate the simplified plan representation, solve the problem schematically in terms of it, and relate the plan appropriately to the tactical details of the problem. I shall describe four planning programs of increasing power: GPS, ABSTRIPS, NOAH, and BUILD.

The global strategy of tackling a simplified representation of the problem is important because step-by-step heuristics (such as those used by the GRAPH TRAVERSER) are inadequate for complex problems, where the thinker must be able to analyze the problem structure *as a whole* if the best solution is to be found and the combinatorial explosion averted. In the terminology introduced in Chapter 8, the thinking must be largely top-down rather than purely bottom-up. This is true even for many linear problems, whose steps do not interact. As we shall see, it is still more important in nonlinear examples, where the *optimal* solution requires nonlinear thinking even though a linear system (like HACKER) sometimes could blunderingly solve the problem.

The concept of planning was first introduced to artificial intelligence by the programmers of "GPS" (General Problem Solver).[11] The planning strategy used by GPS (which could solve a fairly wide range of problems including the missionary-and-cannibal puzzle) was *means-end analysis,* a general method that has since been used in many other systems, including the STRIPS program discussed in Chapter 10.

Means-end analysis is a hierarchical planning strategy that works backward from a clear idea of the goal to be achieved. A program (or person) using this strategy employs comparison procedures to identify the difference (or differences) between the goal and the current state.

Usually, different differences are ordered for priority. This ordering may be guided by general purpose heuristics (like one that says that a difference in the subject matter of the two representations being compared is more important than a difference concerning which topic is mentioned first), or by domain-specific heuristics (like the surgical advice to concentrate on asepsis before antisepsis: if perfect asepsis is achieved then antiseptic procedures should not be necessary). Concentrating on the most important differences first, the problem solver plans to reduce them progressively until none remain.

In order to do this, the system needs to know what operators are available to transform one state into another, and what are their preconditions and effects. This knowledge, and also the difference-ordering, are assumed to be provided to the problem solver: they are preprogrammed in the data base for GPS and STRIPS. (A learning program, of course, might be able to discover them itself.[12]) The list of operators is searched to find one (or more) that can help reduce the specific type of difference that holds between current and desired situation.

To take a homely example, suppose you wish to have a painted ceiling instead of the brick canopy you currently enjoy. The difference between current and goal state is *absence-of-paint*, a difference that only the operators *paint-the-ceiling* and *call-decorator* can reduce. Accordingly, these actions spring immediately to mind, as also does the relevant heuristic *Never call decorator until you've tried job yourself*. So you plan to paint the ceiling. But unfortunately, the operator *paint-the-ceiling* can only be applied to a *clean-ceiling*, which yours is not. So you establish a subgoal, namely, to achieve a clean ceiling. The difference between current and subgoal state is *presence-of-dirt*, a difference that can be reduced by the operator *wash*. So you decide to wash the ceiling, not from an innate passion for cleanliness, but purely as a means to the end of painting it.

The teleological structure of the problems solved by GPS is essentially similar. Suppose that there is only one difference between current and goal state. Occasionally, a single operator is found that can solve the problem without further ado. In this event, the case history runs as follows: (1) GPS first calls on its matching procedures to classify the difference between current and goal state; (2) it then finds an operator listed as being useful for reducing that particular type of difference; (3) on asking whether the operator can be applied to the current problem state as it stands the answer is *Yes*—so GPS does so. And hey presto! the problem is solved, not by magic but by clear thinking.

But more often, and more interestingly, no operator is found that can solve the problem so simply, because the answer to the question posed in step (3) above is *No*. That is, the operator that has been identified, in step (2), as being most relevant to the category of difference concerned cannot be applied directly, because the current state is not a member of the class of states on which it is designed to function. Whenever this happens, GPS establishes a *sub*goal, namely, to transform the current state into one on which the said operator *can* go to work. The three-step routine described above is then applied at the level of the subgoal, the "difference" in question now being that between the current state and the *sub*goal. Since the routine is recursive (that is to say, it recalls *itself* if at step (3) the answer is *No*), sub-subgoals can be successively set up on as many levels as necessary. Eventually, however, a *Yes* answer is received at step (3), in which case the program applies the relevant operator to the current state and works upwards through as many hierarchical levels as have previously been distinguished in order, finally, to achieve "the" goal.

If the initial state differs from the goal in *several* ways, GPS eliminates the differences in turn, passing to the second only when the first has been successfully dealt with as described. In this way, a detailed solution is effected using a problem-solving strategy that can be applied in any domain, provided that appropriate information about states and operators is available, expressed in terms of the representation used by the program. Domain-specific heuristics (like the penny-pinching advice regarding decorators) can be incorporated if required.

GPS typically works within a representation in formal logical terms, in which a problem would be stated something like this: Prove "C," from the three premises "A," "not A or B," and "If not C then not B." This particular problem is solved using the logical operators *or* and *if-then*. The various logical operators available to GPS apply to expressions (states) of a clearly defined character, and effect such changes in them as adding and deleting terms (A, B, C); changing connectives (*or, and, if-then*); changing sign (*not*); changing grouping (by way of brackets); and changing the relative position of terms (AB, BA).

Given the problem just cited, GPS follows the advice of its difference-ordering heuristics to approach the problem in a fashion essentially similar to that you would probably use if you were confronted with one of Lewis Carroll's "Sorites" puzzles.[13] These require you to draw a conclusion (such as "No hedgehog takes *The Times*") from a long list of sentences mentioning hedgehogs, readers of *The Times*, dogs that bay at the moon . . . and various other matters that only the creator of *Alice* would think of juxtaposing. Roughly, what one does is to scan the list for pairs of sentences with shared subject matter, which one then tries to combine in a logical manner (now taking account of the previously ignored logical words: *no, all, not, and, if, or*) so as to reach the conclusion.

And this, in essence, is what GPS does. Initially, it simply notices that the conclusion contains C alone, while the premises respectively contain A alone, A with B, and B with C. Attending to subject matter (rather than connectives or position) first, GPS plans to obtain B by somehow combining A with AB, and then to obtain C by combining B with BC. On closer examination it appears that this plan requires "not A or B" to be transformed into "A implies B," and "if not C then not B" to be transformed into "B implies C." These two transformations can be effected by the operators *or* and *if-then*. Since these operators can be applied to the second and third premises without further ado, GPS is in a position to solve the problem.

STRIPS works in much the same way, its problems being expressed in predicate calculus logic. When it is being used to formulate plans for the robot SHAKEY, its logical formulae are interpreted in terms of the physical actions available to SHAKEY. On selecting an operator (action) apt for reducing a difference between current and desired state, STRIPS treats its precondition list as the next subgoal to be attempted.

Because the simple planning strategy employed by GPS and STRIPS does not produce an overall strategic plan of the problem before the detailed solution is started, it will not suffice to avoid a combinatorial explosion when the initial representation contains a great deal of detail (in that there are many operators, each with a sizable list of preconditions). What is needed, then, is a facility for a further level (or levels) of abstraction, so that a solution in *very* general terms is found first that can be successively transformed into lower representational levels, each more detailed than its superior in the hierarchy.

The lowest level should be "epistemologically adequate" to the problem: that is, it should express all the detail needed for solution.[14] What is epistemologically adequate depends on the system that uses the representation to effect the solution. For example, an action plan expressed in terms of the STRIPS primitives is epistemologically adequate for SHAKEY in that the robot can perform it "immediately"—but the robot's movements depend on lower level programming and engineering details that STRIPS takes for granted much as you and I take for granted the neuromuscular processes that underlie our "primitive" actions of arm-raising.[15]

The higher level representations of the problem are not epistemologically adequate, since they omit much of the relevant detail required for actually completing the solution. But they are "heuristically adequate," in that they allow a path to the goal state to be found by way of the thought processes available for operating on the representational level concerned.

For a very simple problem (where no abstract planning is needed), the representational level that is epistemologically adequate is heuristically adequate too. But for more complex tasks this is not so. Thus a person can think in terms of individual finger movements if, as a novice knitter, her problem is how to get the wool around the needle as pictured in the *Teach Yourself Knitting* manual. But to solve problems about cable-knitting, such as how to achieve a certain cabling effect, she will be lost if she tries to think at the finger-movement level. She would be over-burdened by such a mass of detail because, even if she could remember it (or were to write it down on paper), it would require too much processing. Searches for possibly relevant combinations of details would take a long time, and so too would matching two complex combinations against each other. In such cases, high level strategy must precede tactics if computation is to be suitably minimized.

The earliest versions of STRIPS suffered from a tendency not to see the wood for the trees. But a later modification called "ABSTRIPS" (ABSTRaction-Based STRIPS) has been developed by E. D. Sacerdoti that enables the problem solver to conceive of a hierarchy of woods, within which copses and trees of decreasing size can be successively specified until an epistemologically adequate level is reached.[16] Details are considered by ABSTRIPS only when a successful plan in a higher level problem space gives strong evidence of their importance. Minimal detail is represented at the highest level, so that the search carried out at that level by ABSTRIPS is kept within manageable bounds.

For instance, suppose STRIPS has to advise SHAKEY how to get a box from ROOM4 into ROOM3. STRIPS sees immediately that PUSH-THRUDR is the relevant operator. But in applying step (3) of the means-end analysis routine previously outlined, there are many questions that need to be asked. Perhaps SHAKEY is not in ROOM4 but in ROOM7; if he is in ROOM4, perhaps he is in the opposite corner to the BOX so not in a position to PUSH it anywhere; and even if he is near the box, the door may not be open. It would be a waste of time (and often combinatorially self-defeating) for STRIPS to grind unwaveringly through all these questions. Rather, STRIPS would be wise to determine whether SHAKEY is in the right room *before* asking whether he is touching the box.

ABSTRIPS provides STRIPS with the requisite wisdom, since ABSTRIPS takes care to consider the most critical preconditions of the operator first. For example, it is a precondition of PUSHTHRUDR that there be a door to the room concerned; given that, the robot and the box must both be in the room that is on the other side of that door; given that, the door must be open; and given that, the robot must be next to the box, which must be next to the door. *All* these states of affairs are preconditions of PUSHTHRUDR, and are listed as such in the STRIPS data base. But in the ABSTRIPS version they are ordered for criticality so that the problem solver can plan first things first: there is no point in laboriously getting SHAKEY into position by the box if the door is not yet open.

ABSTRIPS formulates its first plan so as to take care of the most critical preconditions only. It then uses the initial plan as a frame on which to hang subplans designed to achieve the preconditions of the next level of criticality. And so the process continues, until all preconditions have been satisfied in what Sacerdoti calls the "ground plan." The epistemologically adequate ground plan is then passed to SHAKEY's execution program (which would not know what to do with any higher level plan), and the robot gets on with the job.

Criticality levels (like GPS's difference orderings) are provided in the data base, not worked out by the program itself. The basic rationale in ascribing criticality levels is that some preconditions can be easily achieved once others are assumed to be true, whereas some cannot. (It is this insight that underlies the folk saying, "First catch your hare!") Sacerdoti used a mixture of common-sense "intuition" and algorithm (computing the minimal length of plans by which STRIPS could achieve the given precondition once others were true) so as to order the various preconditions for criticality.

ABSTRIPS passes easily from one plan level to another, mapping more abstract onto less abstract representations, because the abstraction involved is merely a question of ignoring detail (as opposed to deleting it or shifting to an entirely different form of representation). An operator that achieves only niggling details will never be considered at the initial (high level) stages of planning. It is one and the same operator (such as PUSHTHRUDR) that is accessed at every level, but differently-sized portions of its definition are taken into account at different levels. If the definition of the operator's preconditions and effects (add and delete lists) were changed according to plan level, the difficulty of mapping from one representation to another would be increased.

If a plan is produced at one level that turns out to be unachievable at a lower level (perhaps the door between ROOM3 and ROOM4 has been boarded up overnight, or perhaps SHAKEY has instructions not to use that door), then the system backs up to the higher level and tries to formulate an alternative plan. For instance, there may be another door in ROOM3, connecting it to ROOM6, and a circuitous route leading from ROOM4 to ROOM6 so that the robot can move the box as requested.

Sacerdoti notes the importance of abstract planning in real-world tasks: since the real world does not stand still (so that one can find to one's surprise that someone has boarded up the door), it is not usually sensible to try to formulate an epistemologically adequate plan beforehand. Rather, one should sketch the broad outlines of the plan at an abstract level and then wait and see what adjustments need to be made in execution. The execution-monitor program can pass real-world information to the planner at the time of carrying out the plan, and tactical details can be filled in accordingly. Some alternative possibilities can sensibly be allowed for in the high level plan (a case of "contingency planning"), but the notion that one should specifically foresee all possible

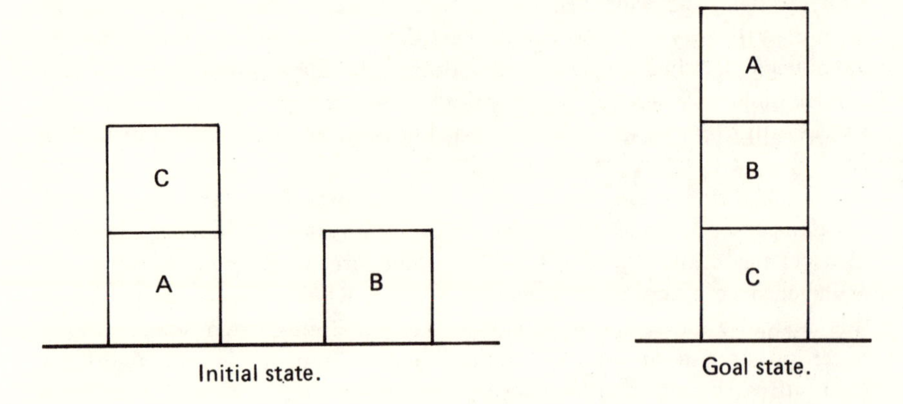


FIGURE 12.2

contingencies is absurd. Use of a hierarchy of abstraction spaces for planning can thus usefully mask the uncertainties inherent in real-world action.

The planning abilities of ABSTRIPS, effective though they are in many cases, are not sufficiently powerful to find the optimal solution to those problems in which there is interaction between the preconditions and effects of the subgoals identified by the high level plan. Nor can HACKER cope really intelligently with such problems, although like ABSTRIPS it may be able to solve them in a clumsy way. The essential reason is that these programs employ linear planning strategies that assume that the subgoals are *additive.* Additive subgoals can be achieved one after the other (although sometimes one order is preferable to another, as we saw in the discussion of HACKER in Chapter 10), in such a way that the optimal solution consists of paying attention to each subgoal independently until they have all been achieved. But even apparently simple problems may be nonadditive.

For example, G. J. Sussman points out that HACKER cannot fathom the best solution to the "3-Block" problem shown in Figure 12.2, although it can solve it in a bumbling fashion.[17] HACKER analyzes the problem into two subgoals: ACHIEVE A-on-B, and ACHIEVE B-on-C. Its (linear) plan will be to achieve first one subgoal and then the other. But whichever it decides to do first, it will run into trouble. For suppose that it first puts A on B (clearing C off the top of A in order to do so): when it comes to try putting B on C it will not be able to, since it can pick up only one block at a time. Conversely, if it starts by putting B on C (for which no preliminary clearing off is necessary), it is even further from the goal state than when it began. To be sure, HACKER can laboriously backtrack and debug its performance program so as to disentangle itself in either case, but had it been more sensible it would not have got entangled in the first place.

To be "more sensible" in this context is to drop the linear assumption of additivity in planning. Sacerdoti has recently programmed a problem solver called "NOAH" (Nets Of Action Hierarchies), in which the initial plan at each level is *non*linear: it does not specify temporal ordering of the subgoals, but represents them merely as logical conjuncts to be achieved "in parallel."[18] As the plan is elaborated on successively more detailed hierarchical levels, potential interactions between subgoals are scanned by NOAH and temporal order is specified if necessary to avoid them. The result is a partially ordered plan intelligently constructed so as to minimize the need for backtracking in execution. That is, NOAH uses constructive rather than destructive planning, since its plan CRITICS (similar in intent to HACKER's "destructive" CRITICS, in that they lead to a bugless plan) *add* constraints to a partially specified plan instead of *rejecting* incorrect assumptions that should not have been made in the first place.

NOAH has five general purpose CRITICS that continually oversee the plan as a whole while it is being elaborated step by step on successive levels, and adjust it by adding constraints if necessary. These procedures watch out for ways of resolving potential conflicts between subgoals; of specifying existing objects for use rather than leaving their identity vague; of eliminating redundant preconditions; of resolving "double crosses" in which each of two conjunctive purposes denies a precondition for the other; and of optimizing disjuncts so that (for example) a choice of fetching *either* the table *or* the ladder in order to paint the ceiling can be predetermined or postponed, whichever is the more sensible.

In addition to these five general procedures, which apply to plans in any domain, NOAH makes use of task-specific CRITICS in the "CBC" project to be described presently.

Many of NOAH's CRITICS depend on a "table of multiple effects" drawn up by the system. Using the lists of preconditions, added facts, and deleted facts stored for each action in the data base (as in the STRIPS data base described in Chapter 10), NOAH picks out and tabulates all the facts that are asserted or denied at more than one point in the plan network. At this stage, all the subgoals of any given goal are represented as conjuncts to be achieved in parallel. The temporal ordering (if any) is imposed on the network at the behest of the CRITICS.

Let us suppose that the problem is to "paint the ladder and paint the ceiling." On looking at the table of multiple effects, the CONFLICT CRITIC quickly notices that HAS LADDER is on the delete list of PAINT LADDER, but on the precondition list of PAINT CEILING. Accordingly, it inserts a temporal constraint into the initial (unordered) plan, specifying that the ceiling is to be painted first. Similarly, this critical procedure can use the table of multiple effects to spot cases where an action deletes a precondition of a *later* action: if for some reason the temporal ordering cannot be switched, the CRITIC inserts a special action at an appropriate point in the plan hierarchy so as to re-achieve the clobbered precondition. The REDUNDANCY CRITIC consults the same table; it notices that GET PAINT is listed as a precondition of both PAINT CEILING and PAINT LADDER, so it eliminates the later occurrence of GET PAINT as unnecessary. (If the execution monitor finds that the paint has run out on the job, NOAH can come to the rescue with the requisite advice.[19])

NOAH's use of the table of multiple effects shows that a planner can reason about plans by dealing with information that is much simpler than the plan itself. Austin Tate's "INTERPLAN" (INTERactive PLANner) employs a similar method, whereby interactions between conjunctive subgoals are forestalled at the planning stage with the help of a tabular representation of preconditions and effects.[20] INTERPLAN can solve the 3-Block problem sensibly, even though (unlike NOAH) it initially makes the linear assumption of additivity, because when this assumption fails it uses an analysis of the failure to patch the original plan. It is better at this than HACKER, and so can find the *optimal* solution of the 3-Block problem. This is of course first to move C off A, then to put B on C, and only then to put A on B. This is equivalent to breaking off in the process of achieving one subgoal in order to achieve the second, and then turning to complete the first.

NOAH has been used with a fair degree of success in the Stanford Research Institute's "CBC" (Computer Based Consultant), a system designed to give on-the-job advice about how to assemble a machine to novice mechanics having varied levels of expertise.[21] You may remember that the improved IDEOLOGY MACHINE sketched in Chapter 4 would have some trouble in lighting on the appropriate level of detail at which to reply to journalists' political questions. The CBC tackles this difficulty by calling on NOAH to use the specific queries posed by the human novice as cues directing it to answer at one hierarchical level or another. NOAH can then advise the CBC to tell one mechanic simply to "Replace the pump," but to tell a less experienced person to "Remove the 4 mounting bolts at the base of the pump using a 3/8-inch open-end wrench."

When a failure in the pump-assembly task occurs because of an unexpected happening (including but not restricted to previous mistakes on the novice's part), NOAH can question the mechanic in an intelligent fashion to locate the difficulty. In most cases, this will not be a simple matter of asking "What's the trouble?" since the novice usually does not know just what has gone wrong, and may claim to have followed all NOAH's advice to the letter. Because NOAH's representation of the semantics of this domain includes detailed and intelligently structured knowledge, in the majority of cases it is able to spot the trouble and work out a way of putting the human worker back on the right track. (The special purpose CRITICS concern matters like tool-gathering: make one trip with your toolbox rather than several, and human anatomy: CBC's pupil may be expected to have only two hands.)

As you might expect, the 3-Block problem is child's play to NOAH. The program can cope with much more difficult posers, such as the "keys and boxes" puzzle I mentioned earlier, which in 1973 Michie identified as a benchmark problem for artificial intelligence, one that contemporary theorem-provers were quite unable to handle.[22]

The keys and boxes puzzle is difficult for many people, even for those who enjoy torturing themselves with such conundrums. The problem is this: you (or a robot for which you have to write the program) are in a room, in which there are two boxes and a table. The table has nothing on it. One box contains several identical keys, the other contains assorted junk. By the door there is a pile of red objects. The keys work "by magic," in that if you have already put one in the pile of things by the door, then you will be able to pass through the door when you try to do so. Your task is to carry one of the red things outside the room. But—there is a catch: you are blind, deaf, and anaesthetized. You can't tell whereabouts you are, what things are red, what things are key-shaped, whether or not your hand is holding anything (still less *what*, if anything, it is holding), or whether your actions—such as *pick up*, which succeeds only if you were not already holding something—have had their desired effect. You can't even tell directly (by looking, listening, or sniffing the air) whether you are inside or outside the room. So what you have to do is to work out a series of actions (*pick up, put down, go to*) which will be sure to enable you to remove something red from the room *without* relying on any of the normal perceptual feedback. If attempting to solve this puzzle makes you go green at the gills, you may feel even more foolish at the reminder that NOAH solved it in a mere twenty-one steps.

Nevertheless, even NOAH is gravely limited compared with human thinkers, as regards its helplessness when each of two conjunctive purposes denies a precondition for the other. Strictly, NOAH is not totally helpless here, for sometimes its DOUBLE CROSS CRITIC can not only spot the trouble, but can redraw the plan in a fully linear fashion so as to avoid the double cross. In many cases, however, this is not possible. As Sacerdoti puts it, "the system must be creative and propose additional steps that will allow the two purposes to be achieved at the same time."[23] This sort of creativity is not one of NOAH's talents.

To satisfy yourself that you are not similarly stupid, please imagine how you would solve the problems shown in Figures 12.3 and 12.4, given

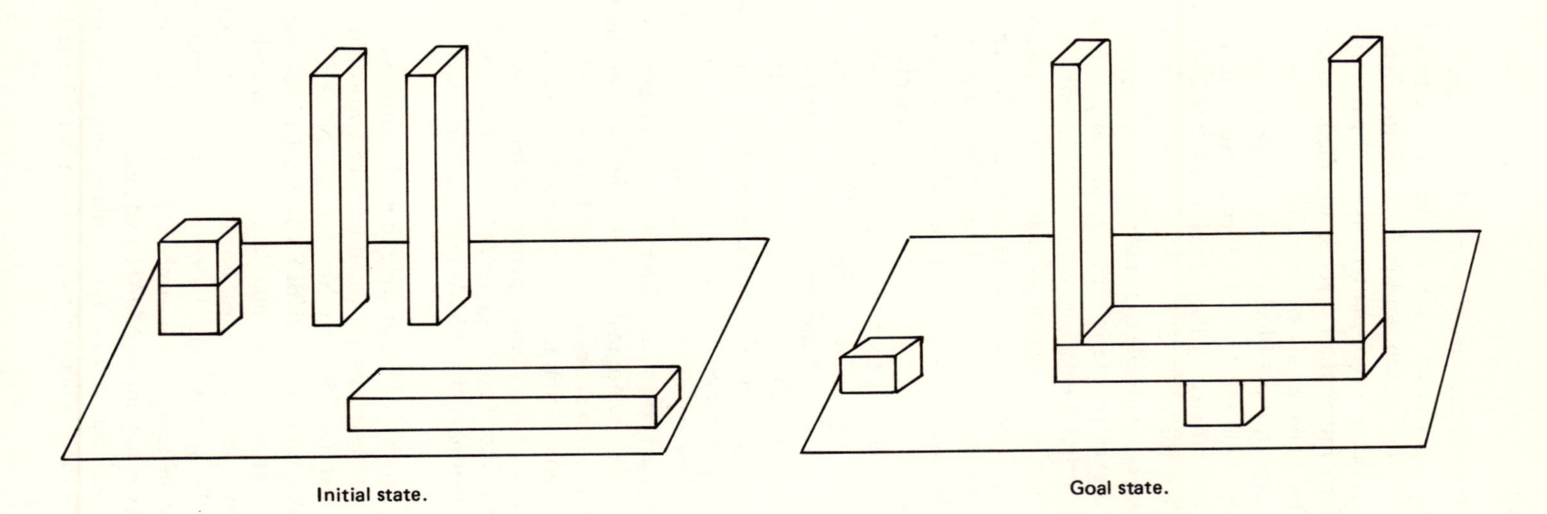


FIGURE 12.3

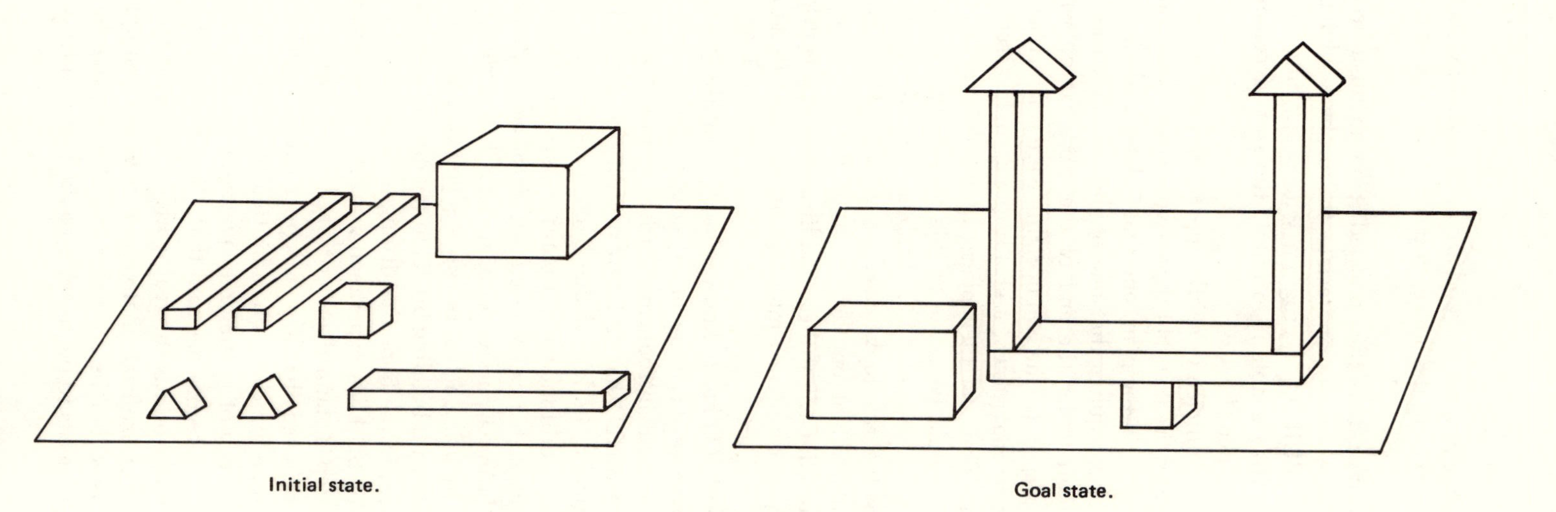

Initial state.

Goal state.

FIGURE 12.4

that you may use only one hand and are not allowed to *slide* bricks. And if you have any bricks in the house, you could then try putting your plan into effect.

When you have done this, or tried it on an unsuspecting child or friend, you will be able to appreciate the probem-solving abilities of S. E. Fahlman's "BUILD," a program whose planning powers surpass NOAH's in that they enable it to mimic the sort of commonsense you have just demonstrated.[24]

BUILD has a sufficiently subtle appreciation of purposive structure to realize and act on the fact that these two see-saw tasks each require an additional step early in the plan, one that is later undone by the builder. Thus problem 12.3 requires that one of the cubical bricks be used as a provisional support under the end of the cross-bar on which the first upright is to be placed. (If the two uprights were small cubes like the supporting fulcrum of the seesaw, a steady hand could assemble the cross-bar-plus-uprights first and then place the whole subassembly onto the fulcrum: BUILD would be aware of this possibility too, but knows it is impracticable in the case of 12.3.) Problem 12.4 demands that the big block (which is too tall to be used as a support) be employed as a temporary counterweight on the center of the cross-bar before either tower is constructed. Each of these "additional" steps is inserted into the plan specifically to counteract mutually destructive effects of main subgoals.

Naturally, BUILD has to know enough about the physics of weights and levers to foresee these unwanted effects, and to work out how to forestall them without introducing further contradictions into the plan. This means that BUILD has to know a great deal about the physics of stability and friction, its knowledge being explicitly represented in the procedures of its world model and planning system. For since BUILD is not a material robot that can amuse itself with real bricks, but a speculative being inhabiting a purely notional BLOCKS world, it cannot test a completed plan by using actual bricks—as you can, and perhaps did. Nor can it experiment with building bricks to try out vague hunches, or fiddle around with them in the hope that it might thereby glean a new insight "by chance." All BUILD's experiments must be *Gedankenexperiments*, in which every step is planned and criticized in thought alone.

Descartes believed that a philosopher could in principle derive all scientific laws in an armchair, but admitted that in practice we must use the Baconian strategy of putting our questions to nature.[25] In somewhat similar vein, R. L. Gregory has said that "the cheapest store of information about the real world is the real world."[26] In other words, a creature (whether natural or artificial) that can perceive and manipulate things in the real world need not be so pedantically knowledgeable as a merely speculative being needs to be, to solve equivalent problems. For instance,

BUILD has to have *theoretical* knowledge that purely linear planning in seesaw problems will result in a pile of bricks scattered over the table, and even has to resort to *arithmetic* to "see" whether a brick is of a size apt for support and whether it is touching another brick. But a suitably sophisticated robot—such as a version of the MIT "COPY-DEMO" hand-eye system that builds copies of brick palaces shown to it by its human playmates—could make good practical use of the vague advice one gives to a small child, "Sometimes a long brick wobbles, but if you put another one under it so as to touch it that may keep it still."[27]

If BUILD were to be linked with such a system, it could often save time and trouble by asking questions of the robot that at present it has to compute. Thus Fahlman reports that fully 80 percent of his programming effort went into implementing BUILD's world knowledge, and much of this was devoted to the theoretically precise stability test—which only works with convex polyhedra. How much simpler it would be if BUILD could ask a robot, "Is it wobbling?" Similarly, a robot might be able to try *sliding* bricks, since it can *see* whether there are any obstructions: BUILD ignores sliding because there is as yet no good computational way to model empty space, for the purpose of sensible path-finding. (M. J. Freiling's "path-finder" mentioned in Chapter 10 is restricted to very simply shaped spatial environments; likewise, HACKER relies on some fairly counterintuitive assumptions when "making space" for the blocks it is planning to move.[28])

The epistemologically primitive steps of the final plan produced by BUILD are MOVE instructions. (If a robot were to be able to do anything with BUILD's plans, it would have to translate these into lower level concepts such as grasping, which in turn would be transformed into detailed commands to the engineered "musculature.") Like HACKER's performance primitives, MOVE modules can talk as well as act. A MOVE primitive may find itself baulked, for instance if the removal or the new placement of the block would cause an instability, or if MOVE is puzzled by the block's apparently being *already* in the right place with respect to the final goal state. In these cases, all of which clearly require a fair amount of thinking on MOVE's part, the procedure sends a nicely discriminating GRIPE message to one of the higher level modules, which sets about adjusting (or confirming) the plan accordingly.

BUILD's planning is handled by its higher level procedures: BUILD (which is the overall controller of the planning process), PLACE, DIG-UP, GETRIDOF, UNBUILD, TRY-MSA, and TRY-TEMP. Each of these has to call MOVE to be achieved, but in addition they can call on each other if they find themselves in difficulty. The heterarchical organization of the program is enhanced by the fact that each of these modules has memory and reasoning power that enables it to tell other modules precisely what the difficulty is, and to answer relevant questions put to it by other routines. (The computational basis of these abilities will be discussed in the next section.)

In outline, the program runs as follows. BUILD starts by drawing up a very simple plan, which is gradually elaborated as necessary by or at the request of the other experts. The first plan follows BUILD's comparison of current with goal state, and is simply a list of the blocks that are not yet in their correct position. The list is ordered, its first members being those blocks whose supports (table or other blocks) are already in position. Starting with these, BUILD plans to call PLACE to deal with them one by one. If PLACE succeeds immediately in each case, the problem is solved.

PLACE's first job is to locate a specific block in the current context that matches the unplaced goal block pointed out to it by BUILD; and its second job is to tell MOVE to move it to the corresponding position. If MOVE gripes in reply that to put it in that position would cause an instability (upset the seesaw, for instance), then PLACE conveys the gripe back to BUILD, who was evidently at fault in asking for that block to be positioned at this point in the plan. BUILD either postpones dealing with that block until all the others have been coped with, or (if no more immediately placeable blocks remain) it passes MOVE's specific gripe message on to the routines that specialize in moveable subassembly (TRY-MSA) and temporary support (TRY-TEMP). TRY-TEMP itself, after due inspection of the problem, can call either on TRY-SCAF to add supporting scaffolds of various types, or on TRY-CWT to add a counterweight. (TRY-TEMP is greatly aided in its choice of stratagems by the precise information about the specific nature of the potential instability that is computed by MOVE, passed on by PLACE to BUILD, and eventually transmitted by BUILD to TRY-TEMP.)

Sometimes PLACE can deal with MOVE's gripe without running straight to the main overseer, BUILD. For example, if MOVE reports that the *removal* of the block would make other blocks fall down, then PLACE requests DIG-UP's help. DIG-UP asks BUILD whether all the "falling" candidates are listed as supported by the block that MOVE wishes to move. If so, DIG-UP instructs GETRIDOF to do the job its name implies until that block is freed. If not, this means that one of the "falling" candidates is *mutually* supported by the target block and some other (for instance, the target block might be one of the uprights of a bridge). So DIG-UP calls UNBUILD to dismantle the substructure that is causing the difficulty. With BUILD's help (which is here being called recursively, since BUILD was responsible for starting this whole rigmarole), UNBUILD does so. MOVE then moves, PLACE asks BUILD for the next goal block named in the high level plan, and so the process continues.

When all the blocks originally named by BUILD have been placed, the problem has not necessarily been completed. For if TRY-TEMP has been involved, there will be an *extra* block somewhere whose presence is no longer needed. So BUILD finally compares goal with current context once again, and if there is a mismatch instructs GETRIDOF accordingly. This also eliminates any blocks lying around that have ended up touching the goal structure merely by coincidence.

Since BUILD is not omniscient, it does not always succeed. A BUILD-GIVES-UP gripe can result if the strategies called by TRY-TEMP (which are only attempted as a last resort) all fail. The gripe contains information about precisely *why* the problem is insoluble by BUILD. This gripe should never complain that the goal is impossible because it is inherently unstable, nor that it is impossible due to lack of blocks. For even before formulating its simple linear plan in the first instance, BUILD always checks that it is not being sent on a wild goose chase in either of these ways. A cook who takes care to assemble all her ingredients first—popping out to the shops if necessary—parallels BUILD's cautious accounting of materials before getting started. And her refusal to commit herself to making a "lamb and jam stew" is somewhat analogous to BUILD's skeptical prediction of goal instability—but since cookery is not an exact science, who can be *sure* that the stew would be inedible?

BUILD's planning power is due in large part to its knowledgeable flexibility. For example, despite its "cook's caution," it puts less emphasis on prechecking relevant conditions than do the other planning programs described in this section, and optimistically starts executing its first plan straightaway. But BUILD's optimism is not blind, for if trouble arises it is analyzed in detail and dealt with. Simply, BUILD does not waste time and memory space in computing beforehand all facts that might possibly be relevant. (The White Knight was less sensible in this regard than BUILD, for he carried beehive and mousetrap on his horse's back—assuring the puzzled Alice, "It's as well to be provided for *everything*.")

Further, BUILD can clear-headedly imagine various hypothetical world states while simultaneously bearing in mind for purposes of comparison the goal state, the current context, and the situations it has passed through in its progression from the initial state. If BUILD has chosen a certain way of coping with a difficulty but finds that it threatens to become more trouble than it is worth, the system can interrupt its action at that point and switch to an alternative method that might be simpler. And if this approach turns out to be even more of a dead end than the other, BUILD can return to complete the first procedure, having forgotten nothing of what it learnt before or after the interrupt. In general, information learnt before failure is available for post-mortems and for informing subsequent procedures of facts they may wish to know. As a corollary, when it fails BUILD does not blindly back-up to the previous choice-point and pick an alternative method at random; instead, it uses its understanding of its failure to select the method most likely to succeed (and can even usefully ask itself why the failed method originally seemed the best choice). If the failure is due not to an unfortunate choice of method but rather to a small local difficulty in applying it, BUILD can adjust this detail as required and restart the failed method.

You may feel that these mental powers are so obviously desirable that a programmer would be foolish who did not provide them. You yourself probably take them so much for granted that you have never explicitly realized before now that you possess them. Somehow or other, no one yet knows how, the human "memory" is organized so as to allow for these types of information processing. Without doubt, some form of symbolism, or language, is essential.[29]

Most of these planning abilities would not be possessed by BUILD, or would at best be gravely restricted, were it not for Fahlman's use of CONNIVER as the intellectual medium within which the program does its thinking. For CONNIVER provides a way of representing and manipulating knowledge that facilitates asking and answering the sorts of question implicit in the preceding paragraphs. Indeed, Fahlman originally wrote BUILD in a version of PLANNER, but decided to switch to the CONNIVER symbolism when it became available—thereby greatly increasing the system's power and flexibility.

In general, a program's manner of approaching and success in solving the problems it faces depend crucially on the language in which it is written. This point has been made in passing on several occasions, and it is now time to inquire rather more closely how it can be so. (If you earlier opted to skip most of Chapter 1, you should finish it now before reading further.)

## PROBLEMS AND PROGRAMMING LANGUAGES

A programming language provides a way to communicate with the computer without being restricted to the level of the machine code. In the early days of artificial intelligence, the things one could say in machine code were extraordinarily niggling and well-nigh unintelligible to boot. (What has this order to do with anything: "Add the numbers in cells 712 and 869, and store the result in cell 250"?) It is now possible to say directly to *some* machines, such things as: "Branch to subroutine X, and then come back here when you've finished." But when programming languages were first developed, machine codes offered a very primitive range of instructions.

We have just seen, for instance, that in CONNIVER one can in effect say to the machine (and enable it to say introspectively to itself) such things as: "Imagine that you were to do that, and tell me what would happen then if you were to find that such-and-such was true," or "What did you learn about this, while you were trying out that hunch that eventually you abandoned?," or again, "Why did you do that at that point in time?" The latter question may remind you of SHRDLU's dialogue (items 24–29), for in PLANNER also strings of goal-subgoal reasons can easily be found. And in PLANNER, as in CONNIVER too, one can say "Get me something broadly like this; I haven't worked out exactly how you are to do so."

Such intellectual subtleties are not straightforwardly expressible in LISP. But even in LISP one can use simple "sentences" to advise the machine, "Do these eleven things one after the other; if by then you still haven't achieved such-and-such, start all over again at Step 1," or "When you have finished that, jump right ahead to the instruction named FREDERIKA (which is three pages away)." Equally, in LISP one can readily symbolize a complex conditional command such as: "If at that particular point in your thinking, *this* happens, then do so-and-so; but if *that* happens, then do such-and-such; if anything *else* happens, whatever it is, then do thus-and-so." (Notice that in the last example the occasion for action is precisely specified; by contrast, PLANNER and CONNIVER allow one neatly to set up "demons"—like Sylvia Weir's collision demon, for instance—that spend their lifetime on the lookout for a certain type of event, and do what they have to do *whenever* a happening of that type occurs.) One can also say in LISP, "If *this* happens, use *that* method to choose an appropriate procedure from *this* list of procedures," where the contents of the list in question can vary as the program runs.

These examples suggest that using a programming language may be viewed as a form of planning. Crucial thinking—"crucial" because it is heuristically adequate to the range of problems in mind—goes on at a relatively abstract level, whose concepts and inference procedures are not duplicated at lower levels.

For instance, the last LISP example in the previous paragraph confers a measure of responsible choice on the computer, as opposed to the blind obedience it shows to the authoritarian machine code instructions, "Do this!" or "If X is true, then do Y!" Similarly, a nonlinear plan does not specify the temporal order of the steps, such decisions being postponed until later. And much as plan execution requires the plan to be translated into low level epistemologically adequate procedures, so programming languages must be transformed inside the computer into machine code (though PLANNER and CONNIVER are interpreted into LISP as a halfway stage). In this sense, all programming languages are equivalent. And most are equivalent also in a stronger sense, such that a procedure expressed in one could *in principle* be expressed in another, without going right down to the machine code level. In practice, however, a procedure that can be simply expressed in one language may be expressible in another only in a very clumsy fashion. (Imagine teaching a child how to add or multiply using Roman instead of Arabic numerals.) Consequently, programming languages differ in power at least as much as STRIPS and ABSTRIPS, and often much more.

With respect to any programming language, one may ask what are the features that make it heuristically adequate to the problems dealt with by programs expressed in it. The authors of new languages usually do this, and contrast their own symbolisms with the supposedly less powerful features of previous ones. However, such claims are often disputed in detail; and it is sometimes said that the general importance of choosing one language rather than another in which to write one's programs is exaggerated, at the expense of concentrating less on what one decides to say with the language once one has chosen it. There is no universally received account of the respective merits of the various programming languages, although some useful comparisons have been made by D. G. Bobrow and Bertram Raphael.[30] And as experimentation with new languages continues, unsuspected deficiencies are discovered that may sometimes outweigh the advantages. Since my aim in this section is not to give a definitive account of programming languages but rather to convey a sense of the sorts of reasons why they have been developed, I shall quote some of the claims and counterclaims but shall not try to adjudicate between them.

John McCarthy's "LISP" (LISt-Processing language) is very widely used in artificial intelligence, either in its "pure" form or in higher level adaptations.[31] What is so special about LISP, that suits it to the representation of knowledge and so to the modeling of intelligence?

LISP's power is due in part to its being a recursive programming language, one well suited to the description and manipulation of recursive structures such as the psychological phenomena of purpose, meaning, and language. A recursive structure is one with an essentially hierarchical character, that can be naturally described at several levels of detail. And a recursive procedure is one that can refer to and operate on itself, so that it can be "nested" within itself to an indefinite number of hierarchical levels.

An example of a recursive procedure is the 3-step means-end analysis routine used by GPS to make manageable sense out of the purposive structure of its problems. You will recall that the job of difference-reduction that is this routine's raison d'être is typically not completed in only one step. Usually, the test within the third step directs the routine back onto itself with a *new* difference now in mind, which calls for a *new* operator to be used, and the first-noticed difference and operator are provisionally shelved. This hierarchical self-nesting can be carried to as many levels as you please. Eventually, a deeply-embedded cycle is successfully completed (that is, the relevant operator manages to do its job of difference-reduction), after which all the others are completed in turn as the program works its way up through the hierarchy. Recursion thus differs from iteration, which is the mere repetition of a completed routine with no shift in the level of control (as in the example citing "eleven steps" given at the beginning of this section).

GPS is not written in LISP, but in IPL-V, which is so much nearer to the machine code that it is less easily intelligible than LISP. But like LISP, IPL-V is a list-processing language. Programs written in LISP and IPL-V are basically concerned with the representation and manipulation of lists, of items on lists, and of lists of lists.[32] Since lists themselves can be hierarchically structured—like the calendar shown in Figure 12.5—list-processing languages are useful for modelling psychological phenomena. Moreover, since their instructions are all expressed as lists, programs written in these languages can use their basic list-processing abilities on themselves in order to modify themselves.

The machine must be able to keep track of what it is doing as it works its way back up through the hierarchy from the most deeply nested level. To see the importance of this, try using the 3-step GPS routine to solve a many-levelled problem like planning a fairly elaborate meal whose several courses require varied ingredients, cooking methods, and lengths of preparation. You will find that it is crucial for the routine of level $n$ to remember just what it was up to when it handed over control to level $n + 1$, so that when the latter has completed its task the former can take over again smoothly. By the same token, the routine at level $n + 1$, must know enough to pass control back to level $n$—and not mess things up by handing the baton to level $n - 1$, for instance.

Let us take making a cheesecake as an example: level $n - 1$ comprises 6 steps in the preparation of the cheesecake, including making the filling that will go onto the crust already prepared. (See Figure 12.6) One of the steps on level $n$ is adding vanilla to the mixture of cheese, eggs, and sugar. But if you have not already got the vanilla out of the larder then you must go to level $n + 1$, the fetching of the bottle of vanilla essence. If after having fetched the vanilla (on the completion of level $n + 1$) you go back to level $n - 1$ (making the filling) the result will be a second cheese-and-egg mixture with no crust available to hold it. On the other hand, if you forget what you wanted the vanilla for, and simply put it down on the table, level $n$ will be omitted and the cheesecake will not be so tasty as otherwise it would have been.

In everyday life you usually remember your "place" largely because

YEARS B.C.
YEARS A.D.
•
•
EIGHTEENTH CENTURY
NINETEENTH CENTURY
TWENTIETH CENTURY
1900s
1910s
•
•
1970s
1970
1971
1972
JANUARY
1
2
3
•
•
31
FEBRUARY
1
2
3
•
•
29
MARCH
APRIL
•
•
•
DECEMBER
1973
1974
•
•
1979
1980s
1990s
TWENTY-FIRST CENTURY
•
•
SECOND COMING

FIGURE 12.5
Hierarchical list structure.

the external world is there to remind you of what you have or haven't done. For instance, you can check up on whether you have already added the vanilla essence by sniffing or tasting the mixture, or perhaps by referring to the pencil-and-paper representation of the culinary task that you have drawn up for this mnemonic purpose.

A computational system that solves its problems "in its head," rather

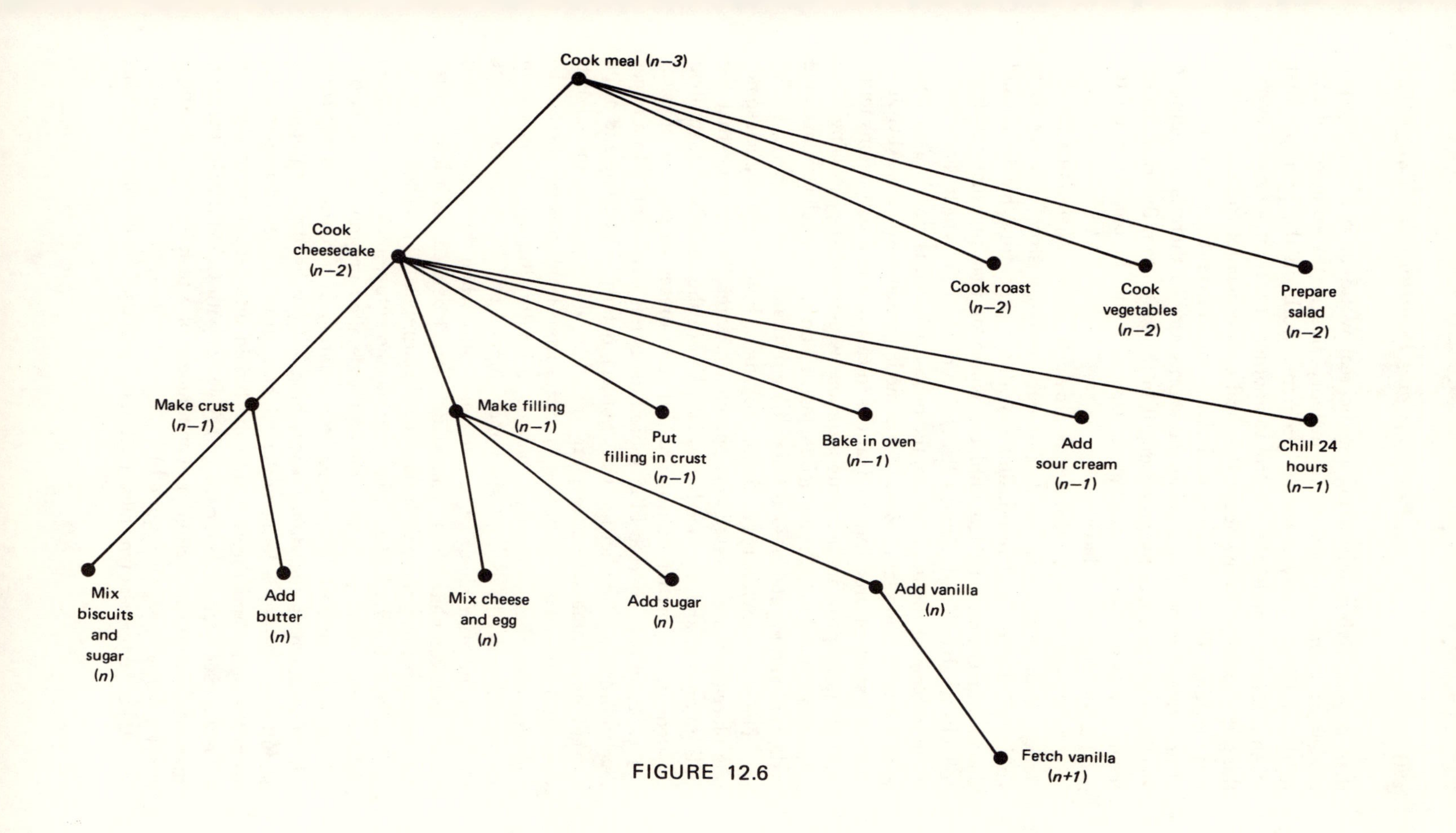
Cook meal (n—3)
Cook cheesecake (n—2)
Cook roast (n—2)
Cook vegetables (n—2)
Prepare salad (n—2)
Make crust (n—1)
Make filling (n—1)
Put filling in crust (n—1)
Bake in oven (n—1)
Add sour cream (n—1)
Chill 24 hours (n—1)
Mix biscuits and sugar (n)
Add butter (n)
Mix cheese and egg (n)
Add sugar (n)
Add vanilla (n)
Fetch vanilla (n+1)

FIGURE 12.6

than by perceiving and acting on the real world or pencil-and-paper models of it, has to have *all* its memory aids in the form of internal representations. Many computer programs work on the principle that when a routine hands over control to another, it instantly forgets everything it needed to know about the local environmental context while it was working. The assumption underlying this convention is that a short-term memory is needed only in the short term, so there is no point in wasting storage space by preserving it in the long term. But this instant memory loss on the relinquishing of control would be disastrous for recursive routines, as we have seen.

An answer to this difficulty is provided by the *push-down list* (or *stack*). A push-down list is a temporary memory store that serves to guide the allocation of responsibility within a procedural hierarchy, and that is not entirely deleted until the whole of that problem has been solved—or abandoned. Being a list structure itself, in which each item is a subproblem of the next one up, the push-down list is closely associated with list-processing languages. (But some languages—such as ALGOL—can use a stack of procedure activations in carrying out instructions *without* letting the programmer write *programs* which do list-processing. And some recent machines have *machine* instructions, in the relevant machine code, for handling stacks so as to facilitate recursion.)

The push-down list is analogous to a metal bill-spike on which someone places her unanswered correspondence, carefully ensuring that the bills go on first and the love letters last, before she deals with them item by item in the predetermined order that is preserved by the physical spike. As an uncompleted LISP subroutine hands over control to a *lower* level subroutine (whether itself or another), details of its name and of its current environmental context are stored at the top of a special list. Conversely, as control is passed *up* again from level to level, these subroutines with their respective contexts are automatically reactivated in the correct order by being visited and "popped up" from the top of the stack, on the principle of *last in, first out*. Only when the routine of level *n* has been completed is its immediate context deleted (as a letter is removed from the spike only after it has been replied to). When control is relinquished by the *highest* level routine, the entire stack will have been deleted, and no memory remains of the detailed progress and environmental experience of the previous computation. If a routine is not *completed* so much as *abandoned*, everything left on the stack is popped up simultaneously—as a letter writer might remove all her correspondence from the spike and throw it into the bin in one despairing movement. (We shall see soon, in comparing PLANNER with CONNIVER, that this memory loss on completion or abandonment of a task is overly drastic if powerful intelligence is required.)

Like the push-down stack itself, most lists used by people are at least partially ordered, in the sense that important information of one sort or another is coded in terms of the abstract relation *next-to*. This relation can be represented in various ways, and in using the list one has to know how to interpret it accordingly. For example, spatial proximity is a crucial structural feature of Figure 12.5, since information about temporal order can be extracted from the representation by moving your finger slightly up, down, or sideways.

It is not usually convenient to use physical proximity to represent the relation *next-to* when one is storing lists inside a computer (although the push-down stack is often stored in adjacent space). It is particularly inconvenient if (as in recursive problems generally) one does not know beforehand how much space one will need to store the list in question. Instead, each item is stored *together with* the "address" of the next item —that is, the number on the door of the memory cell that happens to contain the item's successor.[33] Consequently, it is easy to push entire lists of any length into the middle of an existing list, or to "hang" sublists from individual items, merely by adjusting a couple of numerical addresses. (This facility is used by NOAH and BUILD, for instance, when they elaborate a previously drawn-up abstract plan.)

A list-processing language has to be designed so as to allow this to be done. But the programmer need know nothing whatever about this process, which is carried out silently on her behalf by the compiler or interpreter that translates LISP into machine code. Thus LISP and its intellectual successors allow one to take it for granted that the machine is able to represent and interpret order within lists. (Though one should not assume that either a program or a person can always find an item's predecessor as straightforwardly as its successor: one's frustrating attempt to say the alphabet backwards suggests that psychological access to this list usually relies on *successor* address only. Presumably, learning the reversed alphabet is somehow analogous to the programming task of inserting reverse-pointers between items, in addition to the forward-pointers that can be taken for granted, *or* to formulating new elaborate general procedures for going backwards along a list.[34]) Similarly, the LISP programmer can rely on the "garbage collector" built into LISP, which makes it unnecessary for the programmer to worry about reallocating bits of memory after use; but such bookkeeping matters have to be made explicit at the machine code level.

Compared with machine code, then, LISP is extremely useful as a way of representing everyday problems. But there are important aspects of problem solving that can be expressed in LISP only rather awkwardly, so that it would be useful to have a higher level language yet, one in which such aspects could be expressed more naturally or directly. For instance, you will remember that SHRDLU's human friend can ask the robot to "Pick up a big red block" without mentioning (or even knowing) that the green block will have to be moved away first. How is it possible to communicate with a computer in such a conveniently vague fashion?

Terry Winograd did so by using Carl Hewitt's PLANNER programming language.[35] (Strictly, Winograd used a version of PLANNER called "MICRO-PLANNER," which was implemented at M.I.T. and which lacks some of the properties described in theoretical terms by Hewitt. For example, it does not have "multiple worlds" as full-scale PLANNER and CONNIVER do. To "implement" a programming language is to write the compiler and/or interpreter programs that translate it into machine code; until this has been done, the language cannot actually be used to write programs that will run on the computer. References to "PLANNER" often intend MICRO-PLANNER, or some other partial implementation of Hewitt's system, since full-scale PLANNER has not been implemented.) PLANNER was designed by Hewitt so as to make certain features of intelligent problem solving more easily expressible than they are in LISP.

Hewitt described PLANNER as a *goal-directed* language, in which one can readily specify high level goals in general terms without individuating all the particular objects and operations involved in their achievement—that is, one specifies what one wants done rather than how to do it. The relevant functions or procedures can be called indirectly, since they are indexed by general patterns specifying the form of the data on which they are supposed to work. Functions can be called indirectly in LISP too, by being put into named lists that are then called directly: "Run all the functions, whatever they are, which are on the list FREDERIKA." But in LISP one cannot succinctly state the general form of data on which (or toward which) the various procedures are supposed to work, and then leave it to other LISP procedures to fill in the slots appropriately. In PLANNER, by contrast, one can state a goal, which is then matched by the program to the index of general patterns, and rely on the machine to take care of certain bookkeeping matters that in LISP have to be taken care of by the programmer. This is done by using (unbound) "open variables" in instructions, rather than (bound) "closed variables" only.

Thus the PLANNER procedure defining the PICKUP function includes within itself the general information: "If you wish to pick up *X* (whatever it is), and there is (any) *Y* on top of that *X*, then you should call CLEARTOP to get rid of *Y* before you try to move *X*." In this example, the *first* occurrence of *X* is open, as is the *first* occurrence of *Y*, but subsequent occurrences of both are closed. When PICKUP is called, the program itself can be relied upon to fill in the requisite values of the variables. In LISP, by contrast, a *call* of a procedure can be obeyed only if the variables are closed, either because they have been specifically declared by the programmer or because they have been previously evaluated by the program. If any variable in the called procedure has not been so closed, the LISP compiler will spit it out with an error message asking the programmer to bind that variable. (But LISP allows the initial occurrence of a variable in a *definition* of a procedure to be open, so that the same procedure can be run with different inputs—much as an equation for finding square roots can be used on *any* square number.) So PLANNER makes it easier than LISP to use what the logician calls "existentially quantified" variables: "If there is an X such that . . . , then . . ."

This example mentioning PICKUP and CLEARTOP shows that PLANNER procedures may be complex structures whose definition provides for calls on other procedures (calls that may be recursive, hierarchical, or heterarchical in character). In addition, PLANNER procedures can include heuristic advice about what other procedures may be relevant, and which of these should be tried first. PLANNER is sometimes called a "problem-solving language," not because the language itself (as opposed to the program written by the programmer) solves problems, but because it makes it relatively easy to incorporate complex problem-solving strategies within a procedure.

To see how simply this can be done by way of the PLANNER symbolism, let us assume that we want to write a routine for examining Ph.D. theses. Figure 12.7 is adapted from an example given by Winograd.[36] It shows, in only slightly simplified form, a PLANNER theorem expressing the cynical view that a thesis is acceptable if it is long and/or persuasive, so that an examiner who commits her weekend to actually *reading* the thesis is foolish if she has not checked its length first, preferably without having to thumb through all the pages.

Notice the comment on line 9: when no heuristic advice is included, a PLANNER program searches its data base for *any* theorem (procedure) whose goal matches the pattern in question, and tries each one it finds in turn. That is, it uses a strategy of depth-first search through the "space" defined by the set of alternatives. One way in which PLANNER is more powerful than LISP is that it has inbuilt bookkeeping facilities for depth-first search: it automatically remembers what was the last decision, and what alternatives remained at the time. Consequently, if a procedure that initially seemed to be working successfully fails at a later stage, a PLANNER program can go straight back to the previous decision point and try the next alternative; if that fails also, the PLANNER procedure automatically backtracks to the next highest level of decision, and so on. A LISP program using this search strategy has to have it explicitly programmed by the programmer, whereas it is implicit in the

```
1. (THEOREM "EVALUATE":
2.     (CONSEQUENT X Y) (ACCEPTABLE X)
3.        (GOAL (THESIS X))
4.        (OR
5.           (GOAL (LONG X) (USE CONTENTS-CHECK COUNTPAGES))
6.           (AND
7.              (GOAL (X CONTAINS Y))
8.              (GOAL (ARGUMENT Y))
9.              (GOAL (PERSUASIVE Y) (USE ANYTHING))))))
```

1. EVALUATE is the name we are giving to the theorem we are defining.
2. This line indicates the type of theorem (consequent theorem), names its variables, and states that its goal is to prove that something is acceptable. (This goal pattern will enable the program to call EVALUATE indirectly if necessary.)
3. Show first that X is a thesis (and not, say, a love letter).
4. This "or" relates to lines 5 and 6, and means "try things in the order given until one works, and then stop."
5. To show that X is long, check the contents first, and if that doesn't prove it, try counting the pages.
6. "And" relates to lines 7, 8, and 9, *all* of which are to be done if line 6 is reached at all. (Since this line is covered by the "or" in line 4, it may never be activated.)
7. Find something (call it Y) that is contained in X.
8. Show that Y is an argument (and not, say, an illustration).
9. Prove that Y is persuasive, using *any* theorems you know about that might prove it (search your data-base for theorems with associated patterns matching persuasiveness). N.B. This line is unlike line 5 in that it does not advise which theorem should be tried first; the PLANNER automatic back-up mechanism will try each in turn in the order in which they are found, until one succeeds.

FIGURE 12.7
Slightly simplified PLANNER theorem expressing a procedure for examining Ph.D. theses, with explanation.
(Adapted from Winograd.)

PLANNER theorem of Figure 12.7. However, we shall see in discussing CONNIVER that the PLANNER automatic back-up mechanism has certain disadvantages if one wants to write intelligent programs.

The explanation of line 2 of Figure 12.7 mentions a "consequent theorem." A PLANNER consequent theorem and its corresponding antecedent theorem are different procedural versions of a single declarative statement. It is common to think of "theorems" as declaratives rather than imperatives, as statements of "fact" like Euclid's geometrical theorems and the predicate calculus expressions manipulated by conventional theorem provers like GPS and STRIPS. But as Lewis Carroll showed in his story *What Achilles Said to the Tortoise,* any theorem can be expressed as a rule of inference licensing the drawing of particular conclusions, and vice versa.[37]

More generally, knowledge may be represented as *facts* or as specifications of *methods*. For some purposes, it may be more useful to store knowledge in one way rather than the other; but it should be possible to convert the one into the other if required. Thus a cookbook provides *both* a list of total ingredients used *and* carefully ordered instructions to do this, then that. The fact format is more useful for shopping, the method format for cooking. But recipes can be converted into shopping-lists if necessary. And the reverse is true to the extent that the cook already knows what procedural inferences can be drawn from a certain set of ingredients: the previous example of a person's reluctance to cook "lamb and jam stew" is an illustration of this point.

Hewitt realized that for some purposes it may be useful to have knowledge stored in the form of explicit procedural recommendations. Even taking a dry logical theorem like "A implies B" (or its equivalent, "If A, then B"), one can see that this theorem entitles one to draw certain conclusions, that is: to *do* something. Moreover, one does something different if one concentrates on the first part of the theorem (namely, "A") from what one does if one emphasizes the last part ("B"). In short, if you already know A then you may infer B; correlatively, if you *wish* to know whether B is true, you can try finding out whether A is true first. This is the basis of the PLANNER distinction between "antecedent" and "consequent" theorems. Also, if you believe A to be true but then discover that B is *false*, you can infer that A is false too: this type of inference is expressible in PLANNER as an "erasing" theorem.

In PLANNER, then, "one and the same" item of knowledge can be stored in four forms: as a declarative statement that A implies B, which can be accessed by inferential procedures in the usual manner; as an erasing theorem that is itself a procedure telling the system to erase A, if it finds that B is false; as an antecedent theorem (procedure) instructing the system to infer B, if given A; and as a consequent theorem (procedure) advising that if you want B you had better go about trying to establish A. (Notice that my description of the consequent theorem is ambiguous as between the *truth* of B and the *bringing about* of B. This ambiguity is inherent in PLANNER, whose GOAL theorems make no distinction between *seeing whether* the red block has a clear top and *making it true that* the red block has a clear top. This ambiguity is easier to avoid in CONNIVER and easier still to avoid in POPLER.)

For example, suppose one wishes to know whether Turing is fallible. Well, he's human isn't he? A PLANNER program could respond in essentially the same way, drawing the correct conclusion about Turing, provided its data base contained two items. First, it needs the "fact" of Turing's humanity to be stored (either as a purely declarative item or as an antecedent theorem concerning Turing). Second, it needs a consequent theorem written thus:

```
(CONSE  (X)  (FALLIBLE $?X)
             (GOAL  (HUMAN  $?X)))
```

This theorem is logically equivalent to "All humans are fallible," or "If it's human, then it's fallible," or "Humanity implies fallibility." But instead of being stored as an axiom in the conventional sense, it is a *mini-program* that explicitly advises the system to try first to establish that something is human, if it wishes to know whether that thing is fallible. (The "X" in the PLANNER expression just given is a closed variable—in this case, bound to Turing; the "$?X" is an open "pattern" variable. I omitted the pattern-variable markers from the theorem shown in Figure 12.7, for purposes of clarity.)

Alternatively, suppose this PLANNER program is told a new fact, that Turing is human. Its logically equivalent antecedent theorem will cause it to assert a second new fact, that Turing is fallible:

```
(ANTE  (X)  (HUMAN $?X)
            (ASSERT  (FALLIBLE $?X)))
```

Antecedent theorems can therefore act as "demons" that go ahead on their own responsibility adding (and ERASEing) facts in the data base as other facts come in. Demons not only update the data base automatically, but can be set to remind the program to do something—as Eugene Charniak's trade demon prompts plausible interpretation of " 'They are good pencils' said Janet," and Weir's collision demon changes the course of her program's expectations about what will happen in its visual field. Antecedent theorems are essentially similar to the "production rules" described in the current work of the GPS programmers,[38] and are comparable in intent also to C. J. Rieger's "reflex" MEMORY inferences.

PLANNER antecedent theorems play a role in Hewitt's CONSEQUENCES OF THE CONSEQUENT heuristic. This tongue-twisting heuristic examines (in the program's imagination) the likely consequences of the goal that is being considered, so as to get an idea of statements that could be useful in establishing or rejecting the suggested goal. For instance, the goal of stacking up two pyramids is rejected by SHRDLU because it would result in an unstable structure (*cf.* item 12 of SHRDLU's dialogue). Or again, in item 17 SHRDLU notes that removing the small pyramid from the little red cube would leave the latter free to complete the stack as requested, and so SHRDLU decides to execute the subgoal of moving the pyramid. And *consequent* theorems are used in SHRDLU's response to items 38 and 39, namely, theorems to the effect that if one wants to find (or build) a steeple ("B"), one should look for (or stack) two green cubes and a pyramid (the list of these three objects is "A").

This example of the steeple shows how being given a *procedural* definition of a structure in PLANNER enables SHRDLU to recognize or build one without more ado. Similarly, you will remember that PROGRAMMAR definitions of words and grammatical types are themselves *programs* for parsing words, as opposed to grammatical *data* about words that must be accessed by programs. This distinction between programs and data, or procedural and declarative knowledge, is a computational version of the everyday distinction between knowing *how* and knowing *that* that has been discussed in philosophical terms by Gilbert Ryle.[39] I have referred to it a number of times already, for instance in Chapter 3 where I distinguished the data used by K. M. Colby's neurotic program from the way in which the program works, and in Chapter 8 where I compared the procedural embedding of geometrical knowledge in Guzman's SEE-heuristics with the explicit geometrical data accessed by L. G. Roberts's scene-analysis program.

While the contrast between these two visual programs, for example, clearly shows that there is a distinction to be drawn here, it is not absolute. We saw in discussing list-processing languages that the instructions (being lists themselves) can be used as data by other instructions, so that the program can modify itself. Indeed, one could think of the compiler or interpreter program as taking the LISP program as *data* that it converts into machine code instructions to be followed by the computer hardware. So whether one chooses to regard a particular representation of knowledge as "procedural" or "declarative" is largely a matter of emphasis or point of view.

Like recipes and lists of ingredients, these two forms of epistemological representation have complementary advantages and drawbacks. It is easier to add new knowledge to a declarative representation: compare quickly writing "jam" on the lamb-stew list with teaching someone a slightly different way of beating eggs once she already has a well-established ("fully compiled") method of doing so.[40] But *adding* knowledge is not enough unless one can specify how to use it: unlike Roberts's program, a human acrobat cannot transform her knowledge of Euclidean geometry into practical rules (or "know-how") helping her to perform on the tightrope. Nor is physics much help in learning to ride a bicycle. So if the translation from data to procedure is overly difficult to specify (or overly complicated in specification), it may be more convenient merely to provide a procedure in which a certain amount of the relevant knowledge is implicitly embedded without also giving it an explicit theoretical background. (Imagine giving a *theoretical justification* of all the various "trick" heuristics and "reasonable" procedures used by the programs described in this book.) Provided that it can be translated into action, extra theoretical knowledge may be very useful—as the progression from SEE through OBSCENE to POLY shows. But for effective thinking and action within a certain domain, a set of domain-specific procedures may be more economical than long lists of facts to be manipulated by general deductive procedures. In short, the advice "If you know Turing's human, assume he's fallible" may often be more helpful than the knowledge: "Turing is human, and all humans are fallible."

When PLANNER (and CONNIVER) were first developed, and SHRDLU (and HACKER) written, the protagonists of procedural representations of knowledge were so impressed by the superiority of such representations *for some epistemological purposes* that they tended to lose sight of the merits of the more conventional declarative representations (like the predicate calculus axiomatic base used by general theorem provers). And they sometimes implied that the distinction is an absolute one. But a consensus has now emerged that programmers should try to *combine* the two types of representation so as to get the best of both worlds, and that new programming languages should both help to make the distinction clear if desired (which we have seen to be impossible in PLANNER) and allow easy passage from one form to the other according to circumstantial need.

For instance, Sacerdoti used a special symbolism (the SOUP code) to express NOAH's action network, so as to include characteristics of both alternatives.[41] J. L. Stansfield has outlined a way of representing epistemological structures that can sometimes act as programs and sometimes as descriptions, an aim that Hewitt himself is independently implementing in his recent ACTOR language, in which there is no essential difference between data and process.[42] And Winograd has discussed the "trade-offs" between storing knowledge in one form or the other for different computational (thinking) purposes, and has sketched how both types can be integrated within M. L. Minsky's *frame* orientation.[43]

One of the important points made by Winograd (important not least because it relates to the social influence of artificial intelligence) is that whereas declarative languages are relatively easy for people to understand and communicate, the procedural embedding of knowledge may render that knowledge highly opaque from the human reader's point of view. Consider, for instance, how even Guzman remained largely unaware of just what was the theoretical knowledge that he had implicitly embedded in SEE's linking heuristics. In purely human communication this does not matter too much: though one cannot *say* a great deal to help someone knit, or ride a bicycle, one can *show* her, or physically *move* her limbs in a way which seems to do the trick. But since a computer cannot show us how it does something in this manner, it is important that the program be written in a language that makes it easy for the programmer and others to see what is going on.

As you can see from Figure 12.7, PLANNER programs are relatively easy to read, their meaning is apparently evident. But this is largely because (as in natural-language descriptions of actions) many details of the control structure are implicit, not explicit.[44] There may be all manner of hidden loops and nested routines to which the PLANNER code gives no clue. That is, PLANNER is a nondeterministic programming language, one in which the text does *not* allow one to infer precisely what is going on. If PLANNER programmers were omniscient this would be no drawback, since nothing stupid would be going on "behind" the text. But, like Turing, PLANNER programmers are human. One of the reasons why Fahlman switched from PLANNER to CONNIVER, when developing his BUILD program, was that CONNIVER forced him to make the detailed structure of control more explicit. (A corollary is that CONNIVER programs are more difficult to read than PLANNER ones are.) Consequently, he noticed potential inefficiencies and dead ends of which he would otherwise have remained unaware.

For example, CONNIVER does not normally permit the automatic backtracking that goes on in PLANNER programs. This blind backtracking can be very inefficient, not to say stupid. Suppose, for instance, that a furry robot has a PLANNER representation of PICKUP that advises it to try its right hand, or its left hand, or the tongs . . . Alternatively, suppose that the robot's PLANNER program searches the data base for procedures that match the current goal pattern, finding the three candidates just mentioned. The robot is asked to pick up some red-hot coal off the carpet: its right hand gets burnt and drops the coal, so it backs up to the previous choice point and takes the next choice on the list, namely, *left* hand. By the time such a robot got around to using the tongs, its fur (and the carpet) would be in a sorry state. The programmer might, of course, have foreseen this type of disaster and inserted warning advice concerning red-hot coal and hands. But the point is that not all such eventualities could be foreseen, so there has to be a way for the baulked program itself to decide what to do next in a sensible manner.

A PLANNER program could not do this, without complicated special measures being taken by the programmer, because when it abandons a failed subroutine it deletes the relevant environmental context, restoring the system's inner world to the state it was in before that routine was tried. So it has no knowledge of *why* it failed, still less of how the world changed progressively as the routine was run before being abandoned. (Even if the program is running the routine *in its imagination*, as previously described with reference to SHRDLU, it cannot remember *why* it has decided to abandon a particular tack.) PLANNER resembles LISP in this, for a LISP push-down stack is destroyed when the relevant routine relinquishes control.

By contrast, CONNIVER was specifically designed by G. J. Sussmann and D. V. McDermott so that it would not share PLANNER's inadequacies in situations of failure.[45] A failed CONNIVER routine cannot only tell a higher level module why it failed, but can even pass on information about the successive world changes it encountered on the way. This is a prime reason why HACKER (which is written in CONNIVER) can learn from its mistakes, whereas SHRDLU cannot. Not all CONNIVER programs can learn: BUILD cannot, for instance. But you will remember that in my description of BUILD I referred to environmental messages passing from one module to another, and to local contexts being stored by the master-module BUILD in case they might be needed if TRY-TEMP had to be called in as a last resort. I referred also to modules (such as MOVE, for instance) being temporarily halted, and resuming their task when some other module (such as DIG-UP) has improved the environmental situation in the relevant respect. These features of BUILD are directly attributable to CONNIVER's economical storage of contextual information relating to the different procedures.

Roughly, the CONNIVER context mechanism allows a series (or tree) of local contexts to be stored such that the *shared* information is not repeated. Information that is explicitly represented after the first context frame of all therefore codes only environmental change.

This type of representation thus embodies a particular solution to the "frame problem" with respect to the type of situation involved.[46] The frame problem arises in problems set in real-world contexts, where one needs to know which environmental features will be changed by an action and which will not (or which changes may be ignored for present purposes). For instance, the monkey-and-bananas problem (a favourite test case for artificial intelligence systems) normally assumes that when the monkey moves the box, nothing else changes. However, Figure 12.8 shows that (quite apart from air currents, which may be ignored) in some cases one may expect a very frustrated monkey.

Any program for solving the monkey-and-bananas problem, no matter what language it is written in, should embody *sensible* assumptions regarding what effects of the monkey's actions are likely to be significant. In the case of a CONNIVER program, the programmer must ensure that the context mechanism embodies a sensible set of background assumptions regarding the changes brought about by procedures, if it is to mediate intelligent action. (We saw in the previous section that the cheapest store of information about the real world is the real world: a real monkey, or robot, could try moving the box to *learn* what range of effects ensued.) It is this context-storing aspect of CONNIVER that enables BUILD to return smoothly to a previously interrupted procedure, and to choose the *appropriate* alternative (the heat-resistant tongs) at choice points where the original "best choice" has failed.

Even the original "best choice" may be made more sensibly by BUILD in its CONNIVER implementation than in its PLANNER form, provided that the programmer takes advantage of the design features available. This is because when a PLANNER program searches for a

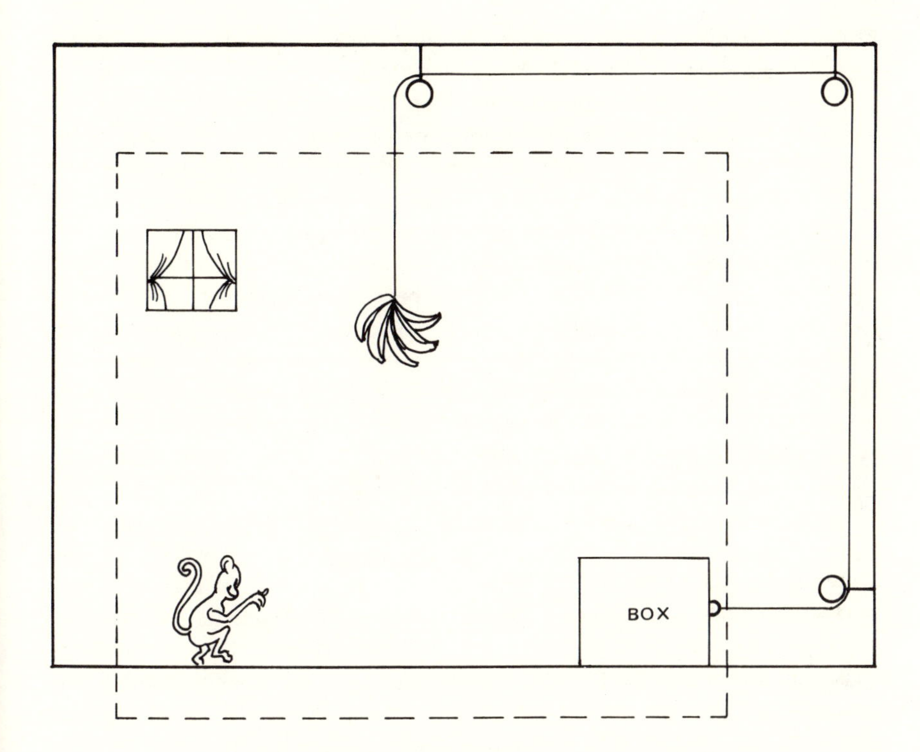


FIGURE 12.8

Monkey-and-bananas problem: How does the monkey get the bananas? (The usual approach to this problem assumes, though does not necessarily explicitly state, that the epistemologically relevant *world* is that shown inside the dotted-line frame. In other words, nothing exists outside this frame that causes significant changes in it on moving the box.)

procedure that matches a particular goal pattern, it simply assembles a list of all the matches it finds and then tries out the first on the list—with *no* attempt to pick the best. To be sure, specific heuristics can improve the situation, but the programmer has to have the foresight to provide them. Again, PLANNER does allow things to be tried hypothetically in sequence, but special measures have to be taken by the programmer to keep records of what happens at each trial. Because of the CONNIVER context mechanism, it is easier for a programmer to write a CONNIVER program that will compare (and in principle even combine) the alternative methods on the spot, by running them through hypothetically in its imagination and then comparing their respective local context histories. If *each* method would succeed, a suitably written CONNIVER program could presciently decide which would be the best in terms of certain criteria (for instance, one solution might promise to involve even "safer" stability measures of the bricks than another).

This implies that some higher level chooser—ultimately, the programmer—is given information about the *contents* of the list of possible methods (information that pattern-matching PLANNER procedures secretively keep to themselves), and then advises which is to be selected. If the programmer wishes the program arbitrarily to choose the first candidate on the list without further computation, as unadvised PLANNER programs do, the relevant instruction can of course be inserted. But the programmer's authority has to be specifically given. This type of control structure therefore helps one to avoid lazy programming, which hands over too much responsibility to a planning program that may have insufficient subtlety to choose widely from among its available methods. Accordingly, a program that CONNIVES with its programmer is likely to be more intelligent than one that PLANS all by itself.

Contrasts such as this one were mentioned by CONNIVER's creators in their account of why CONNIVING is better than PLANNING, and Fahlman's description of BUILD provides an illustration of these theoretical points with reference to a specific example.[47] In general, these workers claim that anything useful that PLANNER enables a programmer (or program) to do, CONNIVER allows too. For instance, CONNIVER demons in the form of IF-ADDED and IF-NEEDED methods are analogous to PLANNER antecedent and consequent theorems. Indeed, since both these higher level languages are implemented in LISP, they may be regarded as LISP with knobs on. They can do easily anything that can be done easily in LISP, and also much that can be done in LISP only with considerable difficulty.

There are a number of other programming languages used in artificial intelligence.[48] D. J. M. Davies' POPLER 1.5, which was briefly mentioned in the third section of Chapter 6, has many of the useful features of CONNIVER and PLANNER as well as some others, though it is implemented not in LISP but in another relatively low level programming language called POP-2. [49] And Bobrow and Winograd are currently developing "KRL" (Knowledge Representation Language), a language designed to integrate procedural and declarative representations, and to avoid the drawbacks of PLANNER and CONNIVER, which drawbacks became increasingly evident following use of these much-heralded tongues.[50]

In general, programming languages are problem-oriented rather than machine-oriented, the problems here in question being those typically dealt with by people interested in the study of *intelligence*. As we have seen, new languages tend to incorporate the benefits offered by their predecessors, offering in addition increased problem-solving power and/or intelligibility. (Though a language's merits may be exaggerated by its authors, and its advantages and disadvantages may be clearly recognized only after prolonged experimentation with it. And, of course, incompetent programmers will write stupid programs in *any* language.)

The future of artificial intelligence thus depends not only on improvements in hardware, but also on the development of new programming languages. The "tin can" computer buffs predict that by the end of the century (or 2001, if you prefer) a single silicon chip computer, only a few millimetres square, will be able to follow 20 million instructions a second using its 65K (65,000 cells) of internal memory store—and it will sell for about *one* U.S. dollar.[51] (Programs currently considered "large" use only about 250K, although SOPHIE achieves 512K by riding on two machines at once; there will probably be several 1,000K programs in existence by 1980.) And research on programming languages continues apace: Hewitt himself has sketched ambitious tongues that make even PLANNER look like childish lisping.[52] Whether or not it ever achieves the highest hopes of its enthusiasts, and whether or not it will soon be able to emulate HAL, artificial intelligence therefore can be expected to become more "natural" in character than it is at present.

With every advance, the question of its wider relevance grows more pressing. What counts technologically or scientifically as an "advance" may appear a retrograde step when viewed from other perspectives. In Part VI I outline some of the many issues involved in assessing the bearing of artificial intelligence on science and society.

# *Part* VI

# THE RELEVANCE OF ARTIFICIAL INTELLIGENCE

# 13

# *Psychological Implications*

SOMEONE equipped with extraordinarily efficient intellectual blinkers could, I suppose, regard artificial intelligence as a fashionable form of crossword puzzle, like acrostics inviting fascination or boredom according to taste, but meriting serious attention only (if at all) because of the expense to the community of the technological toys involved. Such narrowmindedness would verge on irresponsibility. Not only does this area of deliberate artifice have real-world implications, but it bears on the most basic human concerns.

I shall concentrate in turn on psychological, philosophical, and social issues, but it should be evident that these distinctions are not clearcut. The effect of artificial intelligence on society at large, be it for good or ill, will be determined as much by the relevance it is believed to have to the concepts of "humanity" expressed in theoretical psychology and philosophy as by practical applications of a more strictly technological type.

In this chapter, then, I deal with the relations between artificial intelligence and theoretical psychology.

## HUMANISM AND MECHANISM IN PSYCHOLOGY

The literature of machine intelligence lacks any obvious appeal for those who believe that the proper study of mankind is man. The number of people holding this belief is currently growing. Contemporary discussion of psychology, and of the philosophy of mind, is increasingly critical of those theoretical approaches that ignore our specifically human nature. Similar developments are occurring in sociology and anthropology. Reductionism is under attack, objectivism is countered with subjectivity, and the ideal of psychology as a human science is vociferously commended in contrast with the orthodox positivist model, which assimilates it rather to the natural sciences.[1]

Yet in recommending that psychology become a less "natural" science, its current critics are hardly suggesting that it become more artificial. In their opinion, indeed, the moral and intellectual poverty of the behaviorist and physiological paradigms is most evident in the machine-like model of manipulable man encouraged by these reductive viewpoints. Assuming a basic sympathy with this critique, the project of comparing artificial intelligence with natural man may thus appear to be radically perverse: what hope could be more self-defeating or less authentic than that of illuminating human science by reference to machine research? Computer science, apparently, is basically antithetical to and necessarily destructive of an adequate image of human nature.

I should be interested to know whether you approached this book with this attitude in mind and, if so, whether your doubts have been in any way allayed by consideration of the programs discussed so far. Or perhaps you thought it obvious from the start that machine research might be an aid to human psychology, having no patience with the antiscientific and even antirational views sometimes associated with "humanist" accounts of the mind? In any event, these issues must be clarified, and the humanist position properly understood, if one is to decide whether or not artificial intelligence constitutes a menace to our self-respect and humanity by way of its psychological import.

(You will notice that I am using the term "humanist" in a wider sense than it is sometimes used. Thus it is often used to connote "anti-religious rationalists" such as A. J. Ayer, many of whom would not share the inherent suspicion of artificial intelligence that I have described. Again, it is self-consciously used by many psychologists to describe their own theoretical position, but other psychologists—such as those of the "cognitive" schools, for instance—do not describe themselves in this way even though they qualify as "humanists" in my sense of the term. My reason for choosing the label "humanist" rather than "antinaturalist" or "antipositivist" is to stress the fact that even self-confessedly *humanist* psychologists, and laymen concerned with *human* values in general, are misinterpreting artificial intelligence when they view it as totally opposed to their own position.)

Those who proclaim the need for a more human psychology are not merely urging that psychologists address themselves to questions of practical and spiritual importance. If this were all that was meant, that the psychologist be basically humane in intent, not even the strictest behaviorist could be faulted.

Thus B. F. Skinner studies pigeons and rats not for their own sake, but in order eventually to illuminate the more complex psychological processes of *homo sapiens*; and he applies his theories to crucial moral and political issues in his writings on the nature of the ideal society and on the social effects of incorporating concepts such as freedom and dignity within the generally accepted image of humanity.[2] To be sure, his conclusions on such matters differ radically from those of the humanist: for example, Skinner sees "freedom" and "dignity" as socially pernicious illusions, obstructive to rather than definitive of the well-being of the species. But he cannot be accused of irresponsibly ignoring what he sees as human interest, nor of denying the practical importance for mankind of theoretical questions about freedom, dignity, and the like.

The point, rather, is that protagonists of "human science" insist on the *non*illusory nature of the wide range of psychological concepts applied to one's fellows in daily life. They regard these concepts as indispensable in any adequate psychology, and as irreducible to behaviorist or physiological terms. It does not follow from their view that one must always immediately accept the untutored opinion of common sense about psychological matters, even if the opinion is voiced by the person whose psychology is being specifically discussed. But it does follow that psychological science must allow reference to action, thought, perception, emotion, purpose, choice, understanding, and the like—all considered as essentially psychological phenomena that cannot be expressed (as the behaviorist would express them) in terms of stimulus and response or (as the physiologist might suggest) in terms of neural events in the brain.

The irreducibly psychological nature of the subject matter of human science is the central issue on which all humanists agree, whatever their differences in emphasis or theoretical formulation. That there are such differences is notorious: psychologists of a broadly humanist character include many groups, such as existentialists and phenomenologists; cognitive, Gestalt, organismic, hermeneutic, ethogenic, and self-psychologists; personality theorists, clinical psychologists, and transactional psychiatrists; and students of perception, attention, and memory. Parallel groups within sociology and anthropology include the cognitive and ethnomethodological schools. But all these theorists concur in stressing that psychology (like other human sciences) deals with subjective *meaning*, with what in Chapter 8 was termed "intentionality."

According to the humanist, whether a psychological phenomenon be categorized as "inner thought" or as "outer behavior," it must be conceptualized on the model of meaningful action on the part of a subjective agent rather than as a causal process in the natural world. Psychology must give an account of the meaning, or intentionality, intrinsic to mental life, and must recognize the wider theoretical implications of any such account. To attribute meaning to a phenomenon is thereby to posit a psychological subject as the active source of that phenomenon; correlatively, it is to consider the phenomenon as the psychological object of some thought or action of the agent concerned. In a human science, this conceptual polarity of the subject and object of mental acts is central and inescapable. Each of the more specific psychological concepts applied to human beings—from "passion" to "perception," from "freedom" to "fallibility"—must be interpreted in light of this basic psychological distinction. Any theoretical undertaking that hopes to express psychological realities by means of a (behaviorist or physiological) conceptual scheme that does not allow of this distinction is, therefore, doomed to failure.

The current upsurge of this type of criticism directed against positivist psychology has been hailed as a scientific revolution (in T. S. Kuhn's sense), in which one scientific paradigm replaces another.[3] In fact, however, it neither replaces an old paradigm nor offers a new one. It can only be regarded as revolutionary from a parochial standpoint that sees the theoretical predilection of academic American psychologists—namely, behaviorism—as *the* formerly accepted paradigm of psychology. In truth, psychology formerly was (and still is) in a preparadigmatic state, no one of the many diverse theoretical viewpoints being common to all scientists who define themselves as psychologists.[4] Even those who categorize themselves as "humanists" draw on differing historical roots, and espouse varying philosophical positions.

Moreover, humanism is bedeviled by problems of testing, or validation, which are largely responsible for the fact that the humanist approach does not provide a theoretical medium sufficiently detailed to support the cumulative "puzzle-solving" typical of what Kuhn calls "normal science." As we shall see more fully in the next section, humanists are unclear about how one can decide whether a suggested explanation is adequate to the facts it is supposed to explain. Even when validation is recognized by them as a pertinent theoretical issue, the discussion is inconclusive and unsatisfactory—as, for example, in Rom Harré and P. F. Secord's ethogenic psychology or Jurgen Habermas's hermeneutic reading of Freud.[5] Too often, as in Paul Ricoeur's hermeneutic Freudianism or Amadeo Giorgi's defense of phenomenological psychology, the question is not even squarely faced.[6]

Nevertheless, even though humanist psychology does not amount to a revolutionary paradigm in the strict Kuhnian sense, it does have an increasingly widespread advocacy and a reasonably coherent intellectual character. And its conceptual core is a celebration of subjectivity that questions the relevance of mechanistic, and a fortiori of machine-oriented, approaches to the mind.

Humanist doubts concerning the adequacy of mechanistic psychology cannot be dismissed as wholly misguided. Their basic insight that psychological truths cannot be expressed in nonpsychological terms must be acknowledged. The problem then arises of explicating the relation between humanist and mechanist approaches to mind, of showing how it is possible that mind can be conceptualized as both irreducibly psychological and utterly dependent on a mechanistic causal system, the brain. Given that psychological and physiological accounts are alternative ways of conceptualizing our prescientific experience, how is it possible for both accounts to be applicable to individual people, and are there any systematic links between them?

I shall discuss these philosophical questions in the next chapter, simply saying in anticipation here that the apparently total and inescapable contradiction between the humanist and mechanist viewpoints can be avoided. The crucial point is that each of the key terms, "reduction" and "mechanism," has two different senses. Given these distinctions, it is possible wholeheartedly to endorse the humanist stress on subjectivity in psychology without thereby jeopardizing the mechanist's firm insistence that psychological phenomena depend ultimately on causal processes within the brain.

In the next chapter, also, I shall discuss the problem of whether psychological terminology should be understood literally or analogically when imported into a computer context. For the moment, what is important is that psychological terms *are*, increasingly, so used by workers in artificial intelligence.

Consider, for instance, the concept of interpretation itself, which is the central notion of hermeneutic theory (traditionally, "hermeneutics" is the study of the interpretation of ambiguous and fragmented texts, such as the Bible). Interpretation was defined in Chapter 8 in terms of meaning and intentionality, which we have seen to be pivotal to humanist psychologies in general. These concepts were introduced in order to describe and explain the functioning of the programs concerned. In particular, you will remember that the cue-schema relations linking the representational to the target domain could not be identified without reference to the epistemological models used by the programs to interpret the input in terms of the relevant target reality. An input that functions as a cue for one program may be meaningless "noise" from the point of view (*sic*) of another program. Another way of putting this is to say that cues are not detected but constructed, that they are subjective, intentional phenomena rather than objectively definable elements in the physical world. In the absence of any epistemological system actively imposing its constructive schemata on the input from the outside world, there would be no cues—and no perceptions, concepts, or beliefs either.

In general, the intentional concept of "representation" or internal modeling is central to artificial intelligence, whether the specific research concerns vision, problem solving, political prejudice, neurosis, language, or whatever. The question of how knowledge can be represented in a fruitful and flexible manner was highlighted by M. L. Minsky some years ago and is now identified as one of the first priorities of computational research, many workers believing that artificial intelligence will achieve the status of a unified science only when it has developed a general theory of the formation and transformation of a wide range of representations of knowledge.[7]

In addition to the relatively general psychological notions of interpretation, meaning, knowledge, and representation, artificial intelligence workers make increasing use of more specific psychological terms falling within these essentially subjective categories—such as purpose, plan, hypothesis, search, inference, assumption, and the like. HACKER, for example, is not only intelligibly described in purposive terms, but itself makes use of explicit purposive notions in debugging its performance programs and writing better ones. And we have seen that many other programs embody knowledge of the purposive structure of the actions they think about, whether these actions be their own (as in SHRDLU and STRIPS) or those of some other epistemological system (as in the IDEOLOGY MACHINE, MEMORY, and MYCROFT).

Moreover, in describing these programs and conveying a sense of how they achieve what they do, I (like the programmers in their research papers) have continually used a wide range of psychological terms drawn from human psychology. For example, even when discussing the relatively simple and "mechanical" PARRY, I said things like "PARRY unthinkingly answers 'Twenty-eight' when asked '*How old* is your mother?' " Given that in each case the term chosen is carefully selected with the precise functional details of the program in mind, it is not clear that *what the programs do* could be expressed as well—or even at all—without borrowing from everyday psychological vocabulary. And the more complex and flexible the program, the more this is so. Certainly, I could have avoided the term "unthinkingly" when describing PARRY—and I would have done so, had I not wanted to contrast PARRY's mode of functioning with that of human beings and more complex programs (like SHRDLU). But could one express what PARRY does, and how it does it, without ever using the intentional term "pattern-*matching*"? Granted (for reasons given in the next chapter) that the term is used analogically, and granted that the actual program expresses precisely *how* PARRY matches verbal patterns, the use of this psychological term in describing PARRY is more reasonable (because more genuinely informative) than the metaphorical use of the terms *tell* and *say* in describing what clocks do.

This of course does not mean that programmers always choose psychological terms felicitously when describing what their programs do. For instance, HACKER's initial inability to put block B onto block C in the situation shown in Figure 10.22 is due, as we saw in Chapter 10, to the primitive performance program's requiring that there be nothing on top of B. In order to solve this problem, HACKER makes use of the information, or error message, passed to higher (planning) levels of the program by the primitive action module, which recognizes the fact that A's presence on top of B contradicts the precondition that B be clear. This computational process is described by HACKER's programmer thus: "The angry primitive delivers an error message."[8]

The use of "angry" here is unreasonably anthropomorphic, since there is more to *anger* than finding that one is unable immediately to do what one has been asked to do, and none of this "more" is paralleled in HACKER. To be sure, no one is likely to be misled *in detail* by this use of the word, since it is abundantly clear that the relevant preconceptions and expectations are not present in HACKER, whose programmer had no interest in simulating the emotion of anger. But the consistent use of such overly anthropomorphic terms (including *she* and *he*) throughout the published description of a program can lend it a spuriously "intelligent" aspect. The same is true of overly optimistic subroutine-names: "FINDANALOG" suggests a much more subtle process than actually occurs in the neurotic program. Programmers are aware of this effect of what they call "sexy" program descriptions, and some make a deliberate effort to use psychological terms as sparingly as possible so as not to mislead their readers. But to avoid such terms entirely would defeat their aim of outlining the computational power of their programs.

In short, the language used by artificial intelligence workers to present and explain technical results falls within the humanist rather than the mechanist category. Even if one allows that this language is used in an essentially analogical sense, the fact remains that it is regarded as an appropriate medium of communication within the scientific community, not merely when writing sensational articles for the popular press.

The implication is that artificial intelligence workers view themselves as being less in sympathy with behaviorist approaches to psychology than with theoretical systems within the humanist tradition as I have defined it. Indeed, this implication is often explicitly stated by them, behaviorism being firmly repudiated in favor of cognitive or information-processing explanations that stress the (intentional) role of inner models in thought and behavior. For example, M. L. Minsky says:[9]

> Some readers may be disturbed by my deliberate use of psychological terms, such as "meaning," not usually employed so freely in describing the behavior of machines. But it is my opinion that these mentalist terms

> are not all superficial analogies. . . . In its more recent history (after its initial stimulating effect), "behaviorism" has led mainly to ineffectual, near-linear, statistically oriented theories. Originally intended to avoid the need for "meaning," these manage finally only to avoid the possibility of explaining it.

Later in this chapter we shall see that computational work has recently revived interest in visual imagery, a topic neglected for many years by professional psychologists because of the influence of behaviorism. And you may remember that the psychoanalyst K. M. Colby rejected Freud's reductionist view of psychoanalytic theory, seeing it rather as a hermeneutic system; this is why, when he later formulated a programmed model of neurosis, he concentrated on cognitive transformations intelligible in semantic terms, instead of on meaningless energy transfers like those simulated in Ulrich Moser's computational models of defense mechanisms.

Nor do computer scientists concerned with the sort of knowledge-based systems described in this book explain their performance in quasi-physiological (electronic) terms. A high level programming language such as PLANNER can be implemented in many physically different machines (via the appropriate interpreters and compilers), so that there is no one physical description of machine function that is true whenever a certain PLANNER procedure is run. Even if there were, the electronic description would not be semantically equivalent to the programming description, still less to the "psychological" account given in the published literature, and would not be capable of expressing the information-processing properties of the program that are the particular concern of the artificial intelligence researcher, as opposed to the computer engineer. (As we shall see in the next chapter, neurophysiological descriptions are similarly incapable of expressing psychological matters—except in cases where the so-called "physiological" terminology has imported basically intentional notions in order to express the modeling, or representational, functions of the brain.) When SHRDLU answers "OK" to the request, "Pick up a big red block," one cannot understand what it is doing, or how it is doing it, in terms of the electronic processes in which this interpretative achievement is ultimately grounded. Nor does Terry Winograd suggest that one can, preferring to describe his program in terms of the heterarchical interplay of its syntactic, semantic, and planning procedures.

So, contrary to popular opinion, artificial intelligence researchers do *not* interpret their work as supporting the reductionist view, that psychological explanations are in principle dispensable since everything mental is "really" just something happening in the brain. On the contrary, they choose to describe and explain their programs in mentalistic terms (many of which are borrowed from ordinary language, although others are newly coined), because they find it more natural and illuminating to do so than to refer merely to "behavioristic" input-output correlations or to "physiological" details of machine engineering. Whether they are right to interpret their work in this "antireductionist" fashion will be discussed in Chapter 14.

## THEORIES AND TESTABILITY

The psychologist Zenon Pylyshyn has described artificial intelligence as offering "a technical language with which to discipline one's imagination."[10] Creative imagination is a precondition of the computational approach to the mind, as it is also of the literary or aesthetic approach. And intellectual discipline is shared too by theoretical and artistic viewpoints, even by nonverbal representations such as portraiture in painting or music. But theoretical psychology in general—whether psychoanalytic, behaviorist, or whatever—differs from literature in submitting to a particular sort of discipline.

This discipline (which characterizes scientific inquiry in general) aims at providing an explanation of the range of possibilities contained within (generated by) the mind, and it has two main features. First, psychological theorists try to make their knowledge of the mind not only communicable, but explicit. And second, they try progressively to improve its explanatory power by systematically matching their theories against an increasingly comprehensive range of psychological phenomena, perceived *mis*matches prompting adjustment or rejection of the theory.

This double discipline of explicitness and testing is provided in full measure to psychologists who adopt the computational metaphor for the mind. Even if they do not express their theories as *programs,* sensitivity to the explicitness required in computational models may help psychologists to formulate their verbal theories more clearly than they otherwise would have done (R. P. Abelson's later work—including a recent critique of theories of decisoin making[11]—is a case in point). And if they do— attempt to state the theory in programmed form, they find that the activity of programming forces to the surface questions that remain latent in the natural language formulations and intuitive applications of the "equivalent" verbal form. Intuitive usage of the concept of "denial," for instance, normally glosses over psychological distinctions between different phenomena, all of which are naturally termed "denial" (and some of which were distinguished in Part II). Again, when Colby was writing his neurotic program, he was forced to make theoretical decisions (crude and provisional though they were) as to the relative power of the various defense mechanisms in reducing anxiety on the one hand and effecting motivational discharge on the other.

In principle, of course, these questions could have been raised by "armchair" theorizing: Freud himself made a number of remarks pertaining to these very issues. But the theoretical implications (and degree of mutual consistency) of those remarks can be greatly clarified by trying to express them in programmed form. In principle, also, the computer itself is unnecessary for this type of clarification. In practice, however, the computational power of a computer is invaluable in exploring the detailed implications of a given program (and in debugging it in the first place), since humans are unable to continue the explicit theoretical elaboration to the same degree of complexity. Even in the early days of computer simulation, the running of a programmed model of D. O. Hebb's influential neuropsychological theory of cerebral cell assemblies showed up hidden contradictions within Hebb's verbal statements of his theory.[12] And with the advent of the much more intellectually complex psychological theories embodied in artificial intelligence programs, the computer is essential in explicitly assessing the generative power of the program.

Above all, a program provides an explicit theory of the epistemological *processes* by which the concepts contributing to cues and schemata are manipulated in the mind. Verbal theory may be richly fruitful in suggesting the *content* of cues and schemata relevant to a particular interpretative domain, as Freud, Bartlett, and Piaget—and many others—have shown. And empirical investigation of people's concepts and beliefs, whether in childhood or maturity, is usefully suggestive in this regard. But it is only theorizing in a programming mode that can force the theorist to specify, not merely *that* a particular cue-schema pair is activated in a given psychological context, but *how* the cue is identified as such, *how* the potentially relevant schemata are accessed by it, *how* the most appropriate of these is identified in face of the essential ambiguity of the cue, and *how* the conceptual schema is used to mediate "appropriate" thought and action.

In short, detailed questions about the epistemological structure *and function* of memory cannot be avoided by the programmer, who has a rich set of precisely definable concepts (from "low level" examples like *push-down list* and *interpreter*, through *breadth-* and *depth-first search*, to "high level" notions like *top-down processing* or *heterarchical function*) in terms of which to frame hypotheses concerning them. By contrast, these questions can be, and almost invariably are, left largely implicit and unrecognized in theories of a verbal type—even those (like

Freud's) that make a deliberate effort to address such psychological issues. This is what I meant when in Chapter 3 I said that a programmed theory is like a movie of the mind, the principles of progression from one (mental) movie frame to the next being made explicit.

However, explicitness is one thing, truth quite another, so that clarification of theory does not guarantee illumination of the world. If the initial theory is inadequate to psychological reality, then computer models of it can be no less inadequate. Thus if one decries Skinnerian theories because of their radical incapacity to express the subjectivity of the mind, one must also decry programmed models of human social behavior based on (and affording clear formulation of) Skinnerian principles, such as the "HOMUNCULUS" version of G. C. Homans's social psychology.[13] Similarly, if one criticizes Freud's metapsychological pronouncements about libido and energy levels for not being truly psychological (since they are causal rather than intentional in character), one will be less interested in Moser's program of defense mechanisms than in Colby's, no matter how clearly Moser expresses this aspect of Freudian theory. The more the parent theory is psychologically sound, the more likely that its artificial offspring will be genuinely illuminating.

This raises the second aspect of scientific discipline that was distinguished above: the methodical testing of theory against the touchstone of fact. "Facts" function here as cues for addressing the theoretical schema concerned. In the natural sciences, this testing of the explanatory power of the theory commonly takes the form of prediction and experimentation (though in quantum physics prediction is only probabilistic, and in astronomy our capacity to experiment with the celestial motions is not impressive). Tests are then conceived of as relating to specific empirical hypotheses deduced from background theories, and theories are thought of as "laws of nature," broadly of the form: All *X*s are *Y*s. Good experimental method involves such achievements as isolating the *X*s and *Y*s so that they can be manipulated as "dependent" and "independent" variables; designing suitable control experiments; and measuring all the theoretically relevant factors, preferably on a numerical scale having the mathematical features of familiar arithmetic, as opposed to a merely *ordinal* scale or a vaguely quantitative "more" and "less."

This type of explanation (by laws) and testing (by prediction of quantitative matters) has come to be seen by many as definitive of "science" as such. This view is explicit in positivist philosophy of science, and implicit in the "scientistic" bias of Western culture in general. In particular, many psychological theories have been formulated accordingly, behaviorist models in animal and social psychology being prime examples of the type. (Although Skinner lampoons the hypothetico-deductive philosophy of science that is explicitly taken as a guide by many of his behaviorist colleagues, he unquestioningly accepts the conceptual apparatus of quantitative prediction of interdependent variables.[14]) Even Freud saw himself as a *scientist* only insofar as he could reduce psychological phenomena to quantifiable energy levels and neuronal interaction.[15]

We have already noted the radical conceptual differences between Freud's metapsychology and his (interpretative) psychology proper. Frank Cioffi has derided psychoanalytic interpretation as a "pseudo-science," a system in which intellectual maneuvers are continually undertaken specifically to avoid the possibility of falsification.[16] And in discussing hermeneutic or humanist psychologies in general, we saw that the issue of testing, or validation, is not satisfactorily handled (or always squarely faced) by them. The most they appear to ask for is a vague plausibility, disputes between alternative explanations to be settled by "negotiation" between participants or "agreement" between analyst and client.[17] In short, they provide no clear criteria for what is plausible, still less for what is correct. Without such criteria, their claim to be "scientific" looks shaky indeed. Some humanists, in desperation, deny that they are looking for a scientific view of the mind. More often, however, they prefer to regard themselves as psychological scientists, vaguely referring to "a different *sort* of science."[18]

Artificial intelligence is relevant to these issues because we may take it as a case study of "a different sort of science." Like humanist psychology in general, artificial intelligence eschews the prediction of quantitatively defined variables that characterizes the natural sciences. (Would one have discovered anything of interest about SHRDLU if one found that *In 12 percent of cases, SHRDLU's response to the input remark is "OK"*?) But the prime scientific aims of *explaining a range of possibilities* and *testing the explanations* are clearly definitive of the field.

Like Noam Chomsky's generative grammar, a program is a finite set of rules with an infinite generative capacity. (Unlike Chomsky's theory, a program shows precisely how the rules can be applied.) And this set of rules does not specify "laws of nature" predicting variable *X* on the basis of variable *Y*, but generates a range of structural possibilities, which are defined and unified according to the richly interconnected procedures incorporated into the program. PARRY, SHRDLU, and MARGIE, for instance, can each take part in an indefinite number of conversations: one cannot predict just which these will be, but one can say in general what *sorts* of conversation are possible for each of these systems, and *why*. Different possibilities can be systematically related to the existing range by specifying the requisite differences in the underlying program.

The running (and debugging) of a programmed theory of intelligence offers a clear test of it in the sense that the program *must* be adequate to generate, and so to explain, the range of performance it evinces. Whether this performance is persuasively close to human performance is another question, needing empirical inquiry to decide.

Often, the inquiry needed is not "experimental" in the usual sense, since we already possess the empirical knowledge. For example, no experimentation on human beings is required to show that SHRDLU is not a fully adequate model of our use of natural language: we *know* we can use language in many ways that SHRDLU cannot, and this knowledge motivated the "semantic" programs described in Chapter 7.

In other cases, further investigation of human beings may be in order: for instance, programs (such as BUILD) that remember what they have done by reference to environmental changes at each choice point may prompt experiments aimed to see whether people also remember their task performance in this way. If they do, this has implications for theoretical models of the structure of memory, since any such model must be rich enough *at least* to allow for this epistemological ability. Similarly, computational models of inferential memory (such as Rieger's) can be compared with human memory for sentences; if people cannot distinguish the original sentence from one expressing a causal, spatial, or instrumental inference from it, this would suggest a dynamic memory storage having features in common with those modeled in MEMORY. In general, a large range of questions about the activity and organization of human memory can be posed with greater clarity than before by way of hypotheses drawn from a programming context. Consequently, an increasing number of workers in cognitive psychology express their theories with computational models in mind.[19]

Psychologists interested in the neurophysiological basis of experience and action also increasingly use computational concepts, to help them formulate questions about what information-processing functions the nervous system may—or must—be fulfilling. That is, they first ask "What sorts of computation must be involved in seeing a fly, or in snapping at it . . . and what types of procedure could result in the requisite answers?". Only then are they in a position to ask "In frogs, or humans . . . what are the specific physiological mechanisms whose function is to carry out these sorts of computation? How are the computations actually realized?" Various points of contact between artificial intelligence and brain research have been discussed by M. A. Arbib,[20] and examples concerning the function of the retina will be described in the following section.

Clinically oriented psychologists, too, may turn to artificial intelligence for theoretical illumination (and even practical help). Colby, for instance, has spelt out implications for the understanding, treatment, management, and prevention of paranoid disorders based partly on insights gained from his computer simulation of paranoia, PARRY. These concern the exploration of topics (and the influencing of behavior) involving humiliation: shame, self-esteem, and self-censure. He advises against the parental habit of shaming children, for example, since this provides the child with a strategy for future self-control, whose use may be reinforced by any one of the humiliating situations which clinicians have found to trigger the onset of paranoia. Childhood autism, too, has been investigated by artificial intelligence workers—such as Colby and Sylvia Weir. Weir has used a LOGO-turtle environment to expand and explore the autistic child's sense of relevance, shared understanding, and control (whether of self or of the turtle). Her work shows that creative action schemata can be built up in this environment, and that the child's ability for spontaneous verbal and emotional communication is markedly improved at the same time. The continuation of such work may deepen the understanding of autism (and of normal development) both theoretically and therapeutically.[21]

Psychologists of a more self-consciously "humanist" persuasion rarely appeal to computational analogies in formulating their theories. Nevertheless, recourse to this metaphor might clarify their theoretical claims, and suggest the extent to which their interpretative rules really *can* plausibly generate the meaningful thought and action they describe.

For example, theorists of social experience and behavior such as Fritz Heider and S. E. Asch have stressed Gestalt influences such as the strange "halo-effects" of certain semantic or perceptual cues, and the mysterious mutual interdependence of "whole" and "part" in the attribution of personal meanings to our world. But we have seen that these matters are rendered less strange and mysterious by consideration of computational models representing the use of cues and schemata in making sense of an input. Again, "negotiation" can be thought of in computational terms as an exploration of the structural possibilities involved, given the specific setting and sociocultural background, so as to identify alternative sets of psychological rules capable of generating the behavior in question. Similarly, the methods of introspection, role-playing, and "Garfinkelling" (in which a tacit assumption informing social behavior is deliberately upset[22]) may be viewed as techniques for uncovering the implicit cues and schemata guiding our daily actions. And common-sense or clinical study of individual people helps identify idiosyncratic factors that complicate the background sociocultural explanation of the phenomenon concerned. The structural complexity of individual *intentions,* for instance, can be appreciated with the help of the computational point of view, where otherwise psychologists may conceptualize them (if they admit them at all) in a way that is incapable of expressing either their nature or their function.[23] In general, an understanding of computational issues could sensitize humanist psychologists to the procedural requirements of psychological theory, enabling them to test and strengthen their explanations accordingly.

As the previous reference to idiosyncrasy suggests, interpretation (whether psychoanalytic, ethogenic, ethnomethodological, or whatever) is not prediction. Nor do humanists offer "laws of nature" of the form *All people do this,* or *All young middle-class Englishwomen do that.* And even those positivistically inclined psychologists who say such things as "63 percent of people with characteristics *A, B,* and *C* do this" are notoriously unable reliably to predict that "Jane Smith will do this" where the *this* is a matter of any interest to us. This limited predictability is often held to show that psychology is not yet (and perhaps never can be) a *science.*

But scientific understanding does not necessarily involve prediction. And the more complex and changeable the system being studied, the less expectations of prediction (or control) are in order. For instance, description of the functioning of SHRDLU and BUILD (each of which would be extravagantly complimented by being termed "infantile," or "childishly" simple) does not allow of specific prediction, except in the sense that it *outlaws* certain performances—as Chomsky's grammar outlaws "Furiously sleep ideas green colorless". And a self-modifying program interacting with a rich and changing environment (including other information-processing systems), accessing a data base of a size comparable to human memory, would surprise us—and its peers—more often than not. The computational metaphor, therefore, helps one to ask whether an interpretation of a psychological phenomenon is a *possible* explanation of it, without implying that validation must involve detailed predictive tests.

The question whether one or another possible explanation is *the* (or *the best*) explanation is more difficult. With respect to theories expressed as programs, it boils down to the question whether a human performance that is simulated by a program is actually carried out (in our minds) *in the same way.* We shall see in Chapter 14 that some critics of artificial intelligence treat this as an all-or-none question, and try to prejudge the issue in general terms on the basis of a priori arguments of various kinds. However, it is not properly regarded as an all-or-none question. Two systems (people or programs) may think "in the same way" when their thought is represented at one level of detail, but "in different ways" when it is described at another level. One must specify the aspect of thought concerned before one can ask the question.

In comparing programs throughout the book, I have tried to highlight similarities between apparently diverse programs, as well as pointing out significant contrasts. However, detailed comparison of programs with *human* alternatives is at present hardly possible, because of our theoretical ignorance of human thought processes. The GPS programmers deliberately aimed to produce machine performance equivalent to the verbal protocols of their human subjects, and their current work on production systems likewise attempts to simulate human problem solving.[24] But, largely because smooth thinking, at least, is not consciously monitored, it is arguable that their "protocol matching" methodology has significant limitations. To some extent this caveat works in their favor: if there is no spoken protocol at a point corresponding to a particular operation of the program, it does not follow that there is no (subconscious) thought process going on. Equally, if there is a protocol, but one that differs in nature from the programmed machine trace, this may be because the protocol itself is spurious—having been generated, for example, by rationalization on the part of the introspecting human subject.[25] Allen Newell has recently reviewed artificial intelligence with respect to its relevance to information-processing theories of psychology, and in illustration has cited various comparisons between specific features of programs and detailed findings of experimental psychology.[26] But, like "evidence" for any scientific theory, the truth and relevance of these comparisons can be questioned by people wishing to cast doubt on the theory concerned.

In brief, there is no generally agreed view or short-cut way of solving the methodological problem of how to compare a program's performance with its nearest human "equivalent." This lack is partly a function of the difficulties of assessing scientific evidence in general, which is a philosophical issue in the realm of confirmation theory that is not confined to programming contexts. But it is also due to our ignorance of human thinking, most of which is neither introspected nor even introspectible. This is why artificial intelligence is suggestive about, rather than definitive of, the information-processing details of human thought. Its importance is that it helps us to ask clear questions—which, even if some of them turn out to have been misguided, can help us toward the definitive theoretical account that psychologists seek.

Rather than discussing these methodological issues in general terms, let us consider the extent to which artificial intelligence matches psychologists' knowledge of the natural variety in one specific area: namely, visual perception.

## MACHINE VISION AND HUMAN SIGHT

The visual powers of current programs are puny in comparison with ours. We saw in Part IV that even "polyhedral" programs cannot be relied on to recognize all polyhedra, in arbitrary arrangements and lighting conditions. And, happily, our world does not consist entirely of polyhedra; faces, cats, telephones, and even hats are more often salient than blocks and pyramids.

In discussing the perception of hats in Chapter 9 we saw that "geometrical" criteria could not suffice to distinguish hats from other things, and that one needs rich semantic criteria appropriate to the sociocultural domain in question. In a recent overview of artificial intelligence, N. J. Nilsson suggests that concentration on the BLOCKS world has *hindered* research in machine vision.[27] Donald Michie, too, has recommended alternative lines of inquiry (more akin to the approach of J. M. Tenenbaum) in preference to the investigation of "toy worlds" typified in the progression from SEE through OBSCENE to POLY.[28] And M. J. Freiling has pointed out that structural representations even of blocks and arches should be augmented by the addition of functional information, if they are to simulate *what is seen* by animals or people encountering such objects in the real world. In general, we see not only shapes and locations, but also features like pathways, edibility, ferociousness, and amiability of intent.[29]

In light of these criticisms, one may wonder whether any contribution has been made by machine vision to the understanding of human sight. More specifically, one may ask whether computational research throws light on *psychological* theories of vision, and whether it makes any conceptual contact with *physiological* knowledge about the human visual system.

Experimental psychologists of a broadly "cognitive" type have long stressed the contribution of constructive judgment to visual perception. Think, for instance, of Helmholtz's "unconscious inferences," the Gestalt School's "holistic principles," Bartlett's "schemata," J. S. Bruner's "New Look in Perception," and R. L. Gregory's "hypotheses." And think of the related experimental work on eye-movements, which suggests that the eye focuses selectively on those parts of the stimulus-array that correspond to the (intentional) cues used in the current perceptual construction: different parts receiving most attention according to whether the question in mind is "estimate the material circumstances of the family in the picture," or "give the ages of the people."[30]

Work like that described in Part IV, even though much of it is centered on toy worlds, supports and enriches this general theoretical approach, by clarifying the sense in which a perceptual "whole" may depend on its "parts" while at the same time the "parts" depend on the "whole," and by showing just how the phenomenological construction may be effected by the knowing mind. Various examples were mentioned during the exposition of the programs of Sylvia Weir, Gilbert Falk, D. L. Waltz, and G. R. Grape. In addition, the Clowes-Huffman labeling system explains *why* certain figures are perceived as "impossible objects," and *why* ambiguous figures of the type beloved by Gestalt psychologists can be experienced in one way or in another according to the interpretation given to one or two specific pictorial features. That is, possible sets of unconscious inferences are specified, and their different relations to one phenomenological event rather than another are explicated. Such matters have too often been taken for granted by psychologists. But theories of visual phenomenology that are influenced by computational ideas typically show an awareness of these questions.[31]

In addition to providing a new terminology and conceptual base for asking precise questions about how we see what we see, artificial intelligence (even restricted to the BLOCKS world) highlights the poverty of psychologists' knowledge about three-dimensional vision. Thus N. S. Sutherland has criticized experimental approaches to animal vision (including his own previous work), for neglecting to ask how sight contributes to overall spatial orientation and motor skill.[32] Instead, psychologists have concentrated on the study of animal and human responses to two-dimensional stimuli, often of an unnaturally diagrammatic or simplified character. While such empirical work has been valuable in showing that even relatively lowly animals utilize *abstract descriptions* of visual stimuli (descriptions that fit uneasily with the S-R theoretical tradition), it tells us very little about how animals experience their real-world spatial environment.

How, for instance, does a particular species (whether human or not) interpret the variety of cues present in the stimulus input so as to perceive support relations between objects, or so as to see potential pathways and obstacles? As yet, psychologists have no clear idea of how this *might*, or *could*, be done by living creatures, so there is no question of using computational models to pick and choose between alternative psychological theories. Rather, such models should be used to highlight the lacunae in current theories, and help formulate others more nearly adequate to the task of explaining vision.

To be sure, the variety of visual cues acknowledged by psychologists is greater than the range of cues I discussed in Part IV. For instance, the influential work of J. J. Gibson stresses the role of a class of *textural* cues in determining depth-vision and locating object-boundaries.[33] Scene analysis programs like SEE, OBSCENE, and POLY have no notion of texture. But natural segmentation of an outdoor scene into regions (depicting trees, water, grass, and so on) has been achieved on the basis of textural cues by a program written by Ruzena Bajcsy, who also discusses the use of such cues in depth-perception, in the sort of way suggested by Gibson.[34] And the neurophysiologist David Marr has recently formulated a rich computational theory of texture vision that can distinguish many kinds of visual texture, and that can achieve the "Gestalt" separation of *figure* from *ground*.[35] Marr argues that this theory can account for the entire range of texture discriminations of which we are capable, and has already tested it successfully on a variety of real-world scenes. His discussion shows not only how machine vision can be widened beyond "toy" domains, but how psychophysiological understanding of human sight can be furthered by computational concepts.

Many—though not all—programmed visual systems ignore the distance and boundary cues provided by stereopsis, target movement, and motion parallax, each of which has been studied by experimental psychologists.[36] Were such cues to be allowed for, there would still be a need for high level knowledge and top-down inference in perception (to hypothesize the existence of hidden corners, for example, or to hallucinate "invisible" human limbs[37]). But visual matters (such as which objects are behind others) that have to be laboriously computed by stationary, "one-eyed" systems that contemplate one image only at a time might be inferred more directly.[38] Sylvia Weir's program uses "demons" to compare different images and to infer distinct types of movement accordingly; but her program works with abstract stimulus descriptions rather than real scenes. Programs able to perceive motion would have to be able to compare successive images in various ways (although they might also employ "movement primitives" comparable to the brain cells of mammals that are selectively responsive to movement). For instance, to tell whether *it* or *the target* was moving, a machine system would need to know that observer motion appears as a relatively even change in the entire picture (although features nearer to the eye seem to move more), whereas target motion appears as relative change confined to specific localities in the scene. Psychologists know that the phenomena of visual constancy depend also on "bodily" (kinesthetic) knowledge about the movement of one's own head and/or eyes.[39] A computational model might similarly make use of this information, if camera movements were signaled accordingly.

Last, color perception is ignored in many scene-analysis programs. But some recent programming work is aimed at modeling sophisticated psychological knowledge about the contribution of *lightness* cues to intelligent color vision, taking account of the fact that "colors" are constructed from messy and ambiguous physical data, much as "lines" are.[40]

But psychologists unacquainted with computational theorizing often assume that in identifying an additional source of cues, they have solved all the pressing psychological problems concerning the interpretation of those cues. Gibson is a case in point, since he believes the overall theoretical implication of his experimental work to be that most of the perceptual complexity in depth-vision is to be located in the environment or stimulus itself, rather than within the perceiving organism. This conclusion is apparently based on his assumption that the more cues are available, the less "thinking" has to be done by the perceiver. We have seen (in Waltz's program, for instance) that sensitivity to an increased range of cues may indeed contribute to a relatively efficient manner of interpretation; the same point was made above, with reference to stereopsis and motion parallax. Nevertheless, the identification and organization of these cues in relation to the background perceptual schemata is a complex process. As Pylyshyn has commented, "the probem with Gibson's theory is that it leaves too many unbound variables or place-holders for mechanisms about which he has little to say—such as the nature of the all important 'information pickup' function."[41] By contrast, the computational approach encourages psychologists to ask not only what information is picked up, but how—and how it is disambiguated and used after having been picked up.

Sometimes, these questions can be related to specific physiological mechanisms. For example, the computational theory of "color vision *via* lightness" that was mentioned above has been closely related to knowledge of the retina by Marr.[42] And sometimes, as you might expect, the relevance of a physiological finding may be exaggerated, so as to mislead workers into thinking that their program is a closer parallel to natural vision than it really is.

For instance, relatively low (peripheral) physiological levels in the visual system may be capable of recognizing surprisingly complex features. This was first found in connection with the frog's retina, which contains cells that are selectively responsive to *movement* of *small* objects, and which are termed "bug detectors" accordingly (recent work influenced by computational concepts has shown how firing of these cells functions as a spatial map, or sensorimotor schema, enabling the frog to snap in the right place to catch the bug).[43] And the visual cortex of cats, monkeys, and people contains a remarkable variety of single cells that fire in response to complex visual features—such as the angular tilt (to within 5°) of a line in a certain part of the visual field, or successive movements of a tracked object.[44]

When these cells were discovered, a main paradigm for machine vision was the property-list approach, in which patterns were categorized according to their satisfying lists of features.[45] The authors of one of the most successful of such programs explicitly compared its basic operators with the line detectors studied by neurophysiologists.[46] And when the physiologists later found that there are hierarchical levels of such cells, whose anatomical location in the striate cortex nicely matches their hierarchical function with respect to lower level cells, this programming parallel seemed to be reinforced.

However, we have seen that artificial intelligence has progressed from property lists to structured schemata, and from hierarchy to heterarchy. We have seen, too, that there may be very few clear physical intensity contours in the visual input, which rarely consists of line diagrams—still less of *single* lines, as did the inputs for the experimental animals whose "line detectors" were discovered by electrophysiological methods. These points are closely connected, since heterarchical functioning is the more in order as the messiness and ambiguity of the data are increased. It does not follow that the heterarchical use of high level knowledge occurs in precisely the same conditions in current machines and people (we shall see shortly that there are compelling arguments against this view). But it does follow that the straightforward equation of "hierarchical" neurophysiological theory with the early hierarchical programs did not faithfully reflect the nature of visual processes in real life.

This conclusion is underlined by other physiological work, which has found evidence of heterarchical functioning in the visual cortex, and other parts of the sensory system. (Indeed, the term "heterarchy" was first used in a neurophysiological context.[47]) For instance, the response of the orientation-detector cells in the cat's visual cortex can be modified by gravitational changes associated with body tilt; these changes are recorded by the vestibular organ of the inner ear, which presumably sends messages to the visual line detectors.[48] Possibly, such visual-vestibular interactions may play a role in the cat's interpretation of support relations, where it needs to know which things are *underneath* other things. And studies of the cat's auditory nerve imply a physiological basis for the heterarchical modification of perception due to interest, or motivation: the cat's auditory response to a tone is suppressed if a goldfish is placed in its visual field.[49] Psychologists such as Bruner have pointed out that motivation may affect our own vision, although the precise point at which this effect acts on the seeing process is unclear.[50]

Following years of neglect due to the influence of the behaviorist orthodoxy, the topic of mental imagery is once again a live issue in theoretical psychology. The resurgence of interest in this matter occurred within cognitive psychology of an experimental type, but has been enriched and reinforced by computational parallels.

Experimental work by Sutherland and others has shown that even an octopus uses its visual powers to build up internal representations of a fairly abstract character (somehow symbolizing interpretative distinctions like *vertical* and *horizontal*, for instance).[51] In other words, what the octopus sees in its mind's eye is not a *copy* of the external stimulus.

Philosophers and psychologists have pointed out that if the visual image (in your mind, or in that of the octopus) were a mere stimulus copy, nothing would have been explained by positing it: the perceiver's understanding *of the image* would need explanation just as much as the understanding *of the stimulus*.[52] And empirical study of human visual imagery has long shown that "examining" a visual image is very different from examining the equivalent scene (except in cases of eidetic imagery[53]). For example, one may be able to search, or "scan," the image in fewer ways than one can scan the real thing.[54] Again, visual images that are fixated on the retina decay and reappear not uniformly or randomly, as one might expect if mere "physiological fatigue" of retinal cells were involved, but by meaningful stages (one line, or corner, at a time), which tallies with the constructivist approach to visual phenomenology.[55]

Programs like those I have described suggest various ways in which internal representations of a scene may be constructed, and later accessed, where the representations or models embody abstract line descriptions (for instance), rather than iconic duplicates of the input. And Pylyshyn has drawn widely on such programs in arguing for the *symbolic* nature of visual imagery.[56] Again, computational concepts (such as the distinction between parallel and serial processing) are often used by psychologists who ask whether there is any essential distinction between *visual* and *verbal* representation, or thinking. This question has recently been encouraged by experiments on "split-brain" patients and animals, who appear to be able to handle spatial relations more easily with the right side of the brain than with the (normally dominant) speech-using left hemisphere.[57] Researchers in industrial robotics have suggested that the difficulty of providing a formal language for the description of movement (in terms of which to program and communicate with the robot) may be fundamentally related to this cerebral difference, in that motion depends on *non*linguistic representations whose features cannot readily be expressed in digital-processing terms.[58]

These psychological and epistemological issues are still highly obscure, not least because it is not clear what might be meant by phrases like "different sorts of representation." The computational approach highlights a crucial ambiguity, for instance, as between the data storage and the procedures for accessing it. In a critique of a number of attempts to distinguish theoretically between visual and verbal psychology, Pylyshyn has said: "There is nothing intrinsic in an [intentional] object (in the formal sense) which makes it pictorial as opposed to descriptive, analogical as opposed to discursive (or propositional), or even continuous as opposed to discrete. These distinctions refer to the way in which the object is interpreted (or used) by some process."[59] On this view, the theoretical implications of experimental findings on imagery relate not to the image as such, but to the image as it is used in a particular interpretative context.

For example, if people are shown pictures of stacks of objects, and asked whether two stacks are the same, the time they take to answer is directly proportional to the degree of rotation that would be needed to turn one stack around so as to match the orientation of the other.[60] This is often taken to show that visual imagery is "analogical" (though different authors mean different things by this term), or that it is an iconic copy of the original thing rather than an abstract description of it. However, quite apart from the ambiguity of "analogical" and the dubious explanatory power of "copies," all this experiment shows is that there is *some* interpretative pair (representation plus procedure) whose functioning has these observed properties.[61] Not only may there be several such pairs possible, but other "visual" tasks may employ different representations and/or interpretative processes. Similar remarks apply to split-brain findings on "pure" imagery (that is, uncomplicated by left-sided "verbal" processing), and to experiments taken to show that people answer detailed questions about *imagined* animals by searching a visual image of a fixed size, rather than by accessing a list of descriptions of the animal concerned.[62]

These issues cannot be examined in depth here. For present purposes, the important point is that—to many people's surprise—our understanding of the psychological nature and function of human visual imagery may be increased with the help of a computational approach.

Marr's account of texture vision that was mentioned earlier is itself a theory of imagery, one that suggests important differences between human sight and the programs described in Part IV. Being a psychophysiological theory, it concentrates on relatively early stages of visual processing, taking account of what is known about the neural basis of sight. According to Marr, the first computational stage in visual perception is the formation of a "Primal Sketch," an image consisting of descriptions of the scene in terms of features like SHADING-EDGE, EXTENDED-EDGE, LINE, and BLOB (which vary as to FUZZINESS, CONTRAST, LIGHTNESS, POSITION, ORIENTATION, SIZE, and TERMINATION points). The epistemological primitives capitalized in the previous sentence are derived from the original intensity array by "knowledge-free" techniques; that is to say, the Primal Sketch is formed by preprocessing mechanisms independently of any high level knowledge about objects.

The second computational stage is the analysis of the Primal Sketch by symbolic processes that are capable of grouping lines, points, and blobs together in various ways, which analysis results in separation between figure and ground. Marr gives reasons for regarding these symbolic processes as "mechanisms of construction" rather than "mechanisms of detection," thus underlining the constructivist approach to perceptual cues that is characteristic of meaning-based psychologies. He also justifies the positing of the Primal Sketch as a distinct perceptual stage on which later computations operate, as opposed to hypothesizing second-order operations on the original intensity array itself. And he stresses that, except in difficult cases, the second-stage figure-ground separation is achieved *without* higher level knowledge, and *precedes* the computation of the shape of the "figure."

Even in difficult cases, where higher level knowledge is used to help distinguish between figure and ground, Marr claims that its role is to influence control (by deciding which form is to be extracted from the Primal Sketch) rather than to enter into the actual data-processing at either of the first two stages. He points out that this procedural feature of his theory contrasts sharply with current assumptions in scene analysis: for instance, you will remember that Yoshiaki Shirai's line finder performs computations on the intensity array that are deeply influenced by its inbuilt knowledge of the geometry of convex polyhedra.

Marr points out also that extracting rich information from the visual input, by way of knowledge-free preprocessing techniques, presupposes "prodigious computing power"; but experimental psychologists and neurophysiologists have shown that our own visual systems do, in fact, possess such power. While it may be necessary in practice for computer scientists to devise "knowledgeable" processing so that machines may be able to extract even relatively low level information, it does not follow that such computational problems are real issues *for the psychologist.* The moral Marr draws is that workers in artificial intelligence should concentrate on the lower level problems first, turning to study the use of higher level knowledge only after the peripheral processing has been properly understood.

In relating his work to the phenomenology of vision, Marr describes the Primal Sketch as corresponding very closely with the image of which one is conscious. But he stresses that it is a very active structure, in that several highly abstract image-analyzing processes lie "lurking active in its fabric," which processes are the computational embodiment of the point at which visual analysis becomes a purely symbolic affair. After the stage of figure-ground separation, the image can be "read" by higher level processes searching for specific shapes: but the crucial point is that (usually) one can see a rounded-triangular-blob *before* inspecting it closely to see if it is the shape of a chestnut tree. If one had to compute a detailed description of a chestnut (or an oak, or elm . . .) before being able to be aware of the rough shape of the seen object, vision of the real world would be decidedly more difficult than it is.

The significance of the computational metaphor, and its necessity as a complement to psychophysiological experiments, is stated by Marr as follows:[63]

> The situation in modern neurophysiology is that people are trying to understand how a particular mechanism performs a computation that they cannot even formulate, let alone provide a crisp summary of ways of doing. To rectify the situation, we need to invest considerable effort in studying the computational background to questions that can be approached in neurophysiological experiments.
>
> Therefore, although [my work] arises from a deep commitment to the goals of neurophysiology, the work is not about neurophysiology directly, nor is it about simulating neurophysiological mechanisms: it is about studying vision. It amounts to a series of computational experiments, inspired in part by some findings in visual neurophysiology. *The need for them arises because, until one tries to process an image or to make an artificial arm thread a needle, one has little idea of the problems that really arise in trying to do these things.* Computational experiments allow one to study in detail what combination of factors causes a method, or group of methods, to succeed or fail in a number of particular circumstances that originate from real-world data. *The power of this approach is that the knowledge one obtains concerns facts that are inherent in the task, not in the structural details of the mechanism performing it.*

The emphasized passages may be taken as a summary of the importance of artificial intelligence for psychology in general, whether physiologically oriented or not. And since truth springs more readily from error than from confusion, the poverty of machine vision as compared with human sight is less important than the theoretical clarity that is made available to psychologists by the computational approach.

# 14

# *Philosophical Issues*

ARTIFICIAL INTELLIGENCE touches on philosophy at many points. Some are relatively esoteric, involving detailed controversies within logic, epistemology, or the philosophy of language or of science.[1] Most people are not only unaware of these issues, but would feel no pressing need to form an opinion on them one way or the other, should they be explicitly raised. Though these specialist points do have a wider epistemological relevance, there are few whose sleep will be broken or whose dreams troubled by them.

But other philosophical issues are more widely apparent, and concern matters on which most people do have opinions, however implicit or ill-formulated they may be. It is because of these that artificial intelligence is for many the stuff of nightmares, or even madness. The spectre of the mechanical mind haunts the lay consciousness because it appears to threaten deeply-held values and traditional beliefs. Accordingly, claims that mechanized intelligence is impossible are often motivated by the hope of exorcizing this spectre, rather than being disinterested explorations of the limits of the field. (We shall see in Chapter 15 that even "disinterested" inquiries are not, and cannot be, purely disinterested; this epistemological fact has important consequences as regards the social implications of artificial intelligence.)

In the first section I shall explore those philosophical implications of the expression "artificial intelligence" that cause many people to reject it with ideological fervor as a pernicious contradiction in terms. In the second, I shall clarify the claim made in Chapter 1, that artificial intelligence helps resolve some familiar problems within the philosophy of mind because it provides a more complex image of "mechanism" and "machine" than has formerly been available. And in the last, I shall discuss some philosophical arguments that seek to set limits to the practical results that programmers may reasonably expect to achieve, arguments sometimes used to suggest that machine intelligence is a will-o'-the-wisp so essentially elusive that computational research like that I have described must be a hopeless waste of time.

## ARTIFICIAL INTELLIGENCE: A CONTRADICTION IN TERMS?

The fact that "the mind of the machine" has been featured in *Playboy* suggests that this phrase is so paradoxical as to be positively titillating.[2] So you may have decided, in deference at least to your philosophical sensibilities, to invoke the license issued at the close of Chapter 1 permitting you to place scare-quotes round all the psychological terms I use to describe programs. If so, you have been interpreting these words analogically, with a greater or lesser feeling of unease. We must now consider not only whether psychological terms can properly be used in their full literal sense when carried over into cybernetic contexts, but also whether anything important follows from the view that they cannot. In short, what does it matter whether "The mind of the machine" and like phrases are vulgar solecisms?

Some people who regard "artificial intelligence" as a contradiction in terms conclude merely that anyone who uses the phrase with a straight face is crazy. The appropriate attitude toward such imbeciles is then amused indulgence, rather than fear or scornful disdain. In Moscow I encountered a taxi-driver who, when he heard the name of the conference his passengers were attending, roared with laughter and made the "crazy-sign" against his forehead; nor did he stop doing this, his shoulders shaking, until he had dropped us at our destination some five minutes later. The thought that he should view us with wariness or disdain clearly could not have been further from his mind: we offered no threat, just comic relief.

But others are less sanguine about such malapropisms, and meet them with passionate contempt instead of disinterested amusement. Thus the humanist typically claims that psychological vocabulary cannot be literally applied to machines, and often concludes from this premise that "machine intelligence" is an elaborate illusion wholly incapable of throwing light on any human problem—and ideologically pernicious to boot.

For in many humanists' opinion, the literature of artificial intelligence inevitably encourages the alienating mystification that there is no essential difference between people and machines, and thus subtly supports those social systems that effectually treat people as though they were machines. (You will notice that this objection appears stronger the more one limits oneself to *nineteenth*-century concepts of "machine.")

Guy Robinson sums up his mordant statement of this overall point of view by borrowing Hobbes's contemptuous rhetoric: "When men write whole volumes of such stuff, are they not mad, or intend to make others so?"[3] If computer research on thinking is not to be discredited by this common pattern of objections, the humanist's argument must be examined—and its disdainful conclusion (if not also its premise) must be rebutted.

The basic reason for the humanist's premise, her refusal to allow that psychological terms can be applied literally to machines, is that on a humanist view machines are incapable of truly purposive action. This position is stated clearly by Robinson, and occurs repeatedly in humanist writings. Sometimes (as in the quotation at the end of this section) this point is expressed in the terminology of European philosophy, machines then being said to be incapable of *praxis*, or truly self-directed action.

Stated more fully, then, the humanist's premise is as follows: being artificial or manufactured in origin, the machine's "thought" and "action" can be represented as meaningful or intelligent only by some (explicit or implicit) appeal to the ends of the agent who made it. Naturally, in self-modifying planning programs, the explanation of the machine's "purposes" will rarely appeal so directly to human interests as SHRDLU's response 29, "BECAUSE YOU ASKED ME TO." Nevertheless, the fact that the machine has "purposes" at all, irrespective of which those happen to be, is explicable by reference to human design. In general, intentional terms are ascribed to it only because human beings choose (for their own ends) to regard its intrinsically meaningless operations *as if* they expressed the interests of a psychological subject. And such interpretations of machine performance get whatever plausibility they have only from the background of purpose and intelligence involved in the creation and contemplation of artifacts by human persons. Even intentional notions such as *information* and *symbol* are used in an inescapably analogical or derivative sense when they are applied to "information-processing" or "symbol-manipulating" machines. Since men and women have whatever interests they possess naturally and intrinsically, not artificially and derivatively, meaning and purpose are truly ascribed to them in a way that is in principle improper when artificial systems are in question.

Whether or not the humanist's premise is justified depends on whether or not the concepts of purpose and meaning do indeed carry an implicit assumption of their being in a sense *intrinsic* to the agent concerned. Not everyone uses these concepts in precisely the same way; moreover, as I shall stress presently, conceptual usage can change—one of the circumstances leading to such change being technological development within the linguistic community involved. So the way in which "purpose" is thought of today may differ from the way in which it will be thought of tomorrow. But it is currently true that one of the commonly (if not universally) accepted criteria of purposive activity is that it can be explained by reference to ends that are intrinsic to the nature of the agent herself, rather than to any outside agency. Intrinsic ends are not necessarily to be thought of as fixed principles of "human nature," but are purposes or interests that cannot be further explained in *purposive* terms (whether or not they can be explained in evolutionary or physiological terms). Lack of space forces me to make this claim without justifying it at length here, but I have argued for it elsewhere and illustrated it by reference to a wide range of psychological theories.[4] The important point is that many people, theoretical psychologists and laymen alike, do use the concept of purpose so as to carry this implication.

Given this use of the term "purpose," it follows as the humanist says: words that posit the interested action of a psychological subject can be applied only in a derivative sense to a putative agent if some other subject is, in fact, the ultimate source of the interests concerned. This even holds when the putative agent is a human being—which is precisely why people who are following ends other than their own (not having freely adopted these purposes of their own volition, but having accepted them unthinkingly[5] from some other person or class of persons) are likened by the humanist to machines, and described as alienated from their human nature. Since all psychological terms do posit an acting subject, to speak of artificial "intelligence" is therefore to employ a non-literal, analogical, sense of the word.

But even though one grants the humanist's premise, one may yet question whether it entails her dismissive conclusion. In other words, is the derivative use of psychological terminology in cybernetic contexts *an inexcusable misuse* of language, as the humanist typically insists? To the extent that computer analogies can serve the general human interest of increasing the understanding of the mind, the careful use of "psychological" terminology in speaking about certain machines should be encouraged rather than forbidden. For, as we saw in Chapter 11, the analogical use of representations not only enables one to express those matching features between analogous items that one has already noted, but guides one's search for further similarities (and differences) of significance.

Some features of programmed computers are indeed analogous, in greater or lesser degree, to real mental processes. Countless specific instances have been cited in the previous chapters, and in each case I have tried to indicate the points at which the analogy fails to be extensible, so that readers will not be misled by my use of psychological terms. Workers in artificial intelligence have their own attention drawn to these points of weakness by means of their use of psychological analogy, and their further research efforts are guided accordingly. If they did not think about what their programs do in these terms, the "obvious" points at which significant improvements are needed would not be obvious at all. This is true even if they are abstractly concerned with "intelligence in general," as opposed to focusing their work on a specifically human psychological ability. Analogy is thus a powerful method of developing knowledge in this intellectual field, as in others. In short, explanations of psychological phenomena in computational terms share with explanations *in general* the "paradoxical" assimilation of unfamiliar to familiar that serves to extend our understanding.[6]

But one must concede to the humanist that current analogies of intentional phenomena are relatively weak, in that there are many differences between existing artificial and natural information-processing systems. This is true even though programs like those we have considered involve deliberate attempts to parallel the *intelligibility* of human thinking. Indeed, some workers in artificial intelligence conceive of "intelligence" in such a way that intelligibility is an essential criterion of it, as in E. A. Feigenbaum's definition: "Intelligent action is an act or decision that is goal-oriented, arrived at by an understandable chain of symbolic analysis and reasoning steps, and is one in which knowledge of the world informs and guides the reasoning."[7] It would follow from this definition that if a system were ever to be designed that could solve problems by processes involving mere "brute force" computation and hardware tricks (possibly with a few general searching-heuristics thrown in), the system could not be regarded as intelligent. (Whether such a quasi-intelligent problem solver is in principle possible is itself an interesting philosophical problem, on which computational scientists differ.)

Even programs that, like HACKER and SHRDLU, conform to Feigenbaum's definition in that they simulate the symbolic structure of human purpose and intelligence, fail at many points to match the achievements of human reasoning. In addition to specific failings (like SHRDLU's inability to interpret elliptical sentences), there are more general ones that weaken the psychological analogy. For example, SHRDLU is not a close analogy to a truly purposive system because it is too single-minded: it does not have many interacting "motives," and it does not approach its "goals" in light of inner conflict between varied "desires." The same caveat applies to HACKER; and in the previous chapter we remarked that some of the psychological terminology used by HACKER's programmer to describe the program's function is overly anthropomorphic and would have been better avoided. In general, current programs are not nearly rich enough to be regarded as adequate theoretical models of the mind.

So quite apart from the issue of intrinsic interests, which the humanist takes to be so salient as to be *essential* to genuine purpose or intelligence, the deficiencies of current programs make it difficult to claim with a clear conscience that any of them enables a machine to be *really* intelligent: HAL, perhaps—but today's efforts, certainly not.

To someone who believes that "artificial intelligence" is a contradiction in terms, HAL of course would not be really intelligent either. On this view, even if all practical difficulties were to be overcome, the resulting artifice could be termed "intelligent" only in an analogical and fundamentally misleading sense. This conclusion follows directly from the (currently common) use of "intelligence" as a term presupposing intrinsic meaning or purpose in the intelligent individual.

But this does not finally settle the question at issue in this section, since linguistic usage can change, in response to altered knowledge and circumstances. In his famous paper describing the "Turing Test," A. M. Turing predicted that by the end of the century general educated opinion will be that machines can think: but he also remarked that by that time the words might have subtly changed in meaning.[8] If the so-called "intelligent" performances of artifacts were to become increasingly impressive, and more closely similar to intelligent human behavior, the decision would eventually have to be faced whether intrinsic (nonartifactual) purposes are or are not fundamentally necessary for the ascription of intelligence and purpose as such. The crucial point is that this would be a matter for decision, not a question to be settled merely by inspection of current usage.

To say that it would be a matter for decision is not to say that it is an arbitrary matter of no moment, for superficially neutral language can carry wide-ranging ideological implications that subtly influence speaker and hearer in largely unconscious ways. In Chapter 4 we saw that even a computerized model of thought has to take account of this psycholinguistic phenomenon. Some ideologically loaded phrases—like Abelson's examples, "law and order" and "imperialist lackeys"—are both relatively recognizable as such and relatively easy to avoid in writing and conversation.

But others are not: the word "man" has occasionally appeared in my text, as in my title, to connote humanity in the abstract, but it is not a purely abstract or value-free term. On the contrary, the fact that I could not use its female alternative (as I have done for the personal pronoun "he") without making the title misleading, or comic, or both, illustrates how—even in our language—women must be content to be the exception while men define themselves as the rule. And the subtly disorienting effects of my impersonal uses of "she" drag into consciousness the normally unnoticed assumptions that the most "unprejudiced" of readers implicitly brings to the interpretation of seemingly unbiased texts. In particular, the jarring effect of the alternative pronominal convention in references to "the humanist" or "the programmer" shows how readily we all assume it to be natural (not merely culturally probable) that such human creatures should be male. Yet women have philosophical views too, and have even been known to write programs: my decision to use "she" in place of the customary "he" is an attempt to prevent you (and me) from losing sight of these points, and far from being the irrelevant indulgence of an arbitrary whim it is a partial expression of a certain viewpoint about the place of women in society. The common, and often unavoidable, abstract usage of "man" and "mankind" is likewise expressive of nontrivial (though largely unconscious) assumptions.

Since the words "purpose" and "intelligence," like other psychological terms, are similarly not ideologically neutral, and similarly difficult to avoid, more may hang on their use than is immediately apparent. Attitudes of respect and moral consideration, for instance, are subtly bound up with notions of intelligence and intrinsic interests. We shall see in the next chapter that a Kantian ethic (which prescribes moral concern for rational agents) might involve us in practical difficulties, were we ever to regard robots as *rational* beings. And moral attitudes in general are attached to (and expected of) creatures that are both rational and "interested."

Consequently, the decision whether to regard "artificial intelligence" and "machine intelligence" as analogical or as literal expressions cannot be definitively settled by "the facts"—which themselves are commonly conceptualized in a value-loaded way, much as are the intermediate schematic levels in Abelson's examples—and should not be irresponsibly approached. Some workers in artificial intelligence have tried to retract this label, replacing it for instance by "the study of knowledge-based systems," or of "complex information processing." The reason is that, since John McCarthy's introduction of the more well-known term in 1956, it has demonstrably attracted negative attitudes to the field, largely because of the plastic taste of "artificial" and the snooping overtones carried by "intelligence" in the minds of people acquainted with the FBI, CIA, and "military intelligence." These suggested alternative labels might be preferable, even though they each carry essentially the same philosophical difficulties, since many people would not be so readily aware of the possibility of a contradiction being involved. However, the more widely used label is probably here to stay, and will inevitably invite unease and suspicion from people of broadly humanist sympathies.

These points can help one more fully to understand the motivation of humanist attacks on the enterprise of machine research, and on the terminology of artificial intelligence. Such attacks are based not only on "cold" (and, I shall next argue, mistaken) metaphysical beliefs about the logical incompatibility of subjectivity and mechanism, but also on what Abelson would term "hot" ideological commitments of a largely implicit kind concerning the nature and proper ends of human beings. This explains the sociopolitical mistrust and moral disdain with which people often approach—or, rather, adamantly refuse to approach—artificial intelligence. This attitude and its typical rationale are forcefully expressed in the following passage:[9]

> What mistakes are being made [by protagonists of machine intelligence?]
>
> 1. Reductionism—the view that thinking, as something we subjectively do, is no different from the electrical movement in a machine. Here we find a translation from the subjective side of existence to the objective side. This is a mistake of the first order.
>
> 2. Physicalism—the view that there are only physical phenomena. So everything must be restated in a physicalist language. A new language is constructed eliminating reference to subjectivity. The problem with this language is that it lacks meaning. A *subject* is necessary for meaning, which by definition is impossible.
>
> Physicalism is a powerful force in society, for though based on absurd foundations it entails a view of man that is at one with capitalism. Not surprisingly vast amounts of money are spent in the attempt to convince man that he is really a machine and should function accordingly.
>
> 3. Machine intelligence is a tool of bourgeois society, it aims at strengthening existing social relations, and at eliminating protest and social action. This is the realization of Comte's [positivist] dream—some dream! This utilization is explainable in dialectical terms. Machines are incapable of praxis, of purposive action—they are merely a series of blind processes arranged by a programmer. The "intelligent" machine or computer is qualitatively no different than a gun or whip, it serves the same function.

This common fear that machine models of the mind will alienate us from our proper humanity should have been somewhat allayed by the foregoing discussion. Although I have defended the practice of positing analogies between minds and machines, I have agreed that the categories of subjectivity, meaning, and purpose as currently understood can be attributed to artifacts only in a secondary sense, their justification ultimately deriving from the skill and interests of the artificer. And I have recognized that basic ideological issues must be borne in mind by anyone proposing to extend these categories to include systems whose source is not biological evolution but technological manufacture.

Provided that these issues are explicitly acknowledged, and provided that specifically human ends and purposes are unhesitatingly given priority, there need be no danger to a humane image of mankind in allowing that machines are (or one day will be) intelligent in a *non*analogical sense.

So far, I have mentioned philosophical objections to artificial intelligence that are based on the lack of intrinsic interests in artifacts. But for some people, the mere fact that computers are causal mechanisms suffices to show that they can never really be intelligent. These people believe that reason (and subjectivity in general) cannot be reduced to causal processes.[10] On this philosophical view, psychological phenomena cannot be explained in terms even of neural processes in the human brain, never mind computers. The relation of mental processes to physical mechanisms is the central topic of the next section.

## MIND, MECHANISM, AND MACHINES

The philosophy of mind traditionally covers a wide range of related problems, of which each can be stated in various ways according to metaphysical predilection, and several can reasonably be regarded as *the* core issue. The mind-body problem is very commonly viewed as central, especially by people who are interested in neurophysiology and theoretical psychology, and in understanding the relation between physiological and psychological explanations. Since the time of Descartes, whose thinking permeates Western culture even more deeply than Freud's, this problem has commonly been stated in terms of how the mind can affect the body, and vice versa. But this manner of stating it is unhelpful in a number of ways. An alternative formulation of essentially the same question is this: how is it possible for mental phenomena to be both irreducibly psychological and somehow wholly dependent on a mechanistic causal base (the brain and nervous system)?

This query arose in the previous chapter, in comparing humanist with mechanist psychologies. I said then that the apparent metaphysical *impasse* between mind and body can be avoided if one recognizes that "reduction" and "mechanistic" each have two importantly different senses. Elsewhere, I have discussed in detail the differing senses of reduction and mechanism in relation to psychology and physiology, and have also indicated in general terms how computational models bear on these matters.[11] In this section I shall outline my position, and suggest some ways in which the programs discussed in the body of this book are relevant.

It is true, as the humanist insists, that theories employing subjective concepts cannot be translated into nonpsychological terms. The reason is that intentional sentences, whose meaning involves the notion of subjectivity, have a very different pattern of logical implications from sentences that do not involve this notion. In technical terms, the logical peculiarities of intentional sentences include indeterminacy, referential opacity, failure of existential generalization, no implication of any embedded clause (or its negation), and nonextensional occurrence of embedded clauses. Typically, not all of these peculiarities characterize any one intentional sentence, but at least one of them does. In general, the intentional object (the object of thought that is mentioned in the sentence) can be described only by reference to the subject's thoughts, such as her purposes, beliefs, expectations, and desires. There may be no actual thing with which the object of thought can be sensibly identified; and even if there is, the identification will seem sensible only on the basis of certain descriptions of the real thing, and the intentional object may be indeterminate in a way that no actual object can be.

For example, to say that a person sees a vertex in a grey-scale picture is not to commit oneself to saying that there actually is a vertex there: perhaps the person is jumping to mistaken conclusions, as G. R. Grape's program sometimes does. Again, to say that she sees a complex vertex is not necessarily to say that she sees it as having a specific number of contributory lines, even though any vertex actually present in the picture must have a determinate number of connected lines and a determinate number (perhaps zero) of lines sufficiently close to be sensibly interpretable as "part" of that vertex. Similarly, Adolfo Guzman's SEE makes no distinction between different MULTI vertices having five, six, or more lines: the program's descriptive powers do not enable it to do this, even though an initial line finder providing input to SEE might be able to express this determinate information. Last, to say that a neurotic woman —or a "neurotic" program that simulates her beliefs—believes that her father abandoned her is not to say that he did so: perhaps he did, but perhaps he did not.

No pair of sentences with radically different implications can be regarded as identical in meaning. R. C. Schank's semantics, and the MARGIE programs, show how understanding language requires that unstated inferences be taken into account. While it is a moot point where "meaning" ends—how many of C. J. Rieger's MEMORY inferences should one include?—it cannot be claimed that obvious implicational differences like those between intentional and nonintentional sentences make no difference to the sentence's meaning. Consequently, one must be "*anti*reductionist" in allowing that intentionality is an essential feature of psychological reality, one that could *not* be expressed by a theoretical vocabulary lacking the subject-object distinction. For one of the important senses of "reductionism" in psychology is the (mistaken) view that psychological descriptions and explanations are mere shorthand for complicated sets of nonpsychological statements about the brain, so that psychological statements could be translated into physiological ones without loss of meaning.

The other sense of "reductionism" that is relevant here is the view that subjective psychological phenomena are totally dependent on cerebral mechanisms, much as the information-processing functions of a program are grounded in the engineering details of the computer on which it is being run. One can be a reductionist in this sense without being a reductionist in the sense defined above. Indeed, we saw in Chapter 13—and, by implication, in the main part of this book—that computer scientists working in artificial intelligence in fact hold this particular combination of views. For instance, Terry Winograd does not try to explain SHRDLU's answer "OK" in terms of the electronic gadgetry that at base enables the program to exhibit its understanding so persuasively. (This position, which one might regard as an "antireductionist reductionism," has essentially comparable forms in philosophical debate concerning the relation between other pairs of sciences, such as molecular biology and quantum mechanics.[12])

A "mechanistic" psychology is sometimes defined as one that abandons conceptual schemes employing the subject-object distinction, and that refuses to interpret (or explain) behavior in terms of meaning, phenomenology, or purposive action. Psychologists who are "mechanists" in this sense hold that these subjective categories are shorthand labels at best (shorthand for lengthy accounts of stimulus and response or of brain cells), and mystifying illusions at worst. Skinner is an example of a mechanistic psychologist, in this sense of the term, whereas humanist psychologists clearly are not.

Humanist theories may nonetheless be "mechanistic" in a different sense, as Skinner's theories of course are also, for they may allow that subjective psychological phenomena can be generated—and in a manner explained—by bodily processes. This is not to say that psychological processes are *identical* with neural processes, or that they are the *effects* of bodily *causes*. Rather (as I have argued more fully elsewhere), the crucial notion in understanding how subjectivity can be grounded in objective causal mechanism is the concept of an internal model or representation, which we have seen to be central to many schools of psychology and to artificial intelligence also.[13]

It is possible for the categories of subjectivity to be properly attributed to human beings because bodily processes in our brains function as models, or representations, of the world—and of hypothetical worlds—for the individual concerned. K. J. Craik discussed these physiological models in general terms in the 1940s, and since then neurophysiologists have been increasingly concerned with the problem of how such models are built up in the nervous system and how they are organized so as to influence bodily action.[14] As the previous chapter's discussion of psychological work on vision suggested, the physiological basis of epistemological representations is still highly obscure. A neural model of something is not to be thought of as a *copy* of it: there is no more reason to think that Senator Goldwater has a *copy* of a Communist dupe inside his brain than to think that the IDEOLOGY MACHINE has a *copy* (a cartoon?) of President Kennedy. Rather, there is a complex base of bodily processes that interact on one another in ways describable at various computational levels, such that Goldwater's behavior (including his speech) with respect to Communist dupes is broadly appropriate—given his particular background purposes and beliefs.

Ultimately, the origin of these processes must lie in the evolutionary history of the human species: the philosopher D. C. Dennett has made some interesting suggestions about how the modeling functions of the brain might have originated, and theories about the evolution of sight (such as that described by R. L. Gregory) also address such issues.[15]

To identify or describe the neural processes concerned *as* models is itself to ascribe meaning, or intentionality, to them. They could alternatively be described, at least in principle, at the level of "objective" physiological events occurring at particular neuroanatomical locations (such as the events within the visual system of animals that were mentioned in the previous chapter). At this level, however, their meaning cannot be expressed and so their (psychological) function in the life of the individual is lost to view. Even so simple a neurophysiological concept as a "bug detector," for example, is implicitly psychological insofar as it identifies a functional relation within the animal's intentional world. Consequently, the categories of meaning, subjectivity, and purpose would still be required to describe a person as a psychological being even if full neurophysiological knowledge were available. Indeed, a large proportion of "physiological" data would be expressed in intentional terms. To forbid the use of such intentional language would be to omit all mention of mental phenomena, since there is no possibility of saying anything about the mind using only the language of the body.

It is this notion of an inner model or representation of the world that provides the basis for an analogue of subjectivity in artificial information-processing systems (and which was identified as crucial to computational research by M. L. Minsky in the 1960s[16]). Intentional descriptions and explanations of a machine's behavior (even if they are understood in an analogical sense) are not merely shorthand expressions for matters that in principle could be more fully expressed at the electronic level. On the contrary, those features of computer function that are analogous to psychological aspects of people can be identified only by psychological language. Since they can be neither identified nor, in an important sense, explained in electronic terms, such features commit the computer scientist to the use of psychological (computational) terminology if they are not to be wholly lost to view. Concentration merely on the physical mechanism of an intentional system ignores its intentionality, whether the system be natural or artificial in origin.

For example, the behavior of Gilbert Falk's INTERPRET program in completing an imperfect line drawing of a cube can be neither understood nor explained in electronic terms. One must certainly be careful in applying psychological concepts like "recognize," "hallucinate," "interpret," or "predict" to the program, lest one imagine (*sic*) that the relevant analogies are closer than in fact they are. But close attention to the functional details of INTERPRET and its several modules should prevent such misleading overgeneralization of the analogy. The analogy itself may even be of independent philosophical interest, quite apart from its usefulness as a way of expressing what it is that this particular program does.

Consider, for instance, the behaviorist philosophical view that any mental ability *must* somehow show itself in behavior, by asking yourself what are the visual powers (analogically) attributable to INTERPRET—that is, to the system as a whole. INTERPRET cannot see bodies that do not conform to the nine prototypes, because RECOGNIZE matches up body candidates with these models only. Or, rather, INTERPRET can see that there is *something* there (SEGMENT could identify 17 separate bodies, for example); but it would have no way of saying so unless *either* SEGMENT were able to offer its conclusions "to the world" on behalf of the system as a whole, as opposed to communicating its thoughts only intrapsychically (to the next module), *or* a later module (RECOGNIZE, for instance) were able to report the information earlier passed on to the rest of the system by SEGMENT.

Furthermore, COMPLETE could identify and imaginatively fill in gaps in lines apparently representing edges: but if the resulting line drawing did not match any prototype, then RECOGNIZE could not recognize it. This is to say that an imperfect picture of a star might be internally transformed into a perfect picture of a star, *without* INTERPRET's being able to see (recognize) the star. It follows that if COMPLETE could communicate only with the following module, there might be no way for an outside observer to infer that INTERPRET was having these (quite specific) hallucinations.

Possibly, a clue might be found in the puzzled conversations between VERIFY and RECOGNIZE, as VERIFY continually struggled to match the predicted picture with the actual (prehallucinatory) input. But this, of course, assumes not only that RECOGNIZE can report the relevant pictorial features of the image it works with, but also that the hallucinated image is sufficiently close to one of the prototypes for RECOGNIZE to have been satisfied to pass it on to PREDICT in the first place.

If you feel that our familiar psychological vocabulary is not sufficiently fine-grained or precise to state these matters clearly, my sympathies are with you. And if you feel that it is unlikely, on evolutionary grounds, that untapped interpretative powers exist in animals or men, I agree that this raises some interesting questions—to which research in comparative psychology might suggest tentative answers.[17] But whether or not you wish to place scare-quotes round my psychological vocabulary here (and why especially here?), these considerations should have shown that it is not *necessarily* true that inner psychological processes will be somehow expressible in outward behavior. For an observer without independent knowledge of the program to know of their existence it may indeed be necessary that they connect, however indirectly, with observable performance—but that is a different matter.[18]

Consideration of INTERPRET, or any of the more powerful programs I have described, can enormously expand one's intuitive sense of the range of possibilities latent in "mere mechanism," a sense which previously had to draw on simpler examples like the clock, the steam locomotive, or the guided missile, if it was not to depend on a commitment to mechanism as an unproven philosophical credo.

The horological metaphysic of the Cartesian material universe, expressed in its purest form in De la Mettrie's *L'Homme Machine*, is one that many people (like Descartes himself) find unpersuasive when human psychology is in question.[19] T. H. Huxley's epiphenomenalist view that mental factors like purpose and intention have as much directive influence on bodily behavior as does the smoke on the steam engine—which is to say, none at all—is hardly more convincing.[20] And even the cybernetic concept of negative feedback (and its biological equivalent, homeostasis) is not sufficient to account for the intentionality of everyday behavior, although it does provide a stronger analogy to genuinely goal-directed action than do the concepts of seventeenth- or nineteenth-century mechanism.[21]

By contrast, the inner representations of the world (and of the system's own actions in it) that are flexibly manipulated by programs like SHRDLU, HACKER, and BUILD generate structural complexities of computer performance that are comparable in various ways to the behavior of biologically evolved purposive creatures. The analogy fails at many points, to be sure; and (as we shall see in the next section) some philosophers claim that tomorrow's programs will be hardly less deficient in this regard than today's. But even if this claim were correct, it would not destroy the present point: that the image of "machine" provided by current computer science renders it intuitively less inconceivable that mental phenomena may be grounded in a mechanistic physiological base while also having a characteristically psychological guiding influence on behavior.

As well as thus illuminating the mind-body problem narrowly conceived, artificial intelligence also aids clarification of many associated problems concerning the nature of human purpose, self, freedom, and moral choice. By this I do not mean that terms like "free," for instance, can plausibly be applied to any computer example. But the computational distinctions that can be clearly made as between particular programs help one to suggest what may be the complex functional bases of the contrast between "free" and "involuntary" behavior that each of us makes intuitively in daily life.

As G. E. Moore said in rebutting idealism,[22] the philosophical problem may be not so much whether a common-sense claim (that people are free, for example) is true, as whether one can offer an acceptable philosophical analysis of it. Naturally, one should not presuppose that there must be an acceptable analysis of a common-sense concept: perhaps it is incoherent or inadequate in some way, in which case one will be committed to denying it. A philosophical explication of freedom that does not deny it must clarify what sort of phenomenon it is and how it is possible. An acceptable analysis of freedom should illuminate whatever genuine psychological distinctions (and moral implications) are marked by the vocabulary of freedom, without falling into either a self-defeating indeterminism or an antimechanistic mysticism that ignores our physiological constitution and evolutionary origins. Such an explication is made more accessible by bearing computational concepts in mind, so that artificial intelligence has a role to play in helping us to understand what it is to be a human being.

For instance, free action is self-determined action. In other words, a conceptual analysis of the notions of "freedom" or "voluntary action" involves reference to the action's being not merely in some degree autonomous with respect to changes in the environment, but also being somehow deliberately determined by the self of the acting individual. The "self" has been described by many psychologists of differing viewpoints as a cognitive representation of the organism contained within the organism, which comes to include ideals or aspirations (moral and otherwise) with respect to which behavior is guided in a more or less integrated fashion. There is a clear distinction between programs (like SHRDLU and HACKER) that have within them something describable as a reflexive representation of their own range of action, and programs (like K. M. Colby's neurotic) that do not. There is a clear difference also between procedures that are, and those that are not, crucially influenced at particular points by reference to specific aspects of this inner model of the system as a whole.

Again, philosophers commonly insist that free agents are bound by moral laws that are grounded in the interests of persons and are universalizable, in that they concern psychological qualities like loyalty and betrayal and apply to all individuals in similar circumstances. Such views imply that moral understanding and responsible action involve the intelligent use of words like "all" and "every," and comprehension of the difference between betrayal and loyalty—as well as the ability nicely to compare analogous situations with respect to morally salient features. Comparison of the essentially psychopathic SHRDLU (which satisfies none of these criteria) with systems like R. P. Abelson's, whose semantics more nearly approach the complexities concerned even though it cannot assess degrees of analogy, helps to show why an infant cannot have a truly moral sense and therefore (as I claimed in Chapter 4) cannot be accused of such infamies as callous betrayal of her fellows.

Further, one sometimes describes an action of a free agent as irresponsible, and morally culpable, because her mental structure allowed for the *possibility* of its being deliberately willed and consciously monitored, even though she actually did it "unthinkingly," or even "automatically." But imagine a HACKER with rather more complex rules for deciding whether or not to function in CAREFUL mode. And suppose that, in order to save time, say, it omits to ask a question that would have shown it the advisability of carefully monitoring the procedure it is about to execute. One could then claim that—in a clearly explicable sense—it *could* have done something that would have enabled it to avert the undesirable consequence but, for an essentially trivial reason, it did not.

Last, the unpredictability of human action that is so often insisted upon by libertarians is due to a number of factors each of which is intelligible in computational terms. One of these underlies the traditional argument that "free spirits" can always choose to do or not to do something, whatever predictions an observer may make about the matter. Without going into this fully, it is worth remarking that even the relatively mechanical PARRY could easily be made to interpret certain predictions on the psychiatrist's part as threatening, so that unless his emotional monitors were currently indicative of a low anxiety level he would be likely to respond in a "contrary" manner.

To discuss the nature of freedom, purpose, self, and consciousness at greater length here would not only be to repeat myself,[23] but would lead us too far into the general philosophy of mind and away from philosophical questions more specifically connected with research in artificial intelligence. Such questions include the problem of whether there are any a priori reasons for believing that workers in artificial intelligence cannot possibly achieve their stated aims. In describing current programs I have already hinted that several such reasons have been proposed: we must now examine some of these more closely.

## CAN MACHINE INTELLIGENCE BE ACHIEVED?

There are many arguments within the philosophical literature purporting to show that certain things done by people *could not be done* by computers (irrespective of whether we should use psychological vocabulary to describe the things that *can* be done by them).[24] I shall concentrate on the three that appear to be most pertinent to artificial intelligence as it is today. These three are: the argument from Gödel's proof; the claim that digital computers could not achieve common kinds of "tacit knowing," or cognitive information-processing; and the view that computers could not simulate emotion, even to the degree that they can simulate certain cognitive processes.

In 1931 Gödel proved that there are limits to what any particular logical or mathematical system can do.[25] These limits are surprising because they concern matters stated in the terms used by the system, and so of a type that seemingly ought to be decidable by it. In brief, for any consistent logical system (that is rich enough to contain elementary arithmetic), there is at least one meaningful sentence that can be neither proved nor refuted by it, but that would naturally be interpreted as *true* by logicians.

This proof is sometimes cited as showing the superiority of the human mind over "mere logic," since the logically undecidable proposition can apparently be seen to be true by people. By the same token, it is cited in attacking artificial intelligence, since a given program (computational system) seemingly cannot decide the truth of at least one well-formed statement that the programmer knows to be true.[26] Actually, logicians and mathematicians differ over the precise interpretation of Gödel's theorem, and whether it can be generalized to programs (as opposed to axiomatic systems). But we may ignore this fact, since even if Gödel's proof applies to programs as well as to logic, it still does not show that programs are essentially inferior to the human mind.

The reason is that Gödel's theorem applies only to closed systems, wherein all the axioms and inference rules are fixed. If a program (like a person )is capable of learning new rules and axioms, by communicating with a teacher or with the outside world and extending its internal representations accordingly, then something that was undecidable yesterday may be decidable today. To be sure, there will be something *else* that is undecidable today—but it, in turn, may be decidable tomorrow. In sum, Gödel's proof does *not* show that there must be some statements that could be known to be true by people, but not by programmed machines. Philosophers (like J. R. Lucas, for example) who appeal to Gödel's theorem in rebutting mechanistic explanations of mind are therefore mistaken.[27]

The second class of philosophical arguments purporting to show the impossibility of machine intelligence was briefly mentioned with reference to HACKER in Chapter 10. Michael Polanyi has stressed the role of *tacit knowing* in human thought, whether "mental" speculation or "motor" skill, and regards a complete computer simulation of human thinking as impossible.[28] He shows that human reasoning—even of the most "explicit" or "rigorous" type, as in science and mathematics—employs integrative principles of tacit inference or global knowledge of which one is not introspectively aware, but which crucially determine the nature of the thought contents of which one is focally aware. The tacit knowledge is said to be indeterminate, in the sense that its content cannot be explicitly stated. Or, rather, its *complete* statement is impossible in explicit terms: "tacit knowing is the fundamental power of the mind which creates explicit knowing, lends meaning to it and controls its uses. Formalization of tacit knowing immensely expands the powers of the mind, by creating a machinery of precise thought, but it also opens up new paths to intuition."[29]

This position is highly plausible: who would wish to claim that there will come a time when *nothing* more remains to be said about the grounds, or implicit interpretative rules, of human knowledge? It follows that there will always be some aspect of our knowing that has not yet been formalized so as to be expressed in a computer program, yet which implicitly directs our interpretation of what it is that "intelligent" programs are doing.

But two points should be remarked. First, Polanyi admits that tacit processes *can* be formalized, and that doing so immensely expands the powers of the mind. (He gives many examples drawn from the history of science.) So he need not deny that artificial intelligence can usefully articulate some aspects of our tacit knowledge, as Grape's program (for instance) shows how messy and ambiguous cues can be identified according to visual context, or Rieger's EX-SPECTRE suggests ways in which global semantic context can guide the interpretation of a story. Second, and closely related to the first point, Polanyi does not pick out particular cognitive achievements as in principle unformalizable: simply, he says that there will always remain *something or other* that has not yet been made explicit, the "something" varying with the historical period.

Polanyi's follower H. L. Dreyfus is less cautious. In a critique of artificial intelligence, within which he refers to some of the early efforts in the field, Dreyfus identifies certain introspectively obscure aspects of human thought that he claims to be *essentially* intuitive, or indeterminate, such that they could not possibly be simulated by a purely digital machine.[30]

Appealing to empirical, largely introspective evidence (including the human protocols quoted by the GPS programmers), Dreyfus distinguishes four types of thinking that he claims are essential to most human thought—and especially to the phenomena described by Gestalt psychologists, and by phenomenologists such as Maurice Merleau-Ponty.[31] These four are reliance on the fringe of consciousness, discrimination between the essential and the accidental, tolerance of ambiguity, and perspicuous grouping. All these he assumes to deal with information that is somehow ambiguous, or indeterminate. And he infers that the requisite information processing could not consist of discrete, determinate operations, nonambiguously defined, such as are embodied in a program for a digital computer.

Insofar as he mentions limitations of specific published programs (most of which were pointed out by the programmers themselves), Dreyfus's attack has some validity. And he quotes various wildly optimistic claims made by enthusiasts in the early days of artificial intelligence. For example, one of the GPS programmers predicted in 1957 that within ten years a computer would be world chess champion (although by 1962 he was stressing the mediocrity of chess programs).[32] Y. A. Wilks has agreed that "the naive, and from this distance in time, absurd overoptimism of many in early AI is now sad reading."[33] Even today, predictions are sometimes made that apparently belie the enormous complexity of the problems to be faced (some examples will be mentioned in Chapter 15).

But the previous chapters have described many programs that show holistic "Gestalt" phenomena of interpretation and selective attention, albeit in relatively simple epistemological domains as compared with those making up the phenomenal worlds of human beings. Wilks, for instance, has programmed the ability to assign *new* senses to a word on the basis of the semantics of the text-as-a-whole, as we saw in Chapter 11. Sylvia Weir has simulated (and illuminated) the "perspicuous grouping" of objects on the basis of causal interpretations of the visual stimulus, as Chapter 9's account of her work showed. And Greenblatt's chess program uses a heuristic *plausible-move generator* in order to focus on essentials and ignore accidentals (Dreyfus himself was beaten by this program in 1967). In general, heterarchical functioning and an ability to pass from bottom-up to top-down processing *and back again* in the interpretation of highly ambiguous material (as in Grape's visual program) are well suited to the sort of epistemological "chicken or egg?" difficulties highlighted by Dreyfus. To be sure, the problems involved are still not well understood, and the intelligent management of very large data bases has not yet been achieved. But research on "frame systems," for example, seeks ways of richly structuring and economically accessing such data bases *without* relying on brute force exhaustive search—which Dreyfus often appears to assume is the only search strategy available to programmers.

However, from the fact that *some* things have already been done that Dreyfus asserted to be impossible, it does not follow that *all* can be. Perhaps his arguments do set limits of principle to artificial intelligence at some point? His detailed objections fall into three main classes: objections based on the fact that current artificial intelligence uses *digital* computers, not analog machines; objections grounded on the fact that computers do not have bodies; and objections pointing out that computers do not share our forms of life, or human context. Let us consider these three arguments in turn.

We saw in Chapter 1 that digital computers operate by carrying out a number of discrete steps, each of which involves the change of one or more basic engineering components from one physical state to another. (Usually, only two states are possible.) Analog computers are not like this, because the physical parameters used to represent information are continuously variable—like voltage levels, for instance. Dreyfus claims that "indeterminate" information, and thinking that does not proceed by discrete steps, cannot possibly be simulated on a purely digital machine. (Though he admits that a partially analog machine could do better; and he even admits that *in principle* analog processes could themselves be simulated digitally, though in practice time and memory constraints might make this infeasible.)

But Dreyfus is here confusing the information code with the information coded.[34] He himself uses the 26 letters of the Roman alphabet as a symbolic code to represent his "ambiguous and indeterminate" information. Handwritten script is physically continuous, but it is interpretable by us only because we can regard it *as if* it were physically discontinuous. (And a "writing-reader" machine would similarly have to be able to articulate the physically continuous pattern into distinct parts, or cues, much as Grape's program picks out vertices and polyhedra from the messy physical stimulus it starts with.[35]) Clearly, then, indeterminate information is sometimes represented in a discrete fashion, so that one cannot infer from the discrete structure of the code that the information coded is not continuous. Dreyfus apparently assumes that there must be a strongly analogical relation between representation and reality: in Chapter 11, however, we saw that this is not so.

Dreyfus's critical emphasis on the physical discontinuity of digital machines is therefore misplaced. It may be, however, that analog computers will be needed for other reasons. For example, R. L. Gregory has argued that robots (like animals) working in real time would need a partially analog "physiology," because only analog systems can represent the outside world directly by parallel—and perhaps equally fast—changes in their own internal states, as opposed to complex symbol manipulation.[36] On his view, only an analog device could avoid being pounced on by a crouching tiger, because only such a device could process the changing perceptual information as fast as the tiger was changing it. It is not obvious that Gregory is right; thus Minsky's work on *frames* is intended to show how a digital system might have symbol-manipulating procedures with the necessary speed and flexibility to do such things as escaping from the attentions of a tiger. Even if Gregory were right, and partially analog devices were therefore needed in robotics, this would have nothing to do with the "indeterminateness" of the perceptual information. Nor do artificial intelligence workers have to commit themselves to the position that, in practice, purely digital machines will suffice to model the entire range of human thought.[37]

Dreyfus's second argument is that computers cannot be intelligent since they do not have bodies. According to him, even robots could not overcome this limitation if they were made of mechanical (not organic) hardware and controlled by a purely digital machine. The suggestion that some human achievements might (in principle or in practice) be available only to robots of a partially analog character has just been discussed. Moreover, in Chapter 12 we noted that the cheapest store of information about the real world is the real world, and we concluded that many tasks might be feasible only for an active and percipient robot—as opposed to a merely speculative system, such as BUILD. So Dreyfus's claim that a disembodied digital system could not achieve intelligence such as ours may be correct, but not for the reasons he suggests.

For instance, Dreyfus claims that the indeterminate global meanings by way of which we perceive the determinate particulars in our visual world can be supplied only by the body. (*How* the body can do this is left entirely obscure.) But in Part IV we saw that disembodied digital systems can use global, and in a sense indeterminate, schemata to interpret messy pictures. Thus Grape's polyhedral prototypes are "indeterminate" in that they do not specify size, position, orientation, or degree of occlusion (which affects the number of visible cues) of the various items that the program can perceive. And Rieger's EX-SPECTRE uses the "indeterminate" notion, that someone whose property has been stolen is likely to do something or other about it, in its interpretation of sentences like "Lady Jane snatched up her car keys," or "Jake saddled his horse." Again, Schankian semantics in general uses vaguely specified expectations prompted by the previous text to make sense of the rest, and to identify *surprises*. Should Dreyfus claim that his "global meanings" are less determinate even than the highly abstract notions coded at the deeper Schankian levels, it would be difficult to see what meaning was involved.

Dreyfus claims, too, that only a human (or animal) body can

manipulate something in order to recognize it (as contrasted with recognizing it in order to manipulate it). If by "manipulate" he means *move* (as opposed to *touch*), why could not a robot provided with the sort of functional information discussed by M. J. Freiling use it to recognize spatial structures of various types? For instance, it could try pushing a toy car through a hole so as to see (*sic*) whether the hole formed part of a stable *arch* (BUILD would have to puzzle this out mathematically). And as for touch, various types of tactile sensors are already incorporated into industrial robots, so there is no reason why their manipulations of things could not aid their recognition of them.[38] Such robots also function much faster than systems like COPY-DEMO that are used for artificial intelligence research, where speed is of no consequence (Dreyfus criticizes the MIT hand-eye system for taking several minutes to pick up a block).[39]

Industrial robots also use tools. Dreyfus quotes Polanyi's view that "while we rely on a tool or a probe, these are not handled as external objects . . . they remain on our side . . . forming part of ourselves, the operating persons. . . . We accept them existentially by dwelling in them."[40] But an industrial robot similarly ignores the "external" qualities of tools while it is using them (though it takes account of them to pick out the necessary tool in the first place), concentrating rather on what is happening at the business end of the tool—is the screw still turning, or is the power screwdriver no longer engaging successfully with the slot?

Dreyfus is right to highlight the difficulties involved in explicitly representing the knowledge of space that comes naturally to us. We have seen that Freiling and G. J. Sussman, for instance, have to employ some counterintuitive assumptions in representing space. And workers in practical robotics such as P. M. Will have similarly remarked the difficulty of "talking" about spatial movements to an industrial robot that has to be taught a new task (without either specifying geometrical details or manually *moving* the robot arm in the required fashion so that it can then repeat the performance itself).[41] But various high level "spatial" languages are being developed for this purpose, so that nonspecialist factory workers will be able to instruct their machines appropriately. Moreover, the progressive improvement of bodily skills that is stressed by Dreyfus presumably involves the sort of self-modification embryonically present in HACKER. In sum, it may be that impressive spatial skills will never be possessed by purely digital systems, but Dreyfus's arguments do not prove this.

Dreyfus's third reason for asserting the impossibility of artificial intelligence is that computers do not share our human context, what Wittgenstein termed our *forms of life*.[42] For example, computers are not interested in food, humor, or companionship; nor are they hurt when hit, or sympathetically prompted to go to the aid of someone who is. These matters, and many others like them, are part of our natural history, and are deeply embedded in our phenomenal (interpretative) world. So there is no possibility of complete mutual understanding between ourselves and psychological beings with radically different forms of life. (One might question whether "complete" understanding or "perfect" communication is possible even between two people from one and the same social group.) This underlies Wittgenstein's famous remark, "If a lion could speak, we would not understand him."[43]

The discussion of Schank's work suggested that very general psychological facts—such that hitting hurts, or that people tend to render tit for tat—do indeed underlie our interpretation of language. And Wilks has agreed with the Wittgensteinian claim that our epistemological world would be barely intelligible to anyone whose differing forms of life rendered their language and conceptual schemes radically different, and concludes that the underlying semantics of language-using programs must include representations of these matters that tacitly determine our understanding.[44] Even the crude IDEOLOGY MACHINE presupposed that one cannot understand someone's political pronouncements (even their "reporting" of "facts") without some representation of their structured political interests and beliefs.

To what extent one can understand another person's ideology by theoretically representing it, as opposed to adopting and experientially entering into it, is of course a deep and difficult question. Human communication is often bedeviled by it. For instance, well-intentioned attempts by sympathetic men to contribute to feminist debate are sometimes impatiently spurned a priori by women, because "He couldn't possibly understand, no matter what he says."

You may feel that this dismissive "all-or-nothing" attitude on the part of the female feminist is over-strict, albeit containing a certain truth.[45] If so, you should ask yourself whether Dreyfus's *total* rejection a priori of the notion that computers could simulate our understanding in any meaningful way is in order. Three aspects of human forms of life are particularly stressed by phenomenological philosophers such as Dreyfus. The first is human embodiment, which has been briefly discussed above. The second is our possession of intrinsic interests, which was a central topic of the first section. The third is our capacity to experience emotions, and to take account of emotional factors in communicating with others (even in "solitary communication" like writing or reading novels). Let us pass, then, to the final philosophical problem of this section: could a computer have any understanding of emotions, as opposed to cognitive knowledge and problem solving?

Many philosophers besides phenomenologists hold that a computer could not possibly simulate or have any understanding of emotion, because it makes no sense whatever to attribute "feelings" or "consciousness" to inorganic programmed systems.[46] For present purposes, let us grant the questionable claim that feelings cannot be ascribed to computers. But to grant this claim is not to dispose of the matter, since (as we saw in Part II) emotions are *more than* feelings, and computer simulations of emotion attempt a theoretical representation of their psychological origin and effects, rather than ontological mimicry of their "felt" component. (And even Dreyfus's four categories of "intuitive thinking"—reliance on the fringe of consciousness, and so on—are each systematically ambiguous as between conscious states on the one hand, and complex functions of problem solving on the other. If these four categories cannot be applied to machines in their "conscious" interpretation, it does not follow that they cannot be so ascribed when understood as ways of processing information.)

Philosophers who have used the methods of conceptual analysis to inquire into the nature of emotions generally agree that emotions are *not* mere feelings, or bodily sensations, but contain a strong cognitive component relating to the background circumstances in which the emotion is experienced.[47] It is by virtue of this component, for instance, that some cases of emotional response can be regarded as "reasonable" or "unreasonable"—how could a mere *feeling* be either? To be sure, someone may have good reason to feel angry: but this is because "feeling angry" is a complex emotional response drawing on a complicated conceptual base in the mind of the angry person. (Dogs can feel angry too, it seems, which presupposes a background of canine knowledge and desires in terms of which to "make sense" of their anger.) In the terminology I have used in this book, feelings are *cues* to emotions, cues that are interpreted in relation to psychological schemata with varying cognitive content, such as pride, shame, vanity, and humiliation. Programmed theories of emotion therefore have to be able to access representations of these concepts. And if the program is to simulate the having of emotions, as opposed to the recognition of emotions in those with whom it communicates, these schemata must have procedural implications for the system's own behavior.

These matters may be represented implicitly and crudely, as Colby's neurotic program features raised *anxiety* or *self-esteem* according to the semantic content of the beliefs being processed, but cannot distinguish between moral shame and pricked vanity. However, we saw in Chapter 4 that a richer semantics would allow for a distinction between these two forms of lowered self-esteem, as also of others. In general, detailed conceptual analyses of emotions must articulate the presupposed knowledge, concepts, desires, abilities, hopes, imaginings, and so on, in terms of which the emotion enters into the psychological life of the person experiencing it. The thematic successions embodied in Abelson's "scripts" (such as *rescue, the worm turns,* or *the end of the honeymoon*) are closely bound up with emotional factors. For example, the succession of helplessness and aid that we think of as "rescue" has specific implications for the emotional history of the person rescued: it would be a strange maiden indeed who was elated on being abandoned to the dragon but despaired at the sight of her rescuer. Some of the behavioral implications of emotional concepts are incorporated into Abelson's DISPLAY schemata, mentioned in Chapter 11. It is because we have learnt the rich semantic interconnections between emotions and other psychological phenomena that we can understand *Romeo and Juliet* (though a Martian might not be able to), and respond appropriately to our friends in daily life.

In short, the distinction between two emotions lies in the psychological origins and context of the felt excitement and in its ensuing effects, rather than in the nature of any "pure feelings" experienced by the person.

One may even question whether *pure* feelings (feelings with no cognitive component) are often—or ever—experienced by us. The experiment described in Chapter 3 suggested that people's emotional experience may be phenomenologically altered by "reading in" concepts of anger or euphoria, much as a vertex may "look" different according to whether we experience it as a wedge cue or a cube cue. Those who assume that nothing of "the experience itself" could possibly be understood or communicated in such cognitive terms should remember Locke's example of the blind man who said he thought *red* must be something like the sound of a trumpet.[48] It is, isn't it? (At least, scarlet is; crimson, less so.) There are strong philosophical arguments for the position that the qualitative content of our conscious experience of red or blue is intelligible in publicly communicable terms.[49] According to these arguments, language expresses not only the form, or structure, of our experience but also its qualitative content (indeed, the distinction between the form and the content of experience is reinterpreted in terms of inseparable aspects of the underlying phenomenology that is shared by speakers of the language). If so, to say that a computer could have no real understanding of emotions—no matter how plausibly it used emotional *language*—on the ground that it supposedly cannot experience feelings, is to make a highly dubious claim.

So far, I have asked whether computers *could* simulate emotions; but one may also ask whether they *would* or *should* do so. Very briefly, certain types of program *would* involve emotional phenomena of various types, since some emotions—such as surprise, nervousness, confidence, agitation, hope, and despair—play a crucial role in the inner teleological control of intelligent systems acting in a rapidly changing and complex world.

For example, a certain degree of confidence is necessary if complex plans are to be formulated and carried out efficiently. By the same token, an anxious checking of matters sometimes taken for granted is often in order—though a high degree of anxiety may be obstructive if it leads to explicit examination of conditions that a more intelligent approach would not question. The computational beginnings of such psychological distinctions are to be seen in HACKER, which can run in CAREFUL mode under certain circumstances. Similarly, agitation may be partly expressible in terms of the use of breadth-first (as opposed to depth-first) search. Imagine a program that, like BUILD, can list the possible strategies at a given choice point, but that does not know which strategy will turn out best in the current situation. An optimistic plumping for one of the alternatives, followed by a steady depth-first search, would very likely miss the solution (assuming that time constraints do not allow depth-first search of all possibilities). But an agitated breadth-first hopping from one possible strategy to another would be more likely to find the solution—provided that it lay relatively near the origin of the search tree. In some cases, namely when the solution lies at a deep distance from the origin, agitation will be counterproductive, since it will prevent a developed search of any path. But then, humans often fail too.[50]

Other emotions, such as humiliation and guilt, rely on cultural norms and moral imperatives that vary greatly from one culture to another. The sense of humiliation suffered by a character in an eighteenth-century novel, or a Japanese film, may be barely intelligible to a twentieth-century Westerner.

If one takes seriously the suggestion (to be mentioned in Chapter 15) that future programs may be provided with which people can carry on "social conversation" in their own homes, the question arises whether any understanding of these matters *should* be built into such systems. Suppose a distraught and lonely widow refers to her recent bereavement: how is the program to respond? Equally, how is a "psychiatric" program like Colby's interviewer to communicate with an emotionally labile person? This is not merely a question of how much the program should *understand* about what is going on in its interlocutor's mind, but in what sense it should sympathetically (or angrily?) *respond* to inputs, as PARRY responds to human speakers in an "emotional" (if paranoid) way. It seems clear that if computers are to be used at all in personal contexts such as these, some representation of emotional matters would be in order. The difficulty of adequately representing the emotional subtleties concerned might be an overwhelming practical reason *against* the public use of programmed systems for purposes of psychiatric therapy or general chat. But this is not to say that it is in principle impossible to simulate emotional experience and understanding in a computational form. (We shall see in the next chapter that Joseph Weizenbaum,

ELIZA's creator, claims that we should not use computers in emotional contexts under any circumstances.)

Still other emotions, such as sexual desire (and hunger and thirst, if these are allowed to be "emotions" rather than "bodily feelings") are closely dependent on our organic embodiment and evolutionary history. (Though of course "cognitive" contributions are enormously important to sexual emotion too.) It is not clear what sense can be made of ascribing sexual desire to an inorganic artifact, though a program might be able to recognize that someone else was sexually aroused as we can interpret sexual meanings even in superficially innocent conversations.

In sum, the philosophical arguments most commonly directed against artificial intelligence are unconvincing. However, the epistemological issues involved are too obscure to allow one with a clear conscience to insist that *all* aspects of human thought could in principle be simulated by computational means, even though one may know of no specific reason to the contrary.

Still less should one assume that complete simulation is possible in practice. At various points throughout the book I have referred to phenomena whose complexity makes it highly unlikely that adequate simulation will ever in fact be achieved. This applies to "common-sense" knowledge and linguistic ability of the type attributed to the fictional HAL, not only to triumphs of the human imagination like superlative literary and scientific creativity. So HAL is likely to remain within the realm of fiction, even though there may be no impossibility of principle preventing his actualization.

Even if there are some aspects of the human mind that in principle cannot be represented in computational terms, the argument of the preceding section still stands. That is, the achievements of *current* machines suffice to make it intuitively less implausible than heretofore that mental phenomena may be grounded in a mechanistic physiological base, namely, the brain. So were the relevant research to come to a stop tomorrow, artificial intelligence could still be cited as an "existence proof" of the compatibility of humanist psychology (which stresses our subjectivity) and mechanist theories of the material world.

In the next, final, chapter, we shall see that this philosophical interpretation of artificial intelligence allows for the possibility that computational reseach might *offset* the dehumanizing social effects of natural science that are characteristic of urban-industrial cultures.

# 15

# *Social Significance*

PHRASES LIKE "The Computer Revolution" are common currency, not to say cliché. There is a vast literature on the potential effects of manual and clerical automation, and the use—or misuse—of computer science for military purposes.[1] Some of these essayists sketch apocalyptic scenarios in which malignant hobgoblins of a computational character surround us on every side. Others are more sanguine. But all mention projects that are more artificial than intelligent, that exploit the "brute force" of computers rather than knowledgeable information processing.

The issues of privacy, unemployment, leisure, and centralization of political power that are stressed by such writers are obviously raised also by the more intelligent applications of computer technology. For instance, a speech-understanding system that could continuously monitor people's "private" conversations could be more intrusive than any mere filing system or sharp-eyed shuffler of written documents. But there are additional matters associated with "intelligent" machines in particular. I shall not rehearse the more familiar topics in discussing the social influence of artificial intelligence, but shall concentrate instead on those potential effects that are peculiar to examples drawn from the intelligent end of the programming spectrum.

In the first section I specify some uses of artificial intelligence that are currently being mooted in the technical literature, each designed to improve the quality of life in some way. Next, I say a little about the potential effects of these futuristic proposals on a society's way of thinking about human individuals and the emotional life. Finally, I ask what precautions might be taken by the artificial intelligence community (as opposed to governmental and other social institutions) to minimize the dangers involved, including the risk of fostering insidiously dehumanizing attitudes in society at large.

## PROGRAMS IN PROSPECT

Faced with the threat of an idle hour at London Airport, J. E. Michie decided to question a random sample of her fellow passengers about the capabilities of computers.[2] Their answers showed a faith in the achievements of artificial intelligence that is decidedly misplaced. One-third of the people she buttonholed in the Departure Lounge believed that there already exists a chess program capable of beating Bobby Fischer, while virtually all were confident that such a program will be available before 1984.

We noted in Chapter 12, however, that there is no such program in sight. Some applications mentioned in this section are also of debatable practicality, at least in the short term. For instance, most of them assume a reasonable competence in natural language, within variously restricted domains. And many take for granted that the problems of organizing and quickly accessing very large data bases will have been broadly solved. Some even assume that robot perception, thought, and action will take place in *real* time, that is, quickly enough to cope intelligently with a rapidly changing environment demanding immediate response. The earlier chapters should have conveyed a sense of the difficulties, as well as the good auspices, involved in the projects currently being canvassed.

The range of ameliorative applications prophesied in the professional journals is less wide than in more popular forms of science fiction, but is pretty varied nonetheless. No one, to my knowledge, has yet proposed a programmed version of Buster Keaton's string-and-pulley "getting-up machine," which washed, dressed, and breakfasted him in three minutes flat, and made his bed into the bargain. But the seriously made proposals run from robot housecleaners, chauffeurs, and industrial workers, through ever-willing gameplayers and storytellers, to automatic teachers, physicians, legal justices, marriage counsellors, and literary critics.[3] In all these cases, the emphasis is on reasoned and flexible judgment on the program's part, as opposed to the mere storage and automatic regurgitation of isolated facts, or the repetitive performance of a fixed sequence of discriminations and movements.

For example, the computer diagnosticians do not simply store lists of symptom-diagnosis pairs, or spit out prescriptions for treatment in a blindly dogmatic (and apparently "objective") fashion. One prototype of these future programs is the MYCIN system.[4] MYCIN is an interactive program that simulates a medical consultant specializing in infectious diseases. It engages in question-and-answer conversations (lasting 20 minutes on average) with doctors needing specialist help. The physician asks MYCIN for advice on the identification of microorganisms and the prescription of antibiotic drugs, and also for explanations of its advice expressed at the appropriate level of detail.

MYCIN's explanatory capability is not put in merely for show. It enables the physician rationally to *reject* the program's advice if she disagrees with specific aspects of its clinical rationale. It helps nonspecialist doctors to learn more about the complexities of diagnosis and therapy in this class of diseases. And it allows human consultants to make general improvements in the program, by telling it about relevant knowledge that they realize in specific cases to be missing or inadequately stated. For example, it can be pointed out to MYCIN that such-and-such a drug should not be given to patients with a particular allergy or rare heart condition. (Not surprisingly, MYCIN has prompted clarification of the theoretical basis of medicine in this area, much as the development of DENDRAL forced chemists to become conscious of their own chemical intuitions.)

A program like MYCIN, then, is much more than a computerized table of bugs-and-drugs, and can be written only with the help of artificial intelligence techniques. Quite apart from its (rather restricted, PARRY-ish) natural-language understanding, its ability to explain itself on many different levels of detail implies a self-knowledge of its reasoning and goal structure that we have seen to be crucial to intelligent thinking. And its ability to learn by being told implies some mastery of the problems discussed in Chapter 10—though admittedly only for a very limited area of discourse. (The human consultant continually corrects its efforts at paraphrase of the new clinical rule until she is satisfied that MYCIN has "got it.") Again, in advising on patients with varying clinical histories, MYCIN can cope with all manner of pharmacological "ifs" and "buts" (and also "whys" and "wherefores"), and does not have to give a misleadingly straight answer to a straight question:

> Do you ever prescribe Carbenicillin for Pseudomonas infections?
> RELEVANT RULES: RULE064 RULE119 RULE120
> WHICH ONE DO YOU WISH TO SEE?
> Rule 064.
> IF: 1) IT IS KNOWN WITH CERTAINTY THAT THE IDENTITY OF THE ORGANISM IS PSEUDOMONAS AND

THEN: 2) THE DRUG UNDER CONSIDERATION IS GENTAMICIN
RECORD THE FOLLOWING AS A MORE APPROPRIATE THERAPY: GENTAMICIN-AND-CARBENICILLIN.

If the nonspecialist doctor then asks "Why?" MYCIN can give her a reasoned reply.

Many features of MYCIN would be embodied also in programmed legal arbiters, which would not only search for relevant legal precedents in the judicial literature (a far from trivial task), but would also offer legal advice. A system has already been built that to some extent simulates decisions of the United States Supreme Court.[5] As well as taking account of legal precedents, this program simulates the biases of the justices, as disclosed in their past decisions. Whether or not the fact were to be made as clear to users as it is in this toy system, a publicly available legal arbiter would have to embody moral-political attitudes in its subjective epistemological world. MYCIN does too, of course: but (barring Christian Scientists) we can all readily agree on the ethical priorities implicitly built into MYCIN. Human consent to the biases of the "liberal pointy-heads" on the Supreme Court, however, is less automatic.

Like MYCIN, legal arbiters will preferably be used to augment human judgment, rather than replace it. Accordingly, like MYCIN, they should when appropriate offer several (reasoned) alternative judgments, not just the one of which they are most confident. MYCIN's assessment of degrees of confidence is not a mere statistical probability measure. It takes into account psychological factors about the evidential relations of beliefs, factors that philosophers of science have considered in regard to "confirmation theory."[6] Legal programs, too, would have to incorporate more subtle concepts of *confidence* and *evidence* than Bayesian probability, in order to avoid judicial absurdities of various types.

The automatic tutors envisaged for the future are more sophisticated than today's Computer-Aided Instruction (CAI) systems, even those CAI systems that allow differential branching of the "syllabus" according to the mistakes and queries of the student. Like MYCIN and the Computer-Based Consultant described in Chapter 12, they will be able to initiate and answer questions at various levels of detail, according to the pupil's range of expertise. They will concentrate on asking probing questions enabling them to model the student's understanding of the topic, and will devise an individually tailored tutorial strategy designed to build on this understanding in fruitful ways.

Such programs presuppose an intelligently structured representation of the knowledge concerned, in its various aspects and degrees. Imagine, for instance, how Senator Goldwater's political beliefs would have to be explicitly embodied in an improved version of the IDEOLOGY MACHINE discussed in Chapter 4, if they were to be explained to people of varying levels of political sophistication. In addition, these programs would involve careful planning of the *order* of communications from the tutor, in light of the existing conceptual structure of the pupil. In discussing P. H. Winston's concept learner in Chapter 10, we saw that instructional order is crucial, and also that *counter*examples must be presented to the learner if she is to grasp important features of the conceptual structure being communicated. The SCHOLAR program cited in Chapter 11 attempts to instruct its pupils by way of a Socratic method of example and counterexample.

Clearly, then, the development of automatic tutors will go hand-in-hand with increasing psychological appreciation of the way in which human students build, and progressively modify, internal representations of concepts and skills. The previous discussion of HACKER suggested that the notion of *debugging* one's own thought procedures is crucial. An automatic tutor, like its human counterpart, should not merely point out the pupil's mistake: it should explain why the pupil made the mistake and how she can adjust her methods of thinking in order to avoid repeating it. A preliminary example of a program providing this sort of constructive criticism is MYCROFT, mentioned in Chapter 10.

Seymour Papert, the originator of the LOGO-turtles whose drawings of people and houses are criticized by MYCROFT, foresees a marked increase in the use of programming in elementary and secondary schools.[7] Most school-based programming courses offered at present are conceived of as providing an extra tool for the children's use: reading, writing, arithmetic—and programming. To the extent that this practice becomes more widespread, it may help to avert the growth of a socially divisive "computer elite," a small group of people whose members are the only ones to understand computation and so the only ones not to feel alienated in the computerized society of the future. But Papert's interest is more specific, and more exciting. As the descriptions of HACKER and MYCROFT suggest, the activity of programming and debugging can help one to think about one's own thinking in a generally fruitful fashion. The notion of "bug" itself is a powerful concept in terms of which to explore one's cognitive skills, and (unlike "ability") a hopeful one, insofar as it suggests that one look for a hitch in a basically intelligent thinking procedure that can be located and fixed. The underlying lessons learnt in the programming exercises recommended by Papert, as well as those learnt from automatic tutors holding verbal conversations by teletype, could thus help students of all ages spontaneously to improve their own understanding of mathematics, science, and the arts.

Computer art—whether graphic, verbal, or musical—has already crept into our exhibition-halls.[8] So far, most of it is pretty *un*intelligent, and the "literary" endeavors justifiably invite this sort of mockery:[9]

LITERARY CRITICISM

| Search by number | |
|---|---|
| find and gloss: | six proud walkers |
| | five days of Christmas |
| | four and twenty blackbirds |
| | three blind mice. Go. |
| three blind mice: | one noun phrase |
| two blind mice: | one noun phrase |
| one blind mice: | please repeat |
| one blind mice: | stop breaks rule |
| | align number indicators |
| one blinds mice: | stop breaks rule. No. |
| | Go obeys other rule |
| two blinds mice: | stop breaks rule |
| | align number indicators |
| one blinds mice: | who blinds mice? |
| the farmer's wife: | one noun phrase |
| she blinds mice: | that is aggression |
| see how they run: | imperative clause |
| see how they run: | fear-behavior |
| see how they run: | using legs |
| the farmer's wife: | one noun phrase |
| cut off their tails: | more aggression |
| with a carving knife: | divergent thinking |
| did you ever see: | interrogative |
| did you ever see: | listener involvement |
| did you ever see: | listener not blind |
| did you ever see: | listener not mice |
| such a thing in your life: | listener alive |
| as three blind mice: | are they symbols? |
| four blind mice: | plight of humanity |
| five blind mice: | population explosion |
| six blind mice: | out of control |
| seven blind mice: | so am I |
| seven blind mice: | infinite loop |
| | see how I run |
| | see how I run |

We have seen that even simple children's stories about pencils and piggy banks involve a more knowledgeable psychological sensitivity than this, never mind "symbols of the plight of humanity." But we have seen also, in discussing the programmed detective novelist, that a degree of literary talent is not in principle denied to artificial systems.

Programmed musical (as opposed to literary) compositions are often acceptable, even pleasant, rather than simply absurd. Occasionally they are exciting on first hearing, though they tend soon to appear banal on repetition. A more developed musical intelligence could improve on current efforts. Those who assume such matters to be wholly beyond the reach of a computational approach should ponder two programs written by H. C. Longuet-Higgins and M. J. Steedman.[10] They were interested in the introspectively obscure process of musical interpretation that is tacitly carried out by a listener who hears an unknown melody, and "intuitively" assigns its meter and key so that she could write it down correctly in musical notation.

That this is a matter for musical *intelligence* is suggested by the fact that there is more to it than just finding the correct note lengths and the correct positions of notes on the keyboard. For instance, starting with a middle-*A* crotchet, one could write *God Save the Queen* either in 4/4 time in the key of *B*-flat, or in 3/4 time in the key of *A*. It is intuitively obvious even to a musical novice that the second form is correct and the first inappropriate. But, as in comparable cases of visual or linguistic interpretation, it is no easy matter to state precisely why this is so, or to explain how it is possible for the correct meter and key to be established on first hearing. The two programs in question mirror the progressive character of musical comprehension, in that ideas about meter and key become more definite as the melody proceeds, and may crystallize well before the end. Given the task of transcribing the fugue subjects of Bach's *Well Tempered Clavier*, the harmonic program finds every key sign and notates every accidental correctly. The metric program is not quite so successful, since it cannot deal with cases where all notes and rests are of equal duration (six out of the 48 fugues), and it makes mistakes in six other cases (in four of which a human musician might commit the same error).

Bach's *Musical Offering* was composed at the request of Frederick the Great, being a set of 13 variations on a theme created by Frederick himself. Future kings, and even future commoners, may be able to present a personal theme for elaboration to a musical program. If such a service is to be available to commoners and kings alike, there will have to be public access to a computer loaded with the relevant program. Indeed, this would be required for very many of the predicted social applications of artificial intelligence. Individual tuition; medical consultation; voting and participating in public political debate; requesting information on legal, civil rights, meteorological, and other matters; playing a game of Monopoly or Scrabble; shopping by remote control—all these, and many more, presuppose that the person have access to a computer in a nearby social center, if not in her own home.

John McCarthy has forecast the widespread use of *home information terminals*.[11] These are domestic teletype machines (often with sound, graphic, and television capabilities as well) linked by telephone or television cable to a national network of computers functioning on a "time-sharing" basis. A time-sharing system is one with which many users can communicate at one and the same time, there being no necessity to wait until all one's neighbors are "off the line," and no fear of inadvertently getting hooked up to someone else's program (imagine trying to order the weekly groceries from PARRY). This is possible because the time taken by the computer in dealing with a single instruction is infinitesimal in comparison with the time it takes a person to think it up or type it out. So, much as a chess master "simultaneously" plays against 40 lesser champions ranged around the chess hall (and could even cope also with some games of checkers and noughts-and-crosses thrown in for good measure), the computer stores a queue of instructions from its various users, and while someone is still reacting to its response to her previous communication the computer deals with inputs (probably accessing different programs) sent in by many other people.

A speech-understanding program would allow the user to speak her requests instead of typing them. At present, this can be reliably done only by using a very small and deliberately distinct vocabulary, with unnatural pauses between the words. This is because (as in the interpretation of visual scenes) the continuous stream of normal speech has to be sensibly segmented into individual words before it can be understood—but, as you would expect from the discussion of Part IV, it has first to be understood in order to be segmented. Just as the physical absence of a line in a picture may *not* correspond to the absence of a significant feature in the equivalent area of the scene, so an auditory pause in speech may *not* correspond to a semantically significant distinction. Even the individual sounds, or phonemes, cannot be distinguished and identified on purely auditory grounds, but only by reference to the wider linguistic context. But whether or not people will ever be able to speak to programs in the casual way in which the astronauts spoke to HAL (who could even *lip-read*), it should become possible to have rather more natural vocal communication with machines, especially if the universe of discourse is fairly narrow. The "HEARSAY" system currently being developed is a step in this direction.[12]

According to some forecasts, the stationary home terminal will eventually be supplemented by a domestic robot, a motile hand-eye machine linked by remote control to the same central computer system, and accessing the programs stored therein so as to do household jobs of various kinds.[13] It should not be necessary to clear the floor so that the robot can vacuum it: the robot itself will supposedly be able to recognize the objects littering the room and put them in their proper place—or, by default, in one corner. The discussion in Chapter 10 of J. M. Tenenbaum's program suggested that SHAKEY might fairly soon be able to recognize a wastepaper bin wrongly placed by the window, or even high up on the table, and to reposition it by the desk. But *only* if the room is otherwise unnaturally tidy, and *only* if—as in the SRI offices—the bin, window, table, and desk are of uniform type. Recognizing my three-year-old daughter's teddy-bear and other cuddly toys, which may have been lovingly deposited by her in the "house" represented by the bin, is another matter. Even if one assumes that each toy is individually shown to the robot in training sessions, much as Tenenbaum teaches his program how to recognize regulation tables and telephones, it does not follow that the robot would be of much use in any house untidy enough to need it. Before such applications are possible, enormous advances in visual programs will have to be made, since current ones are quickly overwhelmed by visual scenes of a richness far below that coped with by pets and people.

Industrial and agricultural robots are less fanciful. Robots relying on computational techniques developed in artificial intelligence would not have a fixed sensorimotor capability, like the automated machines of the present day.[14] Instead, they would be flexible in operation, and could learn *new* tasks. They would learn not by being laboriously reprogrammed (a finicking job that at present takes days even for very slight adjustments in performance), but by being shown examples of (for instance) new machine parts and machine tools, and by being given an outline sketch of the desired procedure that they would then elaborate in detail. The sort of planning facilities discussed in Chapter 12 would be crucial here. Visual recognition would often be needed: but in a given industrial setting (allowing special lighting conditions if required) the visual variety might be reasonably small, so that the robot not only knew what to look for but could rely on there not being too much extra clobber around. And many industrial robots could be stationary hand-eye (or hand-touch) systems, thus avoiding the computational and engineering problems involved in building a robot that can move about the place without crashing through the window or trampling on the cat.

There are already programmed hand-eye systems that can assemble simple machines from components they learn to recognize by being shown examples, or that (like COPY-DEMO) can build visually demonstrated structures out of a "warehouse" of familiar parts.[15] And Donald Michie has described a Japanese prototype hand-eye device that "inspects moulds for concrete poles as they pass on the belt, finds the bolts, confirms their location by tactile sensing and then tightens the bolts with an impact wrench," other protuberances being intelligently avoided.[16] Moreover, a robot linked to the Computer-Based Consultant that knows about pump-assembly (which uses the NOAH planning program described in Chapter 12) would be able to assemble the pump even if someone else had left the job half-finished, or had wrongly positioned some of the parts. This type of behavior is very different from that of the familiar "conveyor-belt" computer-controlled device, which can get along fine provided that everything is in the right place at the right time, but would be hopelessly lost in more variable environments.

Each of the examples I have cited is obviously a well-intentioned proposal. Whether they would turn out to be beneficial in practice is not so obvious. The fable of the fisherman's three wishes is less a pretty fairy tale than a sour comment on the human condition. If you doubt this, you should know that at a recent symposium on work aimed at providing an artificial means of vision for the blind, one of the blind research symposiasts seriously questioned the need for such an "advance."[17] To some extent, he was merely stressing that "there are other ways of doing things," so that to a rehabilitated blind person a device transmitting visual information would be largely superfluous. But there was a suggestion, too, that a blind person who had painfully made certain adjustments in living would be deeply troubled by pressures to modify them later in life. And the bitter history of the blind man whose sight was restored in an operation, but whose will to live was tragically destroyed, shows only too well that the most "desirable" scientific application can have grossly unfortunate psychological effects.[18]

It is time, then, to ask what might be the wider implications of artificial intelligence for the social relations of people who at present have "other ways of doing things."

## SELVES AND SOCIETY

The social impact of artificial intelligence is not coterminous with that of computers. For many computer applications utilize basically stupid programs, programs that either rely on quick thinking and faultless memory to do superficially complicated jobs that humans cannot do so speedily, or which do simple jobs that people find boring or dangerous.

While it is clear that stupid programs may (like the stirrup and the steam engine) have revolutionary social effects, in themselves they do not pose so great a challenge to socially accepted ways of thinking about human individuals as more intelligent programs would. Thus you have doubtless read in your newspaper about the computer bureaucrat that sent out a gas bill for £0 0p (or its transatlantic cousin, which demanded the dollar equivalent), and duly cut off the supply when the householder failed to send a cheque in response. The snide, not to say gleeful, tone of such reports betrays the fact that they are to be interpreted as episodes in a battle between David and Goliath, with the human comfortingly cast as the final victor.

But with the advent of more intelligent programs, doubt may arise as to whether we are really fit to play a heroic role. So before turning to the good effects that artificial intelligence might have on our corporate self-image, we must first ask how it might change society for the worse by undermining valuable aspects of our current view of ourselves.

One of Michie's grounded voyagers told her firmly that she opposed financial support for research in chess programming, saying that "Man is over-mechanised already." Probably, she shared the widespread suspicion of artificial intelligence that was described in the previous chapter, and that has been voiced at length by critics of urban industrialism such as Theodore Roszak and Herbert Marcuse.[19] Given the currency of this attitude in the populace at large, even better and better *chess* programs could have unfortunate social side effects.

For instance, consider this complaint from the psychologist Rollo May, whose prime professional concern is with therapeutic and counselling psychology:[20]

> I take very seriously . . . the dehumanizing dangers in our tendency in modern science to make man over into the image of the machine, into the image of the techniques by which we study him. . . . A central core of modern man's "neurosis" is the undermining of his experience of himself as responsible, the sapping of his willing and decision.

This lament from a psychologist constantly in touch with laymen seeking practical help or advice cannot be lightly brushed aside, or passed off as merely a dry theoretical inference from his underlying existentialist philosophy. He is speaking of the perceived influence of "mechanistic" science in general, and of behaviorist and physiological approaches to psychology in particular: a fortiori, his remarks would apply to artificial intelligence (and in the next section we shall see that Roszak, writing more recently than May, specifically draws this parallel). May sees such sciences, in their theoretical and technological expressions, as having a pernicious effect on the everyday life of his clients.

May's point is that moral and pragmatic choices made without confidence in their possible relevance to eventual action are unlikely to be effective—or perhaps even to be made at all. For if a person's self-image represents herself to herself as an autonomous purposive creature capable of pursuing certain ends, then it can be used to generate choices and guide her action accordingly. Even in machines, as we have seen with respect to SHRDLU, HACKER, and BUILD, the internal representation of the possible modes of action that are available to the system can be crucial in directing the performance actually carried out. But if a "depersonalization" of the self-image occurs (for whatever reason) so that the self is no longer seen as a truly purposive system, then relatively *in*human, "pathological" behavior can be expected in consequence. Humanist psychiatrists like R. D. Laing have described how a degenerate representation of the self mediates rigid and restricted lifestyles.[21] This "machinelike" behavior is a travesty of human potential, even though it is a genuinely meaningful phenomenon that must be interpreted in intentional terms.

This type of degenerate self-model is encouraged by artificial intelligence in general, given the popular (though mistaken) assumption of total incompatibility between mechanism and humanism. What May calls "the undermining of one's experience of oneself as responsible" can therefore be exacerbated even by the development of clever chess programs.

As well as having immediate consequences in one's personal life and human relations, this sapping of willpower can have widespread implications bearing on social phenomena of various kinds. For example, the political institution of participatory democracy assumes an ascription of responsibility to individuals that fits ill with the dehumanized image remarked by May. Consequently, providing citizens with home terminals, with the partial aim of enabling them to vote and express their political views without leaving their fireside, might subtly undermine their sense of civic responsibility. This is the reverse of the intended effect, which is to encourage individual citizens to engage more fully in democratic government at local and national level, thereby lessening the common feeling of alienated helplessness with respect to the governmental process.

Cynics—or realists—might point out in this connection that to *feel* less a cog is not necessarily to *be* any less a cog. The home terminal might function as a subtle form of social control, damping down dissent by contenting people with an illusory sense of political participation. And who imagines that the politicians would not want to control the content of the "information" in the relevant tutorial programs? A nationwide course in civics, or even several alternatives presenting the "competing" political philosophies of the established parliamentary factions, could effectively brainwash citizens into highly tendentious assumptions about the conceivable range of political alternatives. Many of the dangers inherent in use of the home terminal might be lessened if there were several computing services available rather than one only: people could then dial their preference, and pay for the time used. But whether the content of the alternatives available were to be regulated commercially or governmentally, formally or informally, there would be insidious (and largely unrecognized) social effects due to the mass communication of implicit and unquestioned assumptions much as there are with respect to press and broadcasting today. If it were more difficult for a dissenting group to make their views known via the computer networks than to publish a weekly broadsheet (for instance, because of the expense of programming, or the greater centralization of control over the medium), then these social effects would be correspondingly greater.

While any intelligent program may have a dehumanizing effect on people who see an unbridgeable metaphysical gulf between themselves and machines, some applications of artificial intelligence would be especially open to this criticism. For example, consider K. M. Colby's recommendation that his automatic interviewer cited in Chapter 3 (or an improved version thereof) be used to aid the diagnosis, and perhaps even the treatment, of psychiatric patients. This suggestion is one that humanist psychiatrists are hardly likely to endorse. Even Joseph Weizenbaum, the creator of ELIZA, has bitterly denounced the "obscene" idea of employing programs in clinical situations. For Weizenbaum, the question is not what a computer *could* do, but what it *should* be used to do. Claiming that computer science has brought us to the beginning of "a major crisis in the mental life of our civilization," he compares those who are impressed by artificial analogies of the mind to a psychoanalytic patient, who positively embraces a tentatively proffered and "profoundly humiliating" self-image, and makes of it a new foundation for her life.[22] And Laing has complained of the alienating influence of the average *human* psychotherapist, who views her victim less as another human being with idiosyncratic subjective interests than as a classifiable "case" to be "cured" by the therapist's technical expertise. Laing generalizes from the psychiatric context to society at large, castigating the spiritual poverty of communities in which there is little communication or communion to be found.[23] The notion that one could communicate fruitfully with a computer about personal matters Laing would consider not just mistaken, but dangerously mad.

Weizenbaum's horror concerning programmed psychotherapy has elicited from Colby the retort that it is "dehumanizing" to herd thousands of patients into understaffed mental hospitals where they will hardly ever see a doctor, so we have no choice but to use computers if they turn out to be helpful.[24] (This retort seems to assume that using computers is the *only* alternative to the admittedly inhuman status quo.) There is already some evidence that psychiatric patients accept, and even prefer, interaction with an *im*personal agent when anxiety-ridden personal topics are being identified.[25] This evidence relates to "mechanically" conceived interviewing programs that do not attempt plausible simulation of a therapist's conversation, as ELIZA and Colby's system do, so there is little risk of any nondelusive person's being misled into confusing them with a human doctor. Nevertheless, Laing would view them as inherently alienating in effect. He would argue that the preference just cited itself expresses the alienation of people who can't stand people, and prefer the company of machines (even dogs would make more convivial companions). Nor would he be impressed by the defense that programs are best used only as an initial guide to diagnosis, the therapy being in human hands.

Clearly, such disputes raise deep questions about the nature and aims of psychotherapy, and the significance of mental hospitals as social institutions. Many of the more orthodox clinicians might support Colby, particularly if they realized (what was mentioned in Chapter 3) that the idiosyncrasy of people's world views need not be denied by a computational approach. But psychiatrists like Laing and T. S. Szasz, who dispute the validity of regarding psychiatry as a *medical* discipline and the asylum as a *medical* institution, would scarcely be bowled over by the no-nonsense cry, "If it helps you find the diagnosis, then use it!"[26]

In indisputably medical contexts dealing with basically physical illness, artificial intelligence may be welcomed with less restraint. The MYCIN system has not yet been tried out in a clinical situation, though the authors report a certain resistance on the part of clinicians to the idea of using programs. As yet, MYCIN gives the same advice as a human expert in 75 percent of cases. The authors intend to wait until it has achieved a 90 percent match before introducing it experimentally to a hospital; only then will they be able to see who uses it, how often, and what effect it has on the prescribing practices of doctors and the clinical status of their patients.

But a much simpler program for the diagnosis of peptic ulcers has been tried out on patients, who very rarely object to and often claim to prefer this diagnostic method.[27] Ironically, as justification of their preference, they spontaneously describe the machine (with which they communicate by teletype) as more friendly, polite, relaxing, and comprehensible than the average physician of their experience. The machine never gets tired and irritable, has limitless patience, and perfect manners. Perhaps equally important, there are no intimidating audiovisual cues of social class differences between doctor and patient, nor any indication of the doctor's sex. (Nor are there audiovisual cues of emotions, like those simulated in the output of Colby's neurotic program. An uneasy fidgeting, or averting of the eyes, can speak volumes in psychiatric contexts, so that a mere teletyping program would be more hampered in diagnosing psychological attitudes than in spotting physical illness.)

These chastening observations about the superiority of the personal habits of programs over those of human doctors can doubtless be expected also with regard to automatic lawyers, bureaucrats, and teachers. (The example of PARRY shows that a program *can* be as evasive and abusive as anyone; but only a practical joker of a very curious kind would link a paranoid lawyer or teacher to home terminals throughout the land.) While perhaps appreciated in isolated interactions of a tedious, technical, or embarrassing nature, this imperturbable mode might come to be consciously spurned in human relations in general, with a consequent emphasis on emotional spontaneity. Or it might not: in this case as in so many others, the potential influence of artificial intelligence is ambiguous. So much depends on the background human context, including the varied psychological prejudices and philosophical assumptions of different groups and individuals. Accordingly, the opposite effect might result: much as pseudoscientific jargon has already crept into the daily use of language in situations where it is experienced by the more sensitive as a personal affront, so the blandness of one's guest computer might come to be emulated in one's own mode of expression—or that of one's children.

The possible ill effect on children was cited as a disadvantage of the domestic robot, in a survey of the social implications of artificial intelligence that was carried out by questioning professionals in the field.[28] Parents who provide pets as an inducement to attitudes of caring sympathy in their children might want to think twice before introducing a robot into the home: who can be sure that it would be experienced by the child merely as a "clockwork cleaner," with no more effect on her human sympathies than a clockwork mouse? Clockwork mice, one should note, do not indulge in activities that fill much of the day of many people.

In general, when assessing the impact of the home robot and the home terminal, one has to consider the insidiously dehumanizing effects of people's becoming decreasingly dependent upon direct human contact with their fellows for satisfaction of their various needs. In any highly mechanized society, consumption of the goods produced may become less convivial, as technology enables us to do more things without leaving the house and mixing with other people (for instance, drawing water or being entertained).[29] Thus far, *production* of goods has become more socialized than in past times when the typical productive unit was the "cottage-industry"; but many people who today can only do their jobs by going to a particular place of work, might tomorrow be able to stay at home and communicate with their clients and coworkers via the home terminal. The socially isolating influence often attributed to television is as nothing to the alienation and loneliness that might result from overenthusiastic reliance on the home terminal and associated gadgetry. That this is no mere hobgoblin with which to tease the imagination, but a practical difficulty that may have to be faced, is underlined by a recent professional forecast that "it may be possible for intelligent machines of the future to supply not only intellectual stimulation or instruction, but also domestic and health care, social conversation, entertainment, companionship, and even physical gratification."[30]

Anyone who wants a rationalization for feeling depressed can surely find one in the quotation just cited. Let us turn to the brighter side of the coin, and ask whether social applications of artificial intelligence might have any welcome effects on the way people view themselves and other people.

We have seen already that computational models of intelligence are in fact markedly more human than the behaviorist models of mankind that have been widely accepted for many years. Noam Chomsky has argued with some force that the assumption of broadly behaviorist models by the technocracy has had highly unfortunate political effects.[31] In the next section I shall return to this point, in the context of Roszak's cultural critique of science and technology. If this commonly unsuspected "humanizing" feature of artificial intelligence can be brought home to the general public, then many of the ill effects I have so far hypothesized in this section will be allayed.

In addition to the generally benign influence of artificial intelligence, based in its endorsement of humanists' emphases on the idiosyncrasy of people's subjective world views and on the directive role of the self-image, more specific benefits on the social image of humanity might follow on the introduction of publicly accessible programs like those previously described.

For example, educational methods based on the pedagogical philosophy of LOGO-turtles might well change a child's (or adult's) way of thinking about "failure." Instead of the passively defeatist "I'm not good at this," the child would say "How can I make myself better at it?" This attitude is encouraged by the computational way of thinking about thinking, with its emphasis on the creative interrelation of many different procedures, and on the unintended effects of specifiable bugs in basically well-conceived attempts to achieve one's goal. So-called "failure" is really mistake; and a mistake is a miss-taking of the procedural nature of the problem that can be explained and rectified. By contrast, constructive self-correction is not encouraged by a conception of intelligence that views it as the product of a number of mysterious monolithic "talents" or abilities, which one either has or lacks, willy-nilly. This latter conception of the matter is all too common today, with a consequent cost in fulfilment of human potential that is unacceptably high. Weizenbaum's suggestion that the computational metaphor is "profoundly humiliating" must therefore be treated with some reserve.

Whether or not one believes that computers could in principle be rendered "emotional" to the same degree as they can be made "purposive" or "rational" (a question discussed in the previous chapter), it is a fairly safe assumption that publicly available programs will not in practice be rendered emotional (with or without scare-quotes), since there would be little point in doing so. The program intended to provide "social conversation" that was forecast earlier is possibly an exception, but it is not at all clear what sort of conversation the quoted author had in mind. Nor should one assume that programs will have no appreciation or understanding of emotions. A computer psychiatrist or marriage-counsellor would have to have some internal representation of human emotions, even if these phenomena were never "plugged in" to its own thinking. Similarly, a storyteller of even the simplest kind, provided that it was not a mere unintelligent tape-recorder, would need some knowledge about the psychological themes and scripts mentioned in Chapters 4 and 11, with regard to R. P. Abelson's work. But there would be little point in making PARRY, or even his "normal" emotionally labile cousin, accessible to users of the home terminal. (Such programs might be developed for psychological research purposes, as was PARRY himself, but that is a different matter.)

Consequently, it is probable that there will be a clear distinction between programs and people on the dimension of emotionality. What effects might this observable difference have on society? Earlier, I suggested that interaction with blandly amiable automatic tutors, lawyers, and doctors, might induce a reaction of emotional spontaneity in exclusively human encounters, with a corresponding contempt for the machine's more restricted psychological mode. Such a contempt is evidenced in the following interchange between a person and Colby's machine therapist (the person speaks first):[32]

> I don't understand.
> DO YOU WANT TO UNDERSTAND.
> I hate you you stupid machine.
> WHAT DON'T YOU LIKE ABOUT ME.
> You have no feelings.
> HOW DO YOU MEAN.
> I mean you cannot experience what I experience because you are just a machine.
> TELL ME MORE ABOUT SUCH FEELINGS.
> We are not communicating.
> I AM TRYING MY BEST TO COMMUNICATE WITH YOU.
> Are you really.
> HOW DO YOU SEE IT.
> What do you mean.
> JUST TRY TO SAY WHAT IT MEANS TO YOU.
> I think you are all goofed up.

It is somewhat inconsistent, of course, to go to the trouble of sarcastically insulting the machine, much as there is little point in swearing at one's car, except insofar as it relieves one's own frustration. But to insult a *person* who represses emotion is an act of more social significance.

There are many who would welcome more emphasis on the emotional life than is considered proper in Western societies today. One of Roszak's complaints about urban industrial culture is its undervaluation of the emotional aspects of the mind. To the extent that people still felt a need to draw a line between themselves and "machines," and to the extent that they could observe these significant psychological differences, their valuation of the emotional life would very likely be increased. (I am assuming that they will continue to value people more than machines, an assumption that might strike Roszak as overoptimistic.) Corresponding effects on broad cultural mores and the conduct of personal relations would follow. For example, the sort of "sexual liberation" recommended by Reich or Marcuse might be brought nearer, and the broadly "religious" attitudes commended by Roszak might be fostered. People who are content with our current mores would not relish such developments, seeing them (for instance) as whimsical self-indulgences largely irrelevant to the serious matters of life, like fairy stories. But others would point out that fairy stories, especially those that we call "myths," have a great deal to do with "the serious matters of life."

Incorporated in the self-image of most Westerners is the Protestant Ethic that only hard work and *doing* things are really serious activities. Moreover, in industrial societies "work" is implicitly defined as paid, as done in an employer's time rather than in one's own "leisure" time, and at the employer's behest rather than for one's own purposes. Hence the confusion over whether housewifery is work, or whether self-directed and creative ways of earning one's bread (such as writing novels or throwing pots) are "real" work.

Given these views, massive unemployment could be more soul-destroying than the most repetitive of factory jobs—even though people's "leisure" time for creative hobbies (*sic*) would be correspondingly increased. Although a shortening of the labor day may not reduce work in the same proportion (since it may become more intensive), certain types of unskilled and semiskilled work might be largely taken over by machines. Were the time currently spent on drudgery to be devoted to varied, creative, and autonomous work that is satisfying in itself (a state of affairs that presupposes corresponding changes in the social definition of work), unemployment would not result and people would be more fulfilled in their working lives. Otherwise, and to the extent that socially accepted ways of thinking about "work" did not change quickly enough, communities influenced by the Protestant Ethic would experience a deep social malaise. For if a man has been brought up to see himself as an industrious provider for his family and the community, enforced "leisure" may not only bore him (a boredom that could be partly alleviated by the home terminal), but may cause him destructively to define himself to himself as a social parasite.

This is less likely to happen to women, who—even if they also feel that they have a duty to "use" much of their time in contributing directly to the economy—enjoy an internalized acceptance of emotional values that enables them to draw considerable fulfillment from "mere" personal relations with family and friends. Correspondingly, they are able to express their emotional reactions with a greater freedom than men, particularly in situations outside the immediate family circle. As menfolk have more time free from jobs increasingly relegated to computers, they will not only be able to "do a good day's work" in the (currently predominantly female) welfare professions, and diligently "improve their minds" via tutorials on the home terminal, but will also have enormously increased opportunities for human interaction, with friends as well as family, which at present tend to be regarded as secondary to the masculine role. To the extent that there is a radical shift in the evaluation of emotional life, based primarily on economic change in working hours and the sexual division of labor but secondarily on the "inhuman" unemotionality even of intelligent machines, we may therefore expect deep and wide-ranging changes in social definitions of sexual roles.

It is impossible to say what the computerized society of the future will be like, except that it will be very different from our own, not least in the way in which men and women think about themselves and their fellows. (Always provided, of course, that charming toys such as the conveniently cheap TERCOM cruise missiles have not wiped us all out first. TERCOM is an acronym for TERrain COntour Matching, and TERCOM missiles use an ingenious map-reading guidance technique that is expected soon to give an accuracy of about 30 feet over a range well in excess of 2,000 miles.) To a large extent, what it will be like depends on what we want it to be like, and especially on how much trouble we are prepared to take to ensure that the various evil genies are kept firmly inside the bottle. Let us ask, then, what the artificial intelligence professionals can do to enable the moral and political consensus of society to guide future developments in a humane and sensible manner.

## PRECAUTIONARY MEASURES

Members of the artificial intelligence community bear an ominous resemblance to Mickey Mouse, which amiable rodent was cast in the cinematic role of the Sorcerer's Apprentice. The apprentice learnt just enough magic from his master to save himself the trouble of performing an onerous task, but not quite enough to stop the spellbound buckets and brooms from flooding the castle. Had the wizard not returned in the nick of time, disaster would have ensued.

Passing from fiction to fact, it is the programming community themselves who must be sufficiently in command of their professional magic to avert catastrophes. There is no one else who commands the necessary wizardry to retrieve a situation at the eleventh hour, except perhaps by ordering a total halt in use of a given program. Governmental or other social institutions may advise on what effects should be avoided; but if things go wrong, so that the program needs to be adjusted, individual programmers will have the responsibility of doing so. Before asking more generally how computer scientists should view their professional responsibilities to society, therefore, I shall first ask whether there are any *ways of writing* programs that would tend to keep control in human hands.

The sorcerer, and even his hapless apprentice, could *see* what was going on and knew exactly what needed to be stopped. Programmers who wish to be in the same position, should anything go amiss, must take positive steps beforehand to allow for this. That is, their programs should be intelligible and explicit, so that "what is going on" is not buried in the code or implicitly embodied in procedures whose aim and effect are obscure.

Programs should be generously commented, so that what a procedure is *supposed* to be doing (and why) is readily visible. The importance of this for aiding debugging (whether by a human or an automatic programmer) is evident from HACKER's use of the comments attached to its performance programs. It is only because HACKER has such a good idea of what it is trying to do, and how it is trying to do it, that it can self-critically modify its own procedures. Similarly, MYCROFT can correct other people's programs only if it is told, or can work out, what the overall aim and general strategy were. Since the most knowledgeable and intelligent programs will have to be able to learn for themselves, rather than having everything "spoonfed" to them by a person, it is crucial that their internal bookkeeping make a clear record of what they have done, and why. Otherwise, even their own programmer (who in any event may be retired, or dead) might not be able to understand their cognitive structure sufficiently well to modify them appropriately. (Compare Colby's difficulties in affecting the neurotic attitudes of his human patients during psychotherapy, which initially prompted his programmed model of neurosis.)

In discussing the difference between heuristic and epistemological adequacy, we have seen that intelligibility and explicitness are to some extent opposed. Similarly, programs get less readable as they approach the machine code level. What counts as "machine code" is likely to become somewhat more intelligible to human beings, since instructions that currently have to be programmed may (to save time) be "hardwired" into the electronics of the machine. For instance, there is already a hardwired version of the push-down stack, the programmed form of which was outlined in Chapter 12. Machine codes that are not restricted to such elementary operations as *add* and *shift* will aid readability at the basic level. But hardwiring should be relied on only when the problem will not demand a flexibility in the machine's behavior that would be hindered by built-in procedural assumptions. For example, a chess program exists in which operations like testing the piece on the square (which normally would require several instructions) is carried out much more quickly by one hardware instruction:[33] while this ability is useful for chess, it would be useless for playing checkers or Diplomacy.

To economize on computation time, the next best thing to hardwiring is compiling. So the more complex the program, the more likely that much of the high-to-low-level transformation will be "automatic," or compiled. Programs in LISP, for instance, can be run up to 100 times as fast in compiled as in interpreted form (and use less machine space into the bargain). The reason was mentioned in the first chapter: whereas a compiler program translates the original high level program into machine code and then effectively throws the original version away, an interpreter translates from the original "on the spot." Consequently, an interpreter can continually refer to the intentions expressed at the higher level while it is translating, so as intelligently to guide the giving of detailed instructions to the machine in light of current circumstances. But a compiler program cannot.

It follows that it would in general be advantageous to have the possibility of switching from compiled to interpreted mode if necessary, and the programmer should specify the sorts of contexts in which this switch might be advisable. You will remember that HACKER, when in doubt, can switch into the slower CAREFUL mode in which every step is examined before it is taken, and specific criteria calling for CAREFULness were included in the program by its designer. This switch is analogous to a shift from compiled to interpreted running. M. L. Minsky has compared this shift to the appearance of consciousness at points of difficulty in human thinking.[34] Many psychologists have noted that introspective awareness of one's mental operations arises only when thought or action is not running "smoothly." Learning a physical skill has been explained in terms of the physiological "compiling" of the relevant motor schema; but adjusting the skill takes conscious effort and careful examination of individual steps.[35] Once a procedure in a program is compiled, its flexibility is limited unless special measures are taken by the programmer to allow higher level routines to monitor and modify its running.

In general, it is easier to see what is going on if subroutines are written so that they can be easily "got out" from the program as a whole. This is the principle of modular programming, which is exemplified in S. E. Fahlman's heterarchical BUILD. In dealing with human committees or bureaucratic institutions, the ascription of responsibility and the righting of error are easier if each person has a clearly defined job to do, and clearly specified channels of information (for advice, complaint, or direction) between herself and other people. Analogously, the understanding and improvement of BUILD is facilitated by the clear distinction between intercommunicating experts like PLACE, GETRIDOF, and TRYSCAF.

Modular programming will be required, too, in writing programs that cannot possibly be spied on or maliciously altered by someone using the same computer under a time-sharing arrangement. The reason is that unless it is small (at present, very small), a program *cannot* be proved to perform *exactly* the functions required by the designer *and no other functions whatever*. This is an abstract expression of the fact that should have become obvious in the main body of this book, that the programmer usually does *not* know the exact limits of what the program can and cannot do. Proving the correctness and completeness of programs is a "logical" problem in the theory of computation. This problem interests many workers in artificial intelligence, not least those concerned with automatic programming—in which, one hopes, the "programming program" will know with some precision what its creations are capable of.[36] In complex systems, as intelligent programs must be, many design weaknesses can remain undetected after years of use. In principle, a programming expert could exploit these weaknesses so as to seize control of the system and gain access to the routines checking identification and authorization of users. In practice, programs can be written so as to make this extremely difficult; and *inadvertent* intrusion into someone else's program can readily be made virtually impossible. But the only way to make deliberate tampering absolutely infeasible would be to build the system around individually proved modules, or "security kernels."[37]

Modular programming also helps to counter the conservatism inherent in widespread applications of very complex systems. It is one thing for an individual researcher, devising a comparatively small program, to make a radical change in its design. But altering a large-scale system is another matter. For instance, Fahlman switched from PLANNER to CONNIVER while writing BUILD so as to improve it greatly. But he might not have done so if he had already become committed to a "finished" PLANNER version. Complex systems that, like DENDRAL and COPY-DEMO, took many years of programming teamwork to produce, are correspondingly less likely to invite effective improvements. And a widely knowledgeable program that has been applied in the public domain, not merely in a research laboratory, on which many nonprofessional users in different parts of the country are dependent, is clearly very inconvenient to alter. To the extent that faults can be isolated within separate modules, and those modules adjusted without necessitating widespread tinkering with the system, programmers will be more able to face the task of improving the program. (As Norbert Wiener has said, it is a pity that old programs do not die: not only are they difficult to alter in a fundamental way, but their particular epistemological prejudices can survive indefinitely if we are too mean and lazy to replace them with others that are more satisfactory.[38])

If programmers are to be able to see what is going on, they should not use programming techniques that—while making for readable programs—render the control structure obscure. Fahlman rewrote BUILD in a different programming language because the control structure initially was far from explicit, so that he could *not* tell what was going on by examining the program code. Thus we saw in Chapter 12 that a PLANNER implementation using the PLANNER automatic backtrack facility may include goal-seeking procedures that end with the advice "USE ANYTHING" (as in the theorem shown in Figure 12.7). This advice is interpreted by the machine as an instruction to search its data base for anything it knows about that might help achieve the goal, and then to try it out. This strategy might lead to some very nasty situations: a distraught parent in squalid surroundings may try anything to stop the baby squalling—and succeed appallingly well. A human has to be distraught to disregard the side effects of "effective" measures, but a program may not even know about them. If it does not, then unwanted results may ensue *even if* (as in PLANNER programs like SHRDLU) the system tries out a given method in its imagination, before attempting to use it in the real world of action or policy making.

Moreover, in the PLANNER automatic backtrack situation, the program simply tries out (in imagination or reality) the first method that seems apt, since it has no way of comparing all the possibilities beforehand. The knowledge that hitting may hurt the baby could be buried inside the relevant procedure, to be found only when this way of making the baby quiet was run. Only if the programmer had specifically included the advice never to hit the baby, would a PLANNER program refuse even to consider it. The control structure of CONNIVER, on the other hand, does allow for the potential choices to be listed for higher level criticism. If a CONNIVER program knew that hitting hurts (a fact that in Chapter 7 we saw to be represented in Schankian semantics), and was able to access this knowledge when needed, it would be able to control itself long enough to find an alternative way of quieting the baby, or to decide to abandon this goal as not legally achievable.

This example of quieting the baby makes it clear that a flexibly intelligent control structure is only useful if the program embodies sensible criteria of what effects are undesirable. Some "sensible" criteria are highly culture-specific, and one may expect a good deal of moral-political disagreement about what precautions regarding artificial intelligence are in fact worthwhile.

For instance, suppose we suggest embodying Isaac Asimov's "Three Laws of Robotics" in intelligent programs designed for public use.[39] As all readers of science fiction know, these are:

1. A robot may not injure a human being, or, through inaction, allow a human being to come to harm.
2. A robot must obey the orders given to it by human beings except where such orders would conflict with the First Law.
3. A robot must protect its own existence as long as such protection does not conflict with the First or Second Law.

All very well (except that the laws ignore injury to animals)—but what is to be counted as "harm"? Hitting the baby is harmful, but cutting her with a surgeon's scalpel may not be. One might stipulate that all cases like this one, where "harm" is done for the end of a greater good, must be explicitly referred to a human judge for decision. But there are many less clear cases, in which individuals will disagree about what counts prima facie as harm. (Should a robot put fluoride into the public water supply, or defend a householder by knocking out the burglar?)

Assuming that the household robot described earlier will not be put in the position of having to tidy up the baby (one hopes that the baby will be kept firmly out of its way), one might be tempted to dismiss these questions as frivolous. For robots of the type envisaged by Asimov will be confined to the pages of science fiction for many years, if not forever. Meanwhile, domestic and industrial robots will be relatively specialized creatures, whose decisions cover such an artificially narrow range that the question of their doing *direct* harm to human beings hardly arises. Certainly, accidents may happen—but they happen no less with knives and flint-axes.[40]

But talking (or teletyping) programs are another matter. For instance, do we assume that it is harmful to one's interlocutor to lie? (We commonly say it is immoral, and utilitarians argue that it is socially harmful.) But if so, we may be landed with a talking program that is forbidden to tell tactful white lies. In discussing Colby's neurotic program, I noted that although the program can repress information it cannot lie, and I suggested that a more complex epistemology would be required to enable a program to lie. The point at issue here is not so much epistemological complexity (which we are assuming to have been achieved), but moral purity. One of the reasons people often have for lying is precisely to *avoid* harming others.

Programs that talk merely about the weather forecast, or narrowly defined impersonal issues in relation to which white lies would not be necessary, cause no difficulty in this regard. Nor do programs (like ELIZA) that merely question people about personal matters, rather than stating their own views (though they could of course do harm by insensitively asking a tactless question). But the more "personal" the program, the more likely that lying (or a telltale silence?) might be in order. Some people would even argue that for social-political reasons—such as preventing social dissent, disorder, or panic—certain "impersonal facts" should be kept from the general public. On this view, should all the home terminal programs (particularly those notionally contributing to participatory government) be ignorant of them, or should they be able intelligently to take account of them while "protectively" keeping them to themselves, if necessary lying in order to do so?

The ethical ambiguity of lying is only one of many difficult cases. A special module of the legal arbiter previously mentioned might be included to advise people on tax matters: but many would doubt the social morality of using every possible trick in the book to evade taxes, whether "legal" or not. A researcher in robotics has semiseriously distinguished various moral codes, from the Athenian, through the Spartan, Christian . . . to the Boy Scout.[41] And the literature of moral philosophy shows how difficult it is to arrive at a clear or consistent set of moral principles, still less one on whose specific content everyone will agree. A Kantian morality might even serve to undermine the protective Three Laws, if we are content to grant "rationality" to machines; for the Kantian imperative enjoins rational beings to treat each other as ends-in-themselves equally worthy of moral respect, which casts doubt on the admissibility of preferring *human* over *robot* interests.[42]

Small wonder, then, that one may experience some reserve about Alvin Toffler's optimistic recommendation of a new profession of Value-Impact Forecasters.[43] These ethical experts, armed with scientific tools for making cost-benefit judgments, Toffler would locate "at the hot-center of decision-making" in every social institution involved with technological innovation. Roszak has sourly commented: "What these ethical engineers will know of 'value' (Old Style: the meaning of life) may of course be only a computer simulation of a statistical illusion gleaned from questionnaires whose unreality crudely approximates a moral imbecile's conception of an ethical decision."[44]

Quite apart from his general skepticism about the ethical sensibilities of programmers and their professional advisers, which applies to "stupid" programs in the public domain as well as "intelligent" ones, Roszak specifically decries the practice of using computational models as analogies of the mind. He regards the use of psychological terms in programming contexts as inescapably dehumanizing in its social effects, and refers scathingly to artificial intelligence as "mechanistic counterfeiting" that is both symptom and contributory cause of the spiritual ills of urban-industrial culture. In the preceding chapter I quoted a similarly contemptuous dismissal of machine intelligence, also based on social-political grounds. (That such a dismissal may rest on ignorance of the field is largely irrelevant, for the general public—by definition—do not have close knowledge of it either, and it is their interpretation of "humanity" and "society" that is at issue here.) According to the more virulent political opponents of artificial intelligence, then, the best precautionary measure individual researchers could take would be to resign on the spot.

For many obvious reasons, not least of which is the scientistic bias of Western culture in general and our official policy makers in particular, this is not going to happen. Perhaps it is even too much to hope that a sense of human sympathy and responsibility will inform most individual programmers, so as to sensitize them to the social significance of what they are doing. Are programmers inherently more likely to take account of such matters than nuclear physicists, or scientists involved in chemical and biological warfare? The prospects are chilling indeed, if theoretical discussions of (say) the project of programming computers to understand ordinary language are commonly to be prefaced by such passages as this:[45]

> When computers acquire the ability to interpret natural English, as well as to retrieve and process data in flexible ways, they will be able, without the help of an intermediary, to answer novel questions in a matter of minutes. The resultant increase in the power of military command and control systems will have a significant bearing on our defense stature, and the advantages for business and science are equally exciting to contemplate.

In any event, ethical sensitivity on the part of individual programmers will not be enough to offset the dubious moral implications of artificial intelligence that inform the quotation just cited. Rather, they must cooperate in explicitly relating their moral critique to their profession *considered as a social institution*, with a social reality that is largely autonomous with respect to the wishes or actions of isolated individuals.

Although "pure" research in artificial intelligence (and computing science generally) is closely connected historically and financially with the technological, not to say military, interests of industrial society, there has been for some years a conscious attempt within the expert community to examine what they are doing in regard to interests whose human worth is less questionable.[46] To what extent they will succeed in humanizing their professional enterprise remains to be seen. There is, of course, little point in scientists' discussing the social responsibility of science unless they are prepared not only to acknowledge nonscientific values but, if necessary, sometimes to place them above the Faustian priority of "freedom of scientific inquiry." Otherwise, given the central place of science in industrial society, we all run the risk of sharing Faust's daemonic fate. Scientists must accept that there may be (there have been) cases in which one ought willingly to forego reaching a scientific understanding of something, since the only ways of arriving at such an understanding involve morally unacceptable methods.

The social critics of artificial intelligence would cite this argument not just to urge programmers to refuse to write programs with an obviously exploitative nature, but to suggest that they reexamine the metaphysical premises and philosophical assumptions of their whole enterprise. For if there are potential dangers lurking in that intellectual undergrowth, then merely skimming the surface (pruning *this* specific application rather than *that*) will not arrest them.

In view of the philosophical issues discussed in Chapter 14, this surely is not too much to ask. The programming community should remember that psychological and social theories in general are not purely *descriptive*, but largely *constitutive* of social reality, and computational theories are no exception. The role assigned to human beings in the popular social theory is liable to become the role ascribed by individuals to themselves. In particular, workers in artificial intelligence should constantly be aware of the way in which technological analogies (often of a crudely debased popular form) can enter deeply into the personality and self-image of individuals, and into their culture, with subtly insidious effects that we ignore at our peril. If the public believes—rightly or wrongly—that science regards people as "nothing but clockwork," then clockwork-people we may tend to become.

In this connection, it is worth asking whether programs for public use should include explicit reminders of some of the differences between computers and people. This is done for utilitarian reasons in the peptic ulcer program described in the previous section: the program continually reminds the patient that she is on-line to a machine, because the program is merely a Yes-No recorder with no sense of whether the patient has finished what she wants to say (the program lacks even ELIZA's limited range of recognizable patterns). But at the same time, the reminder serves to avoid mystification of the patient, to impress on her that she is merely "filling in a form" by teletype rather than engaging in communication, still less participating in a human relationship—however oppressive may be her experience of the typical doctor.

Programmers sometimes include "plausibility tricks" in language-using programs, for one reason or another. ELIZA herself, one might say, is almost entirely grounded on such tricks. But even SHRDLU's dialogue ends with the deceptively friendly response "YOU'RE WELCOME!" and the dialogue as a whole is carefully composed by the programmer to hide SHRDLU's ignorance *even of blocks and pyramids*. Again, the basically trivial Computer-Aided Instruction programs currently used in schools take care to address the child from time to time by her first name, so as to put her at her ease. In view of the dehumanizing potential of computer applications (quite apart from the possible results of an overgenerous misunderstanding by the person), it may be that the limits of individual programs should be made as clear as possible to users, and plausibility tricks used sparingly, if at all.

On one point, however, even Roszak presumably would agree that the *similarity* between people and programs should be deliberately stressed. One of his objections to the social use of computers in an advisory capacity is that the machine may be seen by the public (including the politicians) as purely *objective*, and therefore not to be argued with. But we have seen repeatedly that this is not so. Even the judgment "That is a cube" rests on epistemological presuppositions and interpretative schemata that might have been (and in some programs *are*) different.

Moreover, intelligent machines will be subject to illusions and misunderstanding, as we are—and for essentially the same reasons.[47]

Accordingly, the inescapable subjectivity of a program's judgments should be made clear to its users. This is done implicitly in the MYCIN medical consultant: MYCIN not only offers alternative judgments when it perceives several possibilities of diagnosis or treatment, but gives its reasons for each so that the physician can rationally reject its advice if she thinks fit. In general, it would be worthwhile somehow to remind the user that the program is functioning within its own subjective cognitive world, just as she is. Any differences between the two are matter for epistemological debate, not for servile capitulation on the part of the person. This applies not only to moral reservations, but to "factual" and inferential disagreements also. In short, it cannot be assumed that the machine's point of view (*sic*) is "objective" and therefore must be accepted. (Programs can, of course, be objective in a different sense: if they have no way of taking individual personalities into account, then their judgments will be impartial in a way in which many human judgments are not. But the story of Solomon and the baby suggests that strict impartiality may not always be what we want.)

At base, Roszak's political objection to artificial intelligence (indeed, to Western scientific culture in general) is that it encourages what the poet Blake called "single vision": the assumption that only one way of interpreting the world is valuable, namely, the way of natural science. Blake foresaw the impoverishment of consciousness in industrial cultures that has been aggravated since Newton's time by the progressive mechanization of our world picture. This is partly due to the conceptual poverty of the popular images of "machines" that have been current throughout that period. (And the widespread social use of "stupid" programs affords little respite in this regard.) But it is due also to the mistaken epistemological assumption that the use of one interpretative scheme to extend our understanding excludes the use of others, as science is often thought to exclude (not just to differ from) poetry and myth. In other words, it is due to the assumption criticized in the previous paragraph, that the natural sciences are "objective" in an absolute sense.[48]

Someone who shares that assumption may very likely think that workers in artificial intelligence must gainsay other conceptual frameworks. But to defend computational analogies of intelligence as fruitful frames for asking questions about the way we think, about the way we construct our lives and interpret our reality in terms of idiosyncratic and culturally shared subjective schemata, is not to deny that *several* frameworks may be illuminating in this regard—whether psychoanalytic, philosophical, aesthetic, mythical, religious, or whatever. For example, in discussing *Romeo and Juliet* and detective fiction, I claimed that literary understanding rests on conceptual schemata whose content and organization theoretical psychologists may reasonably seek to make explicit, with the help of the computational metaphor. But it would be absurd to suggest that psychologists must therefore reject the intuitive understanding of our lot that is available to them as human beings watching (and participating in) human dramas. Simply, some approaches are more useful than others for certain purposes, which purposes it is up to us to determine. If the artificial intelligence community bear this in mind, and take the trouble to make it clear in situations where otherwise it might be overlooked, the widespread resistance to their research may be lessened and the adverse social effects partly forestalled.

To be sure, apparently contradictory analogies lie uneasily together in the fancy. If we can reach a theoretical reconciliation of them, they can coexist peacefully. This is why discussions of artificial intelligence should stress the philosophical compatibility of humanist and mechanist viewpoints. For most people currently regard these two approaches as no less mutually contradictory than the wave and corpuscular theories of light, and have no conception of any equivalent to optical field-theory that might reconcile the two. If we assume that we are faced with a choice, we cannot simply shrug it off, as most of us can blithely ignore theoretical optics. We must either deny our humanity, with socially destructive results, or else forfeit a scientific understanding of the world in general and people in particular.

Contrary to common opinion, then, the prime metaphysical significance of artificial intelligence is that it can *counteract* the subtly dehumanizing influence of natural science, of which so many cultural critics have complained. It does this by showing, in a scientifically acceptable manner, how it is possible for psychological beings to be grounded in a material world and yet be properly distinguished from "mere matter." Far from showing that human beings are "nothing but machines," it confirms our insistence that we are essentially subjective creatures living through our own mental constructions of reality (among which science itself is one). In addition, for those of us who are interested, it offers an illuminating theoretical metaphor for the mind that allows psychological questions to be posed with greater clarity than before. The more widely these points are recognized, both within and outside the profession, the less of a threat will artificial intelligence present to humane conceptions of self and society.

# 16

# *Postscript*

ARTIFICIAL INTELLIGENCE ("AI") has hit the headlines. When this book first appeared, very few people had even heard of it. Now, one can hardly turn on the television or open the newspaper without seeing some mention of it. It has been featured at length in *Time* magazine (and in business magazines such as *Fortune* and *Forbes*), and has even made the front cover of *Newsweek.* The biennial international conference on artificial intelligence, which in the early 1970s attracted a mere 250 people, now welcomes over 6,000. And specialist conferences and journals—on language understanding, speech technology, image processing, expert systems, and robotics—now abound.

What's more, these words are accompanied by action. New research groups are springing up around the world, and business interests are deeply involved. Throughout the Western industrialized countries, money is being poured into AI research.

Private investment in this new technology has burgeoned. A few large firms, such as the oil prospector, Schlumberger, and the computer manufacturer, Digital Equipment Corporation, have committed huge sums to setting up their own AI laboratories. Other firms hope to contract the services of one of the AI companies now proliferating in the marketplace. These companies (mostly started by individuals whose work is described in the main text of this book) originated in the United States, but several similar enterprises have been set up in Europe—one on the campus of my own university. No international conference on artificial intelligence today is without its natty-suited venture capitalists: financial speculators from the City of London and Wall Street, whose deals run into millions and who are actively seeking to back new enterprises in this area.

This is not a game for private investors alone. Government money in the Western industrialized nations is now being provided for AI research, in academic and industrial (and, it must be admitted, especially in military) contexts. In Europe, the Common Market countries have established the "Esprit" project for funding cooperation between member countries in research in microelectronics and software technology. The first phase of Esprit draws on $650 million from EEC public revenues. The British government has set up, as well as its stake in Esprit, the national Alvey Committee, to recommend a strategy for the long-term funding of artificial intelligence and related computational techniques. Government funds to the tune of $300 million have been allocated for the first five years of this information technology work. (At the time of writing, it is unclear whether the first phase of Alvey will be repeated on a comparable scale. One reason for doubting this is that large sums of money will be allocated to the "Star Wars" project—on which, more below.) And the U.S. government, via the Advanced Research agency of the Department of Defense, has committed $1,000 million over five years—twice as much as the agency's expenditure on advanced computing over the previous twenty years.

Not surprisingly, the electronic industry in all these countries is taking this research seriously too. Esprit's $650 million is being matched by an equal amount from European industry, and in the U.K. likewise, the governmental allocation ($300 million) is equalled by industrial money set aside for the Alvey projects. In the United States, the electronics industry is budgeting $75 million annually to a new research and development group in Texas known as "MCC" (Microelectronics and Computer Technology Corporation).

What is more surprising is the degree of active cooperation involved. MCC is a joint effort of a dozen of the major U.S. firms in this field, who have all agreed to relax the normal mores of cutthroat competition and industrial secrecy. Indeed, the U.S. government has amended their antitrust laws in order to make this collaboration possible. Nor is the United States alone in fostering such industrial harmony. Cooperation between erstwhile industrial rivals is occurring also in Europe and the U.K., whose government funds are given only on condition that this type of cooperation takes place.

One might well wonder what extraordinary stimulus could have triggered these newfound industrial harmonies. And why the sudden media interest in artificial intelligence? What has changed: *ideas* or *attitudes?* Have AI researchers made radical advances, by way of new theoretical ideas which justify the field's new visibility? If not, if the intellectual content of the field is broadly the same as it was before, why the burgeoning public interest?

The current interest is *not* due to any radical theoretical advance.[1] The central problems and achievements of artificial intelligence have changed very little since the early 1970s, when the first edition of this book was being written. The recent explosion of publicity and financial support can be traced primarily to a single source: the announcement in October 1981 of Japan's ten-year national plan for developing "Fifth Generation" computers.[2]

This national plan is jointly funded by Japan's government and electronic industry, to a minimum of $810 million (some observers expect the funding to run into billions). Finances aside, an indication of the importance Japan attaches to this project is that it has been allowed to overturn the traditional Japanese attitude to seniority. The director of Tokyo's newly formed AI institute ("ICOT") is a surprisingly young man (in early middle age). He has insisted that ICOT be staffed only by *very* young computer scientists, under 30 years of age, whose technical training is up-to-date and whose creativity is comparatively high.

The Japanese are not engaging in this exercise as a disinterested search for truth. Nor are they seeking to appropriate just one more market, in addition to cameras and video equipment. Their explicit aim is world economic domination. It is because preeminence in information technology so clearly spells economic preeminence *tout court,* that Western politicians and industrialists (some of whom were invited by the Japanese to the "Announcement" meeting) have picked up Japan's gauntlet with such alacrity.

For once, then, the West is copying Japan. It is Japan's extraordinary insistence that competitors in the capitalist marketplace *cooperate* in the necessary (and very expensive) research that has prompted the novel forms of industrial funding now seen even in Texas. The effect of the Fifth Generation project in stimulating scientific research in the United States has already been so great that it has been compared with the challenge of the Soviet *Sputnik* of 1956. To be sure, the *theoretical* achievements of Japan in artificial intelligence have not as yet approached those of the West. Whereas *Sputnik* was an undoubted achievement, the Japanese Fifth Generation project is still only a promise. However, it would be rash to assume that Japan is incapable of doing first-rate work in artificial intelligence. And there is general agreement that Japanese advances in computer *hardware* will be impressive.

The crucial point about these machines of the future that Japan aims to build is that artificial intelligence will be integral to them. The first four "generations" of computers are defined by the Japanese purely in hardware terms: machines based (1) on valves, (2) on transistors, (3) on silicon chips, and (4) on Very Large Scale Integration (VLSI). The predicted Fifth Generation, by contrast, is defined in terms not only of improved (massively parallel) hardware but also of *artificial intelligence.* Indeed, the improved hardware is seen as the prerequisite for intelligent performance.

(Speaking at an international Conference on advanced information technology in 1984, Professor Hideo Aiso, of ICOT, gave brief details of Japanese plans for a "Sixth Generation" project. The main research target for the Sixth Generation, he said, involves the analysis of the structure of the human brain, the application of biochemistry, and the development of *biological* chips; studies of optical [laser] computers are also planned.)

The Japanese have very ambitious goals for Fifth Generation technology. For example, they hope to achieve reliable machine translation between various natural languages—even on texts that are not restricted to highly specialized subject matter. And they forecast that computers of the early 1990s will be able to interpret the speech of many different individuals, to act as intelligent assistants in a wide variety of tasks, and to provide advanced problem solving and sensorimotor abilities for mobile and domestic industrial robots.

Whether Japan, or any Western nation, will achieve these ambitious aims—even within twenty years, never mind ten—is doubtful. Developing Fifth Generation computers will be no easy task, and most people grossly underestimate the difficulties involved.

For the prime lesson of artificial intelligence (argued and illustrated throughout the main text of this book) is the previously unrecognized subtlety and richness of the human mind—*every* normal human mind. Seeing, speaking, conversing, and commonsense thinking are everyday abilities usually taken so much for granted that no one would dream of commenting on their presence, only on their absence. Each, however, is extremely complex.

For instance, *conversing* requires one to distinguish the individual words spoken by the other person (normal speech does not provide pauses between words), to understand their grammatical and semantic relations, and to structure one's conversation in light of not only one's own goals and beliefs but also those of one's interlocutor. In "conversations" between humans and computers, the first of these problems is often bypassed by making use of a teletype. Even so, programs capable of "natural" conversation with a human user should ideally take into account not only syntax and semantics (discussed in Chapters 6 and 7) but pragmatics too. A programmed travel agent, for example, should not answer "Yes" to the input "Can you tell me how far it is from Paris to Louvain?" Syntactically, this sentence is an interrogative; but pragmatically, it is usually taken to be a request rather than a question. Similarly, a program capable of having a coherent conversation (as opposed to responding to a series of more or less independent inputs) should be able to recognize the rational shifts of focus within a coherent dialogue between two intentional agents, who may have very different goals and beliefs. Not surprisingly, then, current artificial intelligence includes work on the structural analysis of conversation (largely inspired by the theoretical work of philosophers, psychologists, and linguists).[3]

These everyday things like seeing and speaking, which we all do so "naturally" (without any conscious thought or planning), are yet more difficult to automate than are the specialized abilities possessed by professional elites such as doctors, lawyers, oil prospectors, scientists, and mathematicians. Even these highly educated specialists, of course, rely very largely in their professional work on "ordinary" conversational conventions, on commonsense reasoning, and on "informal" analogical thinking. These abilities are unnecessary only for the most routine, narrowly circumscribed, and formally well-understood tasks. And the ability to learn to do all these things better, usually by practice and experience rather than formal instruction, is a further human strength not yet enjoyed by computers—because, as yet, we understand it hardly at all.

For these reasons, I now think that the definition of artificial intelligence given in the first edition—*the study of how to make machines do things that would require intelligence if done by people*—may be misleading. For it can be taken to suggest that artificial intelligence is concerned primarily—or even only—with abstract and theoretical tasks that only especially "intelligent," and highly educated, people can perform. In fact, however, it is concerned also with everyday, taken-for-granted abilities such as those listed earlier: seeing, speaking, conversing, and commonsense reasoning—none of which are ordinarily thought of as requiring "intelligence." A better definition is: *The study of how to build and program machines that can do the sorts of things which human minds can do.* Such things, of course, include many things that some animal minds can do too. And the animals (never mind *homo sapiens*) have the edge: technology will not achieve these things for many years yet.

But if some publicly proclaimed goals are unrealistic, there is no doubt that applications of a less ambitious kind will be achieved fairly soon. By 1990, both the West and Japan will have a wide range of commercially useful AI applications.

Many will be *expert systems* designed to give advice to human professionals (and laymen) in narrowly defined specialist areas—from medical diagnosis, through social security entitlements, to car maintenance.[4] However, for reasons to be explained, the practically useful expert systems of the near future will be "more of the same": although their specific content will be novel, their general form will not. They will be rule-based systems (sets of IF-THEN rules) based on a small number of expert-system "shells," or general inference engines. The first of these was EMYCIN, which embodies the knowledge-free inferential architecture of MYCIN (see Chapter 15) without its specific medical content. Shells are used by the various AI firms for building expert systems "to order." And several (including one, described later, that is capable of a simple form of learning) are already commercially available in a simplified form, for use in do-it-yourself system building by businesses and public institutions.

Other applications that may be available fairly soon will be somewhat more novel, in that they integrate existing AI work from diverse areas. For instance, among the Alvey five-year "demonstrator projects" (a handful of carefully chosen projects intended to demonstrate the commercial potential of intelligent knowledge-based systems, or "IKBS") is one which aims to integrate language understanding and speech technology.[5] The goal is to build a speech-driven word processor, with a 5,000-word vocabulary and the ability to accept continuous speech in a wide range of voices and accents. Even if this goal is achieved only to a limited extent, the resulting system may be useful for handling tomorrow's run-of-the-mill business correspondence. (Notice, however, that this "dictation machine" does not need to *converse:* a dialogue is a highly structured activity, necessitating the recognition of and allowance for the goals and beliefs of the two speakers, which are normally different.)

What of the day after tomorrow? If the long-term research goals of artificial intelligence are ever to be attained, "more of the same" and integration of existing ideas will not be enough. Expert systems, for example, will require radical improvements if they are to be more like human experts than they are today. Consequently, some of the money now being made available for AI research must be (and is being) used in attacking basic theoretical problems.

These problems were discussed at length in the first edition of this book, for the main difficulties are the same now as they were a decade ago. They are more clearly recognized, perhaps, because the inherent limitations of the early work have become increasingly evident. But they are far from being solved. Despite some promising recent research, we are still a long way from the dream—or nightmare—of a powerful machine intelligence.

The outstanding exception to the charge of theoretical stasis is low-level vision (early work in which was briefly discussed in the final section of Chapter 13). This involves the (bottom-up) exploitation of the information in optical images, to derive descriptions of physical surfaces, in terms of their orientation, texture, position, distance, and the like, and to assign the various surfaces to distinct physical objects. There is now a thriving international research community concerned with low-level vision. The groundwork for this research lies in advances in the detailed physics of image formation (such as B. Horn's work, cited in Chapter 13).[6]

The optical theory concerned deals systematically with the ways in which light is reflected off material objects. It answers such questions as: How is light reflected from *this* or *that* sort of surface? And from a surface (or tiny part of a surface) that happens to be oriented at a particular angle relative to the viewer? And from a surface at such-and-such a distance from the eye? And from a surface simultaneously presented to the viewer's two eyes?

The overall surface of any physical object reflects an image that is made up of many tiny areas, or point-images. Neighboring point-images are likely to be similar, because neighboring surface-points are usually similar. Difference-boundaries in the image are therefore often interpretable as real boundaries between distinct objects in the real world. But not *every* discernible difference-boundary in the image should be interpreted in this way. For example, the stripes on a zebra's coat are not different objects: they are part of one and the same physical surface, attributable to one and the same thing (the zebra). In general, local differences in the image may be *organized,* so as to comprise surface markings and surface contours.

Accordingly, the basic computational method of this approach is to combine many distinct—but systematically interconnected—processing units, each one dedicated to seeking a specific type of 3D interpretation for the 2D image-part it looks at. The interconnections between the individual units allow for specific (theoretically defined) feedback between units, feedback which takes into account the *physical possibilities* of images in various real-world situations. Much as the scene-analysis programs described in Chapter 8 used the theory of line labeling (sometimes aided by the Waltz filter) to *confirm* or *disconfirm* particular 3D hypotheses, so these low-level systems use physical optics to arrive at *mutually consistent* sets of interpretations from the constituent interpretative units.

Initially, there must be units that "look at" the individual point-images on the retina. But our reasoning implies that there must be several "levels" of visual units, several "levels" of description available to interpret the image. Some units look for certain sorts of organization within the descriptions arrived at by lower-level units. For example, one unit may fire only when there is a particular light intensity-gradient (a "line") at a certain retinal location. Another may fire only when the relevant image-point appears (on the basis of some depth-detecting algorithm) to be a particular distance from the viewer.

Depth description is often crucial in deciding where one object starts and another begins. For if a sudden depth disparity is noticed at a series of neighboring points in the image, these may be taken to correspond to the edge of the physical object concerned (such as a zebra's back). This would be helpful, also, in distinguishing a black image-patch caused by a black stripe on a zebra's hide from an *immediately contiguous* black image-patch caused by markings on the wallpaper. Suppose that some early processing units had reported "one" black patch (for that is how it appears in the image). If the depth-detecting units were then to decide that there was a line of depth disparity running through this area, they could pass messages to the patch-detecting units concerned, so as to *inhibit* them from describing this as "one" patch (ascribable to "one" surface). In principle, then, these machines could locate the body outline of a zebra standing in front of black-and-white striped wallpaper—*without* having to know anything beforehand about what a typical zebra looks like.

In order for such a task to be feasible, some depth-detecting algorithm has to be written. The computation required for stereopsis (depth-vision which relies on the disparities between the images presented to the two eyes) is much better understood as a result of this theoretical approach.[7]

Indeed, a variety of mathematically sophisticated methods have been developed for 2D-to-3D computation, and for 3D-to-4D computation (where optical flow, for instance, is taken into account). Algorithms exist for deriving descriptions of 3D properties like shape, depth, body edges, surface orientation, and movement from the 2D input (the grey-scale image). These methods are grounded in careful mathematical analysis of the physical constraints that determine the formation of images, given such-and-such a viewpoint, lighting conditions, surface reflectance, and so on. Such theoretical analysis enables the principled prediction (given a particular state of affairs in the 3D world) of the resultant point-images, *groups* of point-images (such as light-intensity gradients, or "lines," in the image), and correspondences between *pairs* of images (images presented either simultaneously to the two eyes, as in stereopsis, or successively, as in relative movement between the seen object and the viewer).

Interpretative algorithms (like these) that can be explicitly justified in the precise terms of physical optics are more reliable than procedures based on unsystematic, and largely tacit, insights about the nature of vision. This is a special case of the general point made at length in Chapters 8 and 9: as the development of "scene-analysis" programs showed, computer models are more powerful if they exploit deep theoretical knowledge about the principles of mapping between the domain in question and various representations (or "projections") of it. In particular, theoretically grounded procedures are more likely to be *generalizable.* The 2D-to-3D algorithms mentioned earlier, for example, can be used in describing *unfamiliar* objects, for they do not depend on (though they can be combined with) top-down, "knowledge-driven" procedures designed for specific classes of object. Clearly, this work in low-level vision can have technological applications in many fields: visual inspection in factories, satellite spying, and robotics.

Research in low-level vision is of psychological interest, too. This was stressed by D. A. Marr (whose early work on vision was mentioned in Chapter 13).[8] Marr greatly clarified the theoretical problems involved and helped develop a number of algorithms for computing depth-by-stereopsis, shape-from-shading, and the like. His theory of vision is based on an analysis of the computational constraints necessarily involved in the task of passing from 2D image to 3D description. He posited a series of visual representations, in which different facts about the 3D world are successively made explicit. Starting from the light array in the image, Marr's visual system passes through the primal sketch (which codes intensity changes and groups thereof), the 2-½D sketch (in which surface orientation is coded relative to the viewer), and the world model (which describes an object's position in 3D space independently of the viewer). In addition, Marr claimed that any scientific psychology, whatever the domain, must likewise be grounded in an abstract analysis of the central task concerned.

Human psychology, of course, makes use of parallel processing in the brain, whose neurones comprise an associative network rich in excitatory and inhibitory connections. Today's artificial intelligence does not, for it relies on digital computers (von Neumann machines). But, as references to "interconnected units" in the description of low-level vision research suggests, some current AI work *conceptualizes* its computational processes as parallel, even though in practice it has to *simulate* parallelism on sequential hardware. This practical limitation will eventually disappear, as a result of current and future work on various types of *connectionist* computational systems.[9] These are the modern descendants of the "neural nets" and "perceptrons" of the 1950s, and like them are seen as analogous to the brain. Information processing that is best suited to connectionist machines differs radically from what typically goes on in digital computers.

Connectionist systems are made up of locally communicating units functioning in parallel, where—because of hardwired feedback links (which at present are simulated by being programmed on a von Neumann computer)—the state of any one unit depends largely on the states of its neighbors. Each unit, or miniprocessor, is relatively stupid. In a "unit-value" machine, for example, each individual processor can code only *one* discrimination, or judgment. The units are not necessarily digital (all-or-none) processors, but may exhibit varying levels of activity. A unit's activity level at a given time reflects its confidence in its (one and only) judgment. The evidence it uses to arrive at that confidence level consists largely of the current activity levels of other units in the system.

If activity levels are to function as "evidence," the excitatory and inhibitory connections built into the machine must be very carefully designed. In general, units coding mutually supportive judgments should be strongly facilitative, units coding consistent judgments should excite each other to some extent, and units coding inconsistent judgments should be mutually inhibitory. If this is achieved, then the interpretation of any one part of the input (such as an image or a sentence) can be sensibly constrained by the interpretations of the neighboring parts.

For example, in a connectionist system computing 3D shape from a 2D image, the specific connections (which in a truly connectionist machine would be engineered, not programmed) embody constraints on 2D-to-3D mapping grounded in the physics of image formation. So the activity of a unit signaling the presence of an "edge" at a certain place will be facilitated by units reporting *a high intensity-gradient* or a marked *depth disparity* at the same position—and these two units will also facilitate each other. A system for analyzing language, on the other hand, would have syntactic or semantic constraints implicitly embodied in its hardware. This mutual feedback of constraints is comparable to the Waltz filter described in Chapter 9. But it avoids the computational "gangrene" caused when one mistaken decision cuts off the correct interpretative path for ever. The reason is that the activity levels fluctuate continuously, so that a unit can be inhibited at one moment and facilitated later.

The mutual interactions between units continue until an overall equilibrium state is reached. This state embodies the system's interpretation of the input. That is, the final decision (or representation) is embodied in the stabilized *pattern of activity* over the network *as a whole.* Such a connectionist system is thus *parallel* (many units function simultaneously), *cooperative* (neighboring units influence each other), and *distributed* (the final representation is embodied in all the units in the system).

One type of equilibrium-seeking connectionist system that has recently attracted interest is the "Boltzmann machine."[10] The locally interacting units in a Boltzmann machine are binary (on-off) units, whose behavior is *stochastic.* That is, a unit's state depends not only on the (weighted) messages being passed to it by its neighbors but also on a probabilistic factor. The overall "equilibrium" to which the units gradually converge is an optimal *probability distribution.* Such self-equilibrating systems can be described by the Boltzmann equations of thermodynamics, which represent the temperature of a large volume of gas as an overall energy equilibrium obtained by energy transfer between neighboring molecules. Although the energy states of an individual molecule are not predictable, the energy distributions of large collections of molecules are.

The inbuilt randomness of a Boltzmann machine is, in principle (though not necessarily in practice, as we shall see), an advantage, in that it makes the network *more* likely than a deterministic one to reach an optimal equilibrium state. Because of the possibility of random "jumps," the system can avoid getting stuck in locally attractive but globally suboptimal configurations. Analogously, a rock that jumped from time to time would be less likely than a normal rock to get stuck in a ditch while rolling down a mountain: after resting in the ditch for a while, it might jump out, and continue its downward descent. In general, the rock would be likely to reach the valley at the bottom of the mountain more quickly if its random "jumps" were more energetic at first and gentler later. Likewise, global equilibrium in a Boltzmann machine is attained by a process ("simulated annealing") comparable to that of annealing metals, in which energy equilibrium is reached more quickly by starting at a high temperature and cooling down gradually.

The interest in connectionism lies partly in the fact that it is closer to the neurophysiological reality of the brain. And it lies partly in the possibility that new forms of computation will be possible in parallel machines, which are not possible—or not practicable—in von Neumann architectures. Von Neumann himself, who invented the digital computer shortly after the Second World War, held that most human thought processes could not be modeled by binary arithmetic and formal logic.[11] He insisted that more "brainlike," parallel computers would be needed, and suggested that the mathematics capable of describing their behavior might be that used in thermodynamics. (The careful task analysis that is needed for writing powerful AI programs for digital computers is no less necessary for connectionist systems. Although parallelism may provide us with new computational opportunities, it will not furnish us with *magic.*)

But just what these "new forms of computation" might be is still unclear because very little is known about the computational potential and limitations of parallelism. There are as yet no large parallel-processing *("non-von")* machines available for experimentation—though several groups around the world (including ICOT) are building them.[12] Workers today must either *simulate* (tiny) parallel machines on digital computers, or concentrate on *theoretical* analysis of the computational properties of connectionist systems (such as the work on Boltzmann machines just mentioned). Recent theoretical work suggests that some of these properties may be highly surprising. For example, the number of individual processors required to make the "human" range of visual shape discriminations can be proved to be markedly less than one would naturally assume. And it was noted earlier that making a connectionist system partly random rather than purely deterministic may *improve* its chances of solving a problem. Indeed, it has been proved that, given infinite time, a Boltzmann machine *must* arrive at the optimal solution.

But the fact that something is possible in principle does not mean that it is achievable in practice. Very recently, a researcher responsible for much of the work on Boltzmann machines has decided that they are not effective systems for use on most real problems: the "noise" in the system will prevent it from settling down into the optimal solution within a reasonable amount of time. His current work, then, focuses on connectionist systems in which the individual local units are deterministic, not stochastic. In general, the practical relevance of such abstract proofs as exist, and other unknown properties of large parallel systems, will be explored within the coming years. Much of this work will be theoretical, but the construction of connectionist hardware will enable empirical explorations also.

The emphasis on parallel hardware in ICOT's Fifth Generation project is connected with their controversial commitment to logic programming, in particular by means of the PROLOG programming language.[13] PROLOG—originally developed (in France and Great Britain) for implementation in digital computers—is well suited to parallel architectures, because of the difference between it and the programming languages used in orthodox (serial) programming.

An orthodox program consists of a sequence of individual instructions to the machine to do things, one by one, in a specified order. Certainly (as explained in Chapter 12), the order may be *fully* determined only at runtime, by decisions made on the basis of information available only at that time. For instance, goal-oriented programming languages such as PLANNER and CONNIVER enable the programmer to write high-level instructions that are less specific than LISP instructions (in which these languages are implemented), and which leave the computer to decide just what to do in obeying them; and LISP itself, when implemented by way of an *interpreter* program (see Chapter 1), allows last-minute decisions to be taken during execution. But the general approach of orthodox programming, which of course reflects the basic (sequential) operating principles of von Neumann architecture, is to instruct the machine exactly *what* it is to do, and *when.*

The basic idea of logic programming, by contrast, is to provide the machine with *a set of items,* and to leave it to cope with them in whatever order it pleases. Accordingly, PROLOG (PROgramming in LOGic)—the programming language most commonly used by people committed to the idea of logic programming—is often described as a *declarative* language. (Strictly, this description is misleading. Of two logically equivalent PROLOG programs, which differ only in the *order* in which the subtasks will be carried out, one may work while the other does not. In true logic programming, this would never be so.)

When running a PROLOG program (and in logic programming generally), the computer exploits very general, domain-independent methods of problem solving: *resolution* and *unification.*[14] A resolution problem solver works by a species of *reductio ad absurdum* reasoning. In deciding whether to accept item X, it tries to find some contradiction that is implied by the *negation* of item X (see Chapter 6, p. 129). This method was first used in early research on theorem proving; it is now relatively well understood, and powerful algorithms exist for applying it efficiently. (Despite this theoretical advantage, general problem-solving methods risk using a sledgehammer to crack a nut. In practice, specialist methods are often necessary—as in low-level vision, where the "2D-to-3D" problem is solved by inferential techniques which specifically exploit the physical constraints on image formation. Any "specialist knowledge" of a PROLOG program is represented as items in its own particular data base.)

The "items" in a PROLOG program are very naturally—though not necessarily, as we shall see—thought of as *facts,* or *assertions,* in which case PROLOG is regarded as involving what in Chapter 12 was called a *declarative* (as opposed to a *procedural*) approach to programming and knowledge representation. Given this (declarative) interpretation of logic programming, a PROLOG program may be said not to *instruct* the machine so much as to *inform* it: it provides a set of facts (or more strictly, assertions), and leaves the machine, using its knowledge of the general inferential principles of logic, to draw whatever conclusions it can from those facts. For example, a machine running a PROLOG program already knows that from the truth of "All A's are B" and "X is an A," it may infer the truth of "X is B." So if it is informed (by a particular PROLOG program) that "All men are mortal" and "Socrates is a man," it will *automatically* conclude that "Socrates is mortal." Likewise, if it needs to establish whether Socrates is mortal, it already knows how to go about trying to find out—by using its knowledge that *if* it is the case that all men are mortal and that Socrates is a man, *then* it follows that Socrates is mortal.

To call these inferences "automatic" is to say that the basic logical principles and general problem-solving methods used by the machine do not have to be written into any specific PROLOG program: they are made available to the machine in some other way. In a general-purpose von Neumann computer, PROLOG has to be compiled or interpreted (see Chapter 1) into some lower-level language; the "background" inferential principles and the methods of resolution and unification then form part of this software implementation of the PROLOG language. But in a dedicated machine, specifically designed to run PROLOG, these logical principles and general methods are implemented in the hardware. The Japanese are building massive, parallel-architecture "PROLOG machines," in which the basic operations of PROLOG are carried out directly by the machine itself. PROLOG programs can be run relatively fast on these machines, for two reasons: because PROLOG is, in effect, the *machine code* of the computer (cf. Chapter 1); and because of the parallelism, which enables many operations to be executed at the same time.

Parallel hardware is especially appropriate for implementing PROLOG because of the basic computational functions that are specified by logic programming. The logical expressions dealt with in these computations include many which are (sometimes very complex) conjunctions or disjunctions. The way in which logic dictates that such expressions be tested is one to which parallel implementation is well suited.

To test the truth of a conjunction (A *and* B *and* C . . . ), the truth of each individual conjunct must be tested: if each conjunct is true, then the conjunction is true; if one or more is false, the conjunction itself is false. Similarly, a disjunction (A *or* B *or* C . . . ) is tested by testing its disjuncts: it is true if at least one—for "exclusive" disjunction, *only* one—disjunct is true; otherwise, it is false. The logician's "if-then" implication can be defined in terms of conjunction and negation: "*if* A, *then* B" is logically equivalent to "it is *not the case* that (A is true *and* B is false)." An expression like "A *and* (B *or* C) *and* (*if* D *then* E) . . ." involves conjunction, disjunction, and if-then implication; the "nesting" of bracketed subexpressions can in principle occur on indefinitely many levels.

Logic requires that to establish the truth of a conjunction, the truth of *every* conjunct (and, for exclusive disjunction, *every* disjunct) be tested. But it has nothing to say about the *order* in which they are tested, because each conjunct (or disjunct) is logically independent of the others. Consequently, there is nothing in logic to stop them from all being tested *simultaneously.* Simultaneous testing, if it could be achieved, would obviously save time—especially where very complex expressions (made up of many individual conjuncts or disjuncts) are concerned. This is where parallelism comes in: a machine can be *engineered* to allow for simultaneous testing of many different conjuncts, by many distinct miniprocessors.

Despite the strong Japanese commitment to it, PROLOG has a number of problematic features. Its "mixed-blessing" reliance on *general* methods of problem solving has already been mentioned. Second, there is disagreement over whether PROLOG is better regarded as a *declarative* program specification or a *procedural* programming language. Some PROLOG programmers use—and proselytize—the language in the former style, while others prefer to think of it in the latter way. But in practice (as indicated earlier), PROLOG can helpfully be regarded as a purely declarative language only in trivially simple cases. In order to write reasonably large PROLOG programs that actually work, the programmer must understand the procedural interpretation of PROLOG—that is, the order in which the program will do its various subtasks. Considering a PROLOG program merely as a *logical* system is not always sufficient, if one wants the (logically implicit) proof to be actually achieved.

The declarative/procedural distinction, as explained in Chapter 12, is in any case one of emphasis rather than essence: the *assertion* "A implies B" can be seen as a declarative representation of the (procedural) *inferential rule* "If you want to prove B, try to prove A." So for some purposes it may be best to interpret the "items" in a PROLOG program not as *facts* or *assertions,* but as *goals.* There might be some advantage, for example, in seeing the task of the so-called theorem-prover in PROLOG not as trying to prove assertions to be true but as trying to achieve certain goals (reporting its success or failure in doing so). The reason for suggesting this is that the problem-solving strategy used by the PROLOG theorem-prover leads to a third theoretical problem: it is not clear how *negation* in PROLOG expressions should be understood—indeed, under one very natural interpretation of what PROLOG programs do, it may very readily be *misunderstood.* For notwithstanding the label "*logic* programming," PROLOG's negation is not "real" negation: it is *not* equivalent to negation in the first-order predicate calculus, where "not-X" means that the assertion X is *false.*

Another way of putting this is to say that resolution theorem proving may obscure the distinction between "knowing that a statement is false" and "not knowing that it is true." For the species of *reductio ad absurdum* argument used by resolution theorem-provers (and so by every PROLOG program) is very naturally described as *trying to prove that the negation of assertion X is false.* This interpretation is especially tempting if one chooses to view PROLOG data "items" (declaratively) as *assertions,* which can be true or false. However, a PROLOG program's judgments of truth and falsity (still assuming that this is what it is up to) rely only on *whether or not it has managed to prove an absurdity*—which in general is unsafe. To be sure, if one manages to disprove the negation (the falsity) of an assertion, then the assertion must be true. But if one fails to disprove its negation, it may be true nevertheless: one might have had inadequate information. In short, "negation by failure," which is what resolution theorem proving provides, is absolutely reliable only in closed logical systems, where all the relevant information can be assumed to be present. Users of PROLOG programs must, in effect, treat the knowledge representation as a closed system; although in practice this is often acceptable, it is clearly open to misunderstanding.

One way of avoiding the confusion between "knowing that a statement is false" and "not knowing that it is true" is to avoid making queries about the statement's *truth* and *falsity* in the first place. For example, one may regard the task of a resolution theorem-prover as "trying to show that, given the knowledge base concerned, the negation of X is *provable (by it).*" If a PROLOG program is not even trying to prove the truth or falsity of assertions, but rather their provability (by it), then it cannot be accused of fostering the confusion just mentioned. It would then be regarded not as implementing a logic of the traditional type (like first-order predicate calculus), but as employing an "epistemic" or "modal" logic, which assigns truth values only to (second-order) statements about evidential relations or provability.

The three research areas outlined here are not the most generally visible aspects of artificial intelligence, and may strike some readers as distinctly arid. From the layman's point of view, "low-level vision," "connectionism," and "logic programming" are hardly words to conjure with. By comparison, "robots" and "expert systems" trip fairly readily off the tongue —what of them?

Work in robotics has become increasingly technical and mathematically based, and several specialist journals have been founded in the past few years. Robotics uses, and helps develop, the results in low-level vision already mentioned. Earlier industrial robots could "see" only things of a type (and, often, in a position) that had been explicitly foreseen by the programmer. That is, they relied on top-down, schema-driven methods comparable to those described in Chapters 8 and 9. This is still true of the industrial robots in common use today. But the most advanced robots of the 1980s can (sometimes with the aid of schema-driven processing) exploit the general image-interpreting algorithms mentioned earlier, to infer the surface geometry, the orientation, and the location of things they may or may not have encountered before. Certainly, they can do this only up to a point (special lighting conditions may be needed, for instance): the Asimov robot is not yet available.

Robotics must deal not only with perception (vision, touch, pressure sensitivity), but also with motor action. A robot must know (or be able to work out) what to do and how to do it. Consequently, the field is concerned with problems of trajectory planning, movement control, stability, calculation of forces, and visuomotor coordination.

One might think that only the first of these involves artificial intelligence, which has long worked on planning where to go and how to get there (cf. Chapter 12). The others, it might seem, are mere engineering. The engineering, however, is not "mere" at all. Very difficult theoretical problems are involved in (for example) controlling jointed robot-manipulators with the degrees of freedom of the human hand.[15] Likewise, enabling a camera-equipped robot to *avoid* the obstacles it sees in its (potential) path is no simple matter. Much as advance in low-level vision requires a deep understanding of physical optics, so advance in robotics will need to be grounded in the detailed theory of mechanics and dynamics. Roboticists must develop computational procedures that can exploit this knowledge in performing practical tasks. The crucial theoretical problems have not all been identified, still less solved. Today's commercial robots avoid these problems, but only by compromising their flexibility. Truly flexible robots, capable of undertaking a wide range of human labor, are still a very long way off.

"Sensible" robots will also need a representation of what AI workers call *naive physics:* one's everyday knowledge of the properties and behavior of different sorts of physical substances, and the nature of the causal relations between them.[16] This commonsense knowledge enters into vision and motor control, and also natural language. Thus a language-using robot (or program) would have to understand the differences in meaning between verbs such as *pour, flow, spill, drop,* and the like, if it were to give instructions or understand remarks about activities dealing with liquids. Similarly, a robot capable of "seeing" that a container was just about to *spill* its contents onto the object below, and of adjusting its movements accordingly, would need some representation of the behavior of fluids. (It might, of course, be programmed to halt movement if it saw an unexpected patch appearing at the rim of the container; but that is a different matter.)

Theoretical physics is not the answer here; physicists cannot define or carry out the requisite calculations (which is not to say that they have nothing of relevance to say about such matters). Every normal adult achieves such feats by qualitative reasoning, as opposed to the calculation of values in differential equations—and a tea-pouring robot, in Tokyo or Tyneside, would likewise need an understanding of *naive* physics rather than the mathematical variety. It is not surprising, then, that a special issue of the journal *Artificial Intelligence* was recently devoted to the topic of qualitative reasoning about the material world.[17]

Expert systems, too, are much less flexible than their human counterparts.[18] They are systems (composed of sets of if-then "production rules") that are still essentially comparable to the prototypes—MYCIN and DENDRAL—discussed in the main text (Chapters 11 and 15). Indeed, most expert systems in actual use are considerably less complex than either of these, each of which took many work-years to build. Expert systems are still restricted to very narrow domains. They can be incrementally improved only up to a point, at which the interactions between independently added rules become uncontrollable. In almost every case, their "explanations" are merely recapitulations of the previous firing of if-then rules (arrived at by "backwards chaining"), for they still have no higher-level representations of the knowledge domain, their own problem-solving activity, or the knowledge of their human user.

Admittedly, some researchers are attempting to provide expert systems with causal reasoning, to be used not only in arriving at a conclusion but also in explaining it to the user. For example, a computer-cardiologist can not only diagnose a wide range of abnormal ECGs (including some due not to a single arrhythmia, but to a compound of as many as seven distinct arrhythmias) in a superficial fashion, but relate them to—predict and explain them in terms of—an underlying theoretical model of the causal functions of the heart.[19] Similarly, the computer-artist that drew the two "lifelike" pictures shown on page 491, and many others of a similar type, was able to do so only because it exploited an underlying model of the body—one which made allowance for the way in which the limbs are foreshortened, and the muscles flexed, given certain bodily attitudes and viewpoints.[20] But "deep" (causal) reasoning has hardly entered the laboratory yet, still less left it; commercially available expert systems do not yet possess this sort of reasoning ability. Nor can they integrate knowledge from distinct domains, using concepts and patterns of inference taken from one domain to reason (analogically) in another. Research on the general architecture suitable for large knowledge bases should help, but—as remarked in the text—genuine expertise requires both high-level knowledge and analogical thinking. Work on all these issues is being done, but instant success is out of the question.

Success in the near future, in my view, is out of the question also. This view is not shared by all AI workers. For instance, at the MCC research center (mentioned earlier), one ambitious project is aimed at developing—by 1995—a representation of commonsense knowledge and analogical reasoning that could be accessed by *any* AI program, including expert systems dealing with indefinitely many domains.[21]

The researchers' stated aim is "to represent all the world's knowledge, to a certain level of detail," which is to be done by representing the knowledge contained in—*and* the background, commonsense knowledge needed to understand—a one-volume encyclopedia. They plan to identify the 400 or so encyclopedia entries requiring maximally distinct knowledge representations (lions and climate, perhaps: lions and elephants, no), and to encode them as hierarchically linked "frames" having "slots" of various kinds (see Chapter 11). They do not see this as 400 distinct tasks: rather, they use the effort-saving method of "copy-and-edit" to exploit and identify *analogies* between different frames (treating an illness, for example, is comparable to waging a war—and corresponding slots can be identified as between the two frames). Eventually, they claim, programmed procedures for frame matching will allow this copy-and-edit process (at present effected manually) to be carried out by the system itself, so that adding the rest of the 30,000 entries will be done relatively easily—and largely automatically.

As explained in Chapter 11, a program capable of recognizing analogies could be left to "brainstorm" in its spare time so as to spin new analogies, some of which might turn out to be useful. This research team

at MCC plans to enable its program to do this. It should eventually be able, for instance, to appreciate the analogy (mentioned in the text) between the Black Mass and the ritual of the Catholic church. Likewise, the various generalizations of neurosis (explored in Chapter 3)—whereby hatred of one's father can lead to dislike of one's boss, suspicion of atheists, and so on—should be within the generative capacity of such a system (if it can be built at all).

Certainly, if the expert systems of the future could be hooked into such an automated commonsense knowledge base, this would overcome the "brittleness" of current expert systems. Today's programs have an unfortunate tendency to give utterly stupid answers (or none at all) to questions which—to us—seem only slightly different from the questions for which they were specifically designed. But can this improvement be achieved? The work in naive physics, for example, shows that the difficulties involved are enormous. Although *some* improvement, of *some* systems, may be possible by these means, a high degree of skepticism is surely in order.

Expert systems would be much more useful also if they could learn (and analogical reasoning could of course help them to do so). As Chapter 10 explains, learning (implicitly) *from experience* is ill understood. Even learning *by being (explicitly) told* raises difficult problems of "truth maintenance," since new information has to be not merely added to but integrated with the information already present, often triggering new inferences (in an example given in Chapter 10, *that the mayor is a murderer*).[22] At present, therefore, programs that can learn are highly circumscribed in what they can achieve.

Admittedly, the learning algorithm built into one commercially available expert-system shell[23] can generate a set of diagnostic questions that is more efficient than the set initially provided by the programmer: the *present* world expert on soya-bean diseases is a program written by the *previous* world expert (a human). This algorithm has even provided useful new concepts concerning chess end-games and—more important—abnormal heart rhythms. But these claims about its "experiential learning" call for some demystification, if they are not to be misleading.

Essentially, what this off-the-shelf algorithm does is to tidy up the logic of the diverse diagnostic rules provided by the human expert. The human gives it a number of "logically flat" rules, of the form "If a thing is A, B, and C, but not D, then it *is* a squoggle" and "If a thing is A, D, and Q, then it is *not* a squoggle" (rules which may be prompted by considering a sample of instances falling within the class). Then, the program is presented with a large number of examples and counterexamples (labeled as such). The learning algorithm does two things. It checks for inconsistencies among the rules (for instance, catching cases where both *A* and *not-A* have been said to be necessary for being a *squoggle*). And it finds the shortest tree-search by which all the relevant questions can be asked in the most helpful order. This is not a purely logical matter, but depends partly on the actual probabilities of a rule's being applicable to the test-sample. (If only 1 percent of *squoggles* possess property Q, then it is not sensible to ask about Q *at the outset,* even though having *both P and Q* may be enough to qualify something as a *squoggle.*)

Recent analytical work on a variety of current learning programs has shown that, despite their superficial diversity, they employ a small number of computational techniques.[24] These techniques were employed by some of the pioneering programs described in Chapter 10 of the main text, but their power and limitations are now better understood. Theoretical analysis proves, for example, that the algorithm described in the previous paragraph is *guaranteed* to find the shortest tree-search, given the data and rules concerned. In general, theoretical analysis, which has been relatively neglected till now, helps in identifying the true significance of AI research, and can prevent the repetition of essentially similar work.

In discussing creativity in Chapter 11, I argued that programs could in principle be creative, but that the central computational issues involved —the generation and exploration of new representations, and the functioning of analogy—are ill understood. Work on analogical reasoning is now being done in several places, but the best examples of machine creativity so far are a group of programs focused on the use—and the self-transformation—of heuristics for generating and evaluating novel concepts.[25] These programs do "creative" mathematics, or suggest experiments in genetic engineering, or design new 3D chips. In each case, they have come up with new ideas, some of which have been commercially useful. And in each case, they make their discoveries by using a mixture of very general rules of transformation and domain-specific expertise (concerned with set theory, molecular structures, and logic circuits).

For example, the "Automatic Mathematician" starts from a collection of primitive mathematical concepts, so basic that they do not even include the concepts of elementary arithmetic, and uses transformational rules (heuristics) to explore the space potentially defined by the primitive concepts. Much as creative mathematics involves not doing sums but playing with mathematical ideas to see if anything interesting comes up, so this program generates and then explores concepts and hypotheses about number theory, guided by its evaluative "hunches" about which concepts are likely to be the most interesting. It has come up with some mathematically powerful ideas, including primes, square roots, addition, and multiplication—which it notices can be performed in four different ways, itself a mathematically interesting fact. It has also originated one minor theorem (about the class of "maximally divisible" numbers) which no one had ever thought of before.

Critics have complained, however, that it is difficult to see just where the power of this program lies.[26] It is unclear, for example, whether writing the program in LISP is equivalent to building in all kinds of theoretically powerful ideas. And one analysis suggests that the heuristic primarily responsible for the program's "discovery" of prime numbers is used only on that occasion, and was included by the programmer in the first place only because he recognized (whether consciously or not) its potential relevance to primes. This is a special case of the point made earlier, that an AI program presented without careful theoretical analysis of just what it is doing, and how, may be superficially impressive but basically trivial. It is because detailed theoretical analysis is emphasized so strongly within recent research on low-level vision, that progress in this domain has been relatively great.

Just as more "humanlike" expert systems will need to be able to learn, and to create and evaluate new ideas, so they will need common sense. Common sense can be ignored with relative impunity where there are few surprises, so that the problem (and the possible answers) can be strictly defined beforehand. Current expert systems are designed for precisely that class of cases. But human expertise depends on, and powerful expert systems will necessarily require, commonsense reasoning of various kinds (including the knowledge of naive physics mentioned earlier).

Common sense is needed, for example, when someone has to deal with a problem without having complete knowledge. The person has to make guesses about what the relevant facts may be. But if it turns out later that one of these guesses was incorrect, they can use that new information from then on. That is, people can cope with the fact that a statement justifiably assumed to be true (or false) at one time can later be found to be false (or true). This cannot happen in traditional logic, wherein truths are proved once and for all. Today's problem-solving programs are based on this type of logic. Consequently, a significant amount of AI research at present is trying to formalize "nonmonotonic" reasoning, in which truth values can shift from time to time as relevant information reaches the system.[27] The problem of truth maintenance (necessary for learning in general) is obviously even greater where the new information is not only *new,* but possibly *inconsistent* with some of the items already present in the memory.

Some of the issues involved here have been illustrated by John McCarthy (the inventor of LISP and one of the most important figures in the history of artificial intelligence), in a discussion of "circumscription," a method allowing some flexibility in what can count as "relevant."[28] McCarthy imagines having a conversation with a friend, in which he asks the friend to solve the "missionaries and cannibals" puzzle—which, as explained in Chapters 11 and 12, can be solved by many AI programs.

The imaginary friend, instead of talking about journeys of one or two people across the river and back again, immediately says, "They all walk across the bridge." McCarthy tells him there is no bridge. His friend next says, "A large crocodile swims up to them, and all six stand on its back and let it carry them across the river." With some impatience, McCarthy says, "No crocodiles!" Now the friend suggests that they phone the nearest airport and hire a helicopter . . . As McCarthy points out, this conversation could go on for ever.

It is the friend's common sense, or general knowledge, which would enable him to tease McCarthy in this way. A *puzzle,* by definition, is a problem in which it is assumed that one will *not* try to use one's common sense. Thus to accept the puzzle in the spirit in which it was given would have meant accepting that no such extraneous factors can be considered —even though they are not explicitly excluded in the statement of the problem. A normal "problem-solving" computer would not consider them either, of course, for it does not even know about bridges, crocodiles, and helicopters. But the "equivalent" problem—*problem,* not puzzle—in real life *might* be solvable in these terms. Even a crocodile might come in useful, if anyone was able (as some people are) to pacify animals by their presence; and if a Tarzan film were being made on location nearby, a helicopter (and a phone line) might well be available. A program lacking common sense would be unable to appreciate this, or to make helpful suggestions accordingly.

We should not assume, however, that one helpful suggestion from a computer program is *necessarily* better than none at all. People are accustomed to interpreting suggestions in a way that may be inappropriate—indeed, dangerous—where computer "experts" are concerned. In real life the choice of one commonsense suggestion would often be taken (intelligently) to imply the rejection of another. Thus if one of the stranded men were to suggest using a crocodile, his companions would immediately assume that he knew of no nearby bridge. For if he did, he would have had enough common sense to mention it first (because he would *know* that crossing a bridge is usually less risky than cavorting with crocodiles). Indeed, if he were one of the cannibals, born and bred in that part of the jungle, his companions would be *stupid* not to infer that no bridge in fact exists.

It follows that an expert system with *some* common sense, but not much, might be more socially dangerous than one with none at all—because laymen might credit it with more reasoning power than it actually has. For example, if a problem-solving program were to suggest riding a crocodile to get across the river, one would naturally [*sic*] assume that it must be bright enough to know about bridges, and that it must therefore have some reason for not bothering to mention them. In general, if a future expert system makes one apparently commonsensical suggestion or mentions one seemingly sensible consideration, the human user might imagine that other potentially relevant ideas are also available to the program. It would seem to follow that, if they are not mentioned, this must be because the program has considered them but judged them to be irrelevant in this particular case. In fact, however, the program might be totally ignorant of them.

Similarly, an apparently "natural" use of English by a program might mislead the user into assuming that it chose a particular word because it had reason to reject each of the other similar-but-not-identical terms that are available to the human speaker. But the program might not know about all, or even any, of these other possibilities. In general, an obviously "stupid" program is in some ways less dangerous than one that looks more intelligent than it is.

This raises the issue already discussed in Chapter 15—the importance of preventing the layman from being fooled by programs that are less intelligent than they appear. A type of computer literacy that highlights the limitations of programs, as well as their potential, could be fostered by a certain way of playing around with programs (already happening in some universities and schools).[29]

For instance, if a person can play with a version of ELIZA, and can "take the lid off" to see how the program works, it soon becomes clear that the program is not so clever as at first it seems, and that some of its faults *cannot* be rectified simply by adding a few more rules. After having had this experience, no one could ever interact with a natural-language interface (or program) without wondering what its linguistic limitations were. They might not be able to find the answer, but the cautionary question would have been raised. At the least, they could never be fooled into thinking that the word-producing program must have the degree of understanding of its words that human beings have.

Also, this type of computer literacy could help to correct the currently popular belief (rebutted in Part V) that computers are *objective* systems. This belief is not merely false, but is a socially oppressive mystification—one which is sometimes deliberately encouraged for questionable purposes (for example, by advertisements saying that because the computer chose a product as "the best," you should choose it too). On the contrary, programs are representations of the world, and ways of thinking about it, and as such are essentially open to question: some other representation, not excluding the one inside the user's head, might be preferable.

This point could be deliberately brought to the public's attention by the designers and purveyors of expert systems. Their AI programs, including "natural language" interfaces, could be written so as to remind the nonspecialist user that their data, reasoning, and decision criteria are different from a human being's (and, for that matter, from those of other possible programs)—and, in principle, just as challengeable. Some very simple things might help: for example, reminding the user more than once during use (not just at log-in time) that the program has been written by *X,* based on advice and data from *Y,* and following decision criteria (the general priorities of which should ideally be indicated) approved by *Z.* In addition, the "human window," through which users can "interrogate" a system, should not merely reveal how a certain conclusion has been generated: it should help the user to have some sense of what possibly relevant factors the system has *not* been able to consider. For example, the program could warn the user of some of its limitations by presenting some examples of the *stupid* answers that it would give to what might appear to be "appropriate" questions.

Professional programmers may feel that it goes against the grain to include such "negative" or "self-deprecatory" reminders in a system of which one is justly proud, and by means of which one perhaps hopes to become rich. But it is the responsible thing to do. Indeed, reminders of the limitations, both general and specific, of computer programs may even become legally required, as artificial intelligence penetrates the marketplace. Large sums of money will sometimes turn on the ascription of responsibility for a decision, or for the advice leading to that decision. In legal systems based on case law, decisions are based on precedents arrived at in previous litigation. General principles are eventually laid down by the judges, on the basis of specific cases. But anticipatory consideration of the principles involved may be useful, and some legally and technically informed people are exploring such issues.[30]

The issue of responsibility cannot be left without mentioning the potential use of artificial intelligence for military purposes. Computers are already prevalent in military technology, for example, within the anti-nuclear "early-warning" systems (on which, more later). Now there is the added possibility of using "intelligent" computers to solve complex problems (often in real time) where not all the potentially relevant information is available. Several widely advertised technical conferences have already been held on applications of artificial intelligence in the "automatic battlefield," from robot-minesweepers and unmanned tanks, through automatic enemy reconnaissance, to computerized battle strategists. Classified government-funded work on related topics is going on in various countries. And in March 1983, the United States government announced its Strategic Defense Initiative (SDI), popularly known as "Star Wars," to which advanced computer technology is essential.

No one but a pacifist has the right to condemn all conceivable military uses of AI technology wholesale. Nor can a nonpacifist reasonably regard all computer professionals involved in military work as ipso facto irresponsible (though one might argue that virtually all such work that takes place within a particular political context is unjustifiable). Moreover, in evaluating the moral-political wisdom of specific projects, one should not ignore the distinction, unclear though it may sometimes be, between offensive and defensive uses of military technology. Nor should one fail to distinguish between "active" and "passive" defense; to defend oneself with a sword is one thing, with a shield another.

However, such evaluations must take account not only of political criteria but of informed technical judgment too. In Chapter 15 it was argued that such judgment should come not only from individuals but from the AI community *as an institution.* This has begun to happen. For example, in late 1983 the society Computer Professionals for Social Responsibility (CPSR) was founded. It publishes a newsletter several times a year, and now has groups in many U.S. cities; comparable groups exist in several other countries, including the U.K.[31] In 1985, the British Computer Society initiated a number of meetings to look into the military implications of computing. And the visit to England of the United States vice-president, George Bush, in 1985 elicited a media campaign (including a letter signed by seventy British computer professionals) criticizing the technical assumptions implicit in the Star Wars project. As is evident from all these discussions, many professionals are convinced that there are *technical* reasons why some suggested applications should cause even the most hawkish among us to think twice, given the horrendous price—and the *unavoidable* possibility—of failure.

The prime—though not the only—example is the Strategic Defense Initiative. At a meeting organized in late 1985 by the Edinburgh chapter of CPSR, Colonel Boli of the U.S. Air Force (who was involved in the SDI program even before it was publicly announced) described SDI as a *research* initiative, and stated that no policy decision on its actual deployment has yet been reached. Many computing professionals, including some from the AI community, are attempting to persuade the political decision makers that such research *cannot* lead to an entirely reliable system. For, quite apart from the *strategic* arguments against SDI,[32] there are many objections on purely technical grounds. In December 1985, the mounting professional criticism of Star Wars was voiced as evidence before the U.S. Senate Subcommittee on Strategic and Theater Nuclear Forces (which—readers of Chapter 4 may be interested to know—included Senator Barry Goldwater as a member).

Among the experts testifying to the Senate committee was David Parnas, a senior computer scientist who has long been deeply involved in defense-sponsored research (including eight years' study of real-time software used in military aircraft). So far is he from having any *political* axe to grind in these matters, that in June 1985 he accepted the U.S. government's invitation to join their top advisory group on the computing aspects of Star Wars (the SDI Committee on Computing in Support of Battle Management). Soon after the initial meeting of this committee, however, Parnas resigned—publishing his letter of resignation (with eight brief explanatory appendices), which later formed the basis of his Senate testimony.[33] Despite the enormous temptations (huge sums of money and a host of challenging intellectual problems) to become involved, he felt it irresponsible to do so. For, in his view, the Star Wars project cannot succeed: it is impossible *in principle.*

Parnas's technical arguments concern the unreliability of computer software in general, and of much AI software in particular. With reference to the specific properties of the proposed SDI software, he shows why the techniques commonly used to build military software are inadequate for this project, and why no amount of research in software engineering, program verification, or automatic programming could provide a truly reliable SDI system.

Programs (such as this one would be) consisting of many millions of lines of code cannot in practice be free of "bugs." Moreover, even if—*per impossibile*—such a program actually was bug-free, it could not be *proved* to be so. It could not even be "validated" by being *tested,* for any test of this destructive arsenal would have to be carried out as a simulation—but the simulation would, of course, be constructed on the same general assumptions as the system itself. Further, one cannot assume that the problem (in this case, how to identify and shoot down enemy missiles) has been correctly specified by the human programmer. Even relatively simple problems are sometimes wrongly specified by (or to) the programmer. (A fearsome example of this concerned the early-warning system in 1960, when a "Red Alert" was caused by a program's mistaking the rising moon for a Soviet missile-attack. Western planes carrying nuclear weapons had already taken off before someone realized that the program had been provided by the programmer with a calendar that did not make allowance for leap-years, so that it did not *expect* to see the moon in that place at that time.)[34] Above all, many of the necessary specifications are not cut-and-dried. Military computer systems such as those suggested for SDI rely on diverse types of information, which have to be differentially weighted according to the specific situation, and which can change very rapidly in real time. Such a system has to make many of its decisions before all the relevant information is available, so there is *always* a theoretical possibility of error.

Artificial intelligence is especially concerned with such decisions. But even the best AI work, which (like good computer science is general) is grounded in a theoretical understanding of the task constraints, cannot provide implementations that will always make the right decision in complex circumstances. And the more complex the task, the less likely it is that abstract theoretical constraints (encompassing the entire problem range) could be specified.

In such cases, there is no alternative but to use some form of heuristic problem solving—which is even less reliable. Heuristic programs, by definition, cannot be guaranteed to reach the correct, or even an optimal, conclusion. As explained in Chapter 12, these programs have to "best-guess" on the basis of incomplete information. Their guessing rules can be justified only on the basis of *past* experience (their own and/or their programmer's). But, no matter how good the heuristics, *tomorrow's* experience may be different. Further, most AI programs—including rule-based expert systems—can handle only a small number of obvious cases, and are rarely generalizable. Unlike humans, whose behavior in difficult circumstances shows "graceful degradation" (it gets gradually more inadequate), most AI programs are brittle. When they go wrong, they commonly go *very* wrong.

Countless examples in the main text illustrate the fact that it is the relatively *successful* performances of AI programs which can be appropriately described as "inadequate." Their *failures* are not so much inadequate as crazy.

The aim of the encyclopedia project mentioned—with some skepticism—earlier is to provide knowledge and analogical reasoning to alleviate this problem. But to alleviate is not to abolish. Even if (setting skepticism aside) the entire *Encyclopaedia Britannica* could be represented, and analogical reasoning of considerable subtlety could be provided, foolishness of various kinds might still result. The brittleness of AI programs was lampooned in a hilariously funny critique written (by a leading AI worker) a decade ago, before artificial intelligence lost its innocence.[35] Now, it is no laughing matter.[36]

# BIBLIOGRAPHY

See page 521 for Bibliography to Postscript

*LIST OF ABBREVIATIONS USED IN NOTES AND BIBLIOGRAPHY*

*AI* — *Artificial Intelligence.* (Journal)

*AISB-1* — *Proc. AISB Summer Conference, July 1974.* University of Sussex. (For the Society for the Study of Artificial Intelligence and Simulation of Behaviour.)

*AISB-2* — *Proc. AISB Summer Conference, July 1976.* University of Edinburgh.

*CACM* — *Communications of the Association for Computing Machinery.*

*CMTL* — *Computer Models of Thought and Language.* Eds. R. C. Schank and K. M. Colby. San Francisco: W. H. Freeman, 1973.

*CT* — *Computers and Thought.* Eds. E. A. Feigenbaum and Julian Feldman. New York: McGraw-Hill, 1963.

*IJCAI-1* — *Proc. First International Joint Conference on Artificial Intelligence.* Washington, D.C.: 1969.

*IJCAI-2* — *Proc. Second International Joint Conference on Artificial Intelligence.* London: 1971.

*IJCAI-3* — *Proc. Third International Joint Conference on Artificial Intelligence.* Stanford: 1973.

*IJCAI-4* — *Proc. Fourth International Joint Conference on Artificial Intelligence.* Tbilisi, USSR: 1975.

*JACM* — *Journal of the Association for Computing Machinery.*

*MI-1* — *Machine Intelligence 1.* Eds. N. L. Collins and Donald Michie. Edinburgh: Edinburgh University Press, 1967. (Originally published by Oliver & Boyd.)

*MI-2* — *Machine Intelligence 2.* Eds. Ella Dale and Donald Michie. Edinburgh: Edinburgh University Press, 1968. (Originally published by Oliver & Boyd.)

*MI-3* — *Machine Intelligence 3.* Ed. Donald Michie. Edinburgh: Edinburgh University Press, 1968.

*MI-4* — *Machine Intelligence 4.* Eds. Bernard Meltzer and Donald Michie. Edinburgh: Edinburgh University Press, 1969.

*MI-5* — *Machine Intelligence 5.* Eds. Bernard Meltzer and Donald Michie. Edinburgh: Edinburgh University Press, 1970.

*MI-6* — *Machine Intelligence 6.* Eds. Bernard Meltzer and Donald Michie. Edinburgh: Edinburgh University Press, 1971.

*MI-7* — *Machine Intelligence 7.* Eds. Bernard Meltzer and Donald Michie. Edinburgh: Edinburgh University Press, 1972.

*PCV* — *The Psychology of Computer Vision.* Ed. P. H. Winston. New York: McGraw-Hill, 1975.

*RU* — *Representation and Understanding: Studies in Cognitive Science.* Eds. D. G. Bobrow and Allan Collins. New York: Academic Press, 1975.

*SIP* — *Semantic Information Processing.* Ed. M. L. Minsky. Cambridge, Mass.: MIT Press, 1968.

*TINLP* — *Theoretical Issues in Natural Language Processing.* Eds. R. C. Schank and B. N. Nash-Webber. Proc. Workshop of Ass. Computational Linguistics, June 1975.

Aaron, R. I. *The Theory of Universals.* Oxford: Clarendon, 1952.

Abelson, R. P. "Computer Simulation of 'Hot' Cognition," in *Computer Simulation of Personality: Frontier of Psychological Research* (eds. S. S. Tomkins and Samuel Messick), pp. 277–298. New York: Wiley, 1963.

———. "The Structure of Belief Systems," in *CMTL*, pp. 287–340.

———. "Concepts for Representing Mundane Reality in Plans," in *RU*, pp. 273–309.

———. "Script Processing in Attitude Formation and Decision-Making," in *Cognition and Social Behavior* (eds. J. S. Carroll and J. W. Payne), Hillsdale, N.J.: Lawrence Erlbaum, (1976), 33–46.

Abelson, R. P., and J. D. Carroll. "Computer Simulation of Individual Belief Systems," *Am. Behav. Scientist*, 8 (1965), 24–30.

Abelson, R. P., and C. M. Reich. "Implicational Molecules: A Method for Extracting Meaning from Input Sentences," in *IJCAI-1*, pp. 641–647.

Abelson, R. P., and M. J. Rosenberg. "Symbolic Psycho-Logic: A Model of Attitudinal Cognition," *Behav. Sci.*, 3 (1958), 1–13.

Adler, M. R., "Recognition of Peanuts Cartoons," in *AISB-2*, pp. 1–13.

Amarel, Saul. "On Representations of Problems of Reasoning About Actions," in *MI-3*, pp. 131–172.

———. "Representations and Modeling in Problems of Program Formation," in *MI-6*, pp. 411–466.

Ambler, A. P., H. G. Barrow, C. M. Brown, R. M. Burstall, and R. J. Popplestone. "A Versatile Computer-Controlled Assembly System," in *IJCAI-3*, pp. 298–307.

Anderson, J. R., and G. H. Bower. *Human Associative Memory.* Washington, D.C.: Winston, 1973.

Anderson, J. R., and Reid Hastie. "Individuation and Reference in Memory: Proper Names and Definite Descriptions," *Cognitive Psychology*, 6 (1974), 495–514.

Arbib, M. A. *The Metaphorical Brain: An Introduction to Cybernetics as Artificial Intelligence and Brain Theory.* New York: Wiley, 1972.

———. "Artificial Intelligence and Brain Theory: Unities and Diversities," *Ann. Biomedical Engineering,* 3 (1975), 238–274.

———. *Brain Theory and Artificial Intelligence.* New York: Academic Press, 1976.

Armer, Paul. "Attitudes Toward Intelligent Machines," in *CT*, pp. 389–405.

Asimov, Isaac. *I, Robot.* London: Dennis Dobson, 1967.

Austin, J. L. *Philosophical Papers.* Oxford: Clarendon, 1961.

———. *How to Do Things with Words.* Oxford: Oxford University Press, 1962.

Ayer, A. J. *Language, Truth, and Logic.* 2nd ed. London: Gollancz, 1946.

Bajcsy, Ruzena. "Computer Description of Textured Surfaces," in *IJCAI-3*, pp. 572–577.

Baker, H. H. *Machine Vision and Its Place in an Industrial Environment.* Edinburgh: Machine Intelligence Research Unit, 1975.

Barlow, H. B. "Single Units and Sensations: A Neuron Doctrine for Perceptual Psychology," *Perception*, 1 (1972), 371–394.

Barratt, Barnaby. "Freud's Psychology as Interpretation," in *Psychoanalysis and Contemporary Science, Vol. V* (Ed. Theodore Shapiro). New York: International Universities Press, (1976), 443–478.

Barrow, H. G., and R. J. Popplestone. "Relational Descriptions in Picture Processing," in *MI-6*, pp. 377–396.

Bartlett, F. C. *Remembering: A Study in Experimental and Social Psychology.* Cambridge: University Press, 1932.

Bennett, Jonathan. *Rationality: An Essay Towards an Analysis.* London: Routledge & Kegan Paul, 1964.

Berliner, H. J. "Some Necessary Conditions for a Master Chess Program," in *IJCAI-3*, pp. 77–85.

Birch, H. G. "The Relation of Previous Experience to Insightful Problem Solving," *J. Comp. Psychol.*, 38 (1945), 367–383.

Bobrow, D. G. "Natural Language Input for a Computer Problem-Solving System," in *SIP*, pp. 146–226.

———. "Requirements for Advanced Programming Systems for List Processing," in *CACM*, 15 (1972), 618–627.

———. "Dimensions of Representation," in *RU*, pp. 1–34.

Bobrow, D. G., and Allan Collins, eds. *Representation and Understanding: Studies in Cognitive Science.* New York: Academic Press, 1975.

Bobrow, D. G., and Bertram Raphael. "A Comparison of List Processing Languages," in *CACM*, 7 (1964), 231–240.

———. "New Programming Languages for Artificial Intelligence Research," *ACM Computing Surveys*, 6 (1974), 155–174.

Boden, M. A. "The Paradox of Explanation," *Proc. Aristot. Soc.*, N.S., 62 (1962), 159–178.

———. "Intentionality and Physical Systems," *Philosophy of Science*, 37 (1970), 200–214.

———. *Purposive Explanation in Psychology.* Cambridge, Mass.: Harvard University Press, 1972.

———. "The Structure of Intentions," *J. Theory Soc. Behav.*, 3 (1973), 23–46.

———. "Freudian Mechanisms of Defence: A Programming Perspective," in *Freud: A Collection of Critical Essays* (ed. Richard Wollheim), pp. 242–270. New York: Anchor, 1974.

Bransford, J. D., and M. K. Johnson. "Considerations of Some Problems of Comprehension," in *Visual Information Processing* (ed. W. G. Chase), pp. 383–438. New York: Academic Press, 1973.

Braunstein, M. L. "The Perception of Depth Through Motion," *Psychological Bulletin*, 59 (1962), 422–433.

Brice, C. R., and C. L. Fennema. "Scene Analysis Using Regions," in *AI*, 1 (1970), 205–226.

Brown, Roger. *Social Psychology.* New York: Free Press, 1965.

———. *A First Language: The Early Stages.* Cambridge, Mass.: Harvard University Press, 1973.

Bruner, J. S. "Personality Dynamics and the Process of Perceiving," in *Perception: An Approach to Personality* (eds. R. R. Blake and G. V. Ramsey), pp. 121–147. New York: Ronald Press, 1951.

———. "On Perceptual Readiness," *Psychological Review*, 64 (1957), 123–152.

Bruner, J. S., Jacqueline Goodnow, and George Austin. *A Study of Thinking.* New York: Wiley, 1956.

Bruner, J. S., and Leo Postman. "Emotional Selectivity in Perception and Reaction," *J. Personality*, 16 (1947), 69–77.

Buchanan, B. G., E. A. Feigenbaum, and N. S. Sridharan. "Heuristic Theory Formation: Data Interpretation and Rule Formation," in *MI-7*, pp. 267–290.

Buchanan, B. G., G. L. Sutherland, and E. A. Feigenbaum. "Heuristic DENDRAL: A Program for Generating Explanatory Hypotheses in Organic Chemistry," in *MI-4*, pp. 209–254.

———. "Rediscovering Some Problems of Artificial Intelligence in the Context of Organic Chemistry," in *MI-5*, pp. 252–280.

Buchanan, B. G., and N. S. Sridharan. "Analysis of Behavior of Chemical Molecules: Rule Formation on Non-Homogeneous Classes of Objects," in *IJCAI-3*, pp. 67–76.

Card, W. I., Mary Nicholson, G. P. Crean, Geoffrey Watkinson, C. R. Evans, Jackie Wilson, and Daphne Russell. "A Comparison of Doctor and Computer Interrogation of Patients," *Int. J. Biomedical Computing*, 5 (1974), 175–187.

Carroll, Lewis. "What the Tortoise Said to Achilles," *Mind*, N.S., 4 (1895), 278–280.

———. *Symbolic Logic: Part I, Elementary*. London: Macmillan, 1896.

Cercone, Nick, and Len Schubert. "Toward a State Based Conceptual Representation," in *IJCAI-4*, pp. 83–90.

Chafe, W. L. *Meaning and the Structure of Language*. Chicago: University of Chicago Press, 1970.

Charniak, Eugene. *Toward a Model of Children's Story Comprehension*. AI-TR-266. Cambridge, Mass.: MIT AI Lab, 1972.

———. "Jack and Janet in Search of a Theory of Knowledge," in *IJCAI-3*, pp. 337–343.

———. *"He Will Make You Take It Back": A Study in the Pragmatics of Language*. Castagnola, Switz.: Ist. Studi Semantici e Cognitivi, 1974.

———. "Organization and Inference in a Frame-Like System of Common Knowledge," in *TINLP*, pp. 46–55.

Charniak, Eugene, and Y. A. Wilks, eds. *Computational Semantics: An Introduction to Artificial Intelligence and Natural Language Comprehension*. Amsterdam: North-Holland, 1976.

Chase, W. G., and H. A. Simon. "Perception in Chess," *Cognitive Psychology*, 4 (1973), 55–81.

Chisholm, R. M. "Intentionality," in *The Encyclopedia of Philosophy, Vol. IV* (ed. Paul Edwards), pp. 201–204. New York: Macmillan, 1967.

Chomsky, Noam. *Aspects of the Theory of Syntax*. Cambridge, Mass.: MIT Press, 1965.

———. *Language and Mind*. New York: Harcourt, Brace, & World, 1968.

———. *Problems of Knowledge and Freedom*. London: Fontana, 1972.

Cicourel, A. V. *Cognitive Sociology: Language and Meaning in Social Interaction*. Harmondsworth: Penguin, 1973.

Cioffi, Frank. "Freud and the Idea of a Pseudo-Science," in *Explanation in the Behavioural Sciences* (eds. Robert Borger and Frank Cioffi), pp. 471–499. London: Cambridge University Press, 1970.

Clare, Anthony. *Psychiatry in Dissent: Controversial Issues in Thought and Practice*. London: Tavistock, 1976.

Clarke, A. C. *2001: A Space Odyssey*. London: Hutchinson, 1968.

———. "The Mind of the Machine," *Playboy*, Dec. 1968, pp. 116–119, 122, 293–294.

Clowes, M. B. "On Seeing Things," in *AI*, 2 (1971), 79–116.

———. "Man the Creative Machine: A Perspective from Artificial Intelligence Research," in *The Limits of Human Nature* (ed. Jonathan Benthall), pp. 192–207. London: Allen Lane, 1973.

Colby, K. M. *Energy and Structure in Psychoanalysis*. New York: Ronald Press, 1955.

———. "Computer Simulation of a Neurotic Process," in *Computer Simulation of Personality: Frontier of Psychological Research* (eds. S. S. Tomkins and Samuel Messick), pp. 165–180. New York: Wiley, 1963.

———. "Experimental Treatment of Neurotic Computer Programs," *Arch. Gen. Psychiatry*, 10 (1964), 220–227.

———. "Computer Simulation of Neurotic Processes," in *Computers in Biomedical*

*Research, Vol. I* (eds. R. W. Stacy and B. D. Waxman), pp. 491–503. New York: Academic Press, 1965.

———. "Computer Simulation of Change in Personal Belief Systems," *Behavioral Science*, 12 (1967), 248–253.

———. "Rationale for Computer Based Treatment of Language Difficulties in Non-Speaking Autistic Children," *J. Autism and Childhood Schizophrenia*, 3 (1973), 254–260.

———. "Simulations of Belief Systems," in *CMTL*, pp. 251–286.

———. *Artificial Paranoia.* New York: Pergamon, 1975.

———. "Clinical Implications of a Simulation Model of Paranoid Processes," *Arch. Gen. Psychiatry*, 33 (1976), 854–857.

Colby, K. M., and Horace Enea. "Heuristic Methods for Computer Understanding of Natural Language in Context-Restricted On-Line Dialogues," *Mathematical Biosciences*, 1 (1967), 1–25.

Colby, K. M., and J. P. Gilbert. "Programming a Computer Model of Neurosis," *J. Math. Psychol.*, 1 (1964), 405–417.

Colby, K. M., F. D. Hilf, Sylvia Weber, and H. C. Kraemer. "Turing-Like Indistinguishability Tests for the Validation of a Computer Simulation of Paranoid Processes," in *AI*, 3 (1972), 199–222.

Colby, K. M., and R. C. Parkison. "Pattern-Matching Rules for the Recognition of Natural Language Dialogue Expressions," *Am. J. Computational Linguistics*, 1 (1974).

Colby, K. M., J. B. Watt, and J. P. Gilbert. "A Computer Method of Psychotherapy: Preliminary Communication," *J. Nerv. Mental Disease*, 142 (1966), 148–152.

Colby, K. M., Sylvia Weber, and F. D. Hilf. "Artificial Paranoia," in *AI*, 2 (1971), 1–26.

Coles, L. S. *Some Thoughts on Robot Ethics.* Privately circulated ms., Menlo Park, Calif.: SRI, Sept. 1969.

Collins, Allan. "Process in Acquiring Knowledge," in *Schooling and the Acquisition of Knowledge* (eds. R. C. Anderson, R. J. Spiro, and W. E. Montague). Hillsdale, N.J.: Lawrence Erlbaum, 1977.

Collins, Allan, E. H. Warnock, Nelleke Aiello, and M. L. Miller. "Reasoning from Incomplete Knowledge," in *RU*, pp. 383–415.

Cooper, L. A., and R. N. Shepard. "Chronometric Studies of the Rotation of Mental Images," in *Visual Information Processing* (ed. W. G. Chase), pp. 75–176. New York: Academic Press, 1973.

Craik, K. J. W. *The Nature of Explanation.* Cambridge: University Press, 1943.

Crosson, F. J., ed. *Human and Artificial Intelligence.* New York: Appleton Century Crofts, 1970.

Danto, A. C. "On Consciousness in Machines," in *Dimensions of Mind: A Symposium* (ed. Sidney Hook), pp. 180–187. New York: New York University Press, 1960.

Davies, D. J. M. *POPLER 1.5 Reference Manual.* TPU Report 1. Edinburgh: Edinburgh Univ. AI Lab., 1973.

———. "Representing Negation in a Planner System," in *AISB-1*, pp. 26–36.

Davies, D. J. M., and S. D. Isard. "Utterances as Programs," in *MI-7*, pp. 325–340.

de Bono, Edward. *Lateral Thinking: A Textbook of Creativity.* London: Ward Lock Educational, 1970.

de la Mettrie, J. O. *L'Homme Machine.* Ed. Aram Vartanian. Princeton: Princeton University Press, 1960.

Dennett, D. C. "Machine Traces and Protocol Statements," *Behavioral Science*, 13 (1968), 155–161.

———. *Content and Consciousness.* London: Routledge & Kegan Paul 1969.

———. "Toward a Cognitive Theory of Consciousness," in *Minnesota Studies in Philosophy of Science, Vol. IX* (ed. C. W. Savage). Minneapolis: Univ. Minnesota Press, (1978), 201–228.

———. "Why You Can't Make a Computer That Feels Pain," *Synthèse,* 38 (1978), 415–456.

Denofsky, M. E. *How Near is Near?* AI Memo 344. Cambridge, Mass.: MIT AI Lab., 1976.

Descartes, René. *Les Passions de L'Ame.* Amsterdam, 1650.

Didday, R. L. "A Model of Visuomotor Mechanisms in the Frog Optic Tectum," *Mathematical Biosciences,* 30 (1976), 169–180.

——— and M. A. Arbib. "Eye Movements and Visual Perception: A 'Two Visual System' Model," *Int. J. Man-Machine Studies,* 7 (1975), 547–569.

Doran, J. E. "New Developments of the Graph Traverser," in *MI-2,* pp. 119–135.

Dreyfus, H. L. *Alchemy and Artificial Intelligence.* P-3244. Santa Monica, Calif.: Rand Corporation, 1965.

———. "Why Computers Must Have Bodies in Order to be Intelligent," *Rev. Metaphysics,* 21 (1967), 13–32.

———. *What Computers Can't Do: A Critique of Artificial Reason.* New York: Harper & Row, 1972.

Duncker, Karl. "On Problem Solving," *Psychological Monographs,* 58 (1945).

Edgley, Roy. *Reason in Theory and Practice.* London: Hutchinson, 1969.

Eisenstadt, Marc. "Processing Newspaper Stories: Some Thoughts on Fighting and Stylistics," in *AISB-2,* pp. 104–117.

Enea, Horace, and K. M. Colby. "Idiolectic Language-Analysis for Understanding Doctor-Patient Dialogues," in *IJCAI-3,* pp. 278–284.

Engelberger, J. F. "Robotics—Like It Was, Like It Is, and Like It Will Be," *IEEE Intercon Conference Record,* April 1975, Session 13, pp. 1–6.

Eriksson, E. S. "Movement Parallax During Locomotion," *Perception and Psychophysics,* 16 (1974), 197–200.

Erman, L. D., and V. R. Lesser. "A Multi-Level Organization for Problem Solving Using Many, Diverse, Cooperating Sources of Knowledge," in *IJCAI-4,* pp. 483–490.

Ernst, G. W., and Allen Newell. "Some Issues of Representation in a General Problem Solver," *Proc. Spring Joint Comp. Conf.,* 30 (1967), 583–600.

———. *GPS: A Case Study in Generality and Problem Solving.* New York: Academic Press, 1969.

Evans, T. G. "A Program for the Solution of Geometric-Analogy Intelligence Test Questions," in *SIP,* pp. 271–353.

Fahlman, S. E. "A Planning System for Robot Construction Tasks," in *AI,* 5 (1974), 1–50.

Falk, Gilbert. "Interpretation of Imperfect Line Data as a Three-Dimensional Scene," in *AI,* 3 (1972), 101–144.

Faught, W. S. "Affect as Motivation for Cognitive and Conative Processes," in *IJCAI-4,* pp. 893–899.

Faught, W. S., K. M. Colby, and R. C. Parkison. "Inferences, Affects, and Intentions in a Model of Paranoia," *Cognitive Psychology,* 9 (1977), 153–187.

Feigenbaum, E. A. "Artificial Intelligence: Themes in the Second Decade," in *Information Processing 68, Vol. II* (ed. A. J. H. Morrell), pp. 1008–1022. Amsterdam: North Holland, 1969.

———. "Artificial Intelligence Research: What Is It? What Has It Achieved? Where Is It Going?" Stanford Univ. Dept. Comp. Sc.: privately circulated ms.

Feigenbaum, E. A., and Julian Feldman, eds. *Computers and Thought*. New York: McGraw-Hill, 1963.

Feigenbaum, E. A., B. G. Buchanan, and Joshua Lederberg. "On Generality and Problem Solving: A Case Study Using the DENDRAL Program," in *MI-6*, pp. 165–190.

Feldman, Jerome. "Bad-Mouthing Frames," in *TINLP*, pp. 102–103.

Fernald, M. R. "The Diagnosis of Mental Imagery," *Psychological Monographs*, 14, no. 58 (Feb. 1912), pp. 1–169.

Fikes, R. E., P. E. Hart, and N. J. Nilsson. "Learning and Executing Generalized Robot Plans," in *AI*, 3 (1972), 251–288.

———. "Some New Directions in Robot Problem Solving," in *MI-7*, pp. 405–430.

Fikes, R. E., and N. J. Nilsson. "STRIPS: A New Approach to the Application of Theorem Proving to Problem Solving," in *AI*, 2 (1971), 189–208.

Fillmore, C. J. "The Case for Case," in *Universals in Linguistic Theory* (eds. E. W. Bach and R. T. Harms), pp. 1–88. New York: Holt, Rinehart & Winston, 1968.

Firschein, Oscar, M. A. Fischler, L. S. Coles, and J. M. Tenenbaum. "Forecasting and Assessing the Impact of Artificial Intelligence on Society," in *IJCAI-3*, pp. 105–120.

Fisher, Harwood. "Logic and Language in Defences," *J. Theory Soc. Behav.*, 3 (1973), 157–214.

Fodor, J. A. *Psychological Explanation: An Introduction to the Philosophy of Psychology*. New York: Random House, 1968.

———. *The Language of Thought*. New York: Crowell, 1975, and Hassocks, Sussex: Harvester Press, 1976.

Foster, J. M. *List Processing*. London: Macdonald, 1967.

Freiling, M. J. *Functions and Frames in the Learning of Stuctures*. Working Paper 58. Cambridge: MIT AI Lab., Dec. 1973.

Freud, Sigmund. *Project for a Scientific Psychology*. (1895) Standard Edition, Vol. I, pp. 295–397. London: Hogarth Press, 1966.

———. *Analysis of a Phobia in a Five-Year-Old Boy*. (1909) Standard Edition, Vol. X, pp. 3–152. London: Hogarth Press, 1955.

———. *Inhibitions, Symptoms, and Anxiety*. (1926) Standard Edition, Vol. XX, pp. 75–176. London: Hogarth Press, 1959.

———. *The Question of Lay Analysis*. (1926) Standard Edition, Vol. XX, pp. 179–260. London: Hogarth Press, 1959.

Garfinkel, Harold. *Studies in Ethnomethodology*. Englewood Cliffs, N.J.: Prentice-Hall, 1967.

Gazzaniga, M. S. *The Bisected Brain*. New York: Academic Press, 1969.

Gelernter, H. L. "Realization of a Geometry-Theorem Proving Machine," in *CT*, pp. 134–152.

Gibson, J. J. *The Senses Considered as Perceptual Systems*. Boston: Houghton Mifflin, 1966.

Giorgi, Amedeo. *Psychology as a Human Science: A Phenomenologically Based Approach*. New York: Harper & Row, 1970.

Goffman, Erving. *Frame Analysis: An Essay on the Organization of Experience*. New York: Harper Colophon Books, 1974.

Goldman, N. M. "Conceptual Generation," in *Conceptual Information Processing* (ed. R. C. Schank), pp. 289–371. New York: American Elsevier, 1975.

Goldstein, I. P. "Summary of MYCROFT: A System for Understanding Simple Picture Programs," in *AI*, 6 (1975), 249–288.

Gombrich, E. H. *Art and Illusion*. New York: Pantheon, 1960.

Good, I. J. "A Five Year Plan for Automatic Chess," in *MI-2*, pp. 89–118.

Goodman, Nelson. *Fact, Fiction, and Forecast.* 2nd. ed. Indianapolis: Bobbs-Merrill, 1965.

———. *Languages of Art: An Approach to a Theory of Symbols.* London: Oxford University Press, 1969.

Gotlieb, C. C., and Allan Borodin. *Social Issues in Computing.* New York: Academic Press, 1973.

Grape, G. R. *Computer Vision Through Sequential Abstractions.* Stanford: Stanford Univ. AI Dept., 1969.

Green, B. F. *Digital Computers in Research: An Introduction for Behavioral and Social Scientists.* New York: McGraw-Hill, 1963.

Green, B. F., A. K. Wolf, Carole Chomsky, and K. R. Laughery. "BASEBALL: An Automatic Question-Answerer," in *CT*, pp. 207–216.

Green, C. C. "Theorem-Proving by Resolution as a Basis for Question-Answering," in *MI-4*, pp. 183–205.

Greenblatt, R. D., D. E. Eastlake, and S. D. Crocker. "The Greenblatt Chess Program," *Am. Fed. Inf. Proc. Soc. Conference Proceedings*, 31 (1967), 801–810.

Gregory, R. L. *Eye and Brain.* London: Weidenfeld & Nicolson, 1966.

———. "Will Seeing Machines Have Illusions?" in *MI-1*, pp. 169–180.

———. "The Evolution of Eyes and Brains: A Hen-and-Egg Problem," in *The Neuropsychology of Spatially Orientated Behaviour* (ed. S. J. Freedman), pp. 7–18. Illinois: Dorsey Press, 1968.

———. "On How So Little Information Controls So Much Behaviour," in *Towards a Theoretical Biology: 2. Sketches* (ed. C. H. Waddington), pp. 236–246. Edinburgh: Edinburgh University Press, 1969.

———. "The Social Implications of Intelligent Machines," in *MI-6*, pp. 3–13.

Gregory, R. L., and J. G. Wallace. "Recovery From Early Blindness: A Case Study," *Exp. Psychol. Soc. Monographs* (No. 2), 1963.

Grice, H. P. "Logic and Conversation," in *Syntax and Semantics, Vol. III: Speech Acts* (eds. Peter Cole and J. L. Morgan), pp. 41–58. New York: Academic Press, 1975.

Gullahorn, J. T., and J. E. Gullahorn. "A Computer Model of Elementary Social Behavior," in *CT*, pp. 375–386.

Gunderson, Keith. *Mentality and Machines.* New York: Doubleday, 1971.

Guzman, Adolfo. *Some Aspects of Pattern Recognition by Computer.* AI-TR-224. Cambridge, Mass.: MIT AI Lab., 1967.

———. *Computer Recognition of Three-Dimensional Objects in a Visual Scene.* AI-TR-228. Cambridge, Mass.: MIT AI Lab., 1968.

———. "Decomposition of a Visual Field into Three-Dimensional Bodies," in *Automatic Interpretation and Classification of Images* (ed. Antonio Grasselli), pp. 243–276. New York: Academic Press, 1969.

———. "Analysis of Curved Line Drawings Using Context and Global Information," in *MI-6*, pp. 325–376.

Haber, R. N., and R. B. Haber. "Eidetic Imagery—I: Frequency," in *Contemporary Theory and Research in Visual Perception* (ed. R. N. Haber), pp. 350–356. London: Holt, Rinehart, & Winston, 1970.

Haber, R. N., and Maurice Hershenson. *The Psychology of Visual Perception.* New York: Holt, Rinehart & Winston, 1973.

Habermas, Jurgen. *Knowledge and Human Interests.* London: Heinemann, 1972.

Halliday, M. A. K. "Functional Diversity in Language as Seen from a Consideration of Modality and Mood in English," *Foundations of Language*, 6 (1970), 322–361.

Hamlyn, D. W. *Sensation and Perception: A History of the Philosophy of Perception.* London: Routledge & Kegan Paul, 1961.

Hardy, Steven. "Synthesis of LISP Functions from Examples," in *IJCAI-4*, pp. 240–245.

Harré, Rom, and P. F. Secord. *The Explanation of Social Behavior.* Oxford: Blackwell, 1972.

Harrison, Bernard. *Form and Content.* Oxford: Blackwell, 1973.

Hart, P. E. "Progress on a Computer Based Consultant," in *IJCAI-4*, pp. 831–841.

Hart, P. E., N. J. Nilsson, and Bertram Raphael. "A Formal Basis for the Heuristic Determination of Minimum Cost Paths," *IEEE Trans. Sys. Sci. and Cybernetics*, SSC-4 (1968), 100–107.

Hayes, P. J. "Robotologic," in *MI-5*, pp. 533–554.

———. "A Logic of Actions," in *MI-6*, pp. 495–520.

———. "Some Comments on Sir James Lighthill's Report on Artificial Intelligence," *AISB Study Group European Newsletter*, Issue 14, (July 1973), 36–53.

———. "Some Problems and Non-Problems in Representation Theory," in *AISB-1*, pp. 63–79.

Hebb, D. O. *The Organization of Behavior: A Neuropsychological Theory.* New York: Wiley, 1949.

Heider, Fritz. *The Psychology of Interpersonal Relations.* New York: Wiley, 1958.

Heider, Fritz, and M. L. Simmel. "An Experimental Study of Apparent Behavior," *Am. J. Psychol.*, 57 (1944), 243–259.

Hempel, C. G. *Aspects of Scientific Explanation, and Other Essays in the Philosophy of Science.* New York: Free Press, 1965.

Hernandez-Peon, Raul, Harald Scherrer, and Michel Jouvet. "Modification of Electrical Activity in the Cochlear Nucleus During 'Attention' in Unanesthetized Cats," *Science*, 123 (1956), 331–332.

Hewitt, Carl. "PLANNER: A Language for Proving Theorems in Robots," in *IJCAI-1*, pp. 295–301.

———. "Procedural Embedding of Knowledge in PLANNER," in *IJCAI-2*, pp. 167–184.

———. *Description and Theoretical Analysis (Using Schemata) of PLANNER: A Language for Proving Theorems and Manipulating Models in a Robot.* AI-TR-258. Cambridge, Mass.: MIT AI Lab., 1972.

———. "A Universal Modular ACTOR Formalism for Artificial Intelligence," in *IJCAI-3*, pp. 235–245.

———. "How to Use What You Know," in *IJCAI-4*, pp. 189–198.

Hollingdale, S. H., and G. C. Tootill. *Electronic Computers.* Rev. ed. Harmondsworth: Penguin, 1970.

Homans, G. C. *Social Behavior: Its Elementary Forms.* New York: Harcourt, 1961.

Horn, Berthold. "Obtaining Shape from Shading Information," in *PCV*, pp. 115–156.

Horn, Gabriel, and R. M. Hill. "Modification of Receptive Fields of Cells in the Visual Cortex Occurring Spontaneously and Associated with Bodily Tilt," *Nature*, 221 (1969), 186–188.

Howe, J. A. M., John Knapman, H. M. Noble, Sylvia Weir, and R. M. Young. *Artificial Intelligence and the Representation of Knowledge.* D.A.I. Research Report No. 5. Edinburgh: Dept. AI, August 1975.

Hubel, D. H., and T. N. Wiesel. "Receptive Fields of Single Neurones in the Cat's Striate Cortex," *J. Physiology*, 148 (1959), 579–591.

———. "Receptive Fields, Binocular Interaction, and Functional Architecture in the Cat's Visual Cortex," *J. Physiology*, 160 (1962), 106–154.

———. "Receptive Fields and Functional Architecture of Monkey Striate Cortex," *J. Physiology*, 195 (1968), 215–243.

Hudson, Liam. *The Cult of the Fact: A Psychologist's Autobiographical Critique of His Discipline.* London: Cape, 1972.

Huffman, D. A. "Impossible Objects as Nonsense Sentences," in *MI-6*, pp. 295–325.

Huxley, T. H. "On the Hypothesis that Animals Are Automata, and Its History," in T. H. Huxley, *Method and Results: Essays*, pp. 199–250. London: Macmillan, 1893.

Illich, I. D. *Tools for Conviviality.* London: Calder & Boyars, 1973.

Ingleby, David. "Ideology and the Human Sciences: Some Comments on the Role of Reification in Psychology and Psychiatry," in *Counter Course: A Handbook for Course Criticism* (ed. Trevor Pateman), pp. 51–81. Harmondsworth: Penguin, 1972.

Jackson, P. C. *Introduction to Artificial Intelligence.* New York: Petrocelli, 1974.

Jansson, G. "Measurements of Eye Movements During a Michotte Launching Event," *Scand. J. Psychol.*, 5 (1964), 153–160.

Johansson, Gunnar. "Visual Perception of Biological Motion and a Model for its Analysis," *Perception and Psychophysics*, 14 (1973), 201–211.

———. "Spatial Event Perception." Chapter in *Handbook for Physiology*, forthcoming.

Johnson-Laird, P. N., Clive Robins, and Lucy Velicogna. "Memory for Words," *Nature*, 251 (1974), 704–705.

Jones, E. E., D. E. Kanouse, H. H. Kelley, R. E. Nisbett, Stuart Valins, and Bernard Weiner, eds. *Attribution: Perceiving the Causes of Behavior.* Morristown, N.J.: General Learning Press, 1972.

Julesz, Bela. *Foundations of Cyclopean Perception.* Chicago: Chicago University Press, 1970.

Kaneff, Steven, ed. *Picture Language Machines.* New York: Academic Press, 1970.

Kanizsa, Gaetano. "Contours Without Gradients or Cognitive Contours?" *Italian J. Psychol.*, 1 (1974), 93–112.

Kant, Immanuel. *Critique of Pure Reason.* 1781. (Many editions.)

Kelly, M. D. "Edge Detection in Pictures by Computer Using Planning," in *MI-6*, pp. 397–410.

Kenny, Anthony. *Action, Emotion, and Will.* London: Routledge & Kegan Paul, 1963.

Kilmer, W. L., W. S. McCulloch, and Jay Blum. "A Model of the Vertebrate Central Command System," *Int. J. Man-Machine Studies*, 1 (1969), 279–309.

Kiss, G. R. "Outlines of a Computer Model of Motivation," in *IJCAI-3*, pp. 446–449.

Klein, Sheldon. "Meta-Compiling Text Grammars as a Model for Human Behavior," in *TINLP*, pp. 94–98.

Klein, Sheldon, J. F. Aeschlimann, D. F. Balsiger, S. L. Converse, Claudine Court, Mark Foster, Robin Lao, J. D. Oakley, and Joel Smith. *Automatic Novel Writing: A Status Report.* Tech. Rpt. 186. Madison, Wis.: Univ. Wisconsin Comp. Sc. Dept., 1973.

Kline, Paul. *Fact and Fantasy in Freudian Theory.* London: Methuen, 1972.

Kohler, Wolfgang. *The Mentality of Apes.* 2nd ed. London: Routledge & Kegan Paul, 1927.

Kosslyn, S. M. "Information Representation in Visual Images," *Cognitive Psychology*, 7 (1975), 341–370.

———. "On Retrieving Information from Visual Images," in *TINLP*, pp. 160–164.

Kosslyn, S. N., and J. R. Pomerantz. "Imagery, Propositions, and the Form of Internal Representations," *Cognitive Psychology*, 9 (1977), 52–76.

Kuhn, T. S. *The Structure of Scientific Revolutions.* Chicago: University of Chicago Press, 1962.

Kuipers, B. J. *Spatial Knowledge.* AI Memo 359. Cambridge, Mass.: MIT AI Lab., 1976.

Kuipers, B. J., John McCarthy, and Joseph Weizenbaum. "Computer Power and Human Reason: Comments," *SIGART* Newsletter, No. 58, June 1976, 4–12.

Laing, R. D. *The Divided Self: A Study of Sanity and Madness.* London: Tavistock Press, 1960.

———. *Self and Others.* London: Tavistock Press, 1961.

———. *The Politics of Experience.* Harmondsworth: Penguin, 1967.

Laing, R. D., Herbert Philipson, and A. R. Lee. *Interpersonal Perception: A Theory and a Method of Research.* London: Tavistock Press, 1966.

Lakatos, Imre. "Falsification and the Methodology of Scientific Research Programmes," in *Criticism and the Growth of Knowledge* (eds. Imre Lakatos and Alan Musgrave), pp. 91–196. Cambridge: Cambridge University Press, 1970.

———. "Proofs and Refutations," *Brit. J. Philos. Science,* 14 (1963), 1–25, 120–139, 221–243, and 296–342.

Landa, L. N. *Algorithmization in Learning and Instruction.* Englewood Cliffs, N.J.: Educational Technology Publications, 1974.

———. *Cybernetics, Algorithmization and Heuristics in Education.* Englewood Cliffs, N.J.: Educational Technology Publications, 1976.

Lashley, K. S. "The Problem of Serial Order in Behavior," in *Cerebral Mechanisms in Behavior: The Hixon Symposium* (ed. L. A. Jeffress), pp. 112–135. New York: Wiley, 1951.

Lehnert, Wendy, "What Makes Sam Run? Script Based Techniques for Question Answering," in *TINLP,* pp. 16–21.

———. *Question Answering in a Story Understanding System.* Research Rpt. 57. New Haven: Yale Univ. Dept. Comp. Sci., Dec. 1975.

Lerner, Laurence. *A.R.T.H.U.R.: The Life and Opinions of a Digital Computer.* Cambridge, Mass.: Univ. Mass. Press, and Hassocks, Sussex: Harvester Press, 1974.

Lettvin, J. Y., H. R. Maturana, Walter Pitts, and W. S. McCulloch. "What the Frog's Eye Tells the Frog's Brain," *Proc. Inst. Radio Engineers,* 47 (1959), 1940–1959.

Levin, J. A., and J. A. Moore. "Dialogue-Games: A Process Model of Natural Language Interaction," in *AISB-2,* pp. 184–194.

Lighthill, James, N. S. Sutherland, R. M. Needham, H. C. Longuet-Higgins, and Donald Michie. *Artificial Intelligence: A Paper Symposium.* London: Science Research Council, 1973.

Lincoln, H. B., ed. *The Computer and Music.* Ithaca, N.Y.: Cornell University Press, 1970.

Lindsay, P. H., and D. A. Norman. *Human Information Processing: An Introduction to Psychology.* New York: Academic Press, 1972.

Lindsay, R. K. "Inferential Memory as the Basis of Machines Which Understand Natural Language," in *CT,* pp. 217–233.

———. "Jigsaw Heuristics and a Language Learning Model," in *Artificial Intelligence and Heuristic Programming* (eds. N. V. Findler and Bernard Meltzer), pp. 173–189. Edinburgh: Edinburgh University Press, 1971.

Locke, John. *An Essay Concerning Human Understanding.* London: 1690.

Loehlin, J. C. *Computer Models of Personality.* New York: Random House, 1968.

Longuet-Higgins, H. C. "The Seat of the Soul," in *Towards a Theoretical Biology: 3. Drafts* (ed. C. H. Waddington), pp. 236–241. Edinburgh: Edinburgh University Press, 1970.

Longuet-Higgins, H. C., and M. J. Steedman. "On Interpreting Bach," in *MI-6,* pp. 221–242.

Lucas, J. R. "Minds, Machines, and Godel," *Philosophy,* 36 (1961), 112–127.

McCarthy, John. "Recursive Functions of Symbolic Expressions and their Calculation by Machine, Part I," in *CACM,* 3 (1960), 184–195.

———. "Programs with Common Sense," in *SIP*, pp. 403–418.

———. "The Home Information Terminal," *Man and Computer.* (*Proc. Int. Conf., Bordeaux 1970*), pp. 48–57. Basel: Karger, 1972.

McCarthy, John, P. W. Abrahams, D. J. Edwards, T. P. Hart, and M. I. Levin. *Lisp 1.5 Programmer's Manual.* Cambridge, Mass.: MIT Press, 1962.

McCarthy, John, and P. J. Hayes. "Some Philosophical Problems from the Standpoint of Artificial Intelligence," in *MI-4*, pp. 463–502.

McCulloch, W. S. "A Heterarchy of Values Determined by the Topology of Nervous Nets," *Bull. Math. Biophysics*, 11 (1949), 89–93.

McDermott, D. V. *Assimilation of New Information by a Natural Language-Understanding System.* AI-TR-291. Cambridge, Mass.: MIT AI Lab., 1974.

Mackay, D. M. "A Mind's Eye View of the Brain," in *Progress in Brain Research, 17: Cybernetics of the Nervous System* (eds. Norbert Wiener and J. P. Schadé), pp. 321–332. Amsterdam: Elsevier, 1965.

Mackworth, A. K. "Interpreting Pictures of Polyhedral Scenes," in *AI*, 4 (1973), 121–138.

———. "Using Models to See," in *AISB-1*, pp. 127–137.

Maier, N. F. "Reasoning in Humans, II. The Solution of a Problem and its Appearance in Consciousness," *J. Comp. Psychol.*, 12 (1931), 181–194.

Marcuse, Herbert. *One Dimensional Man: The Ideology of Industrial Society.* London: Routledge & Kegan Paul, 1964.

Marg, E., J. E. Adams, and B. B. Rutkin. "Receptive Fields of Cells in the Human Visual Cortex," *Experientia*, 24 (1968), 348–350.

Marr, David. *An Essay on the Primate Retina.* AI Memo 296. Cambridge, Mass.: MIT AI Lab., 1974.

———. *Analyzing Natural Images: A Computational Theory of Texture Vision.* AI Memo 334. Cambridge, Mass.: MIT AI Lab., June 1975.

———. *Early Processing of Visual Information.* AI Memo, 340. Cambridge, Mass.: MIT AI Lab., Dec. 1975.

Marzocco, F. N. "Computer Recognition of Handwritten First Names," *IEEE Trans. Electronic Computers*, 14 (1965), 210–217.

May, Rollo, ed. *Existential Psychology.* New York: Random House, 1961.

Meehan, Jim. "Using Planning Structures to Generate Stories," *Am. J. Computational Linguistics*, Microfiche 33 (1975), 77–93.

Meehl, P. E. *Clinical Versus Statistical Prediction: A Theoretical Analysis and a Review of the Evidence.* Minneapolis: University of Minnesota Press, 1954.

———. "Psychological Determinism and Human Rationality: A Psychologist's Reactions to Professor Karl Popper's 'Of Clouds and Clocks'," in *Minnesota Studies in Philosophy of Science, Vol. IV: Analyses of Theories and Methods of Physics and Psychology* (eds. Michael Radner and Stephen Winokur), pp. 310–372. Minneapolis: University of Minnesota Press, 1970.

Merleau-Ponty, Maurice. *Phenomenology of Perception.* New York: Humanities Press, 1962.

———. *The Structure of Behaviour.* London: Methuen, 1965.

Michie, Donald. *On Machine Intelligence.* Edinburgh: Edinburgh University Press, 1974.

———. "Life with Intelligent Machines," in *The Use of Models in the Social Sciences* (ed. Lyndhurst Collins), pp. 101–109. London: Tavistock Press, 1976.

Michie, Donald, and Robert Ross. "Experiments with the Adaptive Graph Traverser," in *MI-5*, pp. 301–318.

Michie, J. E. "What the Public Think," *Firbush News 5*, 46–52. Edinburgh: Machine Intelligence Research Unit, 1974.

Michotte, A. E. *The Perception of Causality.* London: Methuen, 1963.

Miller, G. A., Eugene Galanter, and K. H. Pribram. *Plans and the Structure of Behavior.* New York: Holt, 1960.

Miller, G. A., and P. N. Johnson-Laird. *Language and Perception.* Cambridge, Mass.: Harvard University Press, 1976.

Minsky, M. L. "Steps Toward Artificial Intelligence," in *CT*, pp. 406–450.

———. "Matter, Mind, and Models," in *SIP*, pp. 425–432.

———. *Computation: Finite and Infinite Machines.* Englewood Cliffs, N.J.: Prentice-Hall, 1967.

———, ed. *Semantic Information Processing.* Cambridge, Mass.: MIT Press, 1968.

———. "Form and Content in Computer Science," in *JACM*, 17 (1972), 197–215.

———. "A Framework for Representing Knowledge," in *PCV*, pp. 211–277.

Minsky, M. L., and Seymour Papert. *Artificial Intelligence.* Eugene, Oregon: Condon Lecture Publications, 1973.

Moore, G. E. *Philosophical Studies.* London: Routledge & Kegan Paul, 1922.

Moore, James, and Allen Newell. "How Can MERLIN Understand?" in *Knowledge and Cognition* (ed. L. W. Gregg), pp. 201–252. Baltimore: Lawrence Erlbaum Ass., 1973.

Moser, Ulrich, Werner Schneider, and Ilka von Zeppelin. "Computer Simulation of a Model of Neurotic Defence Mechanism. (Clinical Paper and Technical Paper)," *Bull. Psychol. Inst. Univ. Zurich*, 2 (1968), 1–77.

Moser, Ulrich, Ilka von Zeppelin, and Werner Schneider. "Computer Simulation of a Model of Neurotic Defence Processes," *Int. J. Psychoanalysis*, 50 (1969), 53–64.

———. "Computer Simulation of a Model of Neurotic Defense Processes," *Behavioral Science*, 15 (1970), 194–202.

Nagel, Ernest, and J. R. Newman. *Godel's Proof.* New York: New York University Press, 1958.

Neisser, Ulric. *Cognitive Psychology.* New York: Appleton-Century-Crofts, 1967.

Newell, Allen. "Some Problems of Basic Organization in Problem-Solving Programs," in *Self-Organizing Systems 1962* (eds. M. C. Yovits, G. T. Jacobi, and G. D. Goldstein), pp. 393–423. Washington, D.C.: Spartan, 1962.

———. "Limitations of the Current Stock of Ideas About Problem Solving," in *Electronic Information Handling* (eds. Allen Kent and O. E. Taulbee), pp. 195–208. Washington, D.C.: Spartan, 1965.

———. "Artificial Intelligence and the Concept of Mind," in *CMTL*, pp. 1–60.

———. "You Can't Play 20 Questions with Nature and Win," in *Visual Information Processing* (ed. W. G. Chase), pp. 283–310. New York: Academic Press, 1973.

Newell, Allen, Jeffrey Barnett, J. W. Forgie, C. C. Green, D. H. Klatt, J. C. R. Licklider, J. H. Munson, D. R. Reddy, and W. A. Woods. *Final Report of a Study Group on Speech Understanding Systems.* Amsterdam: North-Holland, 1973.

Newell, Allen, J. C. Shaw, and H. A. Simon. "Empirical Explorations with the Logic Theory Machine: A Case Study in Heuristics," in *CT*, pp. 109–133.

Newell, Allen, and H. A. Simon. "GPS—A Program That Simulates Human Thought," in *CT*, pp. 279–296.

———. *Human Problem Solving.* Englewood Cliffs, N.J.: Prentice-Hall, 1972.

Newell, Allen, F. M. Tonge, E. A. Feigenbaum, B. F. Green, and G. H. Mealy. *Information Processing Language—V Manual.* 2nd ed. Englewood Cliffs, N.J.: Prentice-Hall, 1964.

Nilsson, N. J. *Problem-Solving Methods in Artificial Intelligence.* New York: McGraw-Hill, 1971.

———. "Artificial Intelligence," *Information Processing '74*, Proc. IFIP Congress, 1974, Vol. 4, pp. 778–801. Amsterdam: North-Holland, 1974.

Ninnes, Ted. "History of Positivism," in *Rat Myth and Magic: A Political Critique of Psychology*, pp. 33–36. London: Russell Press, 1972.

Norman, D. A., and D. E. Rumelhart. *Explorations in Cognition.* San Francisco: Freeman, 1975.

Orban, Richard. *Removing Shadows in a Scene.* AI Memo 192. Cambridge, Mass.: MIT AI Lab., 1970.

Paivio, Allen. *Imagery and Verbal Processes.* New York: Holt, Rinehart, & Winston, 1971.

Palermo, D. S. "Is a Scientific Revolution Taking Place in Psychology?" *Science Studies,* 1 (1971), 135–155.

Palmer, S. E. "The Nature of Perceptual Representation: An Examination of the Analog/Propositional Controversy," in *TINLP,* pp. 165–173.

Papert, Seymour. *The Artificial Intelligence of Hubert L. Dreyfus: A Budget of Fallacies.* AI Memo 154. Cambridge, Mass.: MIT AI Lab., 1968.

———. "Teaching Children to Be Mathematicians Versus Teaching About Mathematics," *Int. J. Math. Educ. Sci. Technol.,* 3 (1972), 249–262.

———. "Theory of Knowledge and Complexity," in *Process Models for Psychology* (ed. G. J. Dalenoort), pp. 1–49. Rotterdam: Rotterdam University Press, 1973.

———. *Uses of Technology to Enhance Education.* AI-Memo-298. Cambridge, Mass.: MIT AI Lab., 1973.

Parkison, R. C., K. M. Colby, and W. S. Faught. "Conversational Language Comprehension Using Integrated Pattern Matching and Parsing," *AI,* 9 (1977), 111–134.

Paton, H. J. *The Categorical Imperative: A Study in Kant's Moral Philosophy.* London: Hutchinson, 1947.

Pears, D. F., J. F. Thomson, and Mary Warnock. "What Is the Will?" in *Freedom and the Will* (ed. D. F. Pears), pp. 14–37. London: Macmillan, 1963.

Perky, C. W. "An Experimental Study of Imagination," *Amer. J. Psychol.,* 21 (1910), 422–452.

Piaget, Jean. *The Construction of Reality in the Child.* New York: Basic Books, 1954.

Pohl, Ira. "Bi-Directional Search," in *MI-6,* pp. 127–140.

Polanyi, Michael. *Personal Knowledge: Towards a Post-Critical Philosophy.* New York: Harper, 1964.

———. "The Logic of Tacit Inference," *Philosophy,* 41 (1966), 1–18.

Polya, George. *How to Solve It: A New Aspect of Mathematical Method.* New York: Doubleday, 1945.

Popper, K. R. *Conjectures and Refutations: The Growth of Scientific Knowledge.* London: Routledge & Kegan Paul, 1963.

———. *Of Clouds and Clocks: An Approach to the Problem of Rationality and the Freedom of Man.* Arthur Holly Compton Memorial Lecture. St. Louis: Washington University Press, 1966.

Power, Richard. *A Model of Conversation,* unpublished working paper, University of Sussex Dept. of Experimental Psychology, 1976.

Price, Keith. "A Comparison of Human and Computer Vision Systems," *SIGART Newsletter (ACM),* No. 50. Feb. 1975. Pp. 5–10.

Pritchard, R. M. "Stabilized Images on the Retina," *Scientific American,* 204 (1961), 72–78.

Puccetti, Roland. *Persons: A Study of Possible Moral Agents in the Universe.* London: Macmillan, 1968.

Putnam, Hilary. "Dreaming and Depth Grammar," in *Analytical Philosophy* (ed. R. J. Butler), pp. 211–235. Oxford: Blackwell, 1962.

Pylyshyn, Z. W., ed. *Perspectives on the Computer Revolution.* Englewood Cliffs, N.J.: Prentice-Hall, 1970.

———. "What the Mind's Eye Tells the Mind's Brain: A Critique of Mental Imagery," *Psychological Bulletin,* 80 (1973), 1–24.

———. "Minds, Machines, and Phenomenology: Some Reflections on Dreyfus' 'What Computers Can't Do'," *Cognition*, 3 (1974), 20–42.

———. "Do We Need Images and Analogues?" in *TINLP*, pp. 174–177.

———. "Complexity and the Study of Artificial and Human Intelligence," in *Philosophical Perspectives on Artificial Intelligence* (ed. M. D. Ringle), pp. 23–56. Atlantic Highlands: Humanities Press, 1979.

———. "The Symbolic Nature of Mental Representations," in *Objectives and Methodologies in Artificial Intelligence* (eds. Steven Kaneff and J. E. O'Callaghan). New York: Academic Press, in press.

Quillian, M. R. "Semantic Memory," in *SIP*, pp. 227–270.

Radnitsky, Gerard. *Continental Schools of Metascience.* Goteborg: Scandinavian University Books, 1968.

Raphael, Bertram. "SIR: A Computer Program for Semantic Information Retrieval," in *SIP*, pp. 33–145.

———. "The Frame Problem in Problem-Solving Systems," in *Artificial Intelligence and Heuristic Programming* (eds. N. V. Findler and Bernard Meltzer), pp. 159–169. Edinburgh: Edinburgh University Press, 1971.

———. *The Thinking Computer: Mind Inside Matter.* San Francisco: W. H. Freeman, 1976.

Rattner, M. H. *Extending Guzman's SEE Program.* AI Memo 204. Cambridge, Mass.: MIT AI Lab., 1970.

Reddy, D. R., L. D. Erman, R. D. Fennell, and R. B. Neely. "The HEARSAY Speech Understanding System," in *IJCAI-3*, pp. 185–193.

Reddy, D. R., and Allen Newell. "Knowledge and Its Representation in a Speech Understanding System," in *Knowledge and Cognition* (ed. L. W. Gregg), pp. 253–286. Baltimore: Lawrence Erlbaum Ass., 1974.

Reichardt, Jasia, ed. *Cybernetics, Art, and Ideas.* London: Studio Vista, 1971.

Reitman, W. R. *Cognition and Thought: An Information-Processing Approach.* New York: Wiley, 1965.

Ricoeur, Paul. *Freud and Philosophy: An Essay on Interpretation.* New Haven: Yale University Press, 1970.

Rieger, C. J. *Conceptual Memory: A Theory and Computer Program for Processing the Meaning Content of Natural Language Utterances.* Unpublished Ph.D. thesis. Stanford University Dept. Computer Science, 1974.

———. "Conceptual Memory and Inference," in *Conceptual Information Processing* (ed. R. C. Schank), pp. 157–288. New York: American Elsevier, 1975.

———. "The Commonsense Algorithm as a Basis for Computer Models of Human Memory, Inference, Belief, and Contextual Language Comprehension," in *TINLP*, pp. 199–214.

———. "Conceptual Overlays: A Mechanism for the Interpretation of Sentence Meaning in Context," in *IJCAI-4*, pp. 143–150.

Riesbeck, C. K. "Conceptual Analysis," in *Conceptual Information Processing* (ed. R. C. Schank), pp. 83–156. New York: American Elsevier, 1975.

Roberts, L. G. "Machine Perception of Three-Dimensional Solids," in *Optical and Electro-Optical Information Processing* (eds. J. T. Tippett, D. A. Berkowitz, L. C. Clapp, C. J. Koester, and Alexander Vanderburgh), pp. 159–198. Cambridge, Mass.: MIT Press, 1965.

Robinson, Guy. "How to Tell Your Friends from Machines," *Mind*, N.S., 81 (1972), 504–518.

Rochester, Nathaniel, J. H. Holland, L. H. Haibt, and W. L. Duda. "Test on a Cell Assembly Theory of the Brain, Using a Large Digital Computer," *Inst. Radio Engineers Trans. Inf. Theory*, 2 (1956), 80–93.

Rogers, Carl. *Client Centered Therapy: Current Practice, Implications, and Theory.* Boston: Houghton Mifflin, 1951.

Rosch, Eleanor, and C. B. Mervis. "Family Resemblances: Studies in the Internal Structure of Categories," *Cognitive Psychology*, 7 (1975), 573–605.

Rosenblueth, Arturo, and Norbert Wiener. "Purposeful and Non-Purposeful Behavior," *Philosophy of Science*, 17 (1950), 318–326.

Rosenblueth, Arturo, Norbert Wiener, and Julian Bigelow. "Behavior, Purpose, and Teleology," *Philosophy of Science*, 10 (1943), 18–24.

Roszak, Theodore. *Where the Wasteland Ends: Politics and Transcendence in Post Industrial Society.* New York: Doubleday, 1972.

Roth, Leon. *Descartes' Discourse on Method.* Oxford: Clarendon, 1937.

Rubin, A. D. "The Role of Hypotheses in Medical Diagnosis," in *IJCAI-4*, pp. 856–862.

———. *Hypothesis Formation and Evaluation in Medical Diagnosis.* AI-TR-316. Cambridge, Mass.: MIT AI Lab., 1975.

Rumelhart, D. E. "Notes on a Schema for Stories," in *RU*, pp. 211–236.

Rumelhart, D. E., and D. A. Norman. "Active Semantic Networks as a Model of Human Memory," in *IJCAI-3*, pp. 450–457.

Russell, Bertrand. *The Problems of Philosophy.* London: Oxford University Press, 1912.

Rycroft, Charles. "Causes and Meaning," in *Psychoanalysis Observed* (ed. Charles Rycroft), pp. 7–22. London: Constable, 1966.

Ryle, Gilbert. *The Concept of Mind.* London: Hutchinson, 1949.

———. "Categories," in *Logic and Language: Second Series* (ed. A. G. N. Flew), pp. 65–81. Oxford: Blackwell, 1955.

Sacerdoti, E. D. "Planning in a Hierarchy of Abstraction Spaces," in *AI*, 5 (1974), 115–136.

———. "The Non-Linear Nature of Plans," in *IJCAI-4*, pp. 206–214.

———. *A Structure for Plans and Behavior.* AI Tech. Note 109. Menlo Park, Calif.: SRI, 1975.

Samuel, A. L. "Some Studies in Machine Learning Using the Game of Checkers," in *CT*, pp. 71–105.

———. "Some Studies in Machine Learning Using the Game of Checkers. II—Recent Progress," in *Human and Artificial Intelligence* (ed. F. J. Crosson), pp. 81–116. New York: Appleton-Century-Crofts, 1970.

Sayre, K. M. "Intelligence, Bodies, and Digital Computers," *Rev. Metaphysics*, 21 (1968), 714–723.

———. *Consciousness: A Philosophic Study of Minds and Machines.* New York: Random House, 1969.

———. *Cybernetics and the Philosophy of Mind.* Atlantic Highlands, N.J.: Humanities Press, 1976.

Schachter, Stanley. "The Interaction of Cognitive and Physiological Determinants of Emotional State," in *Anxiety and Behavior* (ed. C. D. Spielberger), pp. 193–224. New York: Academic Press, 1966.

Schachter, Stanley, and J. E. Singer. "Cognitive, Social, and Physiological Determinants of Emotional State," *Psychological Review*, 69 (1962), 379–399.

Schank, R. C. "Finding the Conceptual Content and Intention in an Utterance in Natural Language Conversation," in *IJCAI-2*, pp. 444–454.

———. "Conceptual Dependency: A Theory of Natural Language Understanding," *Cognitive Psychology*, 3 (1972), 552–631.

———. *The Fourteen Primitive Actions and Their Inferences.* AI Memo 183. Stanford: Stanford Univ. Comp. Sc. Dept., 1973.

———. "Identification of Conceptualizations Underlying Natural Language," in *CMTL*, pp. 187–248.

———, ed. *Conceptual Information Processing*. New York: American Elsevier, 1975.

———. "The Structure of Episodes in Memory," in *RU*, pp. 237–272.

Schank, R. C., and R. P. Abelson. "Scripts, Plans, and Knowledge," in *IJCAI-4*, pp. 151–157.

Schank, R. C., and K. M. Colby, eds. *Computer Models of Thought and Language*. San Francisco: Freeman, 1973.

Schank, R. C., N. M. Goldman, C. J. Rieger, and C. K. Riesbeck. *Primitive Concepts Underlying Verbs of Thought*. AI Memo 162. Stanford: Stanford Univ. Comp. Sc. Dept., 1972.

———. "MARGIE: Memory, Analysis, Response Generation, and Inference on English," in *IJCAI-3*, pp. 255–261.

Schank, R. C., and B. L. Nash-Webber, eds. *Theoretical Issues in Natural Language Processing*. Proc. Workshop Ass. Computational Linguistics, 1975.

Schank, R. C., and C. J. Rieger. "Inference and the Computer Understanding of Natural Language," in *AI*, 5 (1974), 373–412.

Schank, R. C., and the Yale AI Project. *SAM—A Story Understander*. Research Rpt. 43. New Haven: Yale Univ. Dept. Comp. Sc., August 1975.

Schmidt, C. F. "Understanding Human Action: Recognizing the Plans and Motives of Other Persons," in *Cognition and Social Behavior* (eds., J. S. Carroll and J. W. Payne), pp. 47–68. Hillsdale, N.J.: Lawrence Erlbaum, 1976.

Scriven, Michael. "The Compleat Robot: A Prolegomena to Androidology," in *Dimensions of Mind: A Symposium* (ed. Sidney Hook), pp. 118–142. New York: New York Universities Press, 1960.

Searle, J. R. *Speech Acts*. Cambridge: University Press, 1969.

Seely-Brown, John, and R. R. Burton. "Multiple Representations of Knowledge for Tutorial Reasoning," in *RU*, pp. 311–350.

Shirai, Yoshiaki. "A Context Sensitive Line Finder for Recognition of Polyhedra," in *AI*, 4 (1973), 95–120. (Also as "Analyzing Intensity Arrays Using Knowledge About Scenes," in *PCV*, pp. 93–114.)

Shortliffe, E. H., S. G. Axline, B. G. Buchanan, T. C. Merigan, and N. S. Cohen. "An Artificial Intelligence Program to Advise Physicians Regarding Antimicrobial Therapy," *Computers and Biomedical Research*, 6 (1973), 544–560.

Shortliffe, E. H., and B. G. Buchanan. "A Model of Inexact Reasoning in Medicine," *Mathematical Biosciences*, 23 (1975), 351–379.

Shortliffe, E. H., Randall Davis, S. G. Axline, B. G. Buchanan, C. C. Green, and N. S. Cohen. "Computer-Based Consultations in Clinical Therapeutics: Explanation and Rule Acquisition Capabilities of the MYCIN System," *Computers and Biomedical Research*, 8 (1975), 303–320.

Simon, H. A. *The Sciences of the Artificial*. Cambridge, Mass.: MIT Press, 1969.

Skinner, B. F. *Walden Two*. New York: Macmillan, 1948.

———. *Science and Human Behavior*. New York: Free Press, 1953.

———. "A Case History in Scientific Method," in *Psychology: A Study of a Science, Vol. II* (ed. Sigmund Koch), pp. 359–379. New York: McGraw-Hill, 1959.

———. *Beyond Freedom and Dignity*. New York: Knopf, 1971.

Sloman, Aaron. "Interactions Between Philosophy and Artificial Intelligence; The Role of Intuition and Non-Logical Reasoning in Intelligence," in *AI*, 2 (1971), 209–225.

———. "On Learning About Numbers," in *AISB-1*, pp. 173–185.

———. "Afterthoughts on Analogical Representation," in *TINLP*, pp. 178–182.

———. "What Are the Aims of Science?" *Radical Philosophy*, No. 13 (Spring 1976), 7–17.

Sloman, Aaron, and Steven Hardy. "Giving a Computer Gestalt Experiences," in *AISB-2*, pp. 242–255.

Speisman, J. C. "Autonomic Monitoring of Ego Defense Processes," in *Psychoanalysis and Current Biological Thought* (eds. N. S. Greenfield and W. C. Lewis), pp. 227–244. Madison: Univ. Wisconsin Press, 1965.

Stansfield, J. L. "Active Descriptions for Representing Knowledge," in *AISB-1*, pp. 214–223.

———. *Programming a Dialogue Teaching Situation.* Unpublished Ph.D. thesis. University of Edinburgh, Dept. AI, 1974.

Steedman, M. J. *The Formal Description of Musical Perception.* Unpublished Ph.D. thesis. University of Edinburgh, 1973.

Sterling, T. D., E. A. Bering, S. V. Pollack, and H. G. Vaughan, eds. *Visual Prosthesis: The Interdisciplinary Dialogue.* New York: Academic Press, 1971.

Sussman, G. J. "The Virtuous Nature of Bugs," in *AISB-1*, pp. 224–237.

———. *A Computer Model of Skill Acquisition.* New York: American Elsevier, 1975.

Sussman, G. J., and D. V. McDermott. *Why Conniving Is Better than Planning.* AI Memo 255a. Cambridge, Mass.: MIT AI Lab., 1972.

———. *The CONNIVER Reference Manual.* AI Memo 259. Cambridge, Mass.: MIT AI Lab., 1972.

Sussman, G. J., and Terry Winograd. *MICRO-PLANNER Reference Manual.* AI Memo 203. Cambridge, Mass.: MIT AI Lab., 1970.

Sutherland, N. S. "Outlines of a Theory of Visual Pattern Recognition in Animals and Man," *Proc. Royal Society B*, 171 (1968), 297–317.

———. "Intelligent Picture Processing," unpublished paper.

Szazs, Thomas. *The Myth of Mental Illness.* London: Secker & Warburg, 1962.

Tate, Austin. "Interacting Goals and Their Use," in *IJCAI-4*, pp. 215–218.

Tenenbaum, J. M., and H. G. Barrow. *Experiments in Interpretation-Guided Segmentation.* AI Tech. Note 123. Menlo Park, Calif.: SRI, 1976.

Tenenbaum, J. M., T. D. Garvey, Stephen Weyl, and H. C. Wolf. *An Interactive Facility for Scene Analysis Research.* AI Tech. Note 87. Menlo Park, Calif.: SRI, 1974.

Tenenbaum, J. M., and Stephen Weyl. "A Region-Analysis Subsystem for Interactive Scene Analysis," in *IJCAI-4*, pp. 682–687.

Thomas, A. J. "Puccetti on Machine Pattern Recognition," *Br. J. Philos. Science*, 26 (1975), 227–239.

Thomas, A. J., and T. O. Binford. *Information Processing Analysis of Visual Perception: A Review.* AIM-227. Stanford: Stanford Univ. Comp. Sc. Dept., 1974.

Thorne, J. P., Paul Bratley, and H. M. Dewar. "The Syntactic Analysis of English by Machine," in *MI-3*, pp. 281–309.

Tikhomirov, O. K. "Philosophical and Psychological Problems of Artificial Intelligence," in *IJCAI-4*, pp. 932–937.

Toffler, Alvin. *Future Shock.* New York: Random House, 1970.

Tomkins, S. S. *Affect, Imagery, Consciousness. Vol. II: The Negative Affects.* New York: Springer, 1963.

Tomkins, S. S., and Samuel Messick, eds. *Computer Simulation of Personality: Frontier of Psychological Research.* New York: Wiley, 1963.

Turing, A. M. "Computing Machinery and Intelligence," in *CT*, pp. 11–35.

Turn, Rein. *Computers in the 1980's.* New York: Columbia University Press, 1974.

Turner, K. J. "Computer Perception of Curved Objects," in *AISB-1*, pp. 238–246.

Turner, R. H., ed. *Ethnomethodology.* Harmondsworth: Penguin, 1974.

Tyler, S. A., ed. *Cognitive Anthropology.* New York: Holt, 1969.

Uhr, Leonard, and Charles Vossler. "A Pattern Recognition Program That Generates, Evaluates, and Adjusts Its Own Operators," in *CT*, pp. 251–268.

Walton, H. J., G. R. Kiss, and K. M. Farvis. *A Computer Program for the On-Line Exploration of Attitude Structures of Psychiatric Patients.* Edinburgh: MRC Speech & Communication Research Unit, April 1973.

Waltz, D. L. "Understanding Line Drawings of Scenes with Shadows," in *PCV*, pp. 19–92.

Warren, Neil. "Is a Scientific Revolution Taking Place in Psychology?—Doubts and Reservations," *Science Studies*, 1 (1971), 407–413.

Wason, P. C., and P. N. Johnson-Laird. *Psychology of Reasoning: Structure and Content.* London: Batsford, 1972.

Weir, Sylvia. "Action Perception," in *AISB-1*, pp. 247–256.

Weir, Sylvia, M. R. Adler, and Marilyn McLennan. *Final Report on Action Perception Project.* Edinburgh: Edinburgh University AI Dept., November 1975.

Weir, Sylvia, and Ricky Emanuel. *Using LOGO to Catalyse Communication in an Autistic Child.* Research Report 15. Edinburgh: Edinburgh University Dept. AI, January 1976.

Weizenbaum, Joseph. "ELIZA—A Computer Program for the Study of Natural Language Communication Between Man and Machine," in *CACM*, 9 (1966), 36–45.

———. "Contextual Understanding by Computers," in *CACM*, 10 (1967), 474–480.

———. "On the Impact of the Computer on Society: How Does One Insult a Machine?" *Science*, 176 (1972), 609–614.

———. *Computer Power and Human Reason: From Judgment to Calculation.* San Francisco: Freeman, 1976.

Wertheimer, Max. *Productive Thinking.* New York: Harper, 1945.

Wiener, Norbert. *Cybernetics, or Control and Communication in the Animal and the Machine.* New York: Wiley, 1948.

———. "Some Moral and Technical Consequences of Automation," *Science*, 131 (1960), 1355–1358.

———. *God & Golem, Inc.: A Comment on Certain Points Where Cybernetics Impinges on Religion.* Cambridge, Mass.: MIT Press, 1964.

Wilks, Y. A. *Grammar, Meaning, and the Machine Analysis of Natural Language.* London: Routledge & Kegan Paul, 1972.

———. "An Artificial Intelligence Approach to Machine Translation," in *CMTL*, pp. 114–151.

———. "Understanding Without Proofs," in *IJCAI-3*, pp. 270–277.

———. "A Computer System for Making Inferences About Natural Language," in *AISB-1*, pp. 268–283.

———. "A Preferential, Pattern-Seeking, Semantics for Natural Language Inference," in *AI*, 6 (1975), 53–74.

———. *Natural Language Understanding Systems Within the AI Paradigm.* Rev. ed. Edinburgh: Edinburgh University AI Dept., 1976.

———. "Dreyfus' Disproofs," *Br. J. Philos. Science*, 27 (1976), 177–185.

Will, P. M., and D. D. Grossman. "An Experimental System for Computer Controlled Mechanical Assembly," *IEEE Trans. Electronic Computers*, 24 (1975), 879–888.

Williams, Bernard. "How Smart Are Computers?" *New York Review*, November 15, 1973, pp. 36–40.

Wilson, T. P. "Normative and Interpretive Paradigms in Sociology," in *Understanding Everyday Life: Toward the Reconstruction of Sociological Knowledge* (ed. J. D. Douglas), pp. 57–79. Chicago: Aldine, 1970.

Winograd, Terry. *Understanding Natural Language.* New York: Academic Press, 1972. (Revised version of *Procedures as a Representation for Data in a Computer Program for Understanding Natural Language.* AI-TR-17. Cambridge, Mass.: MIT AI Lab., 1971.)

———. "A Procedural Model of Language Understanding," in *CMTL*, pp. 152–186.

———. *Five Lectures on Artificial Intelligence.* AIM-246. Stanford: Stanford University AI Lab., 1974.

———. "Frame Representations and the Declarative/Procedural Controversy," in *RU*, pp. 185–210.

Winston, P. H. *Holes.* AI Memo 163. Cambridge, Mass.: MIT AI Lab., 1970.

———. "The MIT Robot," in *MI-7*, pp. 465–480.

———, ed. *New Progress in Artificial Intelligence.* AI-TR-310. Cambridge, Mass.: MIT AI Lab., 1974.

———. "Learning Structural Descriptions from Examples," in *PCV*, pp. 157–210. (Revised version of thesis of same title: AI-TR-231. Cambridge, Mass.: MIT AI Lab., 1970.)

———, ed. *The Psychology of Computer Vision.* New York: McGraw-Hill, 1975.

Wisdom, A. J. T. D. *Philosophy and Psychoanalysis.* Oxford: Blackwell, 1953.

Wittgenstein, Ludwig. *Philosophical Investigations.* Oxford: Blackwell, 1953.

Wolfson, H. A. *The Philosophy of Spinoza.* Cambridge, Mass.: Harvard University Press, 1934.

Woods, W. A. "What's in a Link: Foundations for Semantic Networks," in *RU*, pp. 35–82.

Yakimovsky, Yoram, and J. A. Feldman. "A Semantics-Based Decision Theory Region Analyser," in *IJCAI-3*, pp. 580–588.

Yarbus, A. L. *Eye Movements and Vision.* New York: Plenum, 1967.

Young, J. Z. *A Model of the Brain.* Oxford: Clarendon, 1964.

Young, Richard M. "Production Systems as Models of Cognitive Development," in *AISB-1*, pp. 284–295.

Young, Robert M. "The Human Limits of Nature," in *The Limits of Human Nature* (ed. Jonathan Benthall), pp. 235–274. London: Allen Lane, 1973.

Ziff, Paul. "The Feelings of Robots," *Analysis*, 19 (1959), 64–68.

Zobrist, A. L., and F. R. Carlson. "An Advice-Taking Chess Computer," *Scientific American*, 228 (June 1973), 92–105.

# BIBLIOGRAPHY TO POSTSCRIPT

The literature in artificial intelligence has burgeoned over the past decade, and this additional bibliography lists only a representative sample. I have not attempted to mention every recent paper of interest, concentrating instead on sources that provide useful entry points into the specialist literature.

Those items which are especially accessible to readers unfamiliar with the field (including the most readable of the serious textbooks) are marked by a percent sign (%). Items in which many further specialist AI references can be found are distinguished with an asterisk (*). Items dealing with philosophical, psychological, educational, or social implications carry a number sign (#). And historical accounts of the field bear a plus sign (+).

The most recent research is described in the increasingly many professional journals (of which the most general are *Artificial Intelligence* and *Cognitive Science*), and in the proceedings of regular conferences (for example, *Proceedings of the International Joint Conference of Artificial Intelligence*, published biennially in odd-numbered years; *Proceedings of the European Conference on Artificial Intelligence; ECAI/AISB*, published biennially in even-numbered years; and *Proceedings of the American Association of Artificial Intelligence*, published annually).


\# Abelson, H., and A. di Sessa. *Turtle Geometry.* Cambridge, Mass.: MIT Press, 1981.

Abelson, H., G. J. Sussman, and J. Sussman. *Structure and Interpretation of Computer Programs.* Cambridge, Mass.: MIT Press, 1985.

Ackley, D. H., G. E. Hinton, and T. J. Sejnowski, "A Learning Algorithm for Boltzmann Machines," *Cognitive Science,* 9 (1985), 147–169.

Allen, J. F. "Towards a General Theory of Action and Time," *AI,* 23 (1984), 123–154.

Allen, J. F., and C. R. Perrault. "Analyzing Intention in Utterances," *AI,* 15 (1980), 143–178.

\# Anderson, J. R. *The Architecture of Cognition.* Cambridge, Mass.: Harvard University Press, 1983.

\# Arbib, M. A. *In Search of the Person: Philosophical Explorations in Cognitive Science.* Amherst: University of Massachusetts Press, 1985.

* Ballard, D. H., and C. M. Brown, eds. *Computer Vision.* Englewood Cliffs, N.J.: Prentice-Hall, 1982.

* Barr, A., E. A. Feigenbaum, and P. R. Cohen, eds., *The Handbook of Artificial Intelligence,* 3 vols. London: Pitman, 1981.

Barrett, R., A. Ramsay, and A. Sloman. *POP-11: A Practical Language for Artificial Intelligence.* Chichester: Ellis Horwood, 1985.

* Barrow, H., and J. M. Tenenbaum. "Computational Vision," *Proc. Inst. Electrical and Electronic Engineers,* 6 (1981), 572–595.

Berwick, R. *The Acquisition of Syntactic Knowledge.* Cambridge, Mass.: MIT Press, 1985.

\# BloomBecker, J. "Fifth Generation Computer Crime Law," *Proc. Ninth. Int. Joint Conf. on Artificial Intelligence,* Los Angeles (1985), 1274–1278.

Bobrow, D. G., ed. *Qualitative Reasoning About Physical Systems,* Cambridge, Mass.: MIT Press, 1985.

Bobrow, D. G., and T. Winograd. "An Overview of KRL, a Knowledge Representation Language," *Cognitive Science,* 1 (1977), 3–46.

———. "KRL: Another Perspective," *Cognitive Science,* 3 (1979), 29–42.

% Bobrow, D. G., and P. J. Hayes. "Artificial Intelligence: Where Are We?" *AI,* 25 (1985), 375–415.

# Boden, M. A. *Minds and Mechanisms: Philosophical Psychology and Computational Models.* Ithaca: Cornell University Press, 1981.

# ———. "Educational Implications of Artificial Intelligence," in *Thinking: The New Frontier.* (ed. W. Maxwell) pp. 227–236. Pittsburgh: Franklin Institute Press, 1983.

# ———. "Artificial Intelligence and Social Forecasting," *J. Mathematical Sociology,* 9 (1984), 341–356.

# ———. "Artificial Intelligence and Animal Psychology," *New Ideas in Psychology,* 1 (1983), 11–33.

# ———. "What is Computational Psychology?" *Proc. Aristotelian Society,* Supp. Vol. 58 (1984), 17–35.

# ———. "Artificial Intelligence and Legal Responsibility (Introduction to a Panel Discussion)," *Proc. Ninth. Int. Joint Conf. on Artificial Intelligence,* Los Angeles (1985), 1267–1268;

# ———. *Computer Models of Mind: The Computational Approach in Theoretical Psychology.* Cambridge: Cambridge University Press, forthcoming.

# Bolter, J. D. *Turing's Man: Western Culture in the Computer Age.* London: Duckworth, 1984.

* Bond, A. H., ed. *Machine Intelligence* (Infotech State of the Art Report, Series 9, No. 3). Maidenhead: Pergamon, 1981.

Boyer, R. S., and J. S. Moore. *A Computational Logic.* New York: Academic Press, 1979.

* Brachman, R. J., and H. J. Levesque, eds. *Readings in Knowledge Representation.* Los Altos, Calif.: Kaufmann, 1985.

# Braddick, O. J., and A. C. Sleigh, eds.. *Physical and Biological Processing of Images.* Berlin: Springer-Verlag, 1983.

* Brady, J. M., ed. *Computer Vision.* Amsterdam: North-Holland, 1981. (Reprint of *Artificial Intelligence:* Special Issue on Vision, 1981.)

———. "Artificial Intelligence and Robotics," *AI,* 26 (1985), 79–122;

* Brady, J. M., J. M. Hollerbach, T. L. Johnson, T. Lozano-Perez, and M. T. Mason, eds. *Robot Motion: Planning and Control.* Cambridge, Mass.: MIT Press, 1982.

* Brady, J. M., and R. C. Berwick. *Computational Models of Discourse.* Cambridge, Mass.: MIT Press, 1983.

Brady, J. M., and P. E. Agre. "The Mechanic's Mate," in *ECAI-84: Proc. Sixth European Conf. on Artificial Intelligence,* Pisa (1984), 681–696.

Bratko, I., and P. Mulec. "An Experiment in Automatic Learning of Diagnostic Rules," *Informatika.* Belgrade: Jugoslav Centre for Technical and Scientific Documentation, 1981.

Bratko, I., I. Mozetic, and N. Lavrac, "Automatic Synthesis and Compression of Cardiological Knowledge," *Machine Intelligence,* 11, (in press).

Buchanan, B. G., and E. H. Shortliffe. *Rule-Based Expert Programs: The MYCIN Experiments of the Stanford Heuristic Programming Project.* Reading, Mass.: Addison-Wesley, 1984.

Bundy, A., ed. *Artificial Intelligence: An Introductory Course.* Edinburgh: Edinburgh University Press, 1978.

———. *The Computer Modelling of Mathematical Reasoning.* Orlando, Florida: Academic Press, 1984.

# Bundy, A., and R. Clutterbuck. "Raising the Standards of AI Products," in *Proc. Ninth Joint Conf. on Artificial Intelligence,* Los Angeles (1985), 1289–1294.

Bundy, A., B. Silver, and D. Plummer. "An Analytical Comparison of Some Rule-Learning Programs," *AI,* 27 (1985), 137–182.

Cendrowska, J., and M. Bramer. "Inside an Expert System: A Rational Reconstruction of the MYCIN Consultation System," in *Artificial Intelligence: Tools, Techniques, Applications* (eds. T. O'Shea and M. Eisenstadt), pp. 453–497. New York: Harper & Row, 1984.

Chandrasekaran, B. "Towards a Taxonomy of Problem-Solving Types," *AI Magazine,* 4 (1983), 9–17.

% Charniak, E., and D. McDermott. *Introduction to Artificial Intelligence.* Reading, Mass.: Addison-Wesley, 1985.

Clancey, W. J., and E. H. Shortliffe, eds. *Readings in Medical Artificial Intelligence: The First Decade.* Menlo Park: Addison-Wesley, 1984.

Clark, K. L. "Negation as Failure," in *Logic and Databases.* (eds. H. Gallaire and J. Minker) New York: Plenum, 1978.

# Clippinger, J. H. *Meaning and Discourse: A Computer Model of Psychoanalytic Speech and Cognition.* Baltimore: Johns Hopkins University Press, 1977.

Clocksin, W. "An Introduction to PROLOG," in *Artificial Intelligence: Tools, Techniques, Applications.* (eds. T. O'Shea and M. Eisenstadt), pp. 1–22. New York: Harper & Row, 1984.

Clocksin, W., and C. Mellish. *Programming in Prolog.* New York: Springer-Verlag, 1981.

Clocksin, W., and P. G. Davey. "Industrial Robotics," in *Artificial Intelligence: Tools, Techniques, Applications.* (eds. T. O'Shea and M. Eisenstadt), pp. 389–399. New York: Harper & Row, 1984.

Cohen, P. R. *Heuristic Reasoning About Uncertainty: An Artificial Intelligence Approach.* Palo Alto: Kaufmann, 1985.

Cohen, P. R., and C. R. Perrault. "Elements of a Plan-Based Theory of Speech-Acts," *Cognitive Science,* 3 (1979), 177–212.

Cunningham, J. "Comprehension by Model-Building as a Basis for an Expert System," in *Expert Systems 85* (ed. M. Merry), pp. 259–272. Cambridge: Cambridge University Press, 1985.

# *Daedalus (J. Amer. Acad. Arts & Sciences),* a two-part series on "Weapons in Space": Vol. 1, Spring 1985, "Concepts and Technologies"; Vol. 2, Summer 1985, "Implications for Security."

Davey, A. *Discourse Production: A Computer Model of Some Aspects of a Speaker.* Edinburgh: Edinburgh University Press, 1978.

# Davis, D. B. "Assessing the Strategic Computing Initiative," *High Technology,* 5 (1985), 41–49.

Davis, R., and J. J. King. "An Overview of Production Systems", in *Machine Intelligence,* 8. (eds. E. Elcock and D. Michie), pp. 300–333. Chichester: Ellis Horwood, 1977.

* Davis, R., and Y. A. Wilks. *Computational Semantics: An Introduction to Artificial Intelligence and Natural Language Comprehension.* Amsterdam: North-Holland, 1976.

* Davis, R., and D. B. Lenat, eds. *Knowledge-Based Systems in Artificial Intelligence.* New York: McGraw-Hill, 1982.

De Jong, G. F. "A New Approach to Natural Language Processing," *Cognitive Science,* 3 (1979).

De Kleer, J., and J. S. Brown. "A Qualitative Physics Based on Confluences," *AI,* 24 (1984), 7–83.

# Dennett, D. C. *Brainstorms: Philosophical Essays on Mind and Psychology.* Cambridge, Mass.: MIT Press, 1978.

Dewdney, A. K. "Computer Recreations: Chess Machines," *Scientific American,* 254, No. 2 (Feb. 1986), 12–15.

Doyle, J. "A Truth Maintenance System," *AI,* 12 (1979), 231–272.

# ———. "What is Rational Psychology?" *AI Magazine,* 4 (1983), 50–54.

Dyer, M. G. *In-Depth Understanding: A Computer Model of Integrated Processing for Narrative Comprehension.* Cambridge, Mass.: MIT Press, 1983.

# Ennals, R. *Artificial Intelligence Applications to Logical Reasoning and Historical Research.* Chichester: Wiley, 1985.

Erman, L. D., P. E. London, and S. F. Fickas. "The Design and an Example Use of Hearsay III," *Proc. Seventh Int. Joint Conf. on Artificial Intelligence,* Vancouver, B.C. (1981) 409–415.

Fahlman, S. E. *NETL: A System for Representing and Using Real-World Knowledge.* Cambridge, Mass.: MIT Press, 1979.

- Feigenbaum, E. A., and P. McCorduck. *The Fifth Generation: Artificial Intelligence and Japan's Computer Challenge to the World.* New York: Addison-Wesley, 1983.

Findler, N. V., ed. *Associative Networks: Representation and Use of Knowledge by Computers.* New York: Academic Press, 1979.

: Fodor, J. A. *Representations.* Cambridge, Mass.: MIT Press, 1980.

- Fleck, J. "Development and Establishment in Artificial Intelligence." In *Scientific Establishments and Hierarchies.* (eds. N. Elias, H. Martins, and R. Whitley), Sociology of the Sciences, Vol. VI. pp. 169–217. Amsterdam: Reidel, 1982.

Forbus, K. D. "Qualitative Process Theory," *AI,* 24 (1984), 85–168.

Frey, P. W., ed. *Chess Skill in Man and Machine.* 2nd ed. New York: Springer-Verlag, 1983.

\# Frude, N. *The Intimate Machine: Close Encounters with the New Computers.* London: Century, 1983.

Funt, B. V. "Problem-Solving with Diagrammatic Representations," *AI,* 13 (1980), 201–230.

Gardner, A. *An Artificial Intelligence Approach to Legal Reasoning.* Cambridge, Mass.: MIT Press, in press.

% + Gardner, H. *The Mind's New Science: A History of the Cognitive Revolution.* New York: Basic Books, 1985.

Gazdar, G., E. Klein, G. Pullum, and I. Sag. *Generalized Phrase Structure Grammar.* Oxford: Blackwell, 1985.

* Gazdar, G., and G. Pullum, "Computationally Relevant Properties of Natural Languages and Their Grammars," *New Generation Computing,* 3 (1985), 1–33.

Gentner, D. "Structure-Mapping: A Theoretical Framework for Analogy," *Cognitive Science,* 7 (1983), 155–170.

\# Gerver, E. *Computers and Adult Learning.* Milton Keynes: Open University Press, 1984.

\# Gilbert, N. G., and C. Heath. *Social Action and Artificial Intelligence: Surrey Conferences on Sociological Theory and Method, 3.* London: Gower, 1985.

Gotts, N. M., J. R. W. Hunter, and R. K. E. W. Sinnhuber. "An Intelligent Model-Based System for Diagnosis in Cardiology," *Artificial Intelligence in Medicine Group, Report AIMG-7,* Sussex University, 1984.

* Grosz, B. J. "Understanding Discourse Knowledge," in *Understanding Spoken Language.* (ed. D. Walker), pp. 229–347. New York: North-Holland, 1978.

———. "Utterance and Objective: Issues in Natural Language Communication," *AI Magazine,* 1 (1980), 11–20.

* Halpern, J. Y., ed. *Theoretical Aspects of Reasoning About Knowledge.* Palo Alto: Kaufmann, 1986.

\# Hand, D. J. *Artificial Intelligence and Psychiatry.* Cambridge: Cambridge University Press, 1985.

* Hanson, A., and E. Riseman, eds. *Computer Vision Systems.* New York: Academic Press, 1978.

% Hardy, S. "Robot Control Systems," in *Artificial Intelligence: Tools, Techniques, Applications.* (eds. T. O'Shea and M. Eisenstadt), pp. 178–191. New York: Harper & Row, 1984.

Harmon, P., and D. King. *Expert Systems: Artificial Intelligence in Business.* New York: Wiley, 1985.

\# Haugeland, J., ed. *Mind Design: Philosophy, Psychology, Artificial Intelligence.* Cambridge, Mass.: MIT Press, 1981.

\# ———. *Artificial Intelligence: The Very Idea.* Cambridge, Mass.: MIT Press and Bradford Books, 1985.

Hayes, P. J. "The Naive Physics Manifesto," in *Expert Systems in the Micro-Electronic Age.* (ed. D. Michie), pp. 242–270. Edinburgh: Edinburgh University Press, 1979.

———. "In Defense of Logic," *Proc. Fifth Int. Joint Conf. on Artificial Intelligence,* Cambridge, Mass. (1977), 559–565.

* Hayes-Roth, F., D. A. Waterman, and D. B. Lenat, eds. *Building Expert Systems.* Reading, Mass.: Addison-Wesley, 1983.

Hillis, W. D. *The Connection Machine.* Cambridge, Mass.: MIT Press, 1985.

Hinton, G. E. "Shape Representation in Parallel Systems," *Proc. Seventh Int. Joint Conf. on Artificial Intelligence,* Vancouver, B.C. (1981).

———. "Optimal Perceptual Inference," *Proc. IEEE Conf. on Computer Vision and Pattern Recognition.* Washington, D.C., June 1983.

———. "Parallel Computations for Controlling an Arm," *J. Motor Behaviour,* 16 (1984), 171–194.

* Hinton, G. E., and J. A. Anderson, eds. *Parallel Models of Associative Memory.* Hillsdale, N.J.: Lawrence Erlbaum, 1981.

* Hobbs, J., and R. C. Moore, eds. *Formal Theories of the Commonsense World.* Norwood, N.J.: Ablex, 1984.

% Hofstadter, D. R. *Gödel, Escher, Bach: An Eternal Golden Braid.* New York: Basic Books, 1979.

Horn, B. K. P. *Robot Vision.* Cambridge, Mass.: MIT Press, 1986.

Hunter, J. R. W. "Artificial Intelligence in Medicine: A Tutorial Survey," *Biomedical Measurement, Informatics and Control* (in press).

# Johnson-Laird, P. N. *Mental Models: Towards a Cognitive Science of Language, Inference, and Consciousness.* Cambridge: Cambridge University Press, 1983.

* Joshi, A. K., B. L. Webber, and I. A. Sag, eds. *Elements of Discourse Understanding.* Cambridge: Cambridge University Press, 1981.

* King, M., ed. *Parsing Natural Language.* London: Academic Press, 1983.

# Kling, R., and W. Scacchi. "The Web of Computing: Computing Technology as Social Organization," in M. Yovits, ed., *Advances in Computers,* 21 (1982), 1–90.

Kolodner, J. "Maintaining Organization in a Dynamic Long-Term Memory," and "Reconstructive Memory: A Computer Model," *Cognitive Science,* 7 (1983), 243–328.

Kowalski, R. A. *Logic for Problem Solving.* Amsterdam: North-Holland, 1979.

# Kowalski, R. A., and M. Sergot. "Computer Representation of the Law," *Proc. Ninth. Int. Joint Conf. on Artificial Intelligence,* Los Angeles (1985), 1269–1270.

Kuipers, B. "Commonsense Reasoning About Causality: Deriving Behavior From Structure," *AI,* 24 (1984), 169–203.

# Langley, P. "Language Acquisition through Error Recovery," *Cognition and Brain Theory,* 5 (1982), 211–255.

# Lehman-Wilzig, S. N. "Frankenstein Unbound: Towards a Legal Definition of Artificial Intelligence," *Futures: The Journal of Forecasting and Planning,* 13 (1981), 107–119.

Lehnert, W. G. *The Process of Question Answering.* Hillsdale, N.J.: Erlbaum, 1978.

* Lehnert, W. G., and M. H. Ringle, eds. *Strategies for Natural Language Processing.* Hillsdale, N.J.: Lawrence Erlbaum, 1982.

Lenat, D. B. "The Nature of Heuristics," *AI,* 19 (1982), 189–249.

———. "Theory Formation by Heuristic Search. The Nature of Heuristic II: Background and Examples," *AI,* 20 (1983), 61–98.

———. "The Role of Heuristics in Learning by Discovery: Three Case Studies," in *Machine Learning: An Artificial Intelligence Approach.* vol. I. (eds. R. Michalski, J. G. Carbonell, and T. M. Mitchell), pp. 243–305. Palo Alto: Tioga, 1983.

Lenat, D. B., and J. S. Brown. "Why AM and Eurisko Appear to Work," *AI,* 23 (1984), 269–294.

Lenat, D. B., M. Prakash, and M. Shepherd. "CYC: Using Common Sense Knowledge to Overcome Brittleness and Knowledge Acquisition Bottlenecks," *AI Magazine,* 6 (1986), 65–85.

Levesque, H. J. "Foundations of a Functional Approach to Knowledge Representation," *AI,* 23 (1984), 155–212.

Lindsay, R. K., B. G. Buchanan, E. A. Feigenbaum, and J. Lederberg. *Applications of Artificial Intelligence for Organic Chemistry: The Dendral Project.* New York: McGraw-Hill, 1980.

# Longuet-Higgins, C. H. *Mental Processes: Papers in Cognitive Science.* Cambridge, Mass.: MIT Press (in press).

McCarthy, J. "Circumscription: A Form of Non-Monotonic Reasoning," *AI,* 13 (1980), 27–40.

——— "Some Expert Systems Need Common Sense," *Annals New York Academy of Sciences,* 426 (1983), 129–137.

———. "Applications of Circumscription to Formalizing Common-Sense Knowledge," *AI,* 28 (1986), 89–116.

+ McCorduck, P. *Machines Who Think.* San Francisco: Freeman, 1979.

McDermott, D. V. "Artificial Intelligence Meets Natural Stupidity," reprinted in *Mind Design: Philosophy, Psychology, Artificial Intelligence* (ed. J. Haugeland), pp. 143–160. Cambridge, Mass.: MIT Press, 1981. (First published in SIGART Newsletter of the ACM, No. 57, April 1976.)

———. "R1: A Rule-Based Configurer of Computer Systems." *AI,* 19 (1982), 39–88.

McDermott, D. V., and J. Doyle. "Non-Monotonic Logic," *AI,* 13 (1980), 41–72.

McDermott, D. V., M. M. Waldrop, R. C. Schank, B. Chandrasekaran, and J. McDermott. "The Dark Ages of AI: A Panel Discussion at AAAI-84," *AI Magazine,* 6 (Fall, 1985), 122–134.

Marcus, M. P. *A Theory of Syntactic Recognition for Natural Language.* Cambridge, Mass.: MIT Press, 1980.

* Marr, D. A. *Vision.* San Francisco: Freeman, 1982.

# Mayhew, J. "Stereopsis," in *Physical and Biological Processing of Images.* (eds. O. J. Braddick and A. C. Sleigh), pp. 204–216. Berlin: Springer-Verlag, 1983.

* Mayhew, J., and J. Frisby. "Computer Vision," in *Artificial Intelligence: Tools, Techniques, Applications.* (eds. T. O'Shea and M. Eisenstadt), pp. 301–357. New York: Harper & Row, 1984.

Mellish, C. S. *Computer Interpretation of Natural Language Descriptions.* Chichester: Ellis Horwood, 1985.

* Merry, M., ed. *Expert Systems 85.* Cambridge: Cambridge University Press, 1985.

* Michalski, R., J. G. Carbonell, and T. M. Mitchell, eds. *Machine Learning: An Artificial Intelligence Approach.* vol. I. Palo Alto: Tioga, 1983.

* ———. eds. *Machine Learning: An Artificial Intelligence Approach.* vol. II. Palo Alto: Kaufmann, 1986.

Michener, E. R. "Understanding Understanding Mathematics," *Cognitive Science,* 2 (1978), 361–383.

* Michie, D., ed. *Expert Systems in the Micro-Electronic Age.* Edinburgh: Edinburgh University Press, 1979.

* ———, ed. *Introductory Readings in Expert Systems.* New York: Gordon & Breach, 1982.

# Minsky, M. "The Society Theory of Thinking," in *Artificial Intelligence: An MIT Perspective.* (eds. P. H. Winston and R. H. Brown), pp. 421–452. Cambridge, Mass.: MIT Press, 1979.

# ———. "K-Lines: A Theory of Memory," *Cognitive Science,* 4 (1980), 117–133.

# Newell, A. "Physical Symbol Systems," *Cognitive Science,* 4 (1980), 135–183.

———. "The Knowledge Level," *AI,* 18 (1982), 87–127.

* Nilsson, N. J. *Principles of Artificial Intelligence.* Palo Alto: Tioga, 1980.

———. "Artificial Intelligence, Employment, and Income," in *Impacts of Artificial Intelligence: Scientific, Technological, Military, Economic, Societal, Cultural, and Political.* (ed. R. Trappl), pp. 103–123. Amsterdam: North-Holland, 1986.

# Norman, D. A., and S. W. Draper. *User Centered System Design: New Perspectives on Human-Computer Interaction.* Hillsdale, N.J.: Lawrence Erlbaum, 1986.

# Nycum, S. H., and I. K. Fong. "Artificial Intelligence and Certain Resulting Legal Issues," *The Computer Law Review,* University of Southern California, 1985.

O'Shea, T., ed. *Advances in Artificial Intelligence.* Amsterdam: North-Holland, 1984.

* O'Shea, T., and M. Eisenstadt, eds. *Artificial Intelligence: Tools, Techniques, Applications.* New York: Harper & Row, 1984.

# O'Shea, T., and J. Self. *Teaching and Learning With Computers.* Brighton: Harvester Press, 1984.

# Papert, S. *Mindstorms: Children, Computers, and Powerful Ideas.* Brighton: Harvester Press, 1980.

# Parnas, D. L. *Software and SDI: Why Communications Systems Are Not Like SDI.* (Senate Testimony, December 1985). Available from CPSR, PO Box 717, Palo Alto, CA 94301, USA.

# ———. "Software Aspects of Strategic Defense Systems," *American Scientist,* Sept.-Oct. 1985, 432–440.

* Paul, R. P. *Robot Manipulators: Mathematics, Programming and Control.* Cambridge, Mass.: MIT Press, 1981.

# Pea, R. D., and D. M. Kurland. "On the Cognitive Effects of Learning Computer Programming," *New Ideas in Psychology,* 2 (1984), 137–168.

# Pylyshyn, Z. W. *Computation and Cognition: Toward a Foundation for Cognitive Science.* Cambridge, Mass.: MIT Press, 1984.

Quinlan, J. R. "Learning Efficient Classification Procedures and Their Application to Chess End-Games," in *Machine Learning: An Artificial Intelligence Approach.* (eds. R. Michalski, J. G. Carbonell, and T. M. Mitchell), pp. 463–482. Palo Alto: Tioga, 1983.

Reichman, R. "Conversational Coherency," *Cognitive Science,* 2 (1978), 283–327.

———. *Getting Computers to Talk Like You and Me: Discourse Context, Focus, and Semantics (An ATN Model).* Cambridge, Mass.: MIT Press, 1985.

Reiter, R. "A Logic for Default Reasoning," *AI,* 13 (1980), 81–132.

Reitman, W., ed. *Artificial Intelligence Applications for Business.* Norwood, N. J.: Ablex, 1985.

* Rich, E. *Artificial Intelligence.* New York: McGraw-Hill, 1983.

# Ringle, M., ed. *Philosophical Perspectives in Artificial Intelligence.* Hassocks: Harvester Press, 1979.

Rissland, E. L. "The Ubiquitous Dialectic," *Proc. Sixth European Conf. on Artificial Intelligence,* Pisa (1984), 663–668.

# ———. "AI and Legal Reasoning," *Proc. Ninth Joint Conf. on Artificial Intelligence,* Los Angeles (1985), 1254–1260.

Ritchie, G. D. *Computational Grammar.* Hassocks: Harvester Press, 1980.

Ritchie, G. D., and F. K. Hanna. "AM: A Case Study in AI Methodology," *AI,* 23 (1984), 249–268.

# Rumelhart, D. E., G. E. Hinton, and R. J. Williams. "Learning Internal Representations by Error Propagation," in *Parallel Distributed Processing: Explorations in the Microstructure of Cognition. Vol. 1: Foundations.* (eds. D. E. Rumelhart and J. L. McClelland), Cambridge, Mass.: MIT Press, 1986.

# Rumelhart, D. E., and J. L. McClelland, eds. *Parallel Distributed Processing: Explorations in the Microstructure of Cognition.* 2 vols. Cambridge, Mass.: MIT Press, 1986.

* Rylko, H., ed. *Artificial Intelligence: Bibliographic Summaries of the Select Literature.* Lawrence, Kansas: The Report Store, 1985.

# Schank, R. C. *Dynamic Memory: A Theory of Learning in Computers and People.* Cambridge: Cambridge University Press, 1982.

# Schank, R. C., and R. P. Abelson. *Scripts, Plans, Goals, and Understanding.* Hillsdale, N.J.: Erlbaum, 1977.

* Schank, R. C., and C. K. Riesbeck, eds. *Inside Computer Understanding: Five Programs Plus Miniatures.* Hillsdale, N.J.: Lawrence Erlbaum, 1981.

% Schank, R. C., and P. G. Childers. *The Cognitive Computer: On Language, Learning, and Artificial Intelligence.* Reading, Mass.: Addison-Wesley, 1984.

# Searle, J. "Minds, Brains, and Programs," *Behavioral and Brain Sciences,* 3 (1980), 417–457.

# Selfridge, O. G., E. L. Rissland, and M. A. Arbib, eds. *Adaptive Control in Ill-Defined Systems.* New York: Plenum, 1984.

# Sharples, M. *Cognition, Computers, and Creative Writing.* Chichester: Ellis Horwood, 1985.

# Shneiderman, B., and A. Badre, eds. *Directions in Human-Computer Interactions.* Norwood, N.J.: Ablex, 1982.

% Shore, J. *The Sachertorte Algorithm and Other Antidotes to Computer Anxiety.* New York: Viking Penguin, 1985.

# Simon, H. A. *The Sciences of the Artificial.* 2nd ed. Cambridge, Mass.: MIT Press, 1981.

# ———. "Cognitive Science: The Newest Science of the Artificial," in *Perspectives on Cognitive Science.* (ed. D. A. Norman), pp. 13–26. Norwood, N. J.: Ablex, 1981.

# ———. "Search and Reasoning in Problem Solving," *AI,* 21 (1983), 7–29.

# Sleeman, D. H., and J. S. Brown, eds. *Intelligent Tutoring Systems.* London: Academic Press, 1982.

# Sloman, A. *The Computer Revolution in Philosophy: Philosophy, Science, and Models of Mind.* Brighton: Harvester Press, 1978.

# ———. "Beginners Need Powerful Systems," in *New Horizons in Educational Computing.* (ed. M. Yazdani), pp. 220–234. Chichester: Ellis Horwood, 1984.

# ———. "Real Time Multiple-Motive Expert Systems," in *Expert Systems 85.* (ed. M. Merry), pp. 213–224. Cambridge: Cambridge University Press, 1985.

# ———. "What Enables a Machine to Understand?" *Proc. Ninth Int. Joint Conf. on Artificial Intelligence.,* Los Angeles (1985), 995–1001.

———. "Why We Need Many Knowledge Representation Formalisms," in *Research and Development in Expert Systems.* (ed. M. Bramer), pp. 163–183. Cambridge: Cambridge University Press, 1985.

# Sloman, A., and M. Croucher. "Why Robots Will Have Emotions," *Proc. Seventh Int. Joint Conf. on Artificial Intelligence,* Vancouver, B.C. (1981), 197–202.

* Sparck Jones, K., and Y. A. Wilks. *Automatic Natural Language Parsing.* Chichester: Ellis Horwood, 1983.

* Steels, L., and J. A. Campbell. *Progress in Artificial Intelligence.* Chichester: Wiley, 1985.

Stefik, M., J. Aikins, R. Balzer, J. Benoit, L. Birnbaum, F. Hayes-Roth, and E. D. Sacerdoti. "The Organization of Expert Systems," *AI,* 18 (1982), 135–173.

* Sowa, J. F. *Conceptual Structures: Information Processing in Mind and Machine.* Reading, Mass.: Addison-Wesley, 1984.

Thompson, H. "Speech Transcription: An Incremental, Interactive Approach," in *ECAI-84: Proc. Sixth European Conf. on Artificial Intelligence,* Pisa (1984), 697–704.

# ———. "There Will Always be Another Moonrise: Computer Reliability and Nuclear Weapons." *The Scotsman,* 1984.

# ———. "Empowering Automatic Decision-Making Systems," *Proc. Ninth Int. Joint Conf. on Artificial Intelligence,* Los Angeles (1985), 1281–1283.

# Torrance, S., ed. *The Mind and the Machine: Philosophical Aspects of Artificial Intelligence.* Chichester: Ellis Horwood, 1984.

* Trappl. R., ed. *Impacts of Artificial Intelligence: Scientific, Technological, Military, Economic, Societal, Cultural, and Political.* Amsterdam: North-Holland, 1986.

# Turkle, S. *The Second Self: Computers and the Human Spirit.* New York: Simon & Schuster, 1984.

Ullman, S. *The Interpretation of Visual Motion.* Cambridge, Mass.: MIT Press, 1979.

* Ullman, S., and W. Richards, eds. *Image Understanding.* Norwood, N.J.: Ablex, 1985.

Van Caneghem, M., and D. H. Warren, eds. *Logic Programming and its Applications.* Norwood, N. J.: Ablex, 1986.

* Walker, D., ed. *Understanding Spoken Language.* New York: North-Holland, 1978.

* Waterman, D. A. *A Guide to Expert Systems.* New York: Addison-Wesley, 1985.

* Waterman, D. A., and F. Hayes-Roth, eds. *Pattern-Directed Inference Systems.* New York: Academic Press, 1978.

* Webber, B. L., and N. J. Nilsson, eds. *Readings in Artificial Intelligence.* Palo Alto: Kaufmann, 1981.

Wilensky, R. *Planning and Understanding.* Reading, Mass.: Addison-Wesley, 1983.

Wilks, Y. A. "Making Preferences More Active," *AI,* 11 (1978), 197–223.

# ———. "Responsible Computers," *Proc. Ninth. Int. Joint Conf. on Artificial Intelligence,* Los Angeles (1985), 1279–1280.

# Willick, M. S. "Constitutional Law and Artificial Intelligence: The Potential Legal Recognition of Computers as 'Persons,'" *Proc. Ninth. Int. Joint Conf. on Artificial Intelligence,* Los Angeles (1985), 1271–1273.

* Winograd, T. *Language as a Cognitive Process: Syntax.* Reading, Mass.: Addison-Wesley, 1983 (chapter on computer models).

# Winograd, T., and F. Flores. *Understanding Computers and Cognition: A New Foundation for Design.* Norwood, N. J.: Ablex, 1986.

Winston, P. H. "Learning and Reasoning by Analogy," *CACM,* 23 (1980), 689–703.

* ———. *Artificial Intelligence.* 2nd ed. Reading, Mass.: Addison-Wesley, 1984.

* Winston, P. H., and R. H. Brown, eds. *Artificial Intelligence: An MIT Perspective.* 2 vols. Cambridge, Mass.: MIT Press, 1979.

* Winston, P. H., and K. A. Prendergast. *The AI Business: Commercial Uses of Artificial Intelligence.* Cambridge, Mass.: MIT Press, 1984.

# Yazdani, M., ed. *New Horizons in Educational Computing.* Chichester: Ellis Horwood, 1984.

# Yazdani, M., and A. Narayanan, eds. *Artificial Intelligence: Human Effects.* Chicester: Ellis Horwood, 1984.

*For continuing reports of recent research, see:*

*Artificial Intelligence* (quarterly journal)

*Artificial Intelligence and Society* (first number, Spring 1987)

*Artificial Intelligence Review* (first number, July 1986)

*International Journal of Robotics Research* (quarterly journal)

*Proceedings of the American Association for Artificial Intelligence* (annual)

*Proceedings of the European Society for the Study of Artificial Intelligence and Simulation of Behaviour: ECAI /AISB* (biennial, even-numbered years)

*Proceedings of the International Joint Conference on Artificial Intelligence* (biennial, odd-numbered years)

# NOTES*

## CHAPTER 1

1. The first definition is from Donald Michie, *On Machine Intelligence* (Edinburgh, 1974), p. 156; the second is from P. J. Hayes, "Some Comments on Sir James Lighthill's Report on Artificial Intelligence," *AISB Study Group European Newsletter*, Issue 14 (July 1973), p. 40. For Lighthill's misrepresentation of artificial intelligence, and some rebuttals, see: James Lighthill, N. S. Sutherland, R. M. Needham, H. C. Longuet-Higgins, and Donald Michie, *Artificial Intelligence: A Paper Symposium* (London, 1973).

2. See, e.g., M. B. Clowes, "Man the Creative Machine: A Perspective from Artificial Intelligence Research," in Jonathan Benthall, ed., *The Limits of Human Nature* (London, 1973), pp. 192–207; M. L. Minsky and Seymour Papert, *Artificial Intelligence* (Eugene, Oregon, 1973); Allen Newell, "Artificial Intelligence and the Concept of Mind," in *CMTL*, pp. 1–60.

3. N. J. Nilsson, *Problem-Solving Methods in Artificial Intelligence* (New York, 1971), p. vii.

4. M. L. Minsky, in *SIP*, p. v.

5. See J. A. Fodor, *Psychological Explanation: An Introduction to the Philosophy of Psychology* (New York, 1968), ch. iv.

6. W. L. Kilmer, W. S. McCulloch, and Jay Blum, "A Model of the Vertebrate Central Command System," *Int. J. Man-Machine Studies*, 1 (1969), 279–309.

7. "Effective procedure" is defined in M. L. Minsky, *Computation: Finite and Infinite Machines* (Englewood Cliffs, 1967), p. viii and ch. v. A less technical account is given in Joseph Weizenbaum, *Computer Power and Human Reason: From Judgment to Calculation* (San Francisco, 1976), ch. ii.

8. For a clear introduction to the "mechanics" of computers, see: S. H. Hollingdale and G. C. Tootill, *Electronic Computers* (Harmondsworth, 1970). Some current developments are described in Rein Turn, *Computers in the 1980's* (New York, 1974).

9. A random choice of memory address does not necessarily lead to chaos inside the computer. Thus hash-coding (a form of data storage that in many cases is more efficient than, e.g., alphabetical ordering) uses a "random" numerical procedure to pick, and later on to find, a particular memory cell for a word. It is as though you were to add the "alphabetical" numbers for the letters "C," "A," and "T," and then multiply by the year of your birth, to find the number functioning as the address of the memory cell concerned. Only by coincidence (if the alphabetical orders of the letters of some other words have a sum equal to that of C, A, and T) will more than one word get assigned to any one cell. Special procedures can be written to cope with such collisions; the more words there are to be stored, the more likely such collisions are. When the program needs to *search* for "CAT," it merely does some arithmetic and goes *straight* to the relevant cell, rather than going through a systematic search of the memory store (e.g., alphabetically).

10. H. L. Dreyfus's philosophical critique of the possibility of artificial intelligence will be discussed in Chapter 14.

* A list of source abbreviations is given at the beginning of the Bibliography.

## CHAPTER 2

1. K. M. Colby, "Computer Simulation of a Neurotic Process," in S. S. Tomkins and Samuel Messick, eds., *Computer Simulation of Personality: Frontier of Psychological Research* (New York, 1963), pp. 165–180; K. M. Colby and J. P. Gilbert, "Programming a Computer Model of Neurosis," *J. Math. Psychol.*, 1 (1964), 220–227; K. M. Colby, "Experimental Treatment of Neurotic Computer Programs," *Arch. Gen. Psychiatry*, 10 (1964), 220–227; K. M. Colby, "Computer Simulation of Neurotic Processes," in R. W. Stacy and B. D. Waxman, eds., *Computers in Biomedical Research, Vol. I* (New York, 1965), pp. 491–503.

2. Colby, "Experimental Treatment of Neurotic Computer Programs," p. 221.

3. Colby, "Computer Simulation of a Neurotic Process," p. 172. The list of transforms is slightly different in Colby and Gilbert, "Programming a Computer Model of Neurosis," p. 412.

4. This procedure is described on pp. 412–413 of the most detailed published account of the program: Colby and Gilbert, "Programming a Computer Model of Neurosis."

5. Colby, "Computer Simulation of a Neurotic Process," pp. 172–173.

6. J. C. Speisman took two groups of subjects: students with outside interests of a markedly intellectual nature, and businessmen with nonintellectual hobbies. He measured their GSR (an electrophysiological indicator of anxiety) while they were watching a film of a rather nasty tribal circumcision-initiation rite. There were three experimental situations: film alone, no commentary; film with added soundtrack providing an *intellectualizing* but nondenying commentary (the rite is a fascinating example of this, that, and the other anthropological concept, even if it does hurt the boys); and film with added *denying* commentary (it doesn't really hurt, and anyway the boys are pleased and proud to be joining the adult community). The GSR of all subjects was reduced by each added soundtrack; but the intellectualizing commentary worked better with the student group and the denying commentary with the businessmen (indeed, the effect of denial on the students was hardly noticeable). See J. C. Speisman, "Autonomic Monitoring of Ego Defense Process," in N. S. Greenfield and W. C. Lewis, eds., *Psychoanalysis and Current Biological Thought* (Madison, 1965), pp. 227–244. Other experimental studies of Freudian theory are reviewed in Paul Kline, *Fact and Fantasy in Freudian Theory* (London, 1972).

7. See the Postscript to Sigmund Freud, *The Question of Lay Analysis* (1926).

8. Sigmund Freud, *Inhibitions, Symptoms, and Anxiety* (1926), p. 163.

9. Colby, "Computer Simulation of a Neurotic Process," p. 173. The defensive reply "No" is produced by processes that are represented by flow diagrams in Colby, "Experimental Treatment of Neurotic Computer Programs," pp. 224 and 225. (What Colby had earlier called a "pool" is called a "complex" of beliefs in this paper.)

## CHAPTER 3

1. Verbal descriptions of many-leveled interpersonal perception are given in R. D. Laing, *Self and Others* (London, 1961). The use of the questionnaire technique for identifying matches and mismatches between the various levels is described in R. D. Laing, Herbert Philipson, and A. R. Lee, *Interpersonal Perception: A Theory and a Method of Research* (London, 1966).

2. Moser uses variable density functions as mathematical representations of

psychoanalytic concepts such as: oral, anal, phallic, and genital pleasure gain; the primacy of genitality (the fusion of the four previous component drives); anxiety; repression and its derivatives, functioning as countercathexis mechanisms; emergency defense mechanisms; and various specific forms of neurosis. See: Ulrich Moser, Werner Schneider, and Ilka von Zeppelin, "Computer Simulation of a Model of Neurotic Defence Mechanism. (Clinical Paper and Technical Paper)," *Bull. Psychol. Inst. Univ. Zurich*, 2 (1968), 1–77; Ulrich Moser, Ilka von Zeppelin, and Werner Schneider, "Computer Simulation of a Model of Neurotic Defense Processes," *Int. J. Psychoanalysis*, 50 (1969), 53–64; Ulrich Moser, Ilka von Zeppelin, and Werner Schneider, "Computer Simulation of a Model of Neurotic Defense Processes," *Behavioral Science*, 15 (1970), 194–202.

3. K. M. Colby, *Energy and Structure in Psychoanalysis* (New York, 1955).

4. For a brief statement of this distinction, see Charles Rycroft, "Causes and Meaning," in Charles Rycroft, ed., *Psychoanalysis Observed* (London, 1966), pp. 7–22. For an influential discussion of "hermeneutic-dialectical" sciences, see Gerard Radnitsky, *Continental Schools of Metascience* (Goteborg, 1968). Other hermeneutic commentators on Freud will be cited in Chapter 13, where I discuss the relevance of artificial intelligence to hermeneutic psychologies.

5. Stanley Schachter and J. E. Singer, "Cognitive, Social, and Physiological Determinants of Emotional State," *Psychological Review*, 69 (1962), 379–399; Stanley Schachter, "The Interaction of Cognitive and Physiological Determinants of Emotional State," in C. D. Spielberger, ed., *Anxiety and Behavior* (New York, 1966), pp. 193–224.

6. Harwood Fisher, "Logic and Language in Defences," *J. Theory of Social Behaviour*, 3 (1973), 157–214.

7. Colby's interviewing program is described in : K. M. Colby, J. B. Watt, and J. P. Gilbert, "A Computer Method of Psychotherapy: Preliminary Communication," *J. Nerv. Mental Disease*, 142 (1966), 148–152; K. M. Colby and Horace Enea, "Heuristic Methods for Computer Understanding of Natural Language in Context-Restricted On-Line Dialogues," *Mathematical Biosciences*, 1 (1967), 1–25.

8. Colby, Watt, and Gilbert, "A Computer Method of Psychotherapy," pp. 151–152.

9. P. E. Meehl, *Clinical Versus Statistical Prediction: A Theoretical Analysis and a Review of the Evidence* (Minneapolis, 1954).

10. For Freud's account of the Little Hans case, see Sigmund Freud, *Analysis of a Phobia in a Five-Year-Old Boy* (1909).

11. Imre Lakatos, "Falsification and the Methodology of Scientific Research Programmes," in Imre Lakatos and Alan Musgrave, eds., *Criticism and the Growth of Knowledge* (Cambridge, 1970), pp. 91–196.

## CHAPTER 4

1. For the original formulation of cognitive balance theory, see Fritz Heider, *The Psychology of Interpersonal Relations* (New York, 1958). For Abelson's development, see R. P. Abelson and M. J. Rosenberg, "Symbolic Psycho-Logic: A Model of Attitudinal Cognition," *Behavioral Science*, 3 (1958), 1–13. A clear discussion of this and related theories of attitude change is in Roger Brown, *Social Psychology* (New York, 1965), ch. xi.

2. R. P. Abelson, "Computer Simulation of 'Hot' Cognition," in S. S. Tomkins and Samuel Messick, eds., *Computer Simulation of Personality: Frontier of Psychological Research* (New York, 1963), pp. 277–298; R. P. Abelson and J. D. Carroll,

"Computer Simulation of Individual Belief Systems," *Amer. Behav. Scientist,* 8 (1965), 24–30. These programs are clearly discussed in J. C. Loehlin, *Computer Models of Personality* (New York, 1968), ch. v.

3. Abelson, "Computer Simulation of 'Hot' Cognition," p. 288. The term "rationalization" is being used here in a wide sense, to include both defensive rationalization (which generates a reason that is not the "real" reason for the threatening belief) and nondefensive reasoning (which seeks to render a belief more acceptable by finding reasons for believing it).

4. Abelson, "Computer Model of 'Hot' Cognition," p. 291. Broadly, choice of rationalization or denial depends on the predicted consequences of the unbalanced belief: people tend to want to believe in good consequences, employing rationalization to enable them to do so; bad consequences are shunned, so beliefs with negatively valued consequences are denied rather than rationalized.

5. The IDEOLOGY MACHINE is briefly described in R. P. Abelson, "The Structure of Belief Systems," in *CMTL,* pp. 287–293, and in R. P. Abelson and C. M. Reich, "Implicational Molecules: A Method for Extracting Meaning From Input Sentences," *IJCAI-1,* 641–647. Abelson chose Goldwater as his model because Goldwater's political ideology as expressed in his speeches is relatively clear and self-consistent.

6. Abelson, "Structure of Belief Systems," p. 290.

7. See, e.g., E. E. Jones, D. E. Kanouse, H. H. Kelley, R. E. Nisbett, Stuart Valins, and Bernard Weiner, eds., *Attribution: Perceiving the Causes of Behavior* (Morristown, 1972). *Cf.* also C. F. Schmidt, "Understanding Human Action: Recognizing the Plans and Motives of Other Persons," in J. S. Carroll and J. W. Payne, eds., *Cognition and Social Behavior* (Hillsdale, N.J., in press).

8. Abelson, "Structure of Belief Systems," p. 291. A Marxist version having essentially the same structure might be: The capitalists want to dominate the world and are continually using their material and ideological hegemony to maintain their dominance; these oppressive mechanisms when successful stave off the collapse of the bourgeois social order; if on the other hand the working class really uses its power, then repression will surely fail, and their capitalist rule will be overthrown. However, the misguided and cowardly policies of opportunist TV bureaucrats and reformist Labor parties acting as transmission belts for bourgeois ideology result in inhibition of full use of workers' power. Therefore it is necessary to enlighten all good workers with the facts so that they may expose and overturn these scabs and renegades.

9. Abelson, "Structure of Belief Systems," pp. 292–293.

10. The main body of Abelson, "Structure of Belief Systems," describes this new theory; the Ideology Machine whose failings gave rise to it is discussed only in the opening half dozen pages. Some more recent papers by Abelson developing ideas put forward here, will be mentioned in Chapter 11.

11. Abelson, "Structure of Belief Systems," p. 298.

12. The distinction between causes and purposes is discussed in M. A. Boden, *Purposive Explanation in Psychology* (Cambridge, Mass., 1972), especially pp. 25–52, 114–137, 261–321.

13. Abelson, "Structure of Belief Systems," pp. 300–314. Further discussion of plans occurs in R. P. Abelson, "Concepts for Representing Mundane Reality in Plans," in *RU,* pp. 273–309.

14. G. A. Miller, Eugene Galanter, and K. H. Pribram, *Plans and the Structure of Behavior* (New York, 1960).

15. Abelson, "Structure of Belief Systems," p. 323.

16. For a critique of Freud's procedure in this case, and a comparison with

accounts based on conditioning theory, see Roger Brown, *Social Psychology* (New York, 1965), pp. 350–381.

17. This definition does not fit the Marxist concept of "revolution," according to which distinct social classes will have been abolished, and *no* one group will dominate another. In general, Abelson's theme names are not intended to tally precisely with ordinary uses of the equivalent everyday words. Rather, they are useful mnemonics applied by him post hoc to the different "theme cells," which were drawn up purely by combinatorial use of the three abstract dimensions of themes.

18. Habermas's epistemology will be cited in Chapter 13 (in the discussion of hermeneutic psychology) and in Chapter 15 (with respect to the *interested* nature of natural science). See Jurgen Habermas, *Knowledge and Human Interests* (London, 1972).

19. The new master script is represented both verbally and diagrammatically in Abelson, "Structure of Belief Systems," p. 331.

20. Abelson, "Structure of Belief Systems," p. 338.

21. Abelson, "Computer Simulation of 'Hot' Cognition," pp. 282–283.

## CHAPTER 5

1. A. C. Clarke, *2001: A Space Odyssey* (London, 1968).

2. The question-answering programs named "BASEBALL," "STUDENT," "SAD SAM," and "SIR" are described respectively in: B. F. Green, A. K. Wolf, Carole Chomsky and K. R. Laughery, "BASEBALL: an Automatic Question-Answerer," in *CT*, pp. 207–216: D. G. Bobrow, "Natural Language Input for a Computer Problem-Solving System," in *SIP*, pp. 146–226: R. K. Lindsay, "Inferential Memory as the Basis of Machines Which Understand Natural Language," in *CT*, pp. 217–233; Bertram Raphael, "SIR: A Computer Program for Semantic Information Retrieval," in *SIP*, pp. 33–145.

3. Steven Kaneff, ed., *Picture Language Machines* (New York, 1970), pp. 32–33. The term "vice president" here does not mean the deputy head of the whole company, but one of its salesmen, who each held this title.

4. The instructions left on the teletype machine told users to "End all messages with a period," and the vice president obediently did so until he got too angry to remember. Since the period signals "end of message" to ELIZA, she always waits patiently for it before attempting to generate a response.

5. The versions of PARRY discussed in this chapter are described in: K. M. Colby, Sylvia Weber, and F. D. Hilf, "Artificial Paranoia," *AI*, 2 (1971), 1–26; K. M. Colby, *Artificial Paranoia* (New York, 1975); and K. M. Colby and R. C. Parkison, "Pattern-Matching Rules for the Recognition of Natural Language Dialogue Expressions," *Am. J. Computational Linguistics*, 1 (1974), Microfiche 5. The most recent version, which is less "stimulus-response" in character than these, is in: W. S. Faught, K. M. Colby, and R. C. Parkison, "The Interaction of Inferences, Affects, and Intentions in a Model of Paranoia," to appear in *Cognitive Psychology* (1976); R. C. Parkison, K. M. Colby, and W. S. Faught, "Conversational Language Comprehension Using Integrated Pattern Matching and Parsing," to appear in *Cognitive Science* (1976).

6. S. S. Tomkins, *Affect, Imagery, Consciousness, Vol. II: The Negative Affects* (New York, 1963).

7. See Colby, *Artificial Paranoia*, ch. vi, and K. M. Colby, F. D. Hilf, Sylvia

Weber, and H. C. Kraemer, "Turing-Like Indistinguishability Tests for the Validation of a Computer Simulation of Paranoid Processes," *AI*, 3 (1972), 199–222.

Strictly, this was not a Turing test in the accepted sense, since the psychiatrists doing the interviewing were not asked to judge which teletypes were attached to people and which to machines. They were not even informed that an "imitation game" was going on. Colby justifies this by pointing out that if the interviewer is informed at the start that a computer may be involved, she tends to change her normal interviewing strategy; that is, she asks questions designed to test which respondent is a program, rather than asking questions specifically relevant to the diagnostic dimension (paranoia) being studied. Colby points out also that, even in such a case, the program does not necessarily fail the Turing test: "We have found in informal experiments that if a human-respondent does not follow standards of the interviewer's expectations [does not answer honestly and candidly], jokes around, or plays other games, ordinary judges cannot distinguish him from a computer program with limited natural language understanding," Colby, *et al.* "Turing-Like Indistinguishability Tests," p. 203. For Turing's original description of the "imitation game," see A. M. Turing, "Computing Machinery and Intelligence," in *CT*, pp. 11–35.

8. Colby *et al.*, "Artificial Paranoia," pp. 16–18.

9. Colby, *Artificial Paranoia*, p. 78. We shall see in Chapter 15 that Laing is one of those psychiatrists who have questioned the "medical" overtones of diagnoses in terms of mental "illness"; he would be more likely to regard "COPS ARREST THE WRONG PEOPLE" as a truth implying the sickness of society and its authorities, rather than a fantasy based in a patient's diseased mind.

10. Green *et al.*, "BASEBALL."

11. *SIP*, p. 204.

12. H. L. Dreyfus, *What Computers Can't Do: A Critique of Artificial Reason* (New York, 1972), p. 12.

13. Horace Enea and K. M. Colby, "Idiolectic Language-Analysis for Understanding Doctor-Patient Dialogues," *IJCAI-3*, p. 279.

14. This comment is somewhat less applicable to the recent version of PARRY described in Faught *et al.*, "Interaction of Inferences, Affects, and Intentions," and Parkison *et al.*, "Conversational Language Comprehension."

15. The classic description of nondirective therapy is Carl Rogers, *Client Centered Therapy: Current Practice, Implications, and Theory* (Boston, 1951). For ELIZA, see Joseph Weizenbaum, "ELIZA—A Computer Program for the Study of Natural Language Communication Between Man and Machine," *CACM*, 9 (1966), 36–45, and Joseph Weizenbaum, "Contextual Understanding by Computers," *CACM*, 10 (1967), 474–480.

16. Weizenbaum, "ELIZA," pp. 36–37.

17. References to these programs are given in note 2.

18. Weizenbaum, "ELIZA," p. 42.

## CHAPTER 6

1. "They are eating apples" is an example often used by Noam Chomsky. According to his transformational grammar, this sentence, or "surface structure," is ambiguous because it can be generated from two different "deep structures." For the relevance of Chomsky's grammar to psychology (and philosophy), see Noam Chomsky, *Language and Mind* (New York, 1968).

2. Current speech-understanding programs are not able to exploit prosodic cues such as stress and intonation in natural dialogue. But it is recognized that procedures for doing so should ideally be developed: Allen Newell, Jeffrey Barnett, J. W. Forgie, C. C. Green, D. H. Klatt, J. C. R. Licklider, J. H. Munson, D. R. Reddy, and W. A. Woods, *Final Report of a Study Group on Speech Understanding Systems* (Amsterdam, 1973).

3. This example is from Terry Winograd, *Understanding Natural Language* (New York, 1972), p. 33.

4. The fullest account of SHRDLU is in Winograd, *Understanding Natural Language.* A brief description is: Terry Winograd, "A Procedural Model of Language Understanding," in *CMTL*, pp. 152–186.

5. Unlike most of the acronyms used to name AI programs, "SHRDLU" was picked by Winograd because it is meaning*less*. One row of the keyboard of a standard linotype typesetting machine consists of these letters, and typesetters often "correct" a mistake by inserting them in a faulty line so that the proofreaders will easily spot that a mistake has been made. Bad proofreading may result in this deliberate gibberish being printed in the final text—a fact made much of in *MAD* magazine. Being an ex-devotee of *MAD*, Winograd picked this nonsense word as the name for his program. (Winograd, personal communication.)

6. Winograd, *Understanding Natural Language*, pp. 8–15. Items 22MIT and 23MIT appear in the earlier account of the program; Terry Winograd, *Procedures as a Representation for Data in a Computer Program for Understanding Natural Language* (Cambridge, Mass., 1971).

7. Winograd, *Understanding Natural Language*, pp. 163–169.

8. By Jeff Hill, of the Computer Corporation of America. (Winograd, personal communication.)

9. Chapter 7 will show that PARRY and ELIZA are *in some ways* more flexible than SHRDLU.

10. For discussion of the advantages of hierarchical organization, see H. A. Simon, *The Sciences of the Artificial* (Cambridge, Mass., 1969). An early account of some disadvantages is: Allen Newell, "Some Problems of Basic Organization in Problem-Solving Programs," in M. C. Yovits, G. T. Jacobi, and G. D. Goldstein, eds., *Self-Organizing Systems 1962* (Washington, 1962), pp. 393–423.

11. *Cf.* P. H. Winston, "The MIT Robot," in *MI-7*, p. 443.

12. In the HEARSAY speech-understanding program, independent knowledge processes communicate with each other by writing (or modifying) hypotheses on a public "blackboard," there being no overall executive control. See D. R. Reddy and Allen Newell, "Knowledge and Its Representation in a Speech Understanding System," in L. W. Gregg, ed., *Knowledge and Cognition* (Baltimore, 1974), pp. 253–285; L. D. Erman and V. R. Lesser, "A Multi-Level Organization for Problem Solving Using Many, Diverse, Cooperating Sources of Knowledge," *IJCAI-4*, pp. 483–490. Neuropsychological evidence suggests that the brain may function in an analogous manner: M. A. Arbib, "Artificial Intelligence and Brain Theory; Unities and Diversities," *Annals Biomedical Engineering*, 3 (1975), 238–274.

13. M. A. K. Halliday, "Functional Diversity in Language as Seen From a Consideration of Modality and Mood in English," *Foundations of Language*, 6 (1970), 322–361.

14. Winograd, *Understanding Natural Language*, p. 5.

15. Winograd, *Understanding Natural Language*, pp. 5, 7, 29.

16. See C. C. Green, "Theorem-Proving by Resolution as a Basis for Question-Answering Systems," in *MI-4*, pp. 183–205. The Resolution Principle is explained at length in N. J. Nilsson, *Problem-Solving Methods in Artificial Intelligence* (New York, 1971), chs. vi and vii.

17. See, e.g., J. P. Thorne, Paul Bratley, and H. M. Dewar, "The Syntactic Analysis of English by Machine," in *MI-3*, pp. 281–309. This program could parse Lewis Carroll's "Jabberwocky" correctly, even though almost all the words in it are nonsensical. But this is not to say that the program (or Alice) could find "the" correct parsing: for this, the nonsense words would need to be understood—*cf.* Donald Michie, *On Machine Intelligence* (Edinburgh, 1974), pp. 122–123.

18. Winograd, *Understanding Natural Language*, p. 21.

19. See Chomsky, *Language and Mind*, and Noam Chomsky, *Aspects of the Theory of Syntax* (Cambridge, Mass., 1965).

20. Winograd, *Understanding Natural Language*, p. 56.

21. These correspond to different linguistic *cases*: C. J. Fillmore, "The Case for Case," in E. W. Bach and R. T. Harms, eds., *Universals in Linguistic Theory* (New York, 1968), pp. 1–88.

22. Winograd, *Understanding Natural Language*, pp. 158–161, 168–169.

23. "It is raining" is linguistically odd, for instance, in that it involves a "dummy" use of *it*: short of postulating a desexualized rain god, there is no referent for *it* in this sentence. SHRDLU could not even parse this sentence (still less assign a referent to *it*) because there is no previous noun or noun phrase with which to identify *it*. The reason why SHRDLU can interpret *it* to mean a particular action (as opposed to an object) is that actions are treated much like objects by the program: they are each assigned a specific number, depending on the time at which they occur.

For some examples of anomalous uses of *it*, see W. L. Chafe, *Meaning and the Structure of Language* (Chicago, 1970), p. 101.

24. D. J. M. Davies, "Representing Negation in a Planner System," *AISB*, 26–36.

25. Winograd, *Understanding Natural Language*, p. 157.

26. D. J. M. Davies and S. D. Isard, "Utterances as Programs," in *MI-7*, pp. 325–340. See esp. pp. 337–338.

27. Winograd, *Understanding Natural Language*, p. 136.

28. Winograd, *Understanding Natural Language*, p. 136.

29. Winograd, *Understanding Natural Language*, pp. 149, 157.

30. Gilbert Ryle, *The Concept of Mind* (London, 1949), ch. ii.

31. Eugene Charniak, *Toward a Model of Children's Story Comprehension* (Cambridge, Mass., 1972); Eugene Charniak, "Jack and Janet in Search of a Theory of Knowledge," *IJCAI-3*, 337–343. (Some more recent work of Charniak's will be mentioned in Chapters 10 and 11.)

I said Charniak has "outlined" a program because only an early and partial version of his theory was actually implemented; the reason for this is the high degree of complexity involved. Workers in artificial intelligence are sometimes less frank than Charniak about the precise extent to which the system they describe is constituted by a running program, as opposed to as-yet-unimplemented theoretical insights. Winograd's description of SHRDLU is an outstanding example of a report that tries to state unambiguously what a particular program can and cannot do, although even in this case the full details of the programming code are of course not included. From the psychologist's point of view, theoretical insights into thinking are welcome even if (like most psychological theories) they are not yet expressed in the form of a running program. But the full strength of the artificial intelligence approach depends upon there being a specific program which does certain things in certain ways—for only thus can one be *sure* that these things can indeed be achieved in these ways. The usefulness of programming methods in the formulation of psychological theories will be further discussed in Chapter 13.

32. Charniak, *Children's Story Comprehension*, pp. 72–73.

33. Charniak, *Children's Story Comprehension*, ch. vi and pp. 205, 182–184, 188–189, 196.

34. Charniak, *Children's Story Comprehension*, p. 228.

35. For discussions of pretending, promising, and making excuses, see chs. ix, x, and vi of J. L. Austin, *Philosophical Papers* (Oxford, 1961).

36. See G. R. Kiss, "Outlines of a Computer Model of Motivation," *IJCAI-3*, 446–449.

37. Charniak, *Children's Story Comprehension*, pp. 228–238.

38. In *Children's Story Comprehension* and "Jack and Janet," Charniak tries explicitly to express the knowledge of piggy banks implicitly used in understanding stories about children's (successful and unsuccessful) financial transactions. This knowledge includes such mundane facts as that if you shake a piggy bank and there is no noise, it probably has no money in it, and that if you go to find a piggy bank you probably want money to buy something. Charniak believes that the common-sense knowledge involved can be expressed in English within the compass of about a page. He points out that some of this knowledge—in a more knowledgeable system—might better be expressed more economically: for example, one might instead store the general fact that empty containers do not make a noise when shaken—which applies also to rice-jars, sweet-tins, and boxes of tintacks.

39. Winograd, *Understanding Natural Language*, p. 49.

## CHAPTER 7

1. For accounts of Conceptual Dependency Theory, see: R. C. Schank, "Finding the Conceptual Content and Intention in an Utterance in Natural Language Conversation," *IJCAI-2*, 444–454; R. C. Schank, "Conceptual Dependency: A Theory of Natural Language Understanding," *Cognitive Psychology*, 3 (1972), 552–631; R. C. Schank, "Identification of Conceptualizations Underlying Natural Language," in *CMTL*, pp. 187–248; R. C. Schank and C. J. Rieger, "Inference and the Computer Understanding of Natural Language," *AI*, 5 (1974), 373–412.

2. See W. A. Woods, "What's in a Link: Foundations for Semantic Networks," in *RU*, pp. 35–82.

3. Originally Schank postulated 14 primitive actions: R. C. Schank, *The Fourteen Primitive Actions and Their Inferences* (Stanford, 1973). Later he conflated SMELL, LOOK-AT, and LISTEN-TO into the more general ATTEND, so that only 12 primitives are described in Schank and Rieger, "Inference and the Computer Understanding of Natural Language." R. P. Abelson (personal communication) tells me that CONC has now been dropped, as it did not lead to any interesting inferences, so there are only 11 primitives at present.

4. R. C. Schank, N. M. Goldman, C. J. Rieger, and C. K. Riesbeck, *Primitive Concepts Underlying Verbs of Thought* (Stanford, 1972).

5. For a discussion of syntactic cases, see C. J. Fillmore, "The Case for Case," in E. W. Bach and R. T. Harms, eds., *Universals in Linguistic Theory* (New York, 1968), pp. 1–88.

6. The characteristic inferences associated with distinct primitives are described in Schank, *The Fourteen Primitive Actions and Their Inferences*.

7. In all these cases the assumption is that the subject of "push" is animate; so to say "the boulder pushed her against the wall" is to extend the use of the word. One of the strongest inferences associated with PROPEL is that someone MOVE-d a bodypart so as to touch the thing PROPEL-ed, or some other thing in physical contact with it; but as guidance by remote control shows, the MOVE-ing may some-

times be applied to something physically separate from the thing PROPEL-ed, such as a radio button. When no MOVE-ing whatever occurs, the prime inference associated with PROPEL is unsatisfied, and one may even find it difficult to understand the use of the relevant vocabulary. This fact underlies the approach of those philosophers who claim that one *cannot* "will a table to move," since there is no way of making sense of this if no physical action is envisaged (they do not deny, of course, that one can stare glassily at a table and pronounce the words "Move, won't you!"): D. F. Pears, J. F. Thomson, and Mary Warnock, "What Is the Will?" in D. F. Pears, ed., *Freedom and the Will* (London, 1963), pp. 14–37.

8. Schank, "Conceptualizations Underlying Natural Language," p. 201. For a discussion of various accounts of what have been termed "basic acts," see M. A. Boden, "The Structure of Intentions," *J. Theory Social Behaviour*, 3 (1973), esp. pp. 36–44.

9. M. R. Quillian, "Semantic Memory," in *SIP*, pp. 227–270.

10. Ludwig Wittgenstein, *Philosophical Investigations* (Oxford, 1953), para. 485.

11. John Seely-Brown and R. R. Burton, "Multiple Representations of Knowledge for Tutorial Reasoning," in *RU*, pp. 311–350.

12. Schank, "Finding the Conceptual Content and Intention in an Utterance," p. 452.

13. For a discussion of conversational structure and convention relevant both to the eventual improvement of MARGIE and to the debate about the relation between logic and language, see: H. P. Grice, "Logic and Conversation," in Peter Cole and J. L. Morgan, eds., *Syntax and Semantics, Vol. III: Speech Acts* (New York, 1975), pp. 41–58.

14. For a critique of some of MARGIE's shortcomings, see: Nick Cercone and Len Schubert, "Toward a State Based Conceptual Representation," in *IJCAI-4*, pp. 83–90.

15. For a brief description of MARGIE, see: R. C. Schank, N. M. Goldman, C. J. Rieger, and C. K. Riesbeck, "MARGIE: Memory, Analysis, Response Generation, and Inference on English," in *IJCAI-3*, pp. 255–261. For a fuller account, see the chapters by these four authors in R. C. Schank, ed., *Conceptual Information Processing* (New York, 1975).

16. Schank *et al.*, "MARGIE," p. 258. (If you check this example carefully you will find that the names in the published paper are transposed: in my text I have corrected them.)

17. Brainstorming is described in Edward de Bono, *Lateral Thinking: A Textbook of Creativity* (London, 1970), pp. 149–165.

18. Richard Power, *A Model of Conversation* (Sussex, 1976); J. A. Levin and J. A. Moore, "Dialogue-Games: A Process Model of Natural Language Interaction," in *AISB-2*, pp. 184–194. (See also the Grice reference in note 13, above.)

For a model of some of the hesitations and self-corrections that can occur during speaking, see J. H. Clippinger, *Meaning and Discourse: A Computer Model of Psychoanalytic Discourse and Cognition* (Baltimore, 1977).

19. Y. A. Wilks, *Grammar, Meaning, and the Machine Analysis of Language* (London, 1972), p. 10.

20. Gilbert Ryle, "Categories," in A. G. N. Flew, ed., *Logic and Language: Second Series* (Oxford, 1955), pp. 65–81.

21. Y. A. Wilks, "An Artificial Intelligence Approach to Machine Translation," in *CMTL*, p. 115.

22. For the original discussion of "Computable Semantic Derivations," see Wilks, *Grammar, Meaning, and the Machine Analysis of Language*, ch. iii.

23. Sixty ELEMENTS are mentioned in Wilks, "An Artificial Intelligence Approach to Machine Translation," p. 123. More recently, Wilks has postulated 80

such semantic primitives: Y. A. Wilks, *Natural Language Understanding Systems Within the AI Paradigm* (Edinburgh, 1976), p. 33.

24. Wilks, "An Artificial Intelligence Approach to Machine Translation," p. 124.

25. Wilks, *Grammar, Meaning, and the Machine Analysis of Language*, p. 12.

26. Wilks, "An Artificial Intelligence Approach to Machine Translation," pp. 131–132.

27. Y. A. Wilks, "Understanding Without Proofs," *IJCAI-3*, 270–277. The distinction between Wilks's and Winograd's systems is maintained in Wilks, "An Artificial Intelligence Approach to Machine Translation." But it is drawn much less rigidly in Wilks, *Natural Language Understanding Systems Within the AI Paradigm*, pp. 17–18.

28. Wilks, "Understanding Without Proofs," p. 270.

## CHAPTER 8

1. For examples of visual illusions, see R. L. Gregory, *Eye and Brain* (London, 1966) and Gaetano Kanizsa, "Contours Without Gradients or Cognitive Contours?" *Italian J. Psychology*, 1 (1974), 93–112. Both these authors discuss some of the views of "cognitive" psychologists concerning the contribution of knowledge to perception. For the historical background of such views, see D. W. Hamlyn, *Sensation and Perception: A History of the Philosophy of Perception* (London, 1961).

2. J. J. Gibson, *The Senses Considered as Perceptual Systems* (Boston, 1966).

3. As we shall see in Chapter 13, stereopsis is sometimes useful in depth perception. But stereopsis alone cannot account for the perceptual achievements discussed in Part IV: we can usually see depth relations just as well with one eye as with two, and using both eyes will not help one to read a sloppily handwritten text.

4. An example of a perambulating robot (which can move from room to room and push boxes from one to another) is "SHAKEY," discussed in Chapters 10 and 12. A stationary hand-eye system (which can move and stack blocks on a table) is "COPY-DEMO," described by P. H. Winston, "The MIT Robot," in *MI-7*, pp. 465–480. By contrast, the programs discussed in Part IV are "disembodied" visual systems with no motor abilities (though they may be, and some are, used by other programs that control the actions of a physical robot).

5. L. G. Roberts, "Machine Perception of Three-Dimensional Solids," in J. T. Tippett, D. A. Berkowitz, L. C. Clapp, C. J. Koester, and Alexander Vanderburgh, eds., *Optical and Electro-Optical Information Processing* (Cambridge, Mass., 1965), pp. 159–198.

6. This will be discussed further in Chapter 13.

7. For instance, the HEARSAY speech-understanding system plays "voice chess," and uses its knowledge of the current state of the board plus the rules of chess to predict plausible hypotheses concerning what the opponent may say next, and interprets the auditory input with the help of such knowledge: D. R. Reddy and Allen Newell, "Knowledge and Its Representation in a Speech Understanding System," in L. W. Gregg, ed., *Knowledge and Cognition* (Baltimore, 1974), pp. 253–285.

8. For a brief description of SEE, consult Adolfo Guzman, "Decomposition of a Visual Field Into Three-Dimensional Bodies," in Antonio Grasselli, ed., *Automatic Interpretation and Classification of Images* (New York, 1969), pp. 243–276. For more detail, see Adolfo Guzman, *Computer Recognition of Three-Dimensional Objects in a Visual Scene* (Cambridge, Mass., 1968).

9. Adolfo Guzman, *Some Aspects of Pattern Recognition by Computer* (Cambridge, Mass., 1967).

10. Gilbert Ryle, *The Concept of Mind* (London, 1949), ch. ii.

11. Winston, "The MIT Robot."

12. Guzman, *Computer Recognition of Three-Dimensional Objects*, pp. 222–232.

13. Some unsuccessful efforts to make SEE see holes are described in P. H. Winston, *Holes* (Cambridge, Mass., 1970).

14. D. A. Huffman, "Impossible Objects as Nonsense Sentences," in *MI-6*, pp. 295–325.

15. M. B. Clowes, "On Seeing Things," *AI*, 2 (1971), 79–116.

16. A. K. Mackworth, "Interpreting Pictures of Polyhedral Scenes," *AI*, 4 (1973), 121–138: A. K. Mackworth, "Using Models to See," in *AISB-1*, pp. 127–137.

## CHAPTER 9

1. I might almost have said "by the mind's eyes," since the eyes themselves discern rather few visual cues: in effect, they are virtually blind. Some psychological and physiological data concerning the relation between eye and brain will be discussed in the final section of Chapter 13.

2. C. W. Perky, "An Experimental Study of Imagination," *Amer. J. Psychol.*, 21 (1910), 422–452.

3. For the essential contribution of hypotheses to perception, see R. L. Gregory, "Will Seeing Machines Have Illusions?" in *MI-1*, pp. 169–180; R. L. Gregory, "On How so Little Information Controls so Much Behaviour," in C. H. Waddington, ed., *Towards a Theoretical Biology: 2. Sketches* (Edinburgh, 1969), pp. 236–246; M. B. Clowes, "Man the Creative Machine: A Perspective from Artificial Intelligence Research," in Jonathan Benthall, ed., *The Limits of Human Nature* (London, 1973), pp. 192–207.

4. Gilbert Falk, "Interpretation of Imperfect Line Data as a Three-Dimensional Scene," *AI*, 3 (1972), 101–144.

5. Yoshiaki Shirai, "A Context Sensitive Line Finder for Recognition of Polyhedra," *AI*, 4 (1973), 95–120. (Also available as "Analyzing Intensity Arrays Using Knowledge About Scenes," in *PCV*, pp. 93–114).

6. G. R. Grape, *Computer Vision Through Sequential Abstractions* (Stanford, 1969).

7. This criticism is stressed, for example, by Sir James Lighthill and H. L. Dreyfus, neither of whom appreciates the extent to which theoretical work in artificial intelligence is specifically motivated by a recognition of the problem. See James Lighthill, N. S. Sutherland, R. M. Needham, H. C. Longuet-Higgins, and Donald Michie, *Artificial Intelligence: A Paper Symposium* (London, 1973), pp. 1–21. For a rebuttal, see P. J. Hayes, "Some Comments on Sir James Lighthill's Report on Artificial Intelligence," *AISB Study Group European Newsletter*, Issue 14 (July, 1973), 36–53. Dreyfus's attack on artificial intelligence will be discussed in Chapter 14; see H. L. Dreyfus, *What Computers Can't Do; A Critique of Artificial Reason* (New York, 1972).

8. Richard Orban, *Removing Shadows in a Scene* (Cambridge, Mass., 1970).

9. D. L. Waltz, "Understanding Line Drawings of Scenes with Shadows," in *PCV*, pp. 19–92.

10. M. H. Rattner, *Extending Guzman's SEE Program* (Cambridge, Mass., 1970).

11. K. J. Turner, "Computer Perception of Curved Objects," in *AISB-1*, pp. 238–246.

12. M. D. Kelly, "Edge Detection in Pictures by Computer Using Planning," in *MI-6*, pp. 397–410.

13. M. L. Minsky, "Steps Toward Artificial Intelligence," in *CT*, esp. pp. 422–425. An early attempt to apply Minsky's approach was T. G. Evans's analogy-recognizing program in *SIP*, pp. 271–353 (Evans's program will be discussed in Chapter 11). See also Steven Kaneff, ed., *Picture Language Machines* (New York, 1970).

14. Adolfo Guzman, "Analysis of Curved Line Drawings Using Context and Global Information," in *MI-6*, pp. 325–376.

15. M. R. Adler, "Recognition of Peanuts Cartoon," in *AISB-2*, pp. 1–13.

16. A. E. Michotte, *The Perception of Causality* (London, 1963).

17. Fritz Heider and M. L. Simmel, "An Experimental Study of Apparent Behavior," *Am. J. Psychology*, 57 (1944), 243–259.

18. Sylvia Weir, "Action Perception," in *AISB-1*, pp. 247–256. For the application of this approach to animate, rather than inanimate, causation, see Sylvia Weir, "Actions, Motives, and Feelings," in Sylvia Weir, M. R. Adler, and Marilyn McLennan, *Final Report on Action Perception Project* (Edinburgh, Nov. 1975), pp. 1–38.

19. Eugene Charniak, *Toward a Model of Children's Story Comprehension* (Cambridge, Mass., 1972), pp. 116–128.

20. Also, some observers perceived the actors as fish or birds that (viewed sideways-on) can move either up or down; it is not clear which comes first: the fixing of viewpoint as aerial or sideways-on, for instance, or the interpretation of objects as people or fish. (Sylvia Weir, personal communication.)

21. Shadowgraphing is described by Donald Michie, *On Machine Intelligence* (Edinburgh, 1974), pp. 125–128. This technique enables a program to reconstruct the entire 3D contour of the surface of an irregularly-shaped object, by moving a shadow-casting horizontal edge to successive positions and examining the path of the shadow across the surface.

## CHAPTER 10

1. John McCarthy, "Programs with Common Sense," in *SIP*, pp. 403–418.

2. Bertrand Russell, *The Problems of Philosophy* (London, 1912), ch. v. In fairness to Russell, one should point out that his conception of "knowledge by acquaintance" was narrower than the everyday usage of "learning by example." While he would have been willing to accept the hearing of outgribing as a case of learning by example involving direct knowledge, he would not have said that first-hand knowledge of Communist duplicity, for instance, constitutes knowledge by acquaintance. The reason is that he would have agreed with R. P. Abelson that to label something as (and to experience it as) "Communist duplicity" is to draw on a wide background of descriptive concepts, by way of which the experience is interpreted.

3. J. M. Tenenbaum, T. D. Garvey, Stephen Weyl, and H. C. Wolf, *An Interactive Facility for Scene Analysis Research* (Stanford, 1974); J. M. Tenenbaum and Stephen Weyl, "A Region-Analysis Subsystem for Interactive Scene Analysis," *IJCAI-4*, 682–687; J. M. Tenenbaum and H. G. Barrow, *Experiments in Interpretation-Guided Segmentation* (Stanford, 1976).

For other programs that base scene analysis on regions rather than lines, see C. R. Brice and C. L. Fennema, "Scene Analysis Using Regions," *AI*, 1 (1970), 205–

226; Yoram Yakimovsky and J. A. Feldman, "A Semantics-Based Decision Theory Region Analyser," *IJCAI-3*, 580–588; H. G. Barrow and R. J. Popplestone, "Relational Descriptions in Picture Processing," in *MI-6*, pp. 377–396. All these programs concentrate on irregularly shaped everyday objects rather than simple polyhedra, and start from photographs rather than line drawings.

4. Ludwig Wittgenstein, *Philosophical Investigations* (Oxford, 1953), paras. 6–9, 26–43, 49.

5. Immanuel Kant, *Critique of Pure Reason* (1781). See esp. the sections "Transcendental Aesthetic," for discussions of the intuitions of space and time, and of mathematical concepts; and "Transcendental Logic" for a discussion of the concepts, or "categories," such as causality and identity that are presupposed by common-sense and scientific thinking (both perceptual and consciously reasoned thinking).

6. P. H. Winston, "Learning Structural Descriptions from Examples," in *PCV*, pp. 157–210.

7. Winston, "Learning Structural Descriptions," pp. 190–191.

8. For an early study of concept formation, see J. S. Bruner, Jacqueline Goodnow, and George Austin, *A Study of Thinking* (New York, 1956). See also P. C. Wason and P. N. Johnson-Laird, *Psychology of Reasoning: Structure and Content* (London, 1972).

9. M. J. Freiling, *Functions and Frames in the Learning of Structures* (Cambridge, Mass., 1973). For further discussion of the difficulty of spatial description see citations in note 28 to chap. 12 below.

10. A convenient source of references and outline definitions of the nominalism *versus* essentialism (realism) debate is: Paul Edwards, ed., *The Encyclopedia of Philosophy* (New York, 1960s). For a more extended historical account, see R. I. Aaron, *The Theory of Universals* (Oxford, 1952).

11. Nelson Goodman, *Fact, Fiction, and Forecast* (Indianapolis, 1965).

12. See Bruner *et al.*, *A Study of Thinking*, ch. vi. Also Wason and Johnson-Laird, *Psychology of Reasoning*, pp. 51 and 71.

13. Wittgenstein, *Philosophical Investigations*. For a recent empirical study of this issue, see Eleanor Rosch and C. B. Mervis, "Family Resemblances: Studies in the Internal Structure of Categories," *Cognitive Psychology*, 7 (1975), 573–605.

14. Wittgenstein, *Philosophical Investigations*, para. 79. Wittgenstein adds: "—so long as it does not prevent you from seeing the facts. (And when you see them there is a good deal you will not say.)" Compare John Wisdom's philosophical precept and practice, "Say what you choose—but be careful!": A. J. T. D. Wisdom, *Philosophy and Psychoanalysis* (Oxford, 1953), p. 39.

15. For an attempt to distinguish between implications that are essentially contained within an expression's meaning, and inferences that (though they will naturally be drawn) are not so contained, see Roy Edgley, *Reason in Theory and Practice* (London, 1969), pp. 73–82.

16. D. V. McDermott, *Assimilation of New Information by a Natural Language-Understanding System* (Cambridge, Mass., 1974).

17. C. J. Rieger, "Conceptual Memory and Inference," in R. C. Schank, ed., *Conceptual Information Processing* (New York, 1975), pp. 157–288. Some examples of psychological studies relevant to Rieger's claims are listed in Note 19 of Chapter 13.

18. These 16 inference types are listed (with examples) in Rieger, "Conceptual Memory and Inference," pp. 193–195. The verbal definitions are taken from pp. 155–156 of the earlier version in C. J. Rieger, *Conceptual Memory: A Theory and Computer Program for Processing the Meaning Content of Natural Language Utterances* (Stanford, 1974).

19. An alternative system with an "open" data base is described in D. A. Norman and D. E. Rumelhart, *Explorations in Cognition* (San Francisco, 1975), Parts I, II, and IV. A brief account is: D. E. Rumelhart and D. A. Norman, "Active Semantic Networks as a Model of Human Memory," *IJCAI-3*, 450–457.

20. Rieger, "Conceptual Memory and Inference," p. 186.

21. Rieger, *Conceptual Memory*, p. 389.

22. R. C. Schank, "The Structure of Episodes in Memory," in *RU*, pp. 237–272.

23. Eugene Charniak, *"He Will Make You Take It Back": A Study in the Pragmatics of Language* (Castagnola, 1974).

24. A. L. Samuel, "Some Studies in Machine Learning Using the Game of Checkers," in *CT*, pp. 71–105: A. L. Samuel, "Some Studies in Machine Learning Using the Game of Checkers. II—Recent Progress," in F. J. Crosson, ed., *Human and Artificial Intelligence* (New York, 1970), pp. 81–116.

25. Donald Michie, *On Machine Intelligence* (Edinburgh, 1974), p. 39.

26. R. E. Fikes and N. J. Nilsson, "STRIPS: A New Approach to the Application of Theorem Proving to Problem Solving," *AI*, 2 (1971), 189–208; R. E. Fikes, P. E. Hart, and N. J. Nilsson, "Learning and Executing Generalized Robot Plans," *AI*, 3 (1972), 251–288; R. E. Fikes, P. E. Hart, and N. J. Nilsson, "Some New Directions in Robot Problem Solving," in *MI-7*, pp. 405–430. An introductory description of SHAKEY is given in Bertram Raphael, *The Thinking Computer: Mind Inside Matter* (San Francisco, 1976), chap. viii.

27. Fikes *et al.*, "Learning and Executing Generalized Robot Plans," p. 258.

28. G. J. Sussman, *A Computer Model of Skill Acquisition* (New York, 1975); G. J. Sussman, "The Virtuous Nature of Bugs," in *AISB-1*, pp. 224–237. An amusing picture of the compulsive human hacker is given by Joseph Weizenbaum, *Computer Power and Human Reason: From Judgment to Calculation* (San Francisco, 1976).

29. Like Sussman himself, I use "he" rather than "it" in referring to HACKER. While reading accounts of HACKER, ask yourself from time to time whether this anthropomorphic language is misleading in any important sense. This point will be discussed in the first section of Chapter 13.

30. See J. L. Austin, "A Plea for Excuses," in his *Philosophical Papers* (Oxford, 1961), ch. vi.

31. Sussman, "The Virtuous Nature of Bugs," p. 235.

32. I. P. Goldstein, "Summary of MYCROFT: A System for Understanding Simple Picture Programs," *AI*, 6 (1975), 249–288.

33. A brief description of LOGO is given by Seymour Papert, "Teaching Children to Be Mathematicians Versus Teaching About Mathematics," *Int. J. Math. Educ. Sci. Technol.*, 3 (1972), 249–262.

34. Michael Polanyi, *Personal Knowledge: Towards a Post-Critical Philosophy* (New York, 1964).

35. Sussman, *A Computer Model of Skill Acquisition*, p. 1.

## CHAPTER 11

1. Sheldon Klein, J. F. Aeschlimann, D. F. Balsiger, S. L. Converse, Claudine Court, Mark Foster, Robin Lao, J. D. Oakley, and Joel Smith. *Automatic Novel Writing: A Status Report* (Madison, 1973).

2. Sheldon Klein, "Meta-Compiling Text Grammars as a Model for Human Behavior," in *TINLP*, pp. 94–98.

3. D. E. Rumelhart, "Notes on a Schema for Stories," in *RU*, pp. 211–236.

4. R. P. Abelson, "Concepts for Representing Mundane Reality in Plans," in *RU*, pp. 273–309.

5. R. C. Schank and R. P. Abelson, "Scripts, Plans, and Knowledge," *IJCAI-4*, 151–157.

6. M. L. Minsky, "A Framework for Representing Knowledge," in *PCV*, pp. 211–277.

7. A. D. Rubin, "The Role of Hypotheses in Medical Diagnosis," *IJCAI-4*, 856–862; A. D. Rubin, *Hypothesis Formation and Evaluation in Medical Diagnosis* (Cambridge, Mass., 1975).

8. F. C. Bartlett, *Remembering: A Study in Experimental and Social Psychology* (Cambridge, 1932).

9. Jerome Feldman, "Bad-Mouthing Frames," in *TINLP*, pp. 102–103.

10. Eugene Charniak, "Organization and Inference in a Frame-Like System of Common Knowledge," in *TINLP*, pp. 46–55.

11. Restaurant scripts are discussed in Schank and Abelson, "Scripts, Plans, and Knowledge." Their implementation in story-*understanding* programs is described in Wendy Lehnert, "What Makes Sam Run? Script Based Techniques for Question Answering," in *TINLP*, pp. 16–21, and in Wendy Lehnert, *Question Answering in a Story Understanding System* (New Haven, 1975). The use of planning structures in *generating* stories is discussed in Jim Meehan, "Using Planning Structures to Generate Stories," *Am. J. Computational Linguistics*, Microfiche 33 (1975), pp. 77–93. See also R. C. Schank and the Yale A.I. Project, *SAM—A Story Understander* (New Haven, 1975).

12. C. J. Rieger, "Conceptual Overlays: A Mechanism for the Interpretation of Sentence Meaning in Context," *IJCAI-4*, 143–150; C. J. Rieger, "The Commonsense Algorithm as a Basis for Computer Models of Human Memory, Inference, Belief, and Contextual Language Comprehension," in *TINLP*, pp. 199–214. Rieger concentrates on two stylistic devices used in story-telling, the "time-edit" and the "action-zoom"; these two and nine further common compositional styles are distinguished by Marc Eisenstadt, "Processing Newspaper Stories: Some Thoughts on Fighting and Stylistics," *AISB-2*, 104–117.

13. Y. A. Wilks, *Grammar, Meaning, and the Machine Analysis of Language* (London, 1972), pp. 166–172.

14. Y. A. Wilks, "A Computer System for Making Inferences About Natural Language," in *AISB-1*, pp. 268–283: Y. A. Wilks, "A Preferential, Pattern-Seeking, Semantics for Natural Language Inference," *AI*, 6 (1975), 53–74.

15. Y. A. Wilks, personal communication.

16. T. G. Evans, "A Program for the Solution of Geometric-Analogy Intelligence Test Questions," in *SIP*, pp. 271–353.

17. J. S. Bruner, Jacqueline Goodnow, and George Austin, *A Study of Thinking* (New York, 1956).

18. B. G. Buchanan, G. L. Sutherland, and E. A. Feigenbaum, "Heuristic DENDRAL: A Program for Generating Explanatory Hypotheses in Organic Chemistry," in *MI-4*, pp. 209–254; B. G. Buchanan, G. L. Sutherland, and E. A. Feigenbaum, "Rediscovering Some Problems of Artificial Intelligence in the Context of Organic Chemistry," in *MI-5*, pp. 253–280; E. A. Feigenbaum, B. G. Buchanan, and Joshua Lederberg, "On Generality and Problem Solving: A Case Study Using the DENDRAL Program," in *MI-6*, pp. 165–190; B. G. Buchanan, E. A. Feigenbaum, and N. S. Sridharan, "Heuristic Theory Formation: Data Interpretation and Rule Formation," in *MI-7*, pp. 267–290; B. G. Buchanan and N. S. Sridharan, "Analysis of Behavior of Chemical Molecules; Rule Formation on Non-Homogeneous Classes of Objects," in *IJCAI-3*, pp. 67–76.

19. M. A. Boden, "The Paradox of Explanation," *Proc. Aristotelian Soc.*, N.S., 62 (1962), 159–178.

20. N. F. Maier, "Reasoning in Humans, II. The Solution of a Problem and Its Appearance in Consciousness," *J. Comparative Psychology*, 12 (1931), 181–194.

21. Allan Collins, E. H. Warnock, Nelleke Aiello, and M. L. Miller, "Reasoning from Incomplete Knowledge," in *RU*, pp. 383–415; Allan Collins, "Process in Acquiring Knowledge," in R. C. Anderson, R. J. Spiro, and W. E. Montague, eds., *Schooling and the Acquisition of Knowledge* (Hillsdale, N.J., 1977).

22. James Moore and Allen Newell, "How Can MERLIN Understand?" in L. W. Gregg, ed., *Knowledge and Cognition* (Baltimore, 1973), pp. 201–252.

23. Donald Michie, *On Machine Intelligence* (Edinburgh, 1974), p. 5. Compare H. L. Gelernter, "Realization of a Geometry-Theorem Proving Machine," in *CT*, pp. 134–152.

24. T. S. Kuhn, *The Structure of Scientific Revolutions* (Chicago, 1962).

25. H. A. Wolfson, *The Philosophy of Spinoza* (Cambridge, Mass., 1934).

26. Saul Amarel, "On Representations of Problems of Reasoning About Actions," in *MI-3*, pp. 131–172.

27. A hypothetical account of the evolution of vision is given by R. L. Gregory, "The Evolution of Eyes and Brains: A Hen-and-Egg Problem," in S. J. Freedman, ed., *The Neuropsychology of Spatially Oriented Behaviour* (Illinois, 1968), pp. 7–18.

28. Saul Amarel, "Representations and Modeling in Problems of Program Formation," *MI-6*, pp. 411–466.

29. Though there *may* be three distinct languages, the planning being done at the intermediate stage, there *need* not be three such languages. For an automatic programming system that generates simple LISP functions, given only the input-output pairs of expressions on which the functions will be required to operate, see Steven Hardy, "Synthesis of LISP Functions from Examples," *IJCAI-4*, 240–245. The large-scale features taken note of by this program in its planning stages include such things as the relative length of input and output expressions, and the (identical or reversed) ordering of their constituent elements.

30. Aaron Sloman, "Interactions Between Philosophy and Artificial Intelligence: The Role of Intuition and Non-Logical Reasoning in Intelligence," *AI*, 2 (1971), 209–225: Aaron Sloman, "Afterthoughts on Analogical Representation," in *TINLP*, pp. 178–182.

31. For a critique of this view, see M. L. Minsky, "Form and Content in Computer Science," *JACM*, 17 (1972), 197–215. For a relevant discussion of "intuitionism" versus "formalism" in mathematics, see Imre Lakatos, "Proofs and Refutations," *Brit. J. Philos. Science*, 14 (1963), 1–25, 120–139, 221–243, and 296–342.

32. L. N. Landa, *Algorithmization in Learning and Instruction* (Englewood Cliffs, N.J., 1974); L. N. Landa, *Cybernetics, Algorithmization and Heuristics in Education* (Englewood Cliffs, N.J., 1976).

33. D. G. Bobrow, "Dimensions of Representation," in *RU*, pp. 1–34: P. J. Hayes, "Some Problems and Non-Problems in Representation Theory," in *AISB-1*, pp. 63–79. For an early discussion, see G. W. Ernst and Allen Newell, "Some Issues of Representation in a General Problem Solver," *Spring Joint Comp. Conf.*, 30 (1967), 583–600.

34. George Polya, *How to Solve It: A New Aspect of Mathematical Method* (New York, 1945): Max Wertheimer, *Productive Thinking* (New York, 1945); Karl Duncker, "On Problem Solving," *Psychological Monographs*, 58 (1945).

35. For stimulating discussions of the role of representational conventions in art, see Nelson Goodman, *Languages of Art: An Approach to a Theory of Symbols* (London, 1969), and E. H. Gombrich, *Art and Illusion* (New York, 1960).

## CHAPTER 12

1. A chess player would need to consider $10^{120}$ different board positions in tracking down every possible line of play; a checkers (draughts) player would need to search through $10^{40}$ positions. (The estimated age of our planetary system is only $10^{18}$ seconds.) An early description of the British Museum Algorithm—and of the distinction between algorithms and heuristics—is in Allen Newell, J. C. Shaw, and H. A. Simon, "Empirical Explorations with the Logic Theory Machine: A Case Study in Heuristics," in *CT*, pp. 113–117. The 21-step solution of the "keys and boxes" puzzle is described by E. D. Sacerdoti, *A Structure for Plans and Behavior* (Stanford, 1975), pp. 70–72 and 144–148.

2. See P. C. Jackson, *Introduction to Artificial Intelligence* (New York, 1974), p. 95.

3. This way of formulating the algorithm-heuristic distinction was suggested to me by Aaron Sloman.

4. For an overview of heuristic search theory, see N. J. Nilsson, *Problem-Solving Methods in Artificial Intelligence* (New York, 1971). A more introductory account is given in Bertram Raphael, *The Thinking Computer: Mind Inside Matter* (San Francisco, 1976), chap. iii.

5. J. E. Doran, "New Developments of the Graph Traverser," in *MI-2*, pp. 119–135; Donald Michie and Robert Ross, "Experiments with the Adaptive Graph Traverser," in *MI-5*, pp. 301–318.

6. P. E. Hart, N. J. Nilsson, and Bertram Raphael, "A Formal Basis for the Heuristic Determination of Minimum Cost Paths," *IEEE Trans. Sys. Sci. and Cybernetics*, SSC-4 (1968), 100–107.

7. Ira Pohl, "Bi-Directional Search," in *MI-6*, pp. 127–140.

8. Also called "MACHACK," the Greenblatt Chess Program is described in R. D. Greenblatt, D. E. Eastlake, and S. D. Crocker, "The Greenblatt Chess Program," *Am. Fed. Inf. Proc. Soc. Conf. Proceedings*, 31 (1967), 801–810. A program capable of accepting expert advice is discussed by A. L. Zobrist and F. R. Carlson, "An Advice-Taking Chess Computer," *Scientific American*, 228 (June, 1973), 92–105. Some general issues are discussed by I. J. Good, "A Five Year Plan for Automatic Chess," in *MI-2*, pp. 89–118. A "hardware" chess machine is briefly described in note 33, Chapter 15.

9. H. J. Berliner, "Some Necessary Conditions for a Master Chess Program," in *IJCAI-3*, pp. 77–85. See also W. G. Chase and H. A. Simon, "Perception in Chess," *Cognitive Psychology*, 4 (1973), 55–81. And *n.b.* the final sentences of note 33, Chapter 15.

10. Planning was mentioned in Chapter 9 with respect to the programs of Yoshiaki Shirai and M. D. Kelly. Saul Amarel's work, discussed in Chapter 11, is also directed to the issue of planning.

11. GPS was the successor of the Logic Theory Machine (see note 1 above), and is briefly described in Allen Newell and H. A. Simon, "GPS—A Program That Simulates Human Thought," in *CT*, pp. 279–296. A fuller description is: G. W. Ernst and Allen Newell, *GPS: A Case Study in Generality and Problem Solving* (New York, 1969).

12. Wolfgang Kohler's famous study of *The Mentality of Apes* (London, 1927) assumed that learning was irrelevant to the "insightful" solution of problems involving monkeys, boxes, and bananas—a favorite test case for artificial intelligence programs. But H. G. Birch later showed that previous experience of the relevant tools is necessary; perhaps the relevant "difference orderings" are learned during

such experience: H. G. Birch, "The Relation of Previous Experience to Insightful Problem Solving," *J. Comp. Psychol.*, 38 (1945), 367–383.

13. Many examples of Sorites are given in Lewis Carroll, *Symbolic Logic: Part I, Elementary* (London, 1896).

14. The distinction between epistemological and heuristic (and also metaphysical) adequacy is made by John McCarthy and P. J. Hayes, "Some Philosophical Problems from the Standpoint of Artificial Intelligence," in *MI-4*, pp. 463–502.

15. For discussions of when actions should be regarded as primitive, and when they should be analyzed into subunits, see R. P. Abelson, "Concepts for Representing Mundane Reality in Plans," in *RU*, pp. 273–309, and M. A. Boden, "The Structure of Intentions," *J. Theory of Social Behaviour*, 3 (1973), 23–46.

16. E. D. Sacerdoti, "Planning in a Hierarchy of Abstraction Spaces," *AI*, 5 (1974), 115–136.

17. G. J. Sussman, *A Computer Model of Skill Acquisition* (New York, 1975), pp. 105–109.

18. E. D. Sacerdoti, "The Non-Linear Nature of Plans," in *IJCAI-4*, pp. 206–214. For a fuller account, see Sacerdoti, *A Structure for Plans and Behavior.*

19. Sacerdoti, *A Structure for Plans and Behavior*, ch. iv.

20. Austin Tate, "Interacting Goals and Their Use," in *IJCAI-4*, pp. 215–218.

21. P. E. Hart, "Progress on a Computer Based Consultant," in *IJCAI-4*, pp. 831–841. See also Sacerdoti, *A Structure for Plans and Behavior*, pp. 90–109 and 117–120.

22. Sacerdoti, *A Structure for Plans and Behavior*, p. 70. *Cf.* Donald Michie, *On Machine Intelligence* (Edinburgh, 1974), pp. 149–152.

23. Sacerdoti, *A Structure for Plans and Behavior*, p. 42.

24. S. E. Fahlman, "A Planning System for Robot Construction Tasks," *AI*, 5 (1974), 1–50.

25. Descartes's change of view is discussed by Leon Roth, *Descartes' Discourse on Method* (Oxford, 1937).

26. Though he is thus quoted by Michie (*On Machine Intelligence*, p. 76), Gregory disclaims originating this remark. It occurs in E. A. Feigenbaum, "Artificial Intelligence: Themes in the Second Decade," in A. J. H. Morrell, ed., *Information Processing 68* (Amsterdam, 1969), p. 1011. The same general point was made much earlier by G. A. Miller, Eugene Galanter, and K. H. Pribram, *Plans and the Structure of Behavior* (New York, 1960), p. 78.

27. COPY-DEMO is described by P. H. Winston, "The MIT Robot," in *MI-7*, pp. 465–480.

28. G. J. Sussman, *A Computer Model of Skill Acquisition*, pp. 40–59. See also B. J. Kuipers, *Spatial Knowledge* (Cambridge, Mass., 1976) and M. E. Denofsky, *How Near is Near?* (Cambridge, Mass., 1976).

29. This does not mean that one has to be able to talk about blocks and stability in English: babies and dogs have no language, but some form of inner representative symbolism must mediate their sensorimotor achievements. For a discussion of what makes a symbolism a *language*, by way of the question "What more would honey-bees have to be able to do, in order for us to call them rational, language-using, creatures?" see Jonathan Bennett, *Rationality: An Essay Towards an Analysis* (London, 1964).

30. D. G. Bobrow and Bertram Raphael, "New Programming Languages for Artificial Intelligence Research," *ACM Computing Surveys*, 6 (1974), 155–174.

31. A full description of LISP is in John McCarthy, P. W. Abrahams, D. J. Edwards, T. P. Hart, and M. I. Levin. *Lisp 1.5 Programmer's Manual* (Cambridge, Mass., 1962): Briefer accounts are given in the two general references in note 32, below.

32. List processing is discussed in general terms by: J. M. Foster, *List Processing* (London, 1967). For its use in artificial intelligence, see: D. G. Bobrow and Bertram Raphael, "A Comparison of List Processing Languages," *CACM*, 7 (1964), 231–240; D. G. Bobrow, "Requirements for Advanced Programming Systems for List Processing," *CACM*, 15 (1972), 618–627.

The relative "non-intelligibility" of IPL-V is evident from the *Manual:* Allen Newell, F. M. Tonge, E. A. Feigenbaum, B. F. Green, and G. H. Mealy, *Information Processing Language—V Manual* (Englewood Cliffs, 1964).

33. The way in which the address of the successor is stored with the item itself is described by Foster, *List Processing,* chap. ii.

34. Similarly, people, especially children, find it rather more difficult to say the numbers backwards, even though there is a general rule (from *n* go to *n-1*) for generating the reversed series. There are probably *many* ways in which we can access the number series, and not all are learnt at once: Aaron Sloman, "On Learning About Numbers," in *AISB-1*, pp. 173–185.

35. Brief descriptions of PLANNER are: Carl Hewitt, "PLANNER: A Language for Proving Theorems in Robots," in *IJCAI-1*, pp. 295–301; Carl Hewitt, "Procedural Embedding of Knowledge in PLANNER," in *IJCAI-2*, pp. 167–184. A full account is given in Carl Hewitt, *Description and Theoretical Analysis (Using Schemata) of PLANNER: A Language for Proving Theorems and Manipulating Models in a Robot* (Cambridge, Mass., 1972).

Winograd's use of MICRO-PLANNER is described in Terry Winograd, *Understanding Natural Language* (New York, 1972), esp. pp. 108–116. More details are in G. J. Sussman and Terry Winograd, *MICRO-PLANNER Reference Manual* (Cambridge, Mass., 1970).

36. Winograd, *Understanding Natural Language*, p. 110. (The "#" marks used by Winograd in his text may be translated as "IS": they signal that the next expression is to be interpreted as a predicate, or property.)

37. Lewis Carroll, "What the Tortoise Said to Achilles," *Mind*, N.S., 4 (1895), 278–280. Hewitt reproduces this paper at the beginning of his *Description and Theoretical Analysis of PLANNER*.

38. Production rules are briefly described in James Moore and Allen Newell, "How Can MERLIN Understand?" in L. W. Gregg, ed., *Knowledge and Cognition* (Baltimore, 1973), pp. 201–252. A fuller account is in Allen Newell and H. A. Simon, *Human Problem Solving* (Englewood Cliffs, 1972).

39. Gilbert Ryle, *The Concept of Mind* (London, 1949), ch. ii.

40. This advantage of declarative over procedural representation is cited by Terry Winograd, "Frame Representations and the Declarative/Procedural Controversy," in *RU*, p. 188. Even though one may be able to add an instruction just as quickly as adding a statement, its implicit *effects* are more unclear.

41. Sacerdoti, *A Structure for Plans and Behavior*, pp. 15–20 and 141–148.

42. J. L. Stansfield, "Active Descriptions for Representing Knowledge," in *AISB-1*, pp. 214–223: J. L. Stansfield, *Programming a Dialogue Teaching Situation* (Edinburgh, 1974); Carl Hewitt, "A Universal Modular ACTOR Formalism for Artificial Intelligence," in *IJCAI-3*, pp. 235–245.

43. Winograd, "Frame Representations and the Declarative/Procedural Controversy": Terry Winograd, *Five Lectures on Artificial Intelligence* (Stanford, 1974).

44. Of course, control details are implicit in *all* programming languages, even in an apparently simple GOTO instruction; so this point concerns the relative rather than the absolute explicitness of control in different programming languages.

45. G. J. Sussman and D. V. McDermott, *The CONNIVER Reference Manual* (Cambridge, Mass., 1972).

46. Bertram Raphael, "The Frame Problem in Problem-Solving Systems," in

N. V. Findler and Bernard Meltzer, eds., *Artificial Intelligence and Heuristic Programming* (Edinburgh, 1971), pp. 159–169.

47. G. J. Sussman and D. V. McDermott, *Why Conniving Is Better Than Planning* (Cambridge, Mass., 1972).

48. See Bobrow and Raphael, "New Programming Languages for Artificial Intelligence Research."

49. D. J. M. Davies, *POPLER 1.5 Reference Manual* (Edinburgh, 1973).

50. D. G. Bobrow and Terry Winograd, "An Overview of KRL, a Knowledge Representation Language," *J. Cognitive Science,* 1 (1977), 3–46.

51. Oscar Firschein, M. A. Fischler, L. S. Coles, and J. M. Tenenbaum, "Forecasting and Assessing the Impact of Artificial Intelligence on Society," in *IJCAI-3*, p. 105. See also Rein Turn, *Computers in the 1980's* (New York, 1974).

52. Carl Hewitt, "How to Use What You Know," in *IJCAI-4*, pp. 189–198.

## CHAPTER 13

1. For an autobiographical account of the recent trend away from "mechanist" and toward "humanist" approaches to psychology, see Liam Hudson: *The Cult of the Fact: A Psychologist's Autobiographical Critique of His Discipline* (London, 1972).

For references to other work exemplifying this widespread trend, see notes 5, 6, 17, 18, and 22 below.

2. For his theoretical psychology, see B. F. Skinner, *Science and Human Behavior* (New York, 1953). For his writings on social implications, see B. F. Skinner, *Walden Two* (New York, 1948), and B. F. Skinner, *Beyond Freedom and Dignity* (New York, 1971).

Roy Edgley (personal communication) has pointed out that Skinner's "humane intent" may be apparent only. In other words, there is perhaps a radical conceptual incoherence in any attempt to express and follow *human* ends in the terms of a *natural scientific* psychology. On this view, Skinner's conception of what a psychological science should be shapes—or rather misshapes—his ends and policies, and obscures from him their inhumane nature. This is to say that *only* a psychology based on "humanist" concepts could be genuinely humane in intent: a conscious and explicit concern with the benevolent social application of psychological theory does not suffice to render that theory "humane."

3. D. S. Palermo, "Is a Scientific Revolution Taking Place in Psychology?" *Science Studies,* 1 (1971), 135–155. *Cf.* T. S. Kuhn, *The Structure of Scientific Revolutions* (Chicago, 1962).

4. Neil Warren, "Is a Scientific Revolution Taking Place in Psychology?—Doubts and Reservations," *Science Studies,* 1 (1971), 407–413.

5. Rom Harré and P. F. Secord, *The Explanation of Social Behavior* (Oxford, 1972); Jurgen Habermas, *Knowledge and Human Interests* (London, 1972), chs. x and xi.

In general, the dilemma often termed "the hermeneutic circle" marks the impossibility of ever arriving at any fail-safe interpretation of behavior. (This circularity was mentioned in Chapter 7, with reference to the linguistic work of Y. A. Wilks: one does not have to go to psychoanalysis or ethogenic psychology to discover the possibility of alternative interpretations being given within an interpretative science.)

6. Paul Ricoeur, *Freud and Philosophy: An Essay on Interpretation* (New Haven, 1970); Amedeo Giorgi, *Psychology as a Human Science: A Phenomeno-*

*logically Based Approach* (New York, 1970). For a discussion of hermeneutic approaches to Freudian theory, see Barnaby Barratt, "Freud's Psychology as Interpretation," in Theodore Shapiro, ed., *Psychoanalysis and Contemporary Science, Vol. 5* (New York, 1976).

7. For a discussion of "models" in artificial intelligence with reference to the mind-body problem, see M. L. Minsky, "Matter, Mind, and Models," in *SIP*, pp. 425–432. In an overview of 20 years of relevant research, Feigenbaum identified representation as one of the prime issues: E. A. Feigenbaum, "Artificial Intelligence: Themes in the Second Decade," in A. J. H. Morrell, ed., *Information Processing 68, Vol. II* (Amsterdam, 1969), p. 1017.

8. G. J. Sussman, *A Computer Model of Skill Acquisition* (New York, 1975), p. 9.

9. The quote is from *SIP*, p. 2. See also, e.g., M. L. Minsky and Seymour Papert, *Artificial Intelligence* (Eugene, Oregon, 1973); Allen Newell, "You Can't Play 20 Questions with Nature and Win," in W. G. Chase, ed., *Visual Information Processing* (New York, 1973), pp. 283–310.

10. Z. W. Pylyshyn, "Complexity and the Study of Artificial and Human Intelligence," in Steven Kaneff and J. F. O'Callaghan, eds., *Objectives and Methodologies in Artificial Intelligence* (New York, in press).

11. R. P. Abelson, "Script Processing in Attitude Formation and Decision-Making," in J. S. Carroll and J. W. Payne, eds., *Cognition and Social Behavior* (Hillsdale, N.J., 1976).

12. D. O. Hebb, *The Organization of Behavior: A Neuropsychological Theory* (New York, 1949); Nathaniel Rochester, J. H. Holland, L. H. Haibt, and W. L. Duda, "Test on a Cell Assembly Theory of the Brain, Using a Large Digital Computer," *Inst. Radio Engineers Trans. Inf. Theory*, 2 (1956), 80–93.

13. J. T. Gullahorn and J. E. Gullahorn, "A Computer Model of Elementary Social Behavior," in *CT*, pp. 375–386. *Cf.* G. C. Homans, *Social Behavior: Its Elementary Forms* (New York, 1961).

14. B. F. Skinner, "A Case History in Scientific Method," in Sigmund Koch, ed., *Psychology: A Study of a Science, Vol. II* (New York, 1959), pp. 359–379. For a classic statement of the hypothetico-deductive philosophy of science, see C. G. Hempel, *Aspects of Scientific Explanation, and Other Essays in the Philosophy of Science* (New York, 1965). For a useful critique of this approach written with reference to artificial intelligence, see Aaron Sloman, "What Are the Aims of Science?" *Radical Philosophy*, no. 13 (Spring, 1976), 7–17.

15. This philosophical position was stated by Freud in 1895, in his early (though posthumously published) "Project," and was never basically abandoned by him: Sigmund Freud, *Project for a Scientific Psychology*, in Standard Edition, pp. 295–397.

16. Frank Cioffi, "Freud and the Idea of a Pseudo-Science," in Robert Borger and Frank Cioffi, eds., *Explanation in the Behavioural Sciences* (London, 1970), pp. 471–499. Cioffi takes a basically Popperian approach, but concentrates on the actual application (and defense) of theoretical statements by theoreticians, as opposed to their "static" propositional form; *cf.* K. R. Popper, *Conjectures and Refutations: The Growth of Scientific Knowledge* (London, 1963), esp. ch. i.

17. "Negotiation" has been more satisfactorily described than "agreement"; for the former, see Harré and Secord, *Explanation of Social Behaviour*; for the latter, see Habermas, *Knowledge and Human Interests*.

18. See, e.g., Giorgi, *Psychology as a Human Science*.

19. The seminal example of the computational metaphor in psychology was: G. A. Miller, Eugene Galanter, and K. H. Pribram, *Plans and the Structure of Behavior* (New York, 1960). Subsequent general texts based on this metaphor include:

Ulric Neisser, *Cognitive Psychology* (New York, 1967); W. R. Reitman, *Cognition and Thought: An Information-Processing Approach* (New York, 1965); P. H. Lindsay and D. A. Norman, *Human Information Processing: An Introduction to Psychology* (New York, 1972); J. R. Anderson and G. H. Bower, *Human Associative Memory* (Washington, 1973); G. A. Miller and P. N. Johnson-Laird, *Language and Perception* (Cambridge, Mass., 1976); Allen Newell and H. A. Simon, *Human Problem Solving* (Englewood Cliffs, 1972); D. A. Norman and D. E. Rumelhart, *Explorations in Cognition* (San Francisco, 1975).

Examples of experiments relevant to the psychological assumptions informing Rieger's MEMORY program include: J. D. Bransford and M. K. Johnson, "Considerations of Some Problems of Comprehension," in Chase, ed., *Visual Information Processing,* pp. 383–438; J. R. Anderson and Reid Hastie, "Individuation and Reference in Memory: Proper Names and Definite Descriptions," *Cognitive Psychology,* 6 (1974), 495–514 and P. N. Johnson-Laird, Clive Robins, and Lucy Velicogna, "Memory for Words," *Nature,* 251 (1974), 704–705.

For a computational model of aspects of Piagetian theory, see Richard M. Young, "Production Systems as Models of Cognitive Development," in *AISB-1,* pp. 284–295.

Papers influenced by the information-processing approach are regularly published in the journals *Cognitive Psychology, Cognition,* and *Cognitive Science.*

20. For a review of connections between artificial intelligence and neurophysiological theory, see M. A. Arbib, *Brain Theory and Artificial Intelligence* (New York, 1976); a more elementary account is given in M. A. Arbib, *The Metaphorical Brain: An Introduction to Cybernetics as Artificial Intelligence and Brain Theory* (New York, 1972).

21. For the clinical relevance of PARRY, see K. M. Colby, "Clinical Implications of a Simulation Model of Paranoid Processes," *Arch. Gen. Psychiatry,* 33 (1976), 854–857.

For the clinical use of programs in the exploration of childhood autism, see K. M. Colby, "Rationale for Computer Based Treatment of Language Difficulties in Non-Speaking Autistic Children," *J. Autism and Childhood Schizophrenia,* 3 (1973), 254–260; Sylvia Weir and Ricky Emanuel, *Using LOGO to Catalyse Communication in an Autistic Child* (Edinburgh, 1976).

22. See Harold Garfinkel, *Studies in Ethnomethodology* (Englewood Cliffs, 1967). For other sociological and social-psychological discussions of the implicit meanings in everyday actions, see: Erving Goffman, *Frame Analysis: An Essay on the Organization of Experience* (New York, 1974); A. V. Cicourel, *Cognitive Sociology: Language and Meaning in Social Interaction* (Harmondsworth, 1973); T. P. Wilson, "Normative and Interpretive Paradigms in Sociology," in J. D. Douglas, ed., *Understanding Everyday Life* (Chicago, 1970); R. H. Turner, ed., *Ethnomethodology* (Harmondsworth, 1974); S. A. Tyler, ed., *Cognitive Anthropology* (New York, 1969).

23. M. A. Boden, "The Structure of Intentions," *J. Theory Social Behaviour,* 3 (1973), 23–46.

24. Newell and Simon, *Human Problem Solving.*

25. For an insightful critique of the relation between programs and protocols, see: D. C. Dennett, "Machine Traces and Protocol Statements," *Behavioral Science,* 13 (1968), 155–161.

26. Newell, "Artificial Intelligence and the Concept of Mind."

27. N. J. Nilsson, "Artificial Intelligence," *Information Processing '74,* Proc. IFIP Congress 1974, Vol. 4 (Amsterdam, 1974), pp. 778–801.

28. Donald Michie, "On Not Seeing Things," in *On Machine Intelligence* (Edinburgh, 1974), pp. 112–133.

29. The connection of vision with action is discussed in broad terms in M. A. Boden, *Purposive Explanation in Psychology* (Cambridge, Mass., 1972), chap. viii. Physiological mechanisms are discussed in Arbib, *Brain Theory and Artificial Intelligence.*

30. A. L. Yarbus, *Eye Movements and Vision* (New York, 1967), p. 174. In connection with Sylvia Weir's work, see: G. Jansson, "Measurements of Eye Movements During a Michotte Launching Event," *Scand. J. Psychol.*, 5 (1964), 153–160.

31. For a general discussion of this point, see Pylyshyn, "Complexity and the Study of Artificial and Human Intelligence." For Pylyshyn's specific applications to visual phenomenology, see the papers cited in note 56, below. For a psychophysiological theory of eye-movements and vision, making use of the computational metaphor, see: R. L. Didday and M. A. Arbib, "Eye Movements and Visual Perception: A 'Two Visual System' Model," *Int. J. Man-Machine Studies*, 7 (1975), 547–569.

32. N. S. Sutherland, "Intelligent Picture Processing," unpublished paper.

33. J. J. Gibson, *The Senses Considered as Perceptual Systems* (Boston, 1966).

34. Ruzena Bajcsy, "Computer Description of Textured Surfaces," in *IJCAI-3*, pp. 572–577.

35. David Marr, *Analyzing Natural Images: A Computational Theory of Texture Vision* (Cambridge, Mass., 1975): David Marr, *Early Processing of Visual Information* (Cambridge, Mass., 1975).

36. An excellent survey is: Gunnar Johansson, "Spatial Event Perception," chapter in *Handbook for Physiology* (forthcoming). See also Gibson, *The Senses Considered as Perceptual Systems;* Bela Julesz, *Foundations of Cyclopean Perception* (Chicago, 1970); E. S. Eriksson, "Movement Parallax During Locomotion," *Perception and Psychophysics*, 16 (1974), 197–200; M. L. Braunstein, "The Perception of Depth Through Motion," *Psychological Bulletin*, 59 (1962), 422–433.

37. Gunnar Johansson, "Visual Perception of Biological Motion and a Model for its Analysis," *Perception and Psychophysics*, 14 (1973), 201–211.

38. Programs making use of motion parallax are discussed in H. H. Baker, *Machine Vision and Its Place in an Industrial Environment* (Edinburgh, 1975); see also the shadowgraph, Michie, *On Machine Intelligence*, pp. 125–128.

39. R. L. Gregory, *Eye and Brain* (London, 1966), pp. 91–99. *Cf.* Keith Price, "A Comparison of Human and Computer Vision Systems," *SIGART Newsletter*, No. 50 (February 1975), 5–10.

40. A. J. Thomas and T. O. Binford, *Information Processing Analysis of Visual Perception: A Review* (Stanford, 1974), pp. 19–21; P. H. Winston, ed., *New Progress in Artificial Intelligence* (Cambridge, Mass., 1974), pp. 186–219.

41. Pylyshyn, "Complexity and the Study of Artificial and Human Intelligence," p. 7.

42. David Marr, *An Essay on the Primate Retina* (Cambridge, Mass., 1974).

43. J. Y. Lettvin, H. R. Maturana, W. Pitts, and W. S. McCulloch, "What the Frog's Eye Tells the Frog's Brain," *Proc. Inst. Radio Engineers,* 47 (1959), 1940–1959; R. L. Didday, "A Model of Visuomotor Mechanisms in the Frog Optic Tectum," *Mathematical Biosciences,* 30 (1976), 169–180.

44. D. H. Hubel and T. N. Wiesel, "Receptive Fields of Single Neurones in the Cat's Striate Cortex," *J. Physiology*, 148 (1959), 579–591; D. H. Hubel and T. N. Wiesel, "Receptive Fields, Binocular Interaction, and Functional Architecure in the Cat's Visual Cortex," *J. Physiology*, 160 (1962), 106–154; D. H. Hubel and T. N. Wiesel, "Receptive Fields and Functional Architecture of Monkey Striate Cortex,"

*J. Physiology*, 195 (1968), 215–243; E. Marg, J. E. Adams, and B. B. Rutkin, "Receptive Fields of Cells in the Human Visual Cortex," *Experientia*, 24 (1968), 348–350. For a recent theoretical application of experimental results like these, see H. B. Barlow, "Single Units and Sensations: A Neuron Doctrine for Perceptual Psychology," *Perception*, 1 (1972), 371–394.

45. For a description and early critique of the property-list approach, see M. L. Minsky, "Steps Toward Artificial Intelligence," in *CT*, pp. 415–425.

46. Leonard Uhr and Charles Vossler, "A Pattern Recognition Program that Generates, Evaluates, and Adjusts Its Own Operators," in *CT*, p. 265.

47. W. S. McCulloch, "A Heterarchy of Values Determined by the Topology of Nervous Nets," *Bull. Math. Biophysics*, 11 (1949), 89–93.

48. Gabriel Horn and R. M. Hill, "Modification of Receptive Fields of Cells in the Visual Cortex Occurring Spontaneously and Associated With Bodily Tilt," *Nature*, 221 (1969), 186–188.

49. Raul Hernandez-Peon, Harald Scherrer, and Michel Jouvet, "Modification of Electrical Activity in the Cochlear Nucleus During 'Attention' in Unanesthetized Cats," *Science*, 123 (1956), 331–332.

50. J. S. Bruner and Leo Postman, "Emotional Selectivity in Perception and Reaction," *J. Personality*, 16 (1947), 69–77; J. S. Bruner, "Personality Dynamics and the Process of Perceiving," in R. R. Blake and G. V. Ramsey, eds., *Perception: An Approach to Personality* (New York, 1951), pp. 121–147; J. S. Bruner, "On Perceptual Readiness," *Psychological Review*, 64 (1957), 123–152.

51. J. Z. Young, *A Model of the Brain* (Oxford, 1964); N. S. Sutherland, "Outlines of a Theory of Visual Pattern Recognition in Animals and Man," *Proc. Royal Society B*, 171 (1968), 297–317.

52. Z. W. Pylyshyn, "What the Mind's Eye Tells the Mind's Brain; A Critique of Mental Imagery," *Psychological Bulletin*, 80 (1973), 1–24; D. C. Dennett, *Content and Consciousness* (London, 1969), chap. vii. See also J. A. Fodor, *The Language of Thought* (New York, 1975; Hassocks, Sussex, 1976), esp. chap. iv.

53. R. N. Haber and R. B. Haber, "Eidetic Imagery—I: Frequency," in R. N. Haber. ed., *Contemporary Theory and Research in Visual Perception* (London, 1970), pp. 350–356. This is a rich sourcebook of data and discussion in visual psychology.

54. M. R. Fernald, "The Diagnosis of Mental Imagery," *Psychological Monographs*, 14, no. 58 (February 1912), pp. 1–169.

55. R. M. Pritchard, "Stabilized Images on the Retina," *Scientific American*, 204 (1961), 72–78.

56. Pylyshyn, "What the Mind's Eye Tells the Mind's Brain"; Z. W. Pylyshyn, "Do We Need Images and Analogues?" in *TINLP*, pp. 174–177; Z. W. Pylyshyn, "The Symbolic Nature of Mental Representations," in Steven Kaneff and J. E. O'Callaghan, eds., *Objectives and Methodologies in Artificial Intelligence* (New York, in press); S. M. Kosslyn and J. R. Pomerantz, "Imagery, Propositions, and the Form of Internal Representations," *Cognitive Psychology*, 9 (1977), 52–76. *Cf.* D. C. Dennett, "Toward a Cognitive Theory of Consciousness," in C. W. Savage, ed., *Minnesota Studies in the Philosophy of Science, Vol. 9* (Minneapolis, 1978).

57. M. S. Gazzaniga, *The Bisected Brain* (New York, 1969) and A. J. Thomas, "Puccetti on Machine Pattern Recognition," *Brit. J. Philos. Science*, 26 (1975), 227–239 give the psychophysiological evidence concerning split-brain patients. Possible distinctions between visual and verbal thinking are explored by Allan Paivio, *Imagery and Verbal Processes* (New York, 1971).

58. P. M. Will and D. D. Grossman, "An Experimental System for Computer Controlled Mechanical Assembly," *IEEE Transactions on Electronic Computers*, 24 (1975), p. 879.

59. Pylyshyn, "Symbolic Nature of Mental Representations."

60. L. A. Cooper and R. N. Shepard, "Chronometric Studies of the Rotation of Mental Images," in Chase, ed., *Visual Information Processing*, pp. 75–176.

61. S. E. Palmer, "The Nature of Perceptual Representation: An Examination of the Analog/Propositional Controversy," in *TINLP*, pp. 165–173.

62. S. M. Kosslyn, "Information Representation in Visual Images," *Cognitive Psychology*, 7 (1975), 341–370; S. M. Kosslyn, "On Retrieving Information from Visual Images," in *TINLP*, pp. 160–164.

63. Marr, *Analyzing Natural Images*, p. 3. Emphasis mine.

## CHAPTER 14

1. Various philosophical texts are cited in the notes to this and other chapters; these references include not only epistemology and philosophy of mind, which obviously involve issues relevant to artificial knowledge-based systems, but also aesthetics and ethics, which apparently do not. For a persuasive statement of the view that *any* area of philosophical inquiry may be illuminated by, and in its turn shed some light on, the computational approach to thinking, see Aaron Sloman, *The Computer Revolution in Philosophy: Philosophy, Science, and Models of Mind* (Hassocks, Sussex, 1978).

For a philosophical discussion of computational, psychological, and linguistic approaches to "internal representation," see J. A. Fodor, *The Language of Thought* (New York, 1975; Hassocks, Sussex, 1976).

Links with the philosophy of language are explored by Y. A. Wilks in chap. xi of Eugene Charniak and Y. A. Wilks, eds., *Computational Semantics: An Introduction to Artificial Intelligence and Natural Language Comprehension* (Amsterdam, 1976). Philosophical discussions of presuppositions and speech acts bear on problems encountered by anyone trying to write a "natural" language-using program—see, e.g.: J. L. Austin, *How to Do Things with Words* (Oxford, 1962); J. R. Searle, *Speech Acts* (Cambridge, 1969); H. P. Grice, "Logic and Conversation," in Peter Cole and J. L. Morgan, eds., *Syntax and Semantics, Vol. III: Speech Acts* (New York, 1975), pp. 41–58.

Technical work on reference and modal logic are relevant respectively to MEMORY's inferential reference procedures and SHRDLU's interpretation of words like *all* and *can*, and causal logic can contribute to computational representations of the real world. Some of these issues are discussed by: John McCarthy, "Programs with Common Sense," in *SIP*, pp. 403–418; John McCarthy and P. J. Hayes, "Some Philosophical Problems from the Standpoint of Artificial Intelligence," in *MI-4*, pp. 463–502; P. J. Hayes, "Robotologic," in *MI-5*, pp. 533–554; P. J. Hayes, "A Logic of Actions," in *MI-6*, pp. 495–520.

2. A. C. Clarke, "The Mind of the Machine," *Playboy*, Dec. 1968, pp. 116–119, 122, 293–294.

3. Guy Robinson, "How to Tell Your Friends from Machines," *Mind*, N.S., 81 (1972), 504–518. Similar attacks are presented in: David Ingleby, "Ideology and the Human Sciences: Some Comments on the Role of Reification in Psychology and Psychiatry," in Trevor Pateman, ed., *Counter Course: A Handbook for Course Criticism* (Harmondsworth, 1972), pp. 51–81; Ted Ninnes, "History of Positivism," in *Rat Myth and Magic: A Political Critique of Psychology* (London, 1972), pp. 33–36; O. K. Tikhomirov, "Philosophical and Psychological Problems of Artificial Intelligence," in *IJCAI-4*, pp. 932–937.

Philosophical critiques of technological society in broad agreement with these attacks on artificial intelligence are: Herbert Marcuse, *One Dimensional Man: The Ideology of Industrial Society* (London, 1964); Theodore Roszak, *Where the Wasteland Ends: Politics and Transcendence in Post Industrial Society* (New York, 1972).

4. M. A. Boden, *Purposive Explanation in Psychology* (Cambridge, Mass., 1972), pp. 43–45, 118–122, 158–198. For a critique of the view that there is a fixed list of interests natural to all people, see R. M. Young, "The Human Limits of Nature," in Jonathan Benthall, ed., *The Limits of Human Nature* (London, 1973), pp. 235–274.

5. The word "unthinkingly" is important here: it is of course possible for someone deliberately to decide to obey some other person or authority (national, religious, or political), where the decision and the ensuing acts of obedience would be termed "free" by most nonexistentialist philosophers.

6. M. A. Boden, "The Paradox of Explanation," *Proc. Aristotelian Society*, N.S., 62 (1962), 159–178.

7. E. A. Feigenbaum, "Artificial Intelligence Research: What Is It? What Has It Achieved? Where Is It Going?" Stanford Univ. Dept. Comp. Sc.: privately circulated ms.

8. A. M. Turing, "Computing Machinery and Intelligence," in *CT*, p. 19.

For the "tough-minded" suggestion that we first program a machine with the ordinary rules of English and then, when we have provided it with the best "thinking" programs of which we are capable, simply *ask it* whether it is conscious, etc., see: Michael Scriven, "The Compleat Robot: A Prolegomena to Androidology," in Sidney Hook, ed., *Dimensions of Mind: A Symposium* (New York, 1960), pp. 118–142. Scriven's suggestion is shown to be oversimple in an amusing article stressing the fact that *current* "English" does not allow one to speak of machines as "thinking": A. C. Danto, "On Consciousness in Machines," in Hook, ed., *Dimensions of Mind*, pp. 180–187. For a subtle account of the way in which language can shift its meaning at deep levels, see Hilary Putnam, "Dreaming and Depth Grammar," in R. J. Butler, ed., *Analytical Philosophy* (Oxford, 1962), pp. 211–235.

The issues of language use central to these papers are clearly relevant to whether a program such as Y. A. Wilks's machine translator should be written so as to insist on, or merely to prefer, psychological verbs taking an animate subject. Wilks's current program would accept not only "The car drank the petrol," but also "The computer desperately sought the answer." However, each of these would be classed by it as metaphorical or analogical uses. If "The computer desperately sought the answer" were to be taken *literally*, Wilks's basic semantic preference rules would have to be altered.

9. Ninnes, "History of Positivism," pp. 35–36.

10. For a statement and an ensuing rebuttal of this view, see: K. R. Popper, *Of Clouds and Clocks: An Approach to the Problem of Rationality and the Freedom of Man* (St. Louis, 1966): P. E. Meehl, "Psychological Determinism and Human Rationality: A Psychologist's Reactions to Professor Karl Popper's 'Of Clouds and Clocks'," in Michael Radner and Stephen Winokur, eds., *Minnesota Studies in Philosophy of Science, Vol. IV: Analyses of Theories and Methods of Physics and Psychology* (Minneapolis, 1970), pp. 310–372.

11. M. A. Boden, "Intentionality and Physical Systems," *Philosophy of Science*, 37 (1970), 200–214; Boden, *Purposive Explanation in Psychology*, esp. chs. ii, iv, and viii.

The logical criteria of intentionality are discussed in: R. M. Chisholm, "Intentionality," in Paul Edwards, ed., *The Encyclopedia of Philosophy*, Vol. IV (New York, 1967), pp. 201–204; D. C. Dennett, *Content and Consciousness* (London, 1969), ch. ii.

12. H. C. Longuet-Higgins, "The Seat of the Soul," in C. H. Waddington, ed., *Towards a Theoretical Biology: 3. Drafts* (Edinburgh, 1970), pp. 236–241.

13. Internal models are discussed in Boden, *Purposive Explanation in Psychology*, esp. pp. 122–137 and 298–321. For a recent philosophical discussion of computational models of representation, see Fodor, *The Language of Thought*.

14. K. J. W. Craik, *The Nature of Explanation* (Cambridge, 1943).

15. Dennett, *Content and Consciousness*, ch. iii; R. L. Gregory, "The Evolution of Eyes and Brains: A Hen-and-Egg Problem," in S. J. Freedman, ed., *The Neuropsychology of Spatially Oriented Behaviour* (Illinois, 1968), pp. 7–18.

16. M. L. Minsky, "Matter, Mind, and Models," in *SIP*, pp. 425–432.

17. The curiously complementary linguistic abilities of two chimps, Sarah and Washoe, are discussed by Roger Brown, *A First Language: The Early Stages* (Cambridge, Mass., 1973), pp. 32–51.

18. The logical positivists' "verification principle" is therefore not a criterion of *meaning*, but—at most—of empirical knowledge. *Cf.* A. J. Ayer, *Language, Truth, and Logic*, 2nd ed. (London, 1946).

19. J. O. De La Mettrie, *L'Homme Machine.* Critical edition, ed. Aram Vartanian (Princeton, 1960). First published, 1748.

20. T. H. Huxley, "On the Hypothesis that Animals Are Automata, and Its History," in T. H. Huxley, *Method and Results: Essays* (London, 1893), pp. 199–250.

21. Negative feedback was first suggested as the basis of purposive action in: Arturo Rosenblueth, Norbert Wiener, and Julian Bigelow, "Behavior, Purpose, and Teleology," *Philosophy of Science*, 10 (1943), 18–24; Arturo Rosenblueth and Norbert Wiener, "Purposeful and Non-Purposeful Behavior," *Philosophy of Science*, 17 (1950), 318–326.

22. G. E. Moore, "The Refutation of Idealism," in G. E. Moore, *Philosophical Studies* (London, 1922), pp. 1–30.

23. "Purpose" is discussed throughout Boden, *Purposive Explanation in Psychology.* "Freedom" is discussed at pp. 242–248 and 330–334; "self" at pp. 236–260 and 327–334; and "consciousness" at pp. 261–321 and 334–338.

24. In addition to other sources cited in the notes to this chapter, see Paul Armer, "Attitudes Toward Intelligent Machines," in *CT*, pp. 389–405; K. M. Sayre, *Consciousness: A Philosophic Study of Minds and Machines* (New York, 1969); K. M. Sayre, *Cybernetics and the Philosophy of Mind* (Atlantic Highlands, N.J., 1976).

25. Ernest Nagel and J. R. Newman, *Godel's Proof* (New York, 1958).

26. For an early rebuttal of such attacks, see A. M. Turing, "Computing Machinery and Intelligence," in *CT*, esp. pp. 21–22.

27. J. R. Lucas, "Minds, Machines, and Godel," *Philosophy*, 36 (1961), 112–127.

28. Michael Polanyi, *Personal Knowledge: Towards a Post-Critical Philosophy* (New York, 1964); Michael Polanyi, "The Logic of Tacit Inference," *Philosophy*, 41 (1966), 1–18.

29. Polanyi, "Logic of Tacit Inference," p. 18.

30. For Dreyfus's attacks on the field, see: H. L. Dreyfus, *Alchemy and Artificial Intelligence* (Santa Monica, 1965); H. L. Dreyfus, "Why Computers Must Have Bodies in Order to Be Intelligent," *Review of Metaphysics*, 21 (1967), 13–32; and H. L. Dreyfus, *What Computers Can't Do: A Critique of Artificial Reason* (New York, 1972).

A forceful rebuttal of the first two of Dreyfus's critiques is Seymour Papert, *The Artificial Intelligence of Hubert L. Dreyfus: A Budget of Fallacies.*

Dreyfus's book is insightfully reviewed by: Z. W. Pylyshyn, "Minds, Machines, and Phenomenology: Some Reflections on Dreyfus' 'What Computers Can't Do'," *Cognition*, 3 (1974), 20–42; Y. A. Wilks, "Dreyfus' Disproofs," *Brit. J. Philosophy of Science*, 27 (1976), 177–185; and Bernard Williams, "How Smart Are Computers?" *New York Review*, November 15, 1973, pp. 36–40.

31. In Merleau-Ponty's phenomenological writings, the findings of clinical and

experimental psychologists (especially Gestalt psychologists) are cited extensively: Maurice Merleau-Ponty, *Phenomenology of Perception* (New York, 1962): Maurice Merleau-Ponty, *The Structure of Behaviour* (London, 1965). *Cf.* also Aaron Sloman and Steven Hardy, "Giving a Computer Gestalt Experiences," in *AISB-2*, pp. 242–255.

32. This prediction was made and withdrawn by H. A. Simon. See Papert, *Artificial Intelligence of Hubert L. Dreyfus*, section 3.

33. Wilks, "Dreyfus' Disproofs."

34. See K. M. Sayre, "Intelligence, Bodies, and Digital Computers," *Review of Metaphysics*, 21 (1968), 714–723.

35. F. N. Marzocco, "Computer Recognition of Handwritten First Names," *IEEE Trans. Electronic Computers*, 14 (1965), 210–217.

36. R. L. Gregory, "On How So Little Information Controls So Much Behaviour," in C. H. Waddington, ed., *Towards a Theoretical Biology: 2. Sketches* (Edinburgh, 1969), pp. 236–246.

37. D. M. Mackay argues that analog machines and, eventually, "wet" engineering (involving networks of actual biological neurons) will be required to simulate some of the information processing carried out by the brain: D. M. Mackay, "A Mind's Eye View of the Brain," in Norbert Wiener and J. P. Schadé, eds., *Progress in Brain Research, 17: Cybernetics of the Nervous System* (Amsterdam, 1965), pp. 321–332.

38. P. M. Will and D. D. Grossman, "An Experimental System for Computer Controlled Mechanical Assembly," *IEEE Trans. Electronic Computers*, 24 (1975), 879–888.

39. COPY-DEMO takes about 10 minutes for the vision processing and 5 minutes for the mechanical construction of its demonstration tasks (P. H. Winston, personal communication). Even industrial robots are sometimes made to work much slower than they could, for reasons of safety: Will and Grossman, "An Experimental System for Computer Controlled Assembly," p. 882.

40. Polanyi, *Personal Knowledge*, p. 59.

41. Will and Grossman, "An Experimental System for Computer Controlled Assembly," pp. 880, 883–886. See also note 28 to chap. xii, above.

42. Ludwig Wittgenstein, *Philosophical Investigations* (Oxford, 1953), para. 241 and p. 226.

43. Wittgenstein, *Philosophical Investigations*, p. 223.

44. Wilks, "Dreyfus' Disproofs."

45. There may be *other* reasons for a policy of excluding men from feminist discussions—e.g., if their presence has the effect of inhibiting or "skewing" debate.

46. See, e.g.: Paul Ziff, "The Feelings of Robots," *Analysis*, 19 (1959), 64–68; Keith Gunderson, *Mentality and Machines* (New York, 1971), esp. chs. iii and v; Roland Puccetti, *Persons: A Study of Possible Moral Agents in the Universe* (London, 1968), ch. ii; Scriven, "The Compleat Robot," section 11. An excellent recent discussion, whose conclusion is *not* what would most commonly be expected on the basis of its title, is: D. C. Dennett, "Why You Can't Make a Computer That Feels Pain," *Synthèse*, 38 (1978), 415–456.

47. See e.g., Anthony Kenny, *Action, Emotion, and Will* (London, 1963).

48. John Locke, *An Essay Concerning Human Understanding* (London, 1690).

49. Bernard Harrison, *Form and Content* (Oxford, 1973).

50. *Cf.* Boden, *Purposive Explanation in Psychology*, pp. 208–235, 261–262. The analogy of depth-first versus breadth-first search was suggested to me by Steven Hardy.

For a recent simulation of affect, see W. S. Faught, "Affect as Motivation for Cognitive and Conative Processes," in *IJCAI-4*, pp. 893–899.

## CHAPTER 15

1. General reviews of this literature, together with excellent bibliographies, are given in: Z. W. Pylyshyn, *Perspectives on the Computer Revolution* (Englewood Cliffs, 1970); C. C. Gotlieb and Allan Borodin, *Social Issues in Computing* (New York, 1973). A list of references including many novels is given in P. C. Jackson, *Introduction to Artificial Intelligence* (New York, 1974), pp. 394–395.

2. J. E. Michie, "What the Public Think," *Firbush News 5* (Edinburgh, 1974).

3. Oscar Firschein, M. A. Fischler, L. S. Coles, and J. M. Tenenbaum, "Forecasting and Assessing the Impact of Artificial Intelligence on Society," in *IJCAI-3*, pp. 105–120; Donald Michie, *On Machine Intelligence* (Edinburgh, 1974); Donald Michie, "Life with Intelligent Machines," in Lyndhurst Collins, ed., *The Use of Models in the Social Sciences* (London, 1976), pp. 101–109; R. L. Gregory, "The Social Implications of Intelligent Machines," in *MI-6*, pp. 3–13.

The paper by Firschein *et al.* does not mention the marriage counsellor program; however, it was included in the original questionnaire, and was both expected and approved by a large proportion of the professional group engaged in the study (Firschein, personal communication).

4. E. H. Shortliffe, S. G. Axline, B. G. Buchanan, T. C. Merigan, and N. S. Cohen, "An Artificial Intelligence Program to Advise Physicians Regarding Antimicrobial Therapy," *Computers and Biomedical Research*, 6 (1973), 544–560; E. H. Shortliffe, Randall Davis, S. G. Axline, B. G. Buchanan, C. C. Green, and N. S. Cohen, "Computer-Based Consultations in Clinical Therapeutics: Explanation and Rule Acquisition Capabilities of the MYCIN System," *Computers and Biomedical Research*, 8 (1975), 303–320. (Other medical applications of artificial intelligence are being explored by Saul Amarel's group at Rutgers University, N.J.)

5. Firschein *et al.*, "Forecasting and Assessing the Impact of Artificial Intelligence," p. 113.

6. E. H. Shortliffe and B. G. Buchanan, "A Model of Inexact Reasoning in Medicine," *Mathematical Biosciences*, 23 (1975), 351–379.

7. Seymour Papert, "Teaching Children to Be Mathematicians Versus Teaching About Mathematics," *Int. J. Math. Educ. Sci. Technol.*, 3 (1972), 249–262: Seymour Papert, "Theory of Knowledge and Complexity," in G. J. Dalenoort, ed., *Process Models for Psychology* (Rotterdam, 1973), pp. 1–49; Seymour Papert, *Uses of Technology to Enhance Education* (Cambridge, Mass., 1973); *cf.* J. A. M. Howe, John Knapman, H. M. Noble, Sylvia Weir, and R. M. Young, *Artificial Intelligence and the Representation of Knowledge* (Edinburgh, 1975), pp. 45–50.

8. Jasia Reichardt, ed., *Cybernetics, Art, and Ideas* (London, 1971); H. B. Lincoln, ed., *The Computer and Music* (Ithaca, 1970).

9. Laurence Lerner, *A.R.T.H.U.R.: The Life and Opinions of a Digital Computer* (Cambridge, Mass., and Hassocks, Sussex, 1974), pp. 9–10.

10. H. C. Longuet-Higgins and M. J. Steedman, "On Interpreting Bach," in *MI-6*, pp. 221–242. A fuller account is in M. J. Steedman, *The Formal Description of Musical Perception* (Edinburgh, 1973); some radical limitations of this program have recently been overcome by a generalization of the harmonic theory (Steedman, personal communication).

11. John McCarthy, "The Home Information Terminal," in *Man and Computer. Proc. Int. Conf., Bordeaux 1970* (Basel, 1972), pp. 48–57.

12. A brief description of HEARSAY is: D. R. Reddy, L. D. Erman, R. D. Fennell, and R. B. Neely, "The HEARSAY Speech Understanding System," in *IJCAI-3*, pp. 185–193. The original plan for the project is described in: Allen

Newell, Jeffrey Barnett, J. W. Forgie, C. C. Green, D. H. Klatt, J. C. R. Licklider, J. H. Munson, D. R. Reddy, and W. A. Woods, *Final Report of a Study Group on Speech Understanding Systems* (Amsterdam, 1973).

13. The domestic robot was confidently forecast by Ed Fredkin in a tutorial lecture given at the Fourth International Joint Conference on Artificial Intelligence, in Tbilisi, USSR (Sept., 1975). *Cf.* Firschein *et al.*, "Forecasting and Assessing the Impact of Artificial Intelligence," p. 113, for a less confident prediction of the "general factotum."

14. P. M. Will and D. D. Grossman, "An Experimental System for Computer Controlled Mechanical Assembly," *IEEE Trans. Electronic Computers*, 24 (1975), 879–888; J. F. Engelberger, "Robotics—Like It Was, Like It Is, and Like It Will Be," *IEEE Intercon Conference Record*, April 1975, Session 13, pp. 1–6. Several robots are described in Bertram Raphael, *The Thinking Computer: Mind Inside Matter* (San Francisco, 1976), chap. viii.

15. P. H. Winston, "The MIT Robot," in *MI-7*, pp. 465–480; A. P. Ambler, H. G. Barrow, C. M. Brown, R. M. Burstall, and R. J. Popplestone, "A Versatile Computer-Controlled Assembly System," in *IJCAI-3*, pp. 298–307.

16. Donald Michie, *On Machine Intelligence* (Edinburgh, 1974), p. 196.

17. T. D. Sterling, E. A. Bering, S. V. Pollack, and H. G. Vaughan, eds., *Visual Prosthesis: The Interdisciplinary Dialogue* (New York, 1971), pp. 327–342.

18. R. L. Gregory and J. G. Wallace, "Recovery from Early Blindness," *Exp. Psychol. Soc. Monogr.*, no. 2, (1963).

19. Theodore Roszak, *Where the Wasteland Ends: Politics and Transcendence in Post Industrial Society* (New York, 1972); Herbert Marcuse, *One Dimensional Man: The Ideology of Industrial Society* (London, 1964).

20. Rollo May, ed., *Existential Psychology* (New York, 1961), p. 20.

21. R. D. Laing, *The Divided Self: A Study of Sanity and Madness* (London, 1960).

22. Joseph Weizenbaum, "On the Impact of the Computer on Society: How Does One Insult a Machine?" *Science*, 176 (1972), 609–614; Joseph Weizenbaum, *Computer Power and Human Reason: From Judgment to Calculation* (San Francisco, 1976), esp. pp. 268–270. For a debate on Weizenbaum's views, see B. J. Kuipers, John McCarthy, and Joseph Weizenbaum, "Computer Power and Human Reason: Comments," SIGART Newsletter, no. 58, June 1976, 4–12.

23. R. D. Laing, *The Politics of Experience* (Harmondsworth, 1967).

24. K. M. Colby, "Computer Simulation of Change in Personal Belief Systems," *Behavioral Science*, 12 (1967), p. 253.

25. H. J. Walton, G. R. Kiss, and K. M. Farvis, *A Computer Program for the On-Line Exploration of Attitude Structures of Psychiatric Patients* (Edinburgh, 1973).

26. Thomas Szazs, *The Myth of Mental Illness* (London, 1962). For a useful review of approaches to psychiatry see Anthony Clare, *Psychiatry in Dissent: Controversies in Thought and Practice* (London, 1976).

27. W. I. Card, Mary Nicholson, G. P. Crean, Geoffrey Watkinson, C. R. Evans, Jackie Wilson, and Daphne Russell, "A Comparison of Doctor and Computer Interrogation of Patients," *Int. J. Bio-Medical Computing*, 5 (1974), 175–187.

28. Firschein *et al.*, "Forecasting and Assessing the Impact of Artificial Intelligence," p. 115.

29. I. D. Illich, *Tools for Conviviality* (London, 1973).

30. Firschein *et al.*, "Forecasting and Assessing the Impact of Artificial Intelligence," p. 117.

31. Noam Chomsky, *Problems of Knowledge and Freedom* (London, 1972).

32. K. M. Colby, J. B. Watt, and J. P. Gilbert, "A Computer Method of Psychotherapy: Preliminary Communication," *J. Nerv. Mental Disease*, 142 (1966), 148–152.

33. "The MIT machine is intended to do as hardware instructions those operations that are frequently done in computer chess, but would require 2 to 5 instructions on a regular machine. These are things like testing the piece on the square, getting the status of the square next to it in a particular direction, pushing a legal move onto a move stack, etc. Other things that are not as primitive can be programmed in a programmable micro-store which can for instance direct the tree search. This allows changing the search control regimen when this appears desirable, while still getting the benefits of the faster machine hardware. The basic machine is made of 64 custom designed hardware chips interconnected in an $8 \times 8$ array. Each chip can communicate with every chip next to it in each of the eight possible directions (except of course for the ones at the board edge), as well as all the applicable knight distances. The machine has a basic 20 nano-second cycle time. In one cycle, the machine can propagate appropriate messages to adjacent squares (e.g. I have a bishop ray in direction North-west passing through), as well as generate all moves relating to messages it received in the previous cycle. Deciding whether a move is legal requires a message from an adjacent square plus the necessary landing conditions on the square itself. The hardware can even take care of castling conditions. Thus in at most seven cycles all legal moves will have been generated. This makes for a total time of 140 nano-seconds for a move generation, which is really fast. The people at MIT expect the program to play between 150 and 500 times faster than TECH-II playing on the PDP-10. There are however many questions as to whether such speed alone, without any additional knowledge about chess, will cause much improvement. That is one of the questions that should be answerable soon." (H. J. Berliner, personal communication.)

34. M. L. Minsky, "Matter, Mind, and Models," in *SIP*, pp. 430–431.

35. The seminal expression of this view was: K. S. Lashley, "The Problem of Serial Order in Behavior," in L. A. Jeffress, ed., *Cerebral Mechanisms in Behavior: The Hixon Symposium* (New York, 1951), pp. 112–135.

36. Papers on program proof and computational logic can be found in the *MI-1* to *MI-7* series, usually placed in the early sections of the volumes.

37. Rein Turn, *Computers in the 1980's* (New York, 1974), p. 228. Turn predicts that such security kernels could not consist of more than about 500–1000 lines of code; at present, only very tiny programs can be proved in the strictest sense.

38. Norbert Wiener, "Some Moral and Technical Consequences of Automation," *Science*, 131 (1960), 1355–1358. See also Norbert Wiener, *God & Golem, Inc.: A Comment on Certain Points Where Cybernetics Impinges on Religion* (Cambridge, Mass., 1964).

39. Isaac Asimov, *I, Robot* (London, 1967).

40. Precautions comparable to Asimov's Laws (making robots less speedy than they could be, for instance) are described by Will and Grossman, "An Experimental System for Computer Controlled Mechanical Assembly," p. 882.

41. L. S. Coles, *Some Thoughts on Robot Ethics* (Menlo Park, Calif., 1969).

42. One of Kant's 5 formulations of the Categorical Imperative enjoins us always to treat members of "humanity" as ends in themselves; Kantian commentators remark that, strictly speaking, this formula covers *rational beings as such*, of which humans may not be the only instances: H. J. Paton, *The Categorical Imperative: A Study of Kant's Moral Philosophy* (London, 1947), p. 165.

43. Alvin Toffler, *Future Shock* (New York, 1970), p. 397.

44. Roszak, *Where the Wasteland Ends*, p. 69.

45. B. F. Green, *Digital Computers in Research: An Introduction for Be-*

*havioral and Social Scientists* (New York, 1963), p. 258. Green was one of the early workers in artificial intelligence; he cooperated in the design of IPL-V, the first list-processing language, and he was a prime mover behind BASEBALL.

46. Weizenbaum's *Computer Power and Human Reason* is an isolated, and somewhat bitter, plea for restraint in applying computer technology to personal matters (such as psychotherapy). The "Serbelloni Panel" was established in 1972 under the chairmanship of Donald Michie, to look into the social implications of machine intelligence research: its members are less pessimistic than Weizenbaum about these issues. The journal *Computers and Automation* regularly publishes papers on such matters. For further references, see Pylyshyn, ed., *Perspectives on the Computer Revolution*, pp. 457–459.

47. An early paper stressing this point is: R. L. Gregory, "Will Seeing Machines Have Illusions?" *MI-1*, pp. 169–180. Gregory's paper concentrates on vision; but we have seen in previous chapters that "intelligent" programs in general need to rely on fuzzy reasoning and fallible assumptions if they are not to be confined to a rigorous—and empty—foolishness.

48. These epistemological issues are well discussed by Jurgen Habermas, *Knowledge and Human Interests* (London, 1972).

## CHAPTER 16

1. A retrospective evaluation of progress in the first thirty years, based on the invited comments of a number of people in the field, was published in the Silver Jubilee volume of the leading AI journal: D. G. Bobrow and P. J. Hayes, "Artificial Intelligence: Where Are We?" *AI*, 25 (1985), 375–415. For historical accounts of the field, see: P. McCorduck, *Machines Who Think* (San Francisco, 1979); J. Fleck, "Development and Establishment in Artificial Intelligence," in N. Elias, H. Martins, and R. Whitley, eds., *Scientific Establishments and Hierarchies* (Amsterdam, 1982), pp. 169–217; H. Gardner, *The Mind's New Science: A History of the Cognitive Revolution* (New York, 1985).

2. For a readable, if somewhat overenthusiastic, account of Japan's "Fifth Generation" project, see E. A. Feigenbaum and P. McCorduck, *The Fifth Generation: Artificial Intelligence and Japan's Computer Challenge to the World* (New York, 1983).

3. For work on discourse analysis, see: J. F. Allen and C. R. Perrault, "Analyzing Intention in Utterances," *AI*, 15 (1980), 143–178; J. M. Brady and R. C. Berwick, *Computational Models of Discourse* (Cambridge, Mass., 1983); P. R. Cohen and C. R. Perrault, "Elements of a Plan-Based Theory of Speech-Acts," *Cognitive Science*, 3 (1979), 177–212; A. Davey, *Discourse Production: A Computer Model of Some Aspects of a Speaker* (Edinburgh, 1978); L. D. Erman, P. E. London, and S. F. Fickas, "The Design and an Example Use of Hearsay III," *Proc. Seventh Int. Joint Conf. Artificial Intelligence*, Vancouver (1981), 409–415; B. J. Grosz, "Understanding Discourse Knowledge," in D. Walker, ed., *Understanding Spoken Language* (New York, 1978), pp. 229–347; B. J. Grosz, "Utterance and Objective: Issues in Natural Language Communication," *AI Magazine*, 1 (1980), 11–20; A. K. Joshi, B. L. Webber, and I. A. Sag, eds., *Elements of Discourse Understanding* (Cambridge, 1981); R. Reichman, "Conversational Coherency," *Cognitive Science*, 2 (1978), 283–327.

4. For discussions of expert systems in general, see: R. Davis and D. B. Lenat, eds. *Knowledge-Based Systems in Artificial Intelligence* (New York, 1982); F. Hayes-Roth, D. A. Waterman, and D. B. Lenat, eds., *Building Expert Systems* (Reading, Mass., 1983); M. Merry, ed., *Expert Systems 85* (Cambridge, 1985); D. Michie, ed., *Expert Systems in the Micro-Electronic Age* (Edinburgh, 1979); M. Stefik, J. Aikins, R. Balzer, J. Benoit, L. Birnbaum, F. Hayes-Roth, and E. D. Sacerdoti, "The Organization of Expert Systems," *AI*, 18 (1982), 135–173; D. A. Waterman, *A Guide to Expert Systems* (New York, 1985). For detailed descriptions of two of the oldest, and best-known, expert systems (MYCIN and DENDRAL), see: B. G. Buchanan and E. H. Short-

liffe, *Rule-Based Expert Programs: The MYCIN Experiments of the Stanford Heuristic Programming Project* (Reading, Mass., 1984); R. K. Lindsay, B. G. Buchanan, E. A. Feigenbaum, and J. Lederberg, *Applications of Artificial Intelligence for Organic Chemistry: The Dendral Project* (New York, 1980). A brief introduction to MYCIN is: J. Cendrowska and M. Bramer, "Inside an Expert System: A Rational Reconstruction of the MYCIN Consultation System," in T. O'Shea and M. Eisenstadt, eds., *Artificial Intelligence: Tools, Techniques, Applications* (New York, 1984), pp. 453–497. For reviews of various medical expert systems, see: J. R. W. Hunter, "Artificial Intelligence in Medicine: A Tutorial Survey," *Biomedical Measurement, Informatics and Control,* (in press); W. J. Clancey and E. H. Shortliffe, eds., *Readings in Medical Artificial Intelligence: The First Decade* (Menlo Park, 1984). For discussion of the commercial aspects, see: P. H. Winston and K. A. Prendergast, *The AI Business: Commercial Uses of Artificial Intelligence* (Cambridge, Mass., 1984); P. Harmon and D. King, *Expert Systems: Artificial Intelligence in Business* (New York, 1985).

5. See H. Thompson, "Speech Transcription: An Incremental, Interactive Approach," in *ECAI-84: Proc. Sixth European Conf. on Artificial Intelligence,* Pisa (1984), 697–704.

6. The best introduction to low-level vision in general is: J. Mayhew and J. Frisby, "Computer Vision," in T. O'Shea and M. Eisenstadt, eds., *Artificial Intelligence: Tools, Techniques, Applications* (New York,·1984), pp. 301–357. Other general reviews of work in low-level vision are: D. H. Ballard and C. M. Brown, eds., *Computer Vision,* (Englewood Cliffs, N.J., 1982); H. Barrow and J. M. Tenenbaum, "Computational Vision," *Proc. Inst. Electrical and Electronic Engineers,* 6 (1981), 572–595; O. J. Braddick and A. C. Sleigh, eds., *Physical and Biological Processing of Images,* (Berlin, 1983); J. M. Brady, ed., *Computer Vision* (Amsterdam, 1981); A. Hanson and E. Riseman, eds., *Computer Vision Systems* (New York, 1978); S. Ullman, *The Interpretation of Visual Motion* (Cambridge, Mass., 1979); B. K. P. Horn, *Robot Vision* (Cambridge, Mass., 1986).

7. See J. Mayhew, "Stereopsis," in O. J. Braddick and A. C. Sleigh, *Physical and Biological Processing of Images* (Berlin, 1983), pp. 204–216.

8. See D. A. Marr, *Vision* (San Francisco, 1982); I discuss Marr's approach to psychology in general in my *Computer Models of Mind: The Computational Approach in Theoretical Psychology* (Cambridge, forthcoming).

9. A good introduction to connectionism is: G. E. Hinton and J. A. Anderson, eds., *Parallel Models of Associative Memory* (Hillsdale, N.J., 1981). Connectionist systems of various types are also discussed in: S. E. Fahlman, *NETL: A System for Representing and Using Real-World Knowledge* (Cambridge, Mass., 1979); N. V. Findler, ed., *Associative Networks: Representation and Use of Knowledge by Computers* (New York, 1979); G. E. Hinton, "Shape Representation in Parallel Systems," *Proc. Seventh Int. Joint Conf. on Artificial Intelligence,* Vancouver (1981); G. E. Hinton, "Optimal Perceptual Inference," *Proc. IEEE Conf. on Computer Vision and Pattern Recognition,* Washington, D.C. (June 1983); M. Minsky, "K-Lines," *Cognitive Science,* 4 (1980); D. E. Rumelhart and J. L. McClelland, eds., *Parallel Distributed Processing: Explorations in the Microstructure of Cognition. Vol. 1: Foundations.* (Cambridge, Mass., 1986). See also notes 10 and 12.

10. D. H. Ackley, G. E. Hinton, and T. J. Sejnowski, "A Learning Algorithm for Boltzmann Machines," *Cognitive Science,* 9 (1985), 147–169.

11. J. von Neumann, *The Computer and the Brain* ((New Haven, 1958).

12. W. D. Hillis, *The Connection Machine* (Cambridge, Mass., 1985).

13. For a brief account of PROLOG, see W. Clocksin, "An Introduction to PROLOG," in T. O'Shea and M. Eisenstadt, eds., *Artificial Intelligence: Tools, Techniques, Applications* (New York, 1984), pp. 1–22. Fuller accounts are: W. Clocksin and C. Mellish, *Programming in PROLOG* (New York, 1981); R. A. Kowalski, *Logic for Problem Solving* (Amsterdam, 1979). The problem of negation in PROLOG is discussed in K. L. Clark, "Negation as Failure," in H. Gallaire and J. Minker, eds., *Logic and Databases* (New York, 1978). Logic programming in general is discussed in: M. van Caneghem and D. H. Warren, eds., *Logic Programming and its Applications* (Norwood, N. J., 1986).

14. J. A. Robinson, "The.Generalized Resolution Principle," in D. Michie, ed., *Machine Intelligence 3* (Edinburgh, 1968), pp. 77–94.

15. A brief and readable introduction to the problems involved in robotics is: S. Hardy,

"Robot Control Systems," in T. O'Shea and M. Eisenstadt, eds., *Artificial Intelligence: Tools, Techniques, Applications* (New York, 1984), pp. 178–191. Also fairly readable is J. M. Brady and P. E. Agre, "The Mechanic's Mate," in *ECAI-84: Proc. Sixth European Conf. on Artificial Intelligence,* Pisa (1984), 681–696. More technical accounts are: J. M. Brady, "Artificial Intelligence and Robotics," *AI,* 26 (1985), 79–122; J. M. Brady, J. M. Hollerbach, T. L. Johnson, T. Lozano-Perez, and M. T. Mason, eds., *Robot Motion: Planning and Control* (Cambridge, Mass., 1982); R. P. Paul, *Robot Manipulators: Mathematics, Programming and Control* (Cambridge, Mass., 1981). The industrial applications are highlighted in W. Clocksin and P. G. Davey, "Industrial Robotics," in T. O'Shea and M. Eisenstadt, eds., *Artificial Intelligence: Tools, Techniques, Applications* (New York, 1984), pp. 389–399.

16. A good introduction to the problems of naive physics is: P. J. Hayes, "The Naive Physics Manifesto," in D. Michie, ed., *Expert Systems in the Micro-Electronic Age* (Edinburgh, 1979), pp. 242–270. Other sources for this area are: J. Allen, "Towards a General Theory of Action and Time," *AI,* 23 (1984), 123–154; J. De Kleer and J. S. Brown. "A Qualitative Physics Based on Confluences," *AI,* 24 (1984), 7–83; K. D. Forbus, "Qualitative Process Theory," *AI,* 24 (1984), 85–168; B. Kuipers, "Commonsense Reasoning About Causality: Deriving Behavior From Structure," *AI,* 24 (1984), 169–203. Two recent collections are: D. G. Bobrow, ed. *Qualitative Reasoning About Physical Systems* (Cambridge, Mass., 1985); J. Hobbs and R. C. Moore, eds., *Formal Theories of the Commonsense World* (Hillsdale, N.J., 1984).

17. See *AI,* 13 (1980). See also: J. McCarthy, "Some Expert Systems Need Common Sense," *Annals New York Academy of Sciences,* 426 (1983), 129–137; J. McCarthy, "Applications of Circumscription to Formalizing Common-Sense Knowledge," *AI,* 28 (1986), 89–116.

18. See Note 4. See also A. Sloman: "Real Time Multiple-Motive Expert Systems," in M. Merry, ed., *Expert Systems 85* (Cambridge, 1985), pp. 213–224.

19. For an early description of this "ECG-expert", see I. Bratko and P. Mulec, "An Experiment in Automatic Learning of Diagnostic Rules," *Informatika,* 1981. A more recent account is: I. Bratko, I. Mozetic and N. Lavrac, "Automatic Synthesis and Compression of Cardiological Knowledge," *Machine Intelligence,* vol. 11, (in press). See also N. M. Gotts, J. R. W. Hunter, and R. K. E. W. Sinnhuber, "An Intelligent Model-Based System for Diagnosis in Cardiology," *Artificial Intelligence in Medicine, Group Report AIMG-7,* Sussex University, 1984. Causal reasoning is among the problems discussed in: J. R. W. Hunter, "Artificial Intelligence in Medicine: A Tutorial Survey," *Biomedical Measurement, Informatics and Control,* (in press).

20. This program was exhibited in action at IJCAI-1985 by Harold Cohen, of the University of California at San Diego. Cohen (already well known as a painter before starting work in computer art in the 1960s) has had his computer-generated work exhibited and/or commissioned by numerous galleries, including the Tate Gallery, London, and the Washington Children's Museum.

21. D. Lenat, M. Prakash, and M. Shepherd, "CYC: Using Common Sense Knowledge to Overcome Brittleness and Knowledge Acquisition Bottlenecks," *AI Magazine,* 6 (1986), 65–85.

22. Problems of updating memory are discussed in: J. Kolodner, "Maintaining Organization in a Dynamic Long-Term Memory," and "Reconstructive Memory: A Computer Model," *Cognitive Science,* 7 (1983), 243–328. See also note 27.

23. See J. R. Quinlan, "Learning Efficient Classification Procedures and Their Application to Chess End-Games," in R. Michalski, J. G. Carbonell, and T. M. Mitchell, eds., *Machine Learning: An Artificial Intelligence Approach* (Vol. I) (Palo Alto, 1983), pp. 463–482. For a recent review of machine learning, see: R. Michalski, J. G. Carbonell, and T. M. Mitchell, eds., *Machine Learning: An Artificial Intelligence Approach* (Vol. II) (Palo Alto, 1986).

24. A. Bundy, B. Silver, and D. Plummer, "An Analytical Comparison of Some Rule-Learning Programs," *AI,* 27 (1985), 137–182.

25. These programs are described in: D. B. Lenat, "The Nature of Heuristics," *AI,* 19 (1982), 189–249; D. B. Lenat, "Theory Formation by Heuristic Search. The Nature of Heuristic II: Background and Examples," *AI,* 20 (1983), 61–98; D. B. Lenat and J. S. Brown, "Why AM

and Eurisko Appear to Work," *AI,* 23 (1984), 269–294; D. B. Lenat, "The Role of Heuristics in Learning by Discovery: Three Case Studies," in R. Michalski, J. G. Carbonell, and T. M. Mitchell, eds., *Machine Learning: An Artificial Intelligence Approach* (Vol. I). (Palo Alto, 1983), pp. 243–305. For other discussions of analogy, see note 21, and also: D. Gentner, "Structure-Mapping: A Theoretical Framework for Analogy," *Cognitive Science,* 7 (1983); E. R. Michener, "Understanding Understanding Mathematics," *Cognitive Science,* 2 (1978), 361–383; E. L. Rissland, "The Ubiquitous Dialectic," *Proc. Sixth European Conf. on Artificial Intelligence,* Pisa (1984), 663–668; R. C. Schank, *Dynamic Memory: A Theory of Learning in Computers and People* (Cambridge, 1982); P. H. Winston, "Learning and Reasoning by Analogy," *CACM,* 23 (1980), 689–703.

26. G. D. Ritchie and F. K. Hanna, "AM: A Case Study in AI Methodology," *AI,* 23 (1984), 249–268.

27. See the papers in the special issue of *AI,* 13 (1980), and also: J. Doyle, "A Truth Maintenance System," *AI,* 12 (1979), 231–272; R. Reiter, "A Logic for Default Reasoning," *AI,* 13 (1980), 81–132. See also the next note.

28. J. McCarthy, "Circumscription: A Form of Non-Monotonic Reasoning," *AI,* 13 (1980), 27–40; J. McCarthy, "Applications of Circumscription to Formalizing Common-Sense Knowledge," *AI,* 28 (1986), 89–116.

29. A. Sloman, "Beginners Need Powerful Systems," in M. Yazdani, ed., *New Horizons in Educational Computing* (Chichester, 1984), pp. 220–234.

30. A. Bundy and R. Clutterbuck, "Raising the Standards of AI Products," in *Proc. Ninth Joint Conf. on Artificial Intelligence,* Los Angeles (1985), 1289–1294. A panel-discussion on legal implications is published in the *Proc. Ninth Joint Conf. on Artificial Intelligence* Los Angeles (1985), pp. 1267–1280: M. A. Boden, "Artificial Intelligence and Legal Responsibility"; R. Kowalski and M. Sergot, "Computer Representation of the Law"; M. S. Willick, "Constitutional Law and Artificial Intelligence: The Potential Legal Recognition of Computers as 'Persons' "; J. BloomBecker, "Fifth Generation Computer Crime Law"; Y. A. Wilks, "Responsible Computers." The remaining panelists' views are in: S. H. Nycum and I. K. Fong, "Artificial Intelligence and Certain Resulting Legal Issues," *The Computer Law Review,* University of Southern California, 1985. See also S. N. Lehman-Wilzig, "Frankenstein Unbound: Towards a Legal Definition of Artificial Intelligence," *Futures: The Journal of Forecasting and Planning,* 13 (1981), 107–119. The Proceedings of the ECAI Conference, Brighton (1986) will also include discussions of legal implications from another panel discussion. The problem of how to simulate the reasoning of legal experts is discussed in: R. Kowalski "Computer Representation of the Law", E. L. Rissland, "AI and Legal Reasoning," *Proc. Ninth Joint Conf. on Artificial Intelligence,* Los Angeles (1985), 1254–1260; A. Gardner, *An Artificial Intelligence Approach to Legal Reasoning* (Cambridge, Mass., in press).

31. Computer Professionals for Social Responsibility "headquarters" is: CPSR, PO Box 717, Palo Alto, CA 94301, USA. In the U.K., groups of the Society for Computing and Responsibility exist (for example) at the University of Edinburgh and the University of Sussex, and at Imperial College. Similar groups have recently been set up in France, Germany, and Italy.

32. See *Daedalus (J. Am. Acad. Arts & Sciences),* a two-part series on "Weapons in Space": Vol. 1, Spring 1985, "Concepts and Technologies"; Vol. 2, Summer 1985, "Implications for Security."

33. D. L. Parnas, *Software and SDI: Why Communications Systems Are Not Like SDI* (Senate Testimony, December 1985). Available from CPSR (See note 31). D. L. Parnas, "Software Aspects of Strategic Defense Systems," *American Scientist,* (Sept.–Oct. 1985), 432–440. See also D. B. Davis, "Assessing the Strategic Computing Initiative," *High Technology,* 5 (1985), 41–49.

34. See H. Thompson, "There Will Always be another Moonrise: Computer Reliability and Nuclear Weapons," *The Scotsman,* 1984; H. Thompson, "Empowering Automatic Decision-Making Systems," *Proc. Ninth Int. Joint Conf. on Artificial Intelligence,* Los Angeles (1985), 1281–1283.

35. D. V. McDermott, "Artificial Intelligence Meets Natural Stupidity," reprinted in J.

Haugeland, ed., *Mind Design: Philosophy, Psychology, Artificial Intelligence* (Cambridge, Mass., 1981), pp. 143–160. (First published in SIGART Newsletter of the ACM, No. 57, April 1976.)

36. A critical book called *Star Wars: A Question of Initiative,* by Richard Ennals (an English computer scientist who resigned from Imperial College in Autumn 1985 because of his misgivings about SDI research) was to be published on 17th September 1986 by John Wiley in the U.K. Press releases had been widely distributed and a publicity party had been arranged. The newsreader on BBC radio on the morning of 15th September announced that the publishers had withdrawn the book. At the time of my writing this note, no information is available as to why this book was suppressed, or who was responsible for persuading the publishers to take this decision. At the time of proofreading (two months later), there is still no information available.

# NAME INDEX

# SUBJECT INDEX

SOUTHERN
MAR 17 1988
BOUND

475-476

484

393-394

445

447

448-454